Edited by
Khalid Meksem and Günter Kahl

**The Handbook of
Plant Mutation Screening**

Related Titles

Kahl, G., Meksem, K. (eds.)

The Handbook of Plant Functional Genomics

Concepts and Protocols

2008

ISBN: 978-3-527-31885-8

Meksem, K., Kahl, G. (eds.)

The Handbook of Plant Genome Mapping

Genetic and Physical Mapping

2005

ISBN: 978-3-527-31116-3

Kahl, G.

The Dictionary of Genomics, Transcriptomics and Proteomics

2009

ISBN: 978-3-527-32073-8

Fromme, P. (ed.)

Photosynthetic Protein Complexes

A Structural Approach

2008

ISBN: 978-3-527-31730-1

Stewart, C. N. (ed.)

Plant Biotechnology and Genetics

Principles, Techniques and Applications

2008

ISBN: 978-0-470-04381-3

Ahmad, I., Pichtel, J., Hayat, S. (eds.)

Plant-Bacteria Interactions

Strategies and Techniques to Promote Plant Growth

2008

ISBN: 978-3-527-31901-5

Edited by
Khalid Meksem and Günter Kahl

The Handbook of Plant Mutation Screening

Mining of Natural and Induced Alleles

WILEY-VCH Verlag GmbH & Co. KGaA

The Editors

Prof. Dr. Khalid Meksem
Department of Plant, Soil and
Agricultural Systems
Southern Illinois University
Carbondale, IL 62901-4415
USA

Prof. Dr. Günter Kahl
GenXPro GmbH
Research Innovation Center (FIZ) Biotechnology
Altenhöferallee 3
60438 Frankfurt am Main
Germany

All books published by **Wiley-VCH** are carefully produced. Nevertheless, authors, editors, and publisher do not warrant the information contained in these books, including this book, to be free of errors. Readers are advised to keep in mind that statements, data, illustrations, procedural details or other items may inadvertently be inaccurate.

Library of Congress Card No.: applied for

British Library Cataloguing-in-Publication Data
A catalogue record for this book is available from the British Library.

Bibliographic information published by the Deutsche Nationalbibliothek
The Deutsche Nationalbibliothek lists this publication in the Deutsche Nationalbibliografie; detailed bibliographic data are available on the Internet at http://dnb.d-nb.de.

© 2010 WILEY-VCH Verlag GmbH & Co. KGaA, Weinheim

All rights reserved (including those of translation into other languages). No part of this book may be reproduced in any form – by photoprinting, microfilm, or any other means – nor transmitted or translated into a machine language without written permission from the publishers. Registered names, trademarks, etc. used in this book, even when not specifically marked as such, are not to be considered unprotected by law.

Typesetting   Thomson Digital, Noida, India
Printing and Binding   Strauss GmbH, Mörlenbach
Cover Design   Adam-Design, Weinheim

Printed in the Federal Republic of Germany
Printed on acid-free paper

ISBN: 978-3-527-32604-4

To my families, without their support, I would have been somewhere else!

Khalid Meksem

I appreciate the atmosphere and innovative work of the GenXPro GmbH thinktank in the Frankfurt Innovation Center Biotechnology.

Günter Kahl

Contents

Preface *XV*
List of Contributors *XVII*
List of Abbreviations *XXIII*

Part I   Induced Mutations *1*

1   **Physically Induced Mutation: Ion Beam Mutagenesis** *3*
            Shimpei Magori, Atsushi Tanaka, and Masayoshi Kawaguchi
1.1   Introduction *3*
1.1.1   LET *4*
1.1.2   Mutational Effects of Ion Beams on Plants *5*
1.2   Methods and Protocols *7*
1.2.1   Ion Beam Irradiation *8*
1.2.2   Dose Determination for Ion Beam Irradiation *9*
1.2.3   Plant Radiation Sensitivity *10*
1.2.4   Population Size of the M1 Generation *11*
1.3   Applications *11*
1.3.1   Ion Beams for Forward Genetics *12*
1.3.2   Ion Beams for Plant Breeding *13*
1.3.3   Limitations of Ion Beams *13*
1.4   Perspectives *14*
            References *14*

2   ***Ds* Transposon Mutant Lines for Saturation Mutagenesis of the *Arabidopsis* genome** *17*
            Takashi Kuromori and Takashi Hirayama
2.1   Introduction *17*
2.2   Methods and Protocols *18*
2.3   Applications *26*
2.4   Perspectives *28*
            References *28*

The Handbook of Plant Mutation Screening. Edited by Günter Kahl and Khalid Meksem
Copyright © 2010 WILEY-VCH Verlag GmbH & Co. KGaA, Weinheim
ISBN: 978-3-527-32604-4

3		**Use of Mutants from T-DNA Insertion Populations Generated by High-Throughput Screening** *31*
		Ralf Stracke, Gunnar Huep, and Bernd Weisshaar
3.1		Introduction *31*
3.2		Methods and Protocols *34*
3.2.1		Plant Material and Growth Conditions *34*
3.2.2		Plasmid Design *34*
3.2.3		*Agrobacterium* Culture *35*
3.2.4		Plant Transformation and T1 Seed Harvesting *35*
3.2.5		Sulfadiazine Selection of Transgenic T_1 Plants *36*
3.2.6		DNA Preparation from Sulfadiazine-Selected T1 Plants *36*
3.2.7		FST Production *37*
3.2.8		Sequencing and Computational Sequence Analysis *40*
3.2.9		Genetic Analysis of T-DNA Insertions *41*
3.2.10		DNA-Preparation for Confirmation of FST Predicted Insertion Sites *41*
3.2.11		Confirmation PCR *42*
3.2.12		Sequencing and Computational Sequence Analysis *44*
3.2.13		Seed Donation *45*
3.2.14		Identification of Homozygous Mutants *46*
3.3		Applications and Considerations for Work with T-DNA Insertion Mutants *47*
3.3.1		Unconfirmed T-DNA Insertion Lines *48*
3.3.2		Use of Selectable Marker *48*
3.3.3		Aberrant T-DNA Insertions *48*
3.3.4		Multiple T-DNA Insertions *49*
3.3.5		T-DNA-Induced Dominant Effects *49*
3.3.6		Allelic Series of Mutants *49*
3.3.7		Lethal Knockout Mutants *50*
3.3.8		Search for Knockout Phenotype *50*
3.3.9		Handling of Non-Single-Copy Genes *50*
3.4		Perspectives *51*
		References *52*
4		**Making Mutations is an Active Process: Methods to Examine DNA Polymerase Errors** *55*
		Kristin A. Eckert and Erin E. Gestl
4.1		Introduction *55*
4.2		Methods and Protocols *56*
4.2.1		Overview of the Genetic Assay *56*
4.2.2		Overview of the Biochemical Assay for TLS *67*
4.3		Applications *73*
4.3.1		General Features of the *In Vitro* Genetic Assay *73*
4.3.2		Polymerase Accuracy in the Absence of DNA Damage *74*

4.3.3	Mutational Processing of Alkylation Damage by DNA Polymerases 75
4.3.4	DNA Lesion Discrimination Mechanisms 75
4.4	Perspectives 78
	References 79

5   ***Tnt1* Induced Mutations in *Medicago*: Characterization and Applications** 83
Pascal Ratet, Jiangqi Wen, Viviane Cosson, Million Tadege, and Kirankumar S. Mysore

5.1	Introduction 83
5.2	Methods and Protocols 84
5.2.1	Identification of *Tnt1* Insertion Sites 84
5.2.2	Reverse Genetic Approach 94
5.2.2.1	FST Sequencing 94
5.2.2.2	Screening DNA Pools 94
5.3	Applications 95
5.3.1	Line with a Mutant Phenotype – No FSTs Identified 96
5.3.2	Line with a Mutant Phenotype and FSTs Already Identified 96
5.3.3	FST Sequence in the *Tnt1* Database Matches a Gene of Interest – No Mutant Phenotype is Described in that Line 97
5.3.4	Have a Gene to Work With – No FST or Mutant Phenotype 97
5.4	Perspectives 98
	References 98

Part II   Mutation Discovery 101

6   Mutation Discovery with the Illumina Genome Analyzer 103
Abizar Lakdawalla and Gary P. Schroth

6.1	Introduction 103
6.1.1	Overview of the Illumina Genome Analyzer Sequencing Process 103
6.1.2	Resequencing Strategies 104
6.1.2.1	Resequencing Whole Genomes 105
6.1.2.2	Targeted Genome Selection 105
6.1.2.3	Sequencing Transcriptomes 107
6.2	Methods and Protocols 107
6.3	Applications 116
6.4	Perspectives 118
	References 118

7   Chemical Methods for Mutation Detection: The Chemical Cleavage of Mismatch Method 121
Tania Tabone, Georgina Sallmann, and Richard G.H. Cotton

7.1	Introduction 121
7.2	Methods and Protocols 125

7.3	Applications *127*
7.4	Perspectives *127*
	References *128*

8   **Mutation Detection in Plants by Enzymatic Mismatch Cleavage** *131*
Bradley J. Till

8.1	Introduction *131*
8.2	Methods and Protocols *136*
8.3	Applications *143*
8.4	Perspectives *144*
	References *145*

9   **Mutation Scanning and Genotyping in Plants by High-Resolution DNA Melting** *149*
Jason T. McKinney, Lyle M. Nay, David De Koeyer, Gudrun H. Reed, Mikeal Wall, Robert A. Palais, Robert L. Jarret, and Carl T. Wittwer

9.1	Introduction *149*
9.2	Methods and Protocols *150*
9.2.1	LightScanner Instrument *151*
9.2.2	LightScanner for Variant Scanning *151*
9.2.3	LightScanner for Lunaprobe™ (Unlabeled Probe) Genotyping *156*
9.3	Applications *159*
9.3.1	Sensitivity and Specificity for SNP Heterozygote Detection *159*
9.3.2	Variant Scanning by High-Resolution Melting *160*
9.3.3	Bell Pepper Multiplex Genotyping with Two Unlabeled Probes *161*
9.3.4	Potato Tetraploid Genotyping including Allele Dosage using an Unlabeled Probe *161*
9.4	Perspectives *162*
	References *163*

10   ***In Silico* Methods: Mutation Detection Software for Sanger Sequencing, Genome and Fragment Analysis** *167*
Kevin LeVan, Teresa Snyder-Leiby, C.S. Jonathan Liu, and Ni Shouyong

10.1	Introduction *167*
10.2	Mutation Detection with Sanger Sequencing using Mutation Surveyor *168*
10.3	Mutation Detection with NextGENe™ and Next-Generation Sequence Technologies *175*
10.4	Mutation Detection with DNA Fragments Using GeneMarker® *180*
10.5	Perspectives *182*
	References *182*

Part III	**High-Throughput Screening Methods** 185	

11 **Use of TILLING for Reverse and Forward Genetics of Rice** 187
*Sujay Rakshit, Hiroyuki Kanzaki, Hideo Matsumura, Arunita Rakshit,
Takahiro Fujibe, Yudai Okuyama, Kentaro Yoshida, Muluneh Oli,
Matt Shenton, Hiroe Utsushi, Chikako Mitsuoka, Akira Abe,
Yutaka Kiuchi, and Ryohei Terauchi*

11.1 Introduction 187
11.2 Methods and Protocols 188
11.3 Perspectives 196
References 197

12 **Sequencing-Based Screening of Mutations and Natural
Variation using the KeyPoint™ Technology** 199
Diana Rigola and Michiel J.T. van Eijk

12.1 Introduction 199
12.2 Methods and Protocols 202
12.3 Applications 206
12.3.1 EMS Mutation Screening and Validation 206
12.3.2 Natural Polymorphism Screening and Validation 208
12.4 Perspectives 211
References 211

Part IV	**Applications in Plant Breeding** 215	

13 **Natural and Induced Mutants of Barley: Single Nucleotide
Polymorphisms in Genes Important for Breeding** 217
William T.B. Thomas, Brian P. Forster, and Robbie Waugh

13.1 Brief Review of Barley Mutants 217
13.2 Applications in Breeding 221
13.3 Single Nucleotide Polymorphism Genotyping to Identify Candidate
Genes for Mutants 223
13.3.1 Resources 223
13.3.2 Case Study: Two/Six-Row Locus in Barley 224
13.3.3 Case Study: Graphical Genotyping of a Disease Resistance Locus 227
13.3.4 General Protocol for using High-Throughput Genotyping to
Localize Mutants 227
References 229

14 **Association Mapping for the Exploration of Genetic Diversity
and Identification of Useful Loci for Plant Breeding** 231
André Beló and Stanley D. Luck

14.1 Introduction 231
14.2 Methods and Protocols 233
14.2.1 Population for Association Mapping 233

14.2.2	Genotyping	234
14.2.3	Phenotyping	234
14.2.4	Statistical Procedures	235
14.3	Applications	238
14.3.1	QTL Mapping versus Association Mapping	240
14.3.2	Limitations	241
14.4	Perspectives	242
	References	243

15 Using Mutations in Corn Breeding Programs 247
Anastasia L. Bodnar and M. Paul Scott

15.1	Introduction	247
15.1.1	Factors to Consider Before Starting a Breeding Program	248
15.1.2	Alternatives to Breeding	249
15.2	Methods and Protocols	249
15.2.1	Backcross Breeding	249
15.2.2	Forward Breeding	252
15.2.3	Supplementary Protocols	253
15.2.3.1	Determining How Many Seeds to Plant	253
15.2.3.2	Working with Recessive Mutations	255
15.2.3.3	Intermating	256
15.2.4	Complication: Pleiotropic Effects	257
15.3	Applications	258
15.3.1	Breeding with a Natural Mutation: QPM	258
15.3.2	Breeding with a Transgene: GFP	259
15.4	Perspectives	259
15.4.1	Marker-Assisted Selection	259
15.4.1.1	Marker-Assisted Selection in Backcross Breeding	259
15.4.1.2	Marker-Assisted Selection in Forward Breeding	260
15.4.2	Doubled Haploids	260
	References	261

16 Gene Targeting as a Precise Tool for Plant Mutagenesis 263
Oliver Zobell and Bernd Reiss

16.1	Introduction	263
16.2	Methods and Protocols	266
16.3	Applications	279
16.4	Perspectives	280
	References	280

Part V Emerging Technologies 287

17 True Single Molecule Sequencing (tSMS)™ by Synthesis 289
Scott Jenkins and Avak Kahvejian

17.1	Introduction	289
17.2	Methods, Protocols, and Technical Principles	291

17.2.1	Single Molecule Sequencing Technical Challenges and Solutions	291
17.2.2	Flow Cell Surface Architecture	294
17.2.3	Cyclic SBS	295
17.2.4	Optical Imaging of Growing Strands	297
17.2.5	Mechanical Operation	300
17.2.6	System Components	300
17.2.7	Data Analysis	301
17.3	Applications	301
17.3.1	Single Molecule DGE and RNA-Seq	301
17.3.1.1	DNA Sequencing Applications	303
17.3.2	Single Molecule Sequencing Techniques under Development	303
17.4	Perspectives	304
	References	306

18 High-Throughput Sequencing by Hybridization 307
Sten Linnarsson 307

18.1	Introduction	307
18.2	Methods and Protocol	308
18.3	Discussion	316
18.4	Applications	316
	References	317

19 DNA Sequencing-by-Synthesis using Novel Nucleotide Analogs 319
Lin Yu, Jia Guo, Ning Xu, Zengmin Li, and Jingyue Ju

19.1	Introduction	319
19.2	General Methodology for DNA SBS	321
19.3	Four-Color DNA SBS using CF-NRTs	323
19.3.1	Overview	323
19.3.2	Design, Synthesis, and Characterization of CF-NRTs	324
19.3.3	DNA Chip Construction	326
19.3.4	Four-Color SBS using CF-NRTs	326
19.4	Hybrid DNA SBS using NRTs and CF-ddNTPs	329
19.4.1	Overview	329
19.4.2	Design and Synthesis of NRTs and CF-ddNTPs	331
19.4.3	Four-Color Hybrid DNA SBS	333
19.5	Perspectives	335
	References	336

20 Emerging Technologies: Nanopore Sequencing for Mutation Detection 339
Ryan Rollings and Jiali Li

20.1	Introduction	339
20.1.1	Nanopore Detection Principle	340
20.1.2	Important Parameters and Nanopore Sensing Resolution	341
20.1.3	Biological Nanopore History	341
20.1.4	Solid-State Nanopore History	343

20.1.5	Nanopore Promise *343*	
20.2	Current Developments in Nanopore Sequencing *344*	
20.2.1	Improving Biological Nanopores *344*	
20.2.2	Improving Solid-State Nanopores *345*	
20.2.3	Slowing Translocation and Trapping *347*	
20.2.4	Modification of the DNA *347*	
20.2.5	Resequencing Applications *349*	
20.3	Work Done in Our Lab *350*	
20.4	Perspectives *351*	
	References *352*	

Glossary *355*

Index *427*

Preface

The DNA double helix, as we know it since the seminal paper of Watson and Crick in 1953 (Watson, J.D. and Crick, F.H.C. A structure for deoxyribose nucleic acid. *Nature* 171: 737–738, 1953), probably a special variant of a primordial RNA, has been (and still is) under heavy environmental pressure. It is exposed to chemical and physical agents such as drugs and radiation that altogether have altered nucleic acids since their existence. Therefore, DNA is subject to continuous changes in base composition and structure, generally called mutations (Latin *mutare*: to change). These changes not only shaped the DNA molecule, but eliminated or introduced new information, which was, is, and will be fundamental for evolution. Certainly, evolution would not have been possible without mutation(s). The immense phenotypic variations, which surprise the alert observer, rest upon genetic variation (i.e., mutations). Although large subsets of mutations are beneficial for the survival and evolution of a species, certain mutations can have a detrimental impact on individuals – they may indeed lead to a reduced fitness and diseases down to death. These ambivalent aspects of mutations shape life on Earth.

Mutations range in size from single base exchanges, single base deletions, small deletions, inversions, insertions, and duplications of two or a couple of bases more, to large changes of several kilobases to megabases and chromosome aberrations such as supernumerary chromosomes, missing chromosomes, translocated parts from one chromosome to another one (or others, and vice versa), and aberrant physical appearance of chromosomes (e.g., in various forms of leukemia or in the case of ring chromosomes). This listing is by no means a complete description of all possible mutations, but outlines the variety of potential changes at the molecular and chromosomal level. Keeping in mind that before the discovery of DNA by the pioneer Friedrich Miescher around 1869 in the nuclei of leukocytes from pus on surgical bandages (at that time coined nuclein), mutations were already known in various organisms, but could, of course, not be related to DNA. For example, albinos in plants, animals, and humans were considered exotic and unexplainable erratic forms of life.

Today we know of many chromosomal, genomic, and genic mutations, and as more and more genomes are sequenced and resequenced, more and more mutations are discovered and will serve to identify genetic differences between individuals or species, or the causes underlying the 6000–8000 inherited disorders in humans. It is for this reason that the detection of mutations became an important discipline in biology and medicine, and therefore the techniques of discovery and diagnostics of these changes are rapidly evolving.

This book witnesses the importance of mutations. The detailed description of mutation screening technologies by renowned experts in the field gives a flavor of the past and present state of mutation sciences, but also a sense of what is emerging in the area of biophysics, biocomputing, molecular biology, and genomics.

Carbondale (USA) *Khalid Meksem*
Frankfurt am Main (Germany) *Günter Kahl*
December 2009

List of Contributors

Akira Abe
Iwate Agricultural Research Center
20-1 Narita, Kitakami
Iwate 024-0003
Japan

André Beló
University of Delaware & DuPont
Crop Genetics
Department of Plant and Soil Sciences
152 Townsend Hall
Newark, NJ 19716-2103
USA

Anastasia L. Bodnar
Iowa State University
Agronomy Department
G426 Agronomy Hall
Ames, IA 50011
USA

Viviane Cosson
CNRS, Institut des Sciences du Végétal
1 Avenue de la Terrasse
91198 Gif sur Yvette Cedex
France

Richard G.H. Cotton
Genomic Disorders Research Centre
Level 2, 161 Barry Street
Carlton South, Victoria 3053
Australia

and

University of Melbourne
Department of Medicine
Melbourne, Victoria 3010
Australia

David De Koeyer
Agriculture and Agri-Food Canada
Potato Research Centre
850 Lincoln Road, Fredericton
New Brunswick E3B 4Z7
Canada

Kristin A. Eckert
Pennsylvania State University College
of Medicine
Department of Pathology and
Biochemistry & Molecular Biology
500 University Drive
Hershey, PA 17033
USA

Brian P. Forster
Biohybrids International Ltd
PO Box 4211
Reading RG6 5FY
UK

Takahiro Fujibe
Iwate University
21 Century COE
3-18-8 Ueda, Morioka
Iwate 020-8550
Japan

Erin E. Gestl
West Chester University
Department of Biology
750N. Church Street
West Chester, PA 19383
USA

Jia Guo
Columbia University College
of Physicians and Surgeons
Columbia Genome Center
1150 St. Nicholas Avenue
New York, NY 10032
USA

Takashi Hirayama
RIKEN, Membrane Molecular Biology
Laboratory
1-7-22 Suehiro, Tsurumi
Yokohama 230-0045
Japan

Gunnar Huep
Universität Bielefeld
Fakultät für Biologie/Genomforschung
Universitätsstrasse 27
33615 Bielefeld
Germany

Robert L. Jarret
US Department of Agriculture/
Agricultural Research Service
Plant Genetic Resources Unit
1109 Experiment Street
Griffin, GA 30224
USA

Scott Jenkins
Helicos BioSciences Corporation
One Kendall Square, Building 700
Cambridge, MA 02139
USA

Jingyue Ju
Columbia University College
of Physicians and Surgeons
Columbia Genome Center
1150 St. Nicholas Avenue
New York, NY 10032
USA

Avak Kahvejian
Helicos BioSciences Corporation
One Kendall Square, Building 700
Cambridge, MA 02139
USA

Hiroyuki Kanzaki
Iwate Biotechnology Research Centre
22-174-4 Narita, Kitakami
Iwate 024-0003
Japan

Masayoshi Kawaguchi
University of Tokyo
Department of Biological Sciences
Graduate School of Science
7-3-1 Hongo, Bunkyo-ku
Tokyo 113-0033
Japan

List of Contributors

Yutaka Kiuchi
Iwate Agricultural Research Center
20-1 Narita, Kitakami
Iwate 024-0003
Japan

Takashi Kuromori
RIKEN Plant Science Center
Gene Discovery Research Group
1-7-22 Suehiro, Tsurumi
Yokohama 230-0045
Japan

Abizar Lakdawalla
Illumina, Inc.
Sequencing Applications
25861 Industrial Boulevard
Hayward, CA 94545
USA

Kevin LeVan
SoftGenetics LLC
Suite 235, 200 Innovation Boulevard
State College, PA 16803
USA

Jiali Li
University of Arkansas
Department of Physics
825 West Dickson Street
Fayetteville, AR 72701
USA

Zengmin Li
Columbia University College
of Physicians and Surgeons
Columbia Genome Center
1150 St. Nicholas Avenue
New York, NY 10032
USA

Sten Linnarsson
Karolinska Institutet
Department of Medical Biochemistry
and Biophysics
Laboratory for Molecular Neurobiology
Scheeles väg 1
17177 Stockholm
Sweden

C.S. Jonathan Liu
SoftGenetics LLC
Suite 235, 200 Innovation Boulevard
State College, PA 16803
USA

Stanley D. Luck
University of Delaware & DuPont Crop
Genetics
Department of Plant and Soil Sciences
152 Townsend Hall
Newark, NJ 19716-2103
USA

Shimpei Magori
University of Tokyo
Department of Biological Sciences
Graduate School of Science
7-3-1 Hongo, Bunkyo-ku
Tokyo 113-0033
Japan

Hideo Matsumura
Iwate Biotechnology Research Centre
22-174-4 Narita, Kitakami
Iwate 024-0003
Japan

Jason T. McKinney
Idaho Technology
Genotyping Applications Division
390 Wakara Way
Salt Lake City, UT 84108
USA

Chikako Mitsuoka
Iwate Biotechnology Research Centre
22-174-4 Narita, Kitakami
Iwate 024-0003
Japan

Kirankumar S. Mysore
The Samuel Roberts Noble Foundation
Plant Biology Division
2510 Sam Noble Parkway
Ardmore, OK 73401
USA

Lyle M. Nay
Idaho Technology
Genotyping Applications Division
390 Wakara Way
Salt Lake City, UT 84108
USA

Yudai Okuyama
Iwate Biotechnology Research Centre
22-174-4 Narita, Kitakami
Iwate 024-0003
Japan

Muluneh Oli
Iwate Biotechnology Research Centre
22-174-4 Narita, Kitakami
Iwate 024-0003
Japan

Robert A. Palais
University of Utah
Department of Mathematics
155S. 1400E.
Salt Lake City, UT 84112
USA

Arunita Rakshit
NRC on Plant Biotechnology
Pusa Campus
New Delhi 110 012
India

Sujay Rakshit
National Research Centre on Sorghum
Rajendranagar
Hyderabad 500 030
India

Pascal Ratet
CNRS, Institut des Sciences du Végétal
1 Avenue de la Terrasse
91198 Gif sur Yvette Cedex
France

Gudrun H. Reed
University of Utah
Department of Pathology
50N. Medical Drive
Salt Lake City, UT 84109
USA

Bernd Reiss
Max-Planck-Institut für
Züchtungsforschung
Unabhängige Forschergruppe DNA
Rekombination der Pflanzen
Carl-von-Linne-Weg 10
50829 Köln
Germany

Diana Rigola
Keygene NV
Agro Business Park 90
PO Box 216
6700 AE Wageningen
The Netherlands

Ryan Rollings
University of Arkansas
Department of Physics
825 West Dickson Street
Fayetteville, AR 72701
USA

Georgina Sallmann
Monash University
Department of Medicine
Central and Eastern Clinical School
85 Commercial Road
Melbourne, Victoria 3004
Australia

Gary P. Schroth
Illumina, Inc.
Gene Expression R&D
25861 Industrial Boulevard
Hayward, CA 94545
USA

M. Paul Scott
Iowa State University
Agronomy Department
1407 Agronomy Hall
Ames, IA 50011
USA

Matt Shenton
Iwate Biotechnology Research Centre
22-174-4 Narita, Kitakami
Iwate 024-0003
Japan

Ni Shouyong
SoftGenetics LLC
Suite 235, 200 Innovation Boulevard
State College, PA 16803
USA

Teresa Snyder-Leiby
SoftGenetics LLC
Suite 235, 200 Innovation Boulevard
State College, PA 16803
USA

Ralf Stracke
Universität Bielefeld
Fakultät für Biologie/Genomforschung
Universitätsstrasse 27
33615 Bielefeld
Germany

Tania Tabone
Royal Melbourne Hospital
Ludwig Institute for Cancer Research
Centre for Medical Research
Royal Parade
Parkville, Victoria 3050
Australia

Million Tadege
The Samuel Roberts Noble Foundation
Plant Biology Division
2510 Sam Noble Parkway
Ardmore, OK 73401
USA

Atsushi Tanaka
Japan Atomic Energy Agency
Radiation-Applied Biology Division
Quantum Beam Science Directorate
1233 Watanuki-machi, Takasaki
Gunma 370-1292
Japan

Ryohei Terauchi
Iwate Biotechnology Research Centre
22-174-4 Narita, Kitakami
Iwate 024-0003
Japan

William T.B. Thomas
Scottish Crop Research Institute
Genetics Program
Errol Road
Dundee DD2 5DA
UK

Bradley J. Till
Plant Breeding Unit
FAO/IAEA Agricultural &
Biotechnology Laboratory
International Atomic Energy Agency
Wagramer Strasse 5
1400 Vienna
Austria

Hiroe Utsushi
Iwate Biotechnology Research Centre
Department/Division??
22-174-4 Narita, Kitakami
Iwate 024-0003
Japan

Michiel J.T. van Eijk
Keygene NV
Agro Business Park 90
PO Box 216
6700 AE Wageningen
The Netherlands

Mikeal Wall
Idaho Technology
Genotyping Applications Division
390 Wakara Way
Salt Lake City, UT 84108
USA

Robbie Waugh
Scottish Crop Research Institute
Genetics Program
Errol Road
Dundee DD2 5DA
UK

Bernd Weisshaar
Universität Bielefeld
Fakultät für Biologie/Genomforschung
Universitätsstrasse 27
33615 Bielefeld
Germany

Jiangqi Wen
The Samuel Roberts Noble Foundation
Plant Biology Division
2510 Sam Noble Parkway
Ardmore, OK 73401
USA

Carl T. Wittwer
University of Utah
Department of Pathology
50N. Medical Drive
Salt Lake City, UT 84109
USA

Ning Xu
Columbia University College
of Physicians and Surgeons
Columbia Genome Center
1150 St. Nicholas Avenue
New York, NY 10032
USA

Kentaro Yoshida
Iwate Biotechnology Research Centre
22-174-4 Narita, Kitakami
Iwate 024-0003
Japan

Lin Yu
Columbia University College of
Physicians and Surgeons
Columbia Genome Center
1150 St. Nicholas Avenue
New York, NY 10032
USA

Oliver Zobell
University of Osnabrück
Botany
Barbarastrasse 11
49076 Osnabrück
Germany

Abbreviations

2,4-D	2,4-dichlorophenoxyacetic acid
Ac	*Activator*
AFLP	amplified fragment length polymorphism
ANOVA	ANalysis Of Variance
AOD	altered organ development
AVF	azimuthally varying field
BAC	bacterial artificial chromosome
BaYMV	barley yellow mosaic virus
BOPA	barley oligo pooled array
BSA	bovine serum albumin
Carb	carbenicillin
CaMV	cauliflower mosaic virus
[CCM]	chemical cleavage of mismatch
CJE	celery juice extract
ChIP	chromatin immunoprecipitation
CF	cleavable fluorescent
Cm	chloramphenicol
CTAB	cetyltrimethylammonium bromide
DGAT	diacylglycerol acyltransferase
DGE	digital gene expression
DHPLC	denaturing high-performance liquid chromatography
DMSO	dimethylsulfoxide
Ds	*Dissociator*
DSB	double-strand break
DSBs	double-strand breaks
dsDNA	double-stranded DNA
EELS	electron energy loss spectroscopy
EMS	ethylmethane sulfonate
ERA-PG	European Research Area in Plant Genomics
EST	expressed sequence tag
5-FAM	5-Carboxyfluorescein
6-FAM	6-Carboxyfluorescein
5(6)-FAM	5-(and -6)-Carboxyfluorescein

The Handbook of Plant Mutation Screening. Edited by Günter Kahl and Khalid Meksem
Copyright © 2010 WILEY-VCH Verlag GmbH & Co. KGaA, Weinheim
ISBN: 978-3-527-32604-4

FDR	false discovery rate
FST	flanking sequence tag
FUdR	5-fluoro-2′-deoxyuridine
GD	gapped duplex
GFP	green fluorescent protein
GS	genome sequencer
GSPs	gene-specific primers
GWA	genome-wide association
HEX	hexachlorocarboxyfluorescein
HIMAC	Heavy Ion Medical Accelerator in Chiba
HSV-*tk*	herpes simplex virus type 1 thymidine kinase
I	inverse
IAA	indole-3-acetic acid
Indels	insertion/deletions
INRA	Institut National de la Recherche Agronomique
IRD	infrared dye
JAEA	Japan Atomic Energy Agency
KF polymerase	Klenow fragment of *E. coli* DNA polymerase I
LB	left border
LD	linkage disequilibrium
LD_{50}	median lethal dose
LET	linear energy transfer
LOH	loss of heterozygosity
LTR	long terminal repeat
MALDI-TOF MS	matrix-assisted laser desorption/ionization-time of flight mass spectrometry
MCMC	Markov chain Monte Carlo
m^6G	O^6-methylguanine
MNU	*N*-methyl-*N*-nitrosourea
MS	mutation screening
MSI	microsatellite instability
NA	numerical aperture
NAM	naphthalene acetamide
NCBI	National Center for Biotechnology Information
NIRS	National Institute of Radiological Sciences
NRTs	nucleotide reversible terminators
NVS	natural variation screening
OC	open circle
ORF	open reading frame
PAGE	polyacrylamide gel electrophoresis
PCA	principal component analysis
PCR	polymerase chain reaction
PEB	phenol extraction buffer
PEG	polyethylene glycol
PMSF	phenylmethylsulfonate

PTP	picotiterplate
PVP	polyvinylpyrrolidone
QPM	quality protein maize
QTLs	quantitative trait loci
RB	right border
RIBF	RI beam factory
SBH	sequencing-by-hybridization
SBS	sequencing-by-synthesis
SDS	sodium dodecyl sulfate
SNP	single nucleotide polymorphism
SSC	sodium chloride/sodium citrate
SSC	standard sodium citrate
ssDNA	single-stranded DNA
SSR	short sequence repeat
STM	scanning tunneling microscopy
STRs	short tandem repeats
TAIL	thermal asymmetric interlaced
TCEP	tris(2-carboxy-ethyl) phosphine
TD	transposon display
TEM	transmission electron microscopy
TIARA	Takasaki Ion Accelerators for Advanced Radiation Application
TILLING	targeting induced local lesions in genomes
TIRFM	total internal reflection fluorescence microscopy
TLS	translesion synthesis
T_m	melting temperature
tSMS™	True Single Molecule Sequencing
UTR	untranslated region
VB	Vogel–Bonner
W-MAST	Wakasa Wan Energy Research Center Multi-purpose Accelerator with Synchrotron and Tandem
YAC	yeast artificial chromosome

Part I
Induced Mutations

1
Physically Induced Mutation: Ion Beam Mutagenesis

Shimpei Magori, Atsushi Tanaka, and Masayoshi Kawaguchi

Abstract

Ion beams are novel physical mutagens that have been applied to a wide variety of plant species. Unlike other physical mutagens such as X-rays, γ-rays, and electrons, ion beams have high linear energy transfer, leading to high double-strand break yields and the resulting strong mutational effects. Takasaki Ion Accelerators for Advanced Radiation Application (TIARA) in Japan was established as the first ion beam irradiation facility for biological use. In this facility, positively charged ions are accelerated at a high speed and used to irradiate living materials, including plant seeds and tissue cultures. By utilizing this approach, several novel mutants have been successfully isolated even from *Arabidopsis*, in which thousands of mutants have already been obtained using different mutagens. This demonstrates that ion beams are a powerful alternative mutagen with a mutation spectrum different from other chemical, physical, and T-DNA-based mutagens. The application of such an alternative mutagen is of great importance not only to analyze any gene functions through novel mutant isolation, but also to improve global food situations by providing new crop varieties with beneficial traits. In this chapter, we describe the detailed methods of ion beam irradiation and discuss its applications in genetic research as well as plant breeding.

1.1
Introduction

Mutagenesis is one of the most critical steps for genetic studies as well as selective breeding. Successful mutant isolation largely relies on the use of efficient mutagens. In plant research, a chemical mutagen, ethylmethane sulfonate (EMS), has been commonly used for this purpose. Although this mutagen can be handled easily and applied to any plant, it primarily produces single base substitutions, but not drastic mutations such as large genomic deletions. Therefore, application of more powerful mutagens with different mutation spectra is of great significance in some cases. One good technology for this end is ion beam mutagenesis. The ion beam is a physical mutagen that has just recently come into use for plants. In this type of mutagenesis, positively charged ions are accelerated at a high speed (around 20–80% of the speed of light) and used to irradiate target cells. As a physical mutagen, ion beams are similar to other forms of radiation such as X-rays, γ-rays, and electrons, but it is different from them in that ion beams have much higher linear energy transfer (LET). This characteristic is important to understand the high biological effectiveness of ion beams.

1.1.1
LET

LET is the energy deposited to target material when an ionizing particle passes through it. Once an accelerated particle encounters any substance, it gradually loses its own energy (i.e., the same amount of energy is transferred to the substance causing "damage") and eventually stops at the point where the maximum energy loss is observed (Figure 1.1). LET is usually expressed in kiloelectronvolts per micrometer

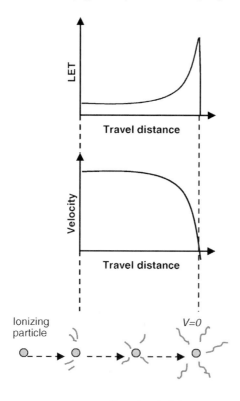

Figure 1.1 Conceptual diagram of LET. An ionizing particle gradually loses its own energy as it slows down in the target material. LET refers to this energy loss, which is deposited to the material. In this cartoon, LET is represented by wavy lines. LET reaches its maximum just before the ionizing particle stops. Immediately after this peak, LET plunges to zero.

(keV/μm), which represents the average amount of energy lost per unit distance. Ion beams have a relatively high LET (around 10–1000 keV/μm or higher), while X-rays, γ-rays, and electrons have low LETs (around 0.2 keV/μm). Therefore, ion beams are able to cause more severe damage to living cells than other forms radiation, resulting in the high relative biological effectiveness [1, 2].

1.1.2
Mutational Effects of Ion Beams on Plants

Biological effects of ion beams have been investigated not only in mammals, but also in plants. For example, studies using *Arabidopsis thaliana* and *Nicotiana tabacum* showed that ion beams were more efficient in decreasing the germination rate and the survival rate than low-LET radiation [3, 4]. More importantly, analysis focusing on *transparent testa* (*tt*) and *glabrous* (*gl*) loci revealed that 113-keV/μm carbon ions induced a 20-fold higher mutation rate per dose than 0.2-keV/μm electrons, thus demonstrating the power of ion beams as a mutagen [5, 6]. The detailed characterization of the carbon ion-induced mutations showed that ion beams can cause large DNA alterations (large deletions, inversions, and translocations) as well as small intragenic mutations and that ion beams frequently, but not always, produce deletions with variable sizes from 1 bp up to 230 kbp, compared to electrons (summarized in Table 1.1) [6]. Since such deletions possibly lead to frameshifts or total gene losses, mutants derived from ion beam mutagenesis can be considered as nulls in many cases. This is a significant difference from the conventional chemical mutagen EMS, which mostly generates point mutations resulting from GC → AT transitions.

These great mutational effects of ion beams are partly due to high double-strand break (DSB) yields induced by ions. The study using tobacco BY-2 protoplasts as a model system showed that initial DSB yields were positively correlated with LET, and that high-LET helium, carbon, and neon ions were more effective in causing DSBs

Table 1.1 Classification of mutations induced by carbon ions and electrons (modified from [6]).

Mutagen (LET)	Intragenic mutation		Large DNA alteration	
Carbon ions (113 keV/μm)	48%		52%	
	deletion	38%	inversion/translocation	21%
	base substitution	7%	total deletion	31%
	insertion	3%		
Electrons (0.2 keV/μm)	75%		25%	
	deletion	33%	inversion/translocation	25%
	base substitution	33%	total deletion	0%
	insertion	8%		

The distributions of the indicated mutation patterns were determined based on the sequence analysis with 29 and 12 mutant alleles produced by carbon ions and electrons, respectively [6]. Note that carbon ions induced large DNA alteration in the tested loci more frequently than electrons. Such large DNA alterations include total deletion, which refers to a complete loss of a gene locus.

than γ-rays [7]. Further, it was found that at least carbon and neon ions produced short DNA fragments more frequently than γ-rays, suggesting that ion particles can act densely and locally on target genomes [7].

It is plausible that DSBs are more difficult for cells to repair than single-strand breaks (i.e., DSB repair can be error-prone), which might partly explain the high mutation rates caused by ion beams. However, the molecular mechanism of ion-mediated mutation induction remains largely unknown. To address this issue, Shikazono *et al.* analyzed the DNA sequences flanking the breakpoints generated by carbon ions and showed that many of the tested sequences contained deletions (1–29 bp), whereas most of the electron-induced breakpoints were flanked by duplications (1–7 bp) [6]. Based on these findings, they hypothesize that unlike electrons, high-LET ions could induce not only DSBs, but also cause severe damage in the broken ends and that such damaged sequences might be eventually excised during the repair processes, resulting in deletion mutations (Figure 1.2) [6].

Although further analysis is necessary to elucidate its precise mode of action, ion beam mutagenesis appears to be a good alternative that can accomplish high mutational effects and a mutation spectrum presumably different from other mutagens such as EMS and low-LET radiation. To date, ion beam mutagenesis has

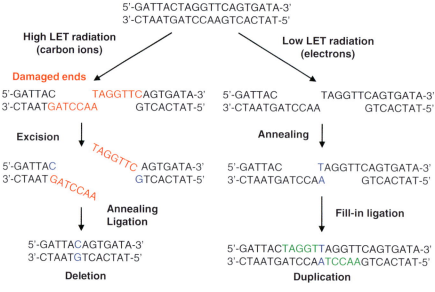

Figure 1.2 Model of mechanisms by which high-LET and low-LET radiation induce mutations (originally proposed by N. Shikazono). High-LET radiation such as carbon ions produce damaged ends of DSBs, which are excised before annealing and ligation of the broken fragments. On the other hand, low-LET radiation such as electrons cause intact ends, which are repaired without any removal of the end sequences. This difference in DSB repair leads to deletions and duplications generated by high-LET and low-LET radiation, respectively. Red letters: bases to be excised; blue letters: bases used for religation; green letters: bases filled in during DSB repair.

been applied to a wide variety of plant species, including *Arabidopsis thaliana*, *Lotus japonicus*, carnations, chrysanthemums, and so on. It is noteworthy that this approach has been successful in the isolation of novel mutants, making a great contribution to plant genetics and breeding (see Section 1.3).

1.2
Methods and Protocols

Currently, there are four facilities available for plant ion beam mutagenesis: Takasaki Ion Accelerators for Advanced Radiation Application (TIARA) of the Japan Atomic Energy Agency (JAEA), RIKEN RI Beam Factory (RIBF), the Wakasa Wan Energy Research Center Multi-purpose Accelerator with Synchrotron and Tandem (W-MAST), and the Heavy Ion Medical Accelerator in Chiba (HIMAC) of National Institute of Radiological Sciences (NIRS). Table 1.2 shows the physical properties of

Table 1.2 Ion beam irradiation facilities and the physical properties of the radiations [modified from the list in The Ion Beam Breeding Society web site (http://wwwsoc.nii.ac.jp/ibbs/)].

Facility	Radiation	Energy (MeV/u)	LET (keV/μm)	Range (mm)
TIARA, JAEA (http://www.taka.jaea.go.jp/index_e.html)	He	12.5	19	1.6
	He	25.0	9	6.2
	C	18.3	122	1.1
	C	26.7	86	2.2
	Ne	17.5	441	0.6
RIBF, RIKEN Nishina Center (http://www.rarf.riken.go.jp/Eng/index.html)	C	135	23	43
	N	135	31	37
	Ne	135	62	26
	Ar	95	280	9
	Fe	90	624	6
W-MAST, The Wakasa Wan Energy Research Center (http://www.werc.or.jp/english/index.htm)	H	200	0.5	256
	C	41.7	52	5.3
HIMAC, National Institute of Radiological Sciences (http://www.nirs.go.jp/ENG/index.html)	C	290	13	163
	Ne	400	30	165
	Si	490	54	163
	Ar	500	89	145
	Fe	500	185	97

Listed are representative ion radiations that have been used in each facility. The energy, LET, and effective range for each ion species are shown.

the ion beams frequently used in these facilities. Here, we describe the protocol of ion beam irradiation in TIARA, which was originally described elsewhere [3, 8].

1.2.1
Ion Beam Irradiation

In general, a variety of ion species, from protons to uranium ions, can be utilized for ion beam applications. In the case of carbon ions, they are produced by an electron cyclotron resonance ion source and accelerated by an azimuthally varying field (AVF) cyclotron to obtain 18.3 MeV/u $^{12}C^{5+}$ ions. At the target surface, the energy of the carbon ions slightly decreases to 17.4 MeV/u, resulting in the estimated 122 keV/μm mean LET in the target material (0.25 mm thick) as water equivalent. In this case, the effective range of the carbon ions is about 1.1 mm. These physical properties can be predicted by the ELOSSM code program [8]. ELOSSM requires the elemental composition and density of the specified substance to determine the potential LET of ion beams. As shown in Figure 1.3, ion beams scan a field of more than 60×60 mm^2 in a vacuum chamber and exit it through a 30-μm titanium foil in the beam window. The samples to be irradiated are placed in the air at a distance of 10 cm below the beam window. In the case of *Arabidopsis* or tobacco seeds, for example, 100–3000 seeds are sandwiched between two Kapton films (7.5 μm in thickness; Toray-Dupont) to make a monolayer of seeds for homogeneous irradiation. As for rice or barley seeds, the embryo sides should be kept facing toward the beam window. On the other hand, when calli or explants cultured in a Petri dish need to be irradiated, the lid of the Petri dish should be replaced by a Kapton film cover to

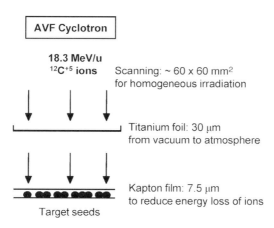

Figure 1.3 Schematic diagram of ion beam irradiation. Ion beams such as carbon ions accelerated by the AVF cyclotron first scan the irradiation field (greater than 60×60 mm^2) in a vacuum chamber. Then, the accelerated ion beams exit through a titanium foil into the atmospheric conditions. Finally, the ion particles attack thinly prepared target samples. Here, plant seeds kept between two Kapton films are shown as an example of target biological materials.

minimize the energy loss of ion beams. The target samples are irradiated for less than 3 min for any dose.

1.2.2
Dose Determination for Ion Beam Irradiation

Determining an optimal irradiation dose of ion beams is the most important and laborious step before irradiating your samples. In principal, the ideal irradiation dose would be a dose at which ion beams show the highest mutation rate at any loci of interest; therefore, you might want to figure out your own favorite irradiation doses by testing different doses at a time and screening all of the resulting samples for your desired mutants. However, such an approach is not practical because plenty of time and effort need to be taken. Alternatively, survival rate, growth rate, chlorophyll mutation, and so on, can be the good indicators to determine appropriate doses for mutation induction.

Figure 1.4 shows the survival curves of *Arabidopsis* dry seeds against several ion beams in comparison with low-LET electrons. The effect of ion beams on the survival rate is higher than that of electrons, but it varies by energy and species of ions. Until now, 18.3 MeV/u carbon ions have been widely used, leading to high mutation rates and efficient novel mutant isolations. However, it has not been fully understood which kind of ions with how high energy would be the most effective for mutation induction. Supposedly, the optimal ion radiation might depend on plant species and materials as well as genome size, ploidy, water content, and also what kind of mutation a researcher wants to produce. Based on several results up to date, it has been suggested that the effectiveness of ion beams as a mutagen might not be

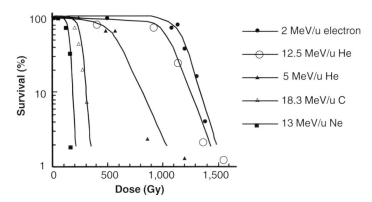

Figure 1.4 Survival curves of *Arabidopsis* dry seeds after irradiation of ion beams (modified from [3]). Dry seeds of the Columbia ecotype of *Arabidopsis* were irradiated with different kinds of ion beams as well as electrons for a low-LET radiation control. Survival responses are shown as a function of irradiation dose. A dose at the shoulder end of each survival curve (e.g., 200 Gy for carbon ions) or less than this dose is supposed to be the most efficient for mutation induction.

determined by the species of ions, but mostly by the LET of ions. So far, ion beams with LET of around 10–500 keV/u appear to be suitable.

As for doses, the median lethal dose (i.e., LD$_{50}$) has been thought to be the best dose for mutation induction using X-ray or γ-ray irradiation. Recent studies have shown that the dose at the shoulder end of the survival curves (200 and 1000 Gy for carbon ions and electrons, respectively, in Figure 1.4) or less than these doses is more efficient for ion beams as well as low-LET radiation (unpublished data). In fact, we are currently using 150 Gy with 18.3 MeV/u carbon ions for *Arabidopsis* dry seeds. In the case of plantlets, we usually irradiate ion beams at such doses that show 100–80% growth rate (around the shoulder end of the growth curve). Also, when tissue culture is concerned, we favor doses that lead to more than around 80% regeneration or growth rate of calli compared to unirradiated controls.

1.2.3
Plant Radiation Sensitivity

In order to determine irradiation doses, it is very useful to understand general radiation sensitivities of plants against radiation. Radiation sensitivities of plants differ greatly among not only plant species, but also plant materials (seeds, plantlets, tissues, etc.). Table 1.3 shows a comparison of the D$_{50}$s of representative plant materials. Basically, the radiation sensitivity of living cells depends on the genome size (i.e., the nuclear contents per cell). With increasing genome size of plant species,

Table 1.3 Effective irradiation dose on plant materials.

Plant material	Radiation		
	18.3 MeV/u C	12.5 MeV/u He	Low-LET radiations
(a) Dry seeds (genome size)			
Arabidopsis (130 Mb)	300	1100	1200 (electrons)
Rice (430 Mb)	40–50	200	350
Tomato (950 Mb)	70	240	—
Barley (4.8 Gb)	10–20	—	—
Wheat (16 Gb)	25	—	—
(b) Tissue culture			
Chrysanthemum var. Taihei	15	10–20	∼60–80
Chrysanthemum var. Jimba	3	2–3	∼10
Carnation	15	40	60

Listed are D$_{50}$s (Gy), the irradiation doses that lead to 50% lethality (a. dry seeds) or growth rate (b. tissue culture). D$_{50}$ is a good indicator to know general sensitivity of plants against radiations. Here, carbon and helium ions with the indicated energy were used for high-LET radiation. For low-LET radiation comparison, γ-rays were used, unless otherwise noted. Note that D$_{50}$ decreases as genome size increases (a) (i.e., plant species with larger genomes are more sensitive to radiation). In addition, even in the same species, D$_{50}$ varies among different varieties (b). Data were extracted from experiments performed at TIARA, the electron beam facilities in JAEA, and the γ-ray irradiation facilities in Institute of Radiation Breeding (unpublished data).

the sensitivity against radiation increases. Occasionally, radiation sensitivities vary significantly even among different varieties of the same plant species. In the case of "Jimba," which is a major variety of chrysanthemum in Japan, its sensitivity is more than 5 times higher than that of a variety "Taihei," of which the sensitivity is considered as a standard level in chrysanthemum. Radiation sensitivities also differ among plant organs. This difference is thought to be due to DNA content, water content, and so on. Cells in S phase of the cell cycle are the most sensitive to radiation because in this stage, the DNA content increases and the chromosomal DNA molecules are unpacked, leading to a cell status that is readily attacked by radiation and the secondary radical products. Radicals such as hydroxyl radicals are a major cause of DNA damage. It is well known that these radicals are generated by reactions between water and radiation. Therefore, plant materials such as dry seeds, in which the water content is very low, tend to show high resistance to radiations.

In conclusion, irradiation dose should be carefully determined according to the kinds of ion species and energies, plant species, plant varieties, plant state of materials such as cell cycle, and water content.

1.2.4
Population Size of the M1 Generation

Apparently, it is preferable to prepare as much of the target samples as possible because mutations basically happen at random and therefore under the laws of probability. When the mutation frequency of a particular locus is known, the minimum size of irradiation treatment samples can be roughly estimated. In the case of 18.3 MeV/u carbon ions, the mutation rate at *tt* and *gl* loci is 1.9×10^{-6} per locus per dose [5]. As the irradiation was performed with a dose of 150 Gy, the mutation rate was about 2.85×10^{-4} (roughly 1/3500) per locus, indicating that about 3500 seeds are necessary on average to obtain at least one mutant for a certain locus.

In practice, the minimum population size to isolate one phenotypic mutation (not one gene) is likely to be around 2000–5000 M1 seeds for *Arabidopsis* [9–11], rice, and other crops (unpublished results). However, it is not fully understood how many seeds will be required for plants with different genome sizes, gene numbers, and ploidies. On the other hand, it seems that a smaller population size would be sufficient for mutation induction from explants or tissue cultures. Moreover, several phenotypes, such as flower colors and shapes, chlorophyll mutations, waxes, and so on, have been obtained even in the M1 generation, although the mutation mechanisms are still unclear [12–15].

1.3
Applications

Considering its high mutation rate and its mutation spectrum that potentially differs from other chemical and physical mutagens, ion beam mutagenesis can be a powerful and useful technique to induce novel mutants. In fact, ion beam mutagenesis has been employed in many plant species and several novel mutants have been produced. Identification of such novel mutants will bring about a better understanding of any biological process of interest, and also a dramatic improvement in agriculture and horticulture. Here, we describe the effectiveness of ion beams by citing recent studies using ion beam radiation.

1.3.1
Ion Beams for Forward Genetics

In forward genetics, isolation of mutants is merely the first step, yet it is a very critical procedure that enables us to analyze any relevant gene functions and gain a new insight into any developmental/physiological event. The new technique of ion beam mutagenesis has contributed significantly to plant research in this respect. For example, a novel mutant, *antiauxin resistant1-1* (*aar1-1*), was identified by screening the M2 progeny of carbon-ion-irradiated *Arabidopsis* seeds for plants resistant to *p*-chlorophenoxyisobutyric acid – a chemical that inhibits the auxin signaling pathway [16]. Further characterization of *aar1-1* showed that this mutant exhibits attenuated response specifically to a synthetic auxin 2,4-dichlorophenoxyacetic acid (2,4-D), but not to the native auxin, indole-3-acetic acid (IAA) [16]. This finding is quite surprising because it has been believed that 2,4-D and IAA have similar effects on auxin signaling despite differences in their stability. It was revealed that the *aar1-1* mutation is a 44-kb deletion encompassing eight annotated genes [16]. Among them, a gene encoding a small acidic protein (SMAP1) was shown to be solely responsible for the *aar1* phenotype [16]. Further molecular analysis of SMAP1 is necessary to dissect the previously underestimated 2,4-D-specific auxin signaling pathway.

Ion beam mutagenesis has also been applied to the model legume *Lotus japonicus*. Leguminous plants develop symbiotic root nodules to confine soil bacteria called rhizobia, which provide the host plants with ammonia produced through bacterial nitrogen fixation. Since this organogenesis is energetically expensive, the host plants should tightly regulate the development and number of nodules. For this purpose, legumes have evolved a long-distance signaling pathway that inhibits unfavorable overproduction of nodules. This systemic regulation requires at least a *CLAVATA1*-like receptor kinase gene and the mutations of this gene lead to the hypernodulation phenotype [17–20]. However, the precise molecular mechanism have been unclear, partly due to the absence of any other hypernodulating mutants, in spite of many attempts to isolate such plants from *L. japonicus* using EMS or T-DNA mutagenesis ([18, 21, 22] and N. Suganuma, personal communication). To circumvent this problem, helium ions were utilized as an alternative mutagen and a novel *Lotus* hypernodulating mutant, *klavier* (*klv*), was readily produced [23]. Grafting experiment using *klv* mutants showed that *KLV* is necessary in the shoots rather than in the roots, indicating that *KLV*, together with a *CLV1*-like receptor kinase gene, constitutes a long-distance signaling control of the nodule number control [23]. This successful identification of the *klv* mutant indicates that ion beams can be a relatively efficient mutagen, possibly having a different mutation spectrum from EMS and T-DNA.

1.3.2
Ion Beams for Plant Breeding

The problem of food shortages is one of the most crucial global challenges that we have ever faced. For this concern, production of new crop varieties with beneficial traits such as drought tolerance is important to fulfill a stable food supply. Moreover, industrialization of these induced varieties could have a great economical impact on societies.

Kirin Agribio in collaboration with the JAEA has generated many varieties of ornamental plants including carnations, chrysanthemums, and petunias by utilizing ion beams [12, 24, 25]. In the case of carnations, the parental leaf tissues were irradiated with carbon ions and then the plants were regenerated from them [24]. Using this approach, a great number of flower mutants including unprecedented round-petal carnations were obtained and some of the new varieties have been commercialized as "Ion Series" varieties (Figure 1.5) [12, 25].

1.3.3
Limitations of Ion Beams

We have shown that ion beam mutagenesis has been applied to a wide variety of plant species in many research fields and it has been successful for novel mutant production. The effectiveness of ion beams can be attributed largely to their high-LET characteristics, which lead to high DSB yields, strong mutational effects, and a

Figure 1.5 Carnation varieties codeveloped by Kirin Agribio and the JAEA using ion beams. The flower on the upper-left corner is the parent carnation flower (var. "Vital") and the others are mutant flowers produced by carbon ions. Note that ion beams successfully induced many flower color and shape mutants.

unique mutation spectrum compared to other chemical and physical mutagens. However, some limitations of ion beams also need to be taken into consideration. For example, ion beam-induced mutations are mostly deletions that can cause frameshifts or total gene losses; therefore, ion beams may not be favorable for hypomorphic mutant isolation. In addition, ion beam irradiation results in various kinds of mutations such as small intragenic deletions, large deletions (greater than 100 kb), translocations, inversions, and chromosomal aberrations. Although this broad mutational effect of ion beams is advantageous with respect to novel mutant induction, the unpredictability of the mutation patterns could potentially hinder the subsequent molecular cloning of the relevant genes in some cases.

1.4
Perspectives

A mutagenesis technique – ion beam irradiation – has been exerting a huge impact on plant basic and applied research. Given that only a small fraction of the annotated genes have been analyzed for their functions even in *Arabidopsis*, the presence of such an alternative mutagen will become increasingly important. Further, application of ion beams in plant biotechnology will be more and more valuable to tackle global issues like food and environmental problems. However, some improvements are still necessary to make this mutagen a more reliable tool. For example, at present, the size of deletions generated by ion beams is variable from 1 bp to over 6 Mbp [26]. In this regard, development of techniques that enable us to control the deletion size will provide us with more efficient gene knockout approaches that can delete only a single gene at a time or sometimes tandem-duplicated multiple genes altogether if necessary. To achieve such an improvement, the precise molecular mechanism by which ion beams induce mutations needs to be elucidated.

Acknowledgments

We would like to thank N. Shikazono for sharing his original model of an ion-mediated mutation induction mechanism and M. Okamura for the generous gift of the photograph of the carnation varieties produced by ion beams. We would also like to thank N. Suganuma for providing us with information about his symbiotic mutant screening in *L. japonicus*.

References

1 Blakely, E.A. (1992) Cell inactivation by heavy charged particles. *Radiat. Environ. Biophys.*, **31**, 181–196.
2 Lett, J.T. (1992) Damage to cellular DNA from particulate radiations, the efficacy of its processing and the radiosensitivity of mammalian cells. Emphasis on DNA double strand breaks and chromatin breaks. *Radiat. Environ. Biophys.*, **31**, 257–277.

3 Tanaka, A., Shikazono, N., Yokota, Y., Watanabe, H., and Tano, S. (1997) Effects of heavy ions on the germination and survival of *Arabidopsis thaliana*. *Int. J. Radiat. Biol.*, **72**, 121–127.

4 Hase, Y., Shimono, K., Inoue, M., Tanaka, A., and Watanabe, H. (1999) Biological effects of ion beams in *Nicotiana tabacum* L. *Radiat. Environ. Biophys.*, **38**, 111–115.

5 Shikazono, N., Yokota, Y., Kitamura, S., Suzuki, C., Watanabe, H., Tano, S., and Tanaka, A. (2003) Mutation rate and novel *tt* mutants of *Arabidopsis thaliana* induced by carbon ions. *Genetics*, **163**, 1449–1455.

6 Shikazono, N., Suzuki, C., Kitamura, S., Watanabe, H., Tano, S., and Tanaka, A. (2005) Analysis of mutations induced by carbon ions in *Arabidopsis thaliana*. *J. Exp. Bot.*, **56**, 587–596.

7 Yokota, Y., Yamada, S., Hase, Y., Shikazono, N., Narumi, I., Tanaka, A., and Inoue, M. (2007) Initial yields of DNA double-strand breaks and DNA Fragmentation patterns depend on linear energy transfer in tobacco BY-2 protoplasts irradiated with helium, carbon and neon ions. *Radiat. Res.*, **167**, 94–101.

8 Fukuda, M., Itoh, H., Ohshima, T., Saidoh, M., and Tanaka, A. (1995) New applications of ion beams to material, space, and biological science and engineering, in *Charged Particle and Photon Interactions with Matter: Chemical, Physiochemical, and Biological Consequences with Applications* (eds A. Mozumder and Y. Hatano), Marcel Dekker, New York.

9 Tanaka, A., Sakamoto, A., Ishigaki, Y., Nikaido, O., Sun, G., Hase, Y., Shikazono, N., Tano, S., and Watanabe, H. (2002) An ultraviolet-B-resistant mutant with enhanced DNA repair in *Arabidopsis*. *Plant Physiol.*, **129**, 64–71.

10 Sakamoto, A., Lan, V.T., Hase, Y., Shikazono, N., Matsunaga, T., and Tanaka, A. (2003) Disruption of the *AtREV3* gene causes hypersensitivity to ultraviolet B light and gamma-rays in *Arabidopsis*: implication of the presence of a translation synthesis mechanism in plants. *Plant Cell*, **15**, 2042–2057.

11 Hase, Y., Tanaka, A., Baba, T., and Watanabe, H. (2000) *FRL1* is required for petal and sepal development in *Arabidopsis*. *Plant J.*, **24**, 21–32.

12 Okamura, M., Yasuno, N., Ohtsuka, M., Tanaka, A., Shikazono, N., and Hase, Y. (2003) Wide variety of flower-color and -shape mutants regenerated from leaf cultures irradiated with ion beams. *Nucl. Instrum. Methods B.*, **206**, 574–578.

13 Nagatomi, S., Tanaka, A., Kato, A., Watanabe, H., and Tano, S. (1996) Mutation induction on chrysanthemum plants regenerated from *in vitro* cultured explants irradiated with C ion beam. *TIARA Annu. Rep.*, **5**, 50–52.

14 Ishii, K., Yamada, Y., Hase, Y., Shikazono, N., and Tanaka, A. (2003) RAPD analysis of mutants obtained by ion beam irradiation to Hinoki cypress shoot primordial. *Nucl. Instrum. Methods B*, **206**, 570–573.

15 Takahashi, M., Kohama, S., Kondo, K., Hakata, M., Hase, Y., Shikazono, N., Tanaka, A., and Morikawa, H. (2005) Effects of ion beam irradiation on the regeneration and morphology of *Ficus thunbergii* Maxim. *Plant Biotechnol.*, **22**, 63–67.

16 Rahman, A., Nakasone, A., Chhun, T., Ooura, C., Biswas, K.K., Uchimiya, H., Tsurumi, S., Baskin, T.I., Tanaka, A., and Oono, Y. (2006) A small acidic protein 1 (SMAP1) mediates responses of the *Arabidopsis* root to the synthetic auxin 2,4-dichlorophenoxyacetic acid. *Plant J.*, **47**, 788–801.

17 Krusell, L., Madsen, L.H., Sato, S., Aubert, G., Genua, A., Szczyglowski, K., Duc, G., Kaneko, T., Tabata, S., de Bruijn, F., Pajuelo, E., Sandal, N., and Stougaard, J. (2002) Shoot control of root development and nodulation is mediated by a receptor-like kinase. *Nature*, **420**, 422–426.

18 Nishimura, R., Hayashi, M., Wu, G.J., Kouchi, H., Imaizumi-Anraku, H., Murakami, Y., Kawasaki, S., Akao, S., Ohmori, M., Nagasawa, M., Harada, K., and Kawaguchi, M. (2002) *HAR1* mediates systemic regulation of symbiotic organ development. *Nature*, **420**, 426–429.

19 Searle, I.R., Men, A.E., Laniya, T.S., Buzas, D.M., Iturbe-Ormaetxe, I., Carroll, B.J., and Gresshoff, P.M. (2003) Long-distance signaling in nodulation directed by a

CLAVATA1-like receptor kinase. *Science*, **299**, 109–112.

20 Schnabel, E., Journet, E.P., de Carvalho-Niebel, F., Duc, G., and Frugoli, J. (2005) The *Medicago truncatula SUNN* gene encodes a *CLV1*-like leucine-rich repeat receptor kinase that regulates nodule number and root length. *Plant Mol. Biol.*, **58**, 809–822.

21 Schauser, L., Handberg, K., Sandal, N., Stiller, J., Thykjaer, T., Pajuelo, E., Nielsen, A., and Stougaard, J. (1998) Symbiotic mutants deficient in nodule establishment identified after T-DNA transformation of *Lotus japonicus*. *Mol. Gen. Genet.*, **259**, 414–423.

22 Szczyglowski, K., Shaw, R.S., Wopereis, J., Copeland, S., Hamburger, D., Kasiborski, B., Dazzo, F.B., and de Bruijn, F.J. (1998) Nodule organogenesis and symbiotic mutants of the model legume *Lotus japonicus*. *Mol. Plant-Microbe. Interact.*, **11**, 684–697.

23 Oka-Kira, E., Tateno, K., Miura, K., Haga, T., Hayashi, M., Harada, K., Sato, S., Tabata, S., Shikazono, N., Tanaka, A., Watanabe, Y., Fukuhara, I., Nagata, T., and Kawaguchi, M. (2005) *klavier* (*klv*), a novel hypernodulation mutant of *Lotus japonicus* affected in vascular tissue organization and floral induction. *Plant J.*, **44**, 505–515.

24 Okamura, M., Ohtsuka, M., Yasuno, N., Hirosawa, T., Tanaka, A., Shikazono, N., Hase, Y., and Tanase, M. (2001) Mutation generation in carnation plants regenerated from *in vitro* leaf cultures irradiated with ion beams. *JAERI Rev.*, **39**, 52–54.

25 Okamura, M., Tanaka, A., Momose, M., Umemoto, N., Teixeira da Silva, J.A., and Toguri, T. (2006) Advances of mutagenesis in flowers and their industrialization, in *Floriculture, Ornamental and Plant Biotechnology*, vol. I (ed. J.A. Teixeira da Silva), Global Science Books, Isleworth.

26 Naito, K., Kusaba, M., Shikazono, N., Takano, T., Tanaka, A., Tanisaka, T., and Nishimura, M. (2005) Molecular analysis of radiation-induced DNA lesions in *Arabidopsis* genome by the pollen irradiation system. *Genetics*, **169**, 881–889.

2
Ds Transposon Mutant Lines for Saturation Mutagenesis of the *Arabidopsis* genome
Takashi Kuromori and Takashi Hirayama

Abstract

Analysis of genetic mutations is one of the most effective techniques for investigating gene function. We now have methods that allow for mass production of mutant lines and cells created by gene disruption or silencing in model organisms, and great progress is being made in the use of those tools for comprehensive phenotypic analysis. In plants, insertion mutations can be produced using T-DNA or transposons, making it possible to monitor the effects of a defect in a single gene. Through bulk storage of mutations in the form of seeds, which is not an option in animal models, it is now feasible to use insertion mutations to analyze every gene in a model plant genome, especially *Arabidopsis*. This makes *Arabidopsis* useful not only as a model organism for plant research, but also as the only multicellular organism in which it is currently possible to perform "saturation mutagenesis" to create knockout strains for every gene. Transposon-tagged lines are generated as a gene knockout mutant resource for a wide variety of functional genomics studies in model plants. This chapter introduces a mutagenesis method using the transposon *Ac*/*Ds* system.

2.1
Introduction

Insertional mutagenesis is a useful method for constructing mutants to investigate gene function because those mutations are tagged with DNA fragments of known sequence [1]. There are two major types of tagged lines in plant research – T-DNA-tagged lines and transposon-tagged lines [2, 3]. In *Arabidopsis*, T-DNA-tagged lines have become an important resource for studies of gene function, as these lines are readily generated in large numbers [2]. However, it has been reported that T-DNA-tagged lines may contain multiple T-DNA insertions into the *Arabidopsis* genome [4]. In contrast, using a transposon *Activator* (*Ac*)/*Dissociator* (*Ds*) system, it is possible to generate mutants with a high proportion of single-copy transposon insertions. This system

The Handbook of Plant Mutation Screening. Edited by Günter Kahl and Khalid Meksem
Copyright © 2010 WILEY-VCH Verlag GmbH & Co. KGaA, Weinheim
ISBN: 978-3-527-32604-4

requires the production of a large number of mutated lines for genome coverage, but the single insertion site in each line is easily determined, thereby simplifying the production and subsequent genetic analysis of a single gene knockout series [5].

This system has other advantages as well.

(i) The *Ds* insertion can be remobilized in the presence of the transposase under appropriate conditions [5]. The recovery of phenotypic revertants by *Ds* remobilization can confirm whether the gene under study is indeed responsible for the phenotype and can make complementation experiments by transgenes unnecessary [6].

(ii) The generated mutants may also be useful as gene trap or enhancer trap lines using *Ds* transposable elements harboring various types of reporter genes and cis-acting elements [7, 8].

With these advantages, transposon-tagged lines are a powerful resource for molecular genetic analysis. Although T-DNA insertion lines are a major tagging resource in *Arabidopsis*, *Ds* transposon-tagged lines serve to complement T-DNA-tagged lines by providing knockout mutants for saturation mutagenesis.

In the last decade, many research groups have determined the flanking sequences of the insertion sites of abundant independent lines of transposon-tagging resources and have compiled the data in a database for effective use [9–17] (Table 2.1). There are several methods of amplifying the DNA fragment derived from the flanking regions of the tags. Among them, the thermal asymmetric interlaced polymerase chain reaction (thermal asymmetric interlacedTAIL-polymerase chain reactionPCR) method is popular [9, 17–20]. Here, however, the adaptor-ligation method is utilized to amplify the sequence flanking the transposon insertion site. The adaptor-ligation method is fairly stable and independent of the quality of template DNA, while the TAIL-PCR method is prone to the generation of multiple PCR products or amplification failure [10, 21]. Adoption of the adaptor-ligation method allows for use of a high-throughput analysis system to promptly analyze many samples.

Using the methods described here, the transposon *Ac/Ds* system developed by Dr. N.V. Fedoroff and colleagues, and the adaptor-ligation method, we generated a total of 18 000 independent transposon-tagged lines with insertion site information within the genome for each tagged plant line [12–14]. A searchable database based on the insertion site information has been developed (rarge.gsc.riken.jp) [22]. This database, containing more than 10 000 insertional mutants and their flanking sequences, provides a useful resource for not only reverse genetic studies, but also genome-wide comprehensive analyzes of plant gene function.

2.2
Methods and Protocols

Development of the mutant resource occurs in three stages: (i) construction of the *Ds*-transposed lines (Figure 2.1), (ii) high-throughput analysis of flanking sequences (Figure 2.2), and (iii) sequence analysis (Figure 2.3).

Table 2.1 Databases for transposon-tagged mutant resources in *Arabidopsis* and rice.

Plant species	Resource name	Institute	Transposon	Numbers mapped to genome[a]	URL
Arabidopsis	CSHL	Cold Spring Harbor Laboratory, USA	*Ds*	21661	genetrap.cshl.org
Arabidopsis	EXOTIC	11 laboratories from eight EU countries	*Ds, Spm*	23573	http://www.jic.bbsrc.ac.uk/science/cdb/exotic/index.htm
Arabidopsis	IMA	National University of Singapore, Singapore	*Ds*	822	http://www.arabidopsis.org/abrc/ima.jsp
Arabidopsis	RIKEN	RIKEN, Japan	*Ds*	18566	http://rarge.gsc.riken.jp/dsmutant/index.pl
Rice	TOS17	National Institute of Agrobiological Sciences, Japan	TOS17	17937	http://www.dna.affrc.go.jp/database
Rice	UCD	University of California Davis, USA	*Ds, Spm*	13666	http://www-plb.ucdavis.edu/Labs/sundar/Rice_Genomics.htm
Rice	CSIRO	Commonwealth Scientific and Industrial Research Organisation, Australia	*Ds*	589	http://www.pi.csiro.au/fgrttpub
Rice	GSNU	Gyeongsang National University, South Korea	*Ds*	1046	—
Rice	EU-OSTID	seven laboratories from five EU countries	*Ds*	1301	orygenesdb.cirad.fr

a) Mapped numbers are quoted from the SIGnAL web site of the Salk Institute (http://signal.salk.edu/) as of the end of April 2009.

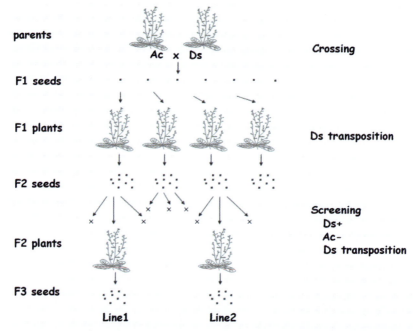

Figure 2.1 Construction of the *Ds*-transposed lines. *Ac* (transposase) parents and *Ds* (transposon) parents are crossed in the parental generation. F1–3 are the generations resulting from the parental cross. In the F1 generation, the transposon would be relocated in the genome. In the F2 generation, plants that are carrying the *Ds* transposon that has moved once from the parental site and are not carrying the transposase gene are selected. F3 seeds are the original stocks of developed lines. The plants in the F3 generation may still be segregating for the transposon – homozygous insertion, heterozygous insertion, or no insertion.

Construction of the *Ds*-Transposed Lines

This section describes how to select *Ds*-transposed plants from parental lines and how to systematically generate transposon-tagged lines (Figure 2.1).

Plant Material

The *Ds* transposon is mobilized by genetic crossing between homozygous *Ds* donor lines and homozygous *Ac* donor lines using, for example, NaeAc-380-16 as a source of the transposase [5, 7]. (*Note:* These donor lines are available from the *Arabidopsis* Biological Resource Center (http://aims.cps.msu.edu/aims) [11].)

Selection of the *Ds*-Transposed Plants

1. Usually, the *Ds* donor lines are the pollen parent and the *Ac* donor lines are the egg parent. The success of the cross is determined by selecting F1 progeny on

MS medium containing hygromycin, resistance to which is conferred by the *aph4* gene on the *Ds* transposon.
2. F2 seeds are obtained by selfing F1 plants. Seeds are harvested separately from each plant.
3. About 100 F2 seeds derived from a single F1 plant are sown on agar-solidified MS medium containing 1% (w/v) sucrose, hygromycin (20 µg/ml), chlorsulfuron (6 µg/ml), and naphthalene acetamide (NAM, 5 µM). (*Note:* Plants in which transposition has occurred but which do not carry the transposase gene should be identified. NAM is a negative selectable compound for the agrobacterial indole acetamide hydrolase gene (*tms2*) that is located near the *Ac* transposase gene [5]. Chlorsulfuron is a selectable marker for *Ds* relocated from parental sites [5].)
4. After imbibition at 4 °C for 1 week to induce germination, plants are incubated for 3 weeks under a 16-h light/8-h dark cycle at 22 °C.
5. One triple-resistant F2 plant derived from a single F1 plant is transferred to soil and grown under the same conditions. (*Note:* It is possible that F2 plants derived from a single F1 plant may have different *Ds* insertions dependent on the timing and the tissue of the transposition. Therefore, several F2 plants can be selected for the next step, if there is enough space. The frequency of different *Ds* insertions might be varied with the parental lines. A pilot test to determine the frequency of different insertions would be worthy to determine the number of lines to be collected.)

Seed Sterilization by Chlorine Gas

Do these steps in a ventilated safety hood, as the chlorine gas produced is toxic.

1. Put the seed samples in a hermetic chamber placed in the hood.
2. >The gas is produced by mixing 200 ml chlorine bleach and 8 ml HCl in a 500-ml beaker in the chamber.
3. Expose the seed samples in the chamber for 3 h.
4. Open the chamber and let the samples air-ventilate in a clean cabinet for more than 10 h.

Note: This method is useful for the sterilization of many seed samples at a time. However, chlorine gas is very toxic. We recommend other methods for seed sterilization unless safety concerns can be met.

High-Throughput Analysis of Flanking Sequences

This section describes a high-throughput scheme for determining the transposon insertion site within the genome for each tagged line. The procedure consists of three steps: (i) genomic DNA purification, (ii) restriction enzyme-mediated adaptor ligation, and (iii) PCR amplification of the *Ds*/genomic DNA junction (Figure 2.2). The Promega MagneSil® DNA purification kit was used for purification of genomic DNA, but other similar kits can be used as well.

Materials

One rosette leaf from
each 3-week-old plant was
sampled using 96-well plates.

↓

DNA extraction

Release DNA by homogenization.
Purify DNA using extraction kits.

↓

Amplification of the franking sequences

Restriction enzyme treatment
Adapter ligation
1st PCR
2nd PCR
PEG precipitation
Sequencing reaction

A liquid-handling
robot can be used
for semi-
automated
performance.

96-well plates are used
for high-throughput
analysis of massive
amounts of routinely
collected samples.

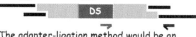

The adapter-ligation method would be an
effective way to amplify fragments near the
insertion sites.

Figure 2.2 High-throughput analysis of flanking sequences. A dark square box represents Ds elements (DS) inserted in the genome, light areas represent the restriction enzymes-treated genomic DNA fragment split by the Ds insertion, solid lines represent adaptors ligated to the genomic DNA fragment, and dark brown and black arrows represent Ds-specific primers and adaptor-specific primers, respectively.

Genomic DNA Purification

1. Selected plants of the F2 generation are grown in soil.
2. One rosette leaf per plant is placed in a well of a deep 96-well plate (Greiner Bio-one).
3. Approximately five ceramic balls (diameter 2.3 mm) and 180 µl Buffer A (MagneSil) are placed into each well of the plate after sampling.
4. The sample plates are shaken for 2 min using a 96-well-plate shaker such as the Shake Master (Bio Medical Science) to crush the leaf samples.
5. Sample plates are centrifuged briefly with a swing-out rotor to spin down the samples.
6. 90 µl Buffer B (MagneSil) is added to each well and mixed for 1 min with a vortex machine (e.g., Titec MicroMixer E-36).
7. Sample plates are centrifuged at $1970 \times g$ (4000 rpm) for 20 min at room temperature with a swing-out rotor (e.g., Kubota 5400).
8. After centrifugation, the supernatant fluid is transferred to a new plate.
9. Genomic DNA is isolated with a MagneSil kit according to the manufacturer's instructions. All further procedures up through sample sequencing can be performed by a liquid-handling robot, such as a Biomek 2000 (Beckman Coulter) (Figure 2.2).

Restriction Enzyme-Mediated Ligation of an Adaptor to the Genomic DNA

The adaptor-ligation method has been described in detail by O'Malley et al. [23].

1. An adaptor is made by mixing 100 µl each of high-performance liquid chromatography-purified 20 µM oligonucleotides, Adaptor Long, Adaptor Short1, Adaptor Short2, and Adaptor Short3 (Table 2.2). Three short oligonucleotides (indicated by "-NH2") need 3'-terminal C7 amino modification to prevent DNA synthesis from the 3' end [23].
2. Incubate the mixture for 5 min at 85 °C in a thermal block, cool slowly from 85 °C to less than 30 °C from 3 h to overnight. Do not freeze and thaw more than 2 times.
3. Digest 10 µl of the 50 µl of extracted genomic DNA from the plant samples with three restriction enzymes, BamHI/EcoRI/HindIII (1.5–10 U each), in a 30-µl reaction mixture using NEB buffer number 4 (or TaKaRa buffer K) for 2 h at 37 °C in a 96-well multititer plate.
4. Add the following materials to each well: 5 µl of the ligation mix, 0.5 µl NEB buffer number 4 (or TaKaRa buffer K), 0.5 µl of 10 mM ATP, 0.125 µl of ligase (0.5 U; Toyobo), 1 µl of adaptor, and 2.875 µl of water. Incubate at room temperature for 1.5 h. The appropriate adaptor to be ligated in this step depends on the restriction enzyme used.

Note: For selecting the restriction enzymes, two issues should be taken into account: (i) the restriction enzymes must not have any cut sites between the *Ds*-specific primer binding sites and the *Ds* junction with the genomic DNA, and (ii) the number of restriction enzyme cut sites and the product length

Table 2.2 Oligonucleotides used in the high-throughput analysis of flanking sequences.

Adaptor sources	
Long	5'-GTAATACGACTCACTATAGGGCACGCGTGGTCGACGGCCCGGGCTGGT-3'
Short1 for *Bam*HI (*Bgl*II in Option)	5'-GATCACCAGCCC-NH2-3'[a]
Short2 for *Eco*RI	5'-AATTACCAGCCC-NH2-3'[a]
Short3 for *Hin*dIII	5'-AGCTACCAGCCC-NH2-3'[a]
Short4 for *Sal*I/*Xho*I	5'-TCGAACCAGCCC-NH2-3'[a]
Sequencing primers	
AP1	5'-GTAATACGACTCACTATAGGGC-3'
AP2"	5'-ATAGGGCACGCGTGGTCGA-3'
Ds-specific primers	
Ds5-1a	5'-ACGGGATCCCGGTGAAACGGT-3'
Ds5-2a	5'-TCCGTTCCGTTTTCGTTTTTTAC-3'
Ds3-1a	5'-CTTCTTATGTTAGCCAAGAGC-3'
Ds3-2a	5'-CCGGATCGTATCGGTTTTCG-3'
Sequencing primers	
Ds5-3	5'-TACCTCGGGTTCGAAATCGAT-3'
Ds5-4	5'-CCGTCCCGCAAGTTAAATATG-3'

a) "-NH2" corresponds to 3'-terminal C7 amino modification [21].

should be considered. Theoretically, using three different restriction enzymes that each recognizes a 6-bp sequence should result in an average of one cut site for every 1300 bp, which is suitable for PCR amplification [23]. However, the frequency of cut sites differs among enzymes and genomes.

Option: A *BglII/SalI/XhoI* (1.5–10 U each) triple digestion can be used when good results cannot be obtained with *BamHI/EcoRI/HindIII* (1.5–10 U each). In this case, an adaptor is made by mixing 100 μl of 20 μM Adaptor Long and Adaptor Short1, and 200 μl of 20 μM Adaptor Short4 (Table 2.2). The reaction buffer is NEB buffer number 3 (or TaKaRa buffer H).

PCR Amplification of the *Ds*/genomic DNA Junction with Primers Specific to the Adaptor and *Ds*

1. PCR is performed twice using different primers and cycling conditions. The primer pairs for the first PCR are Ds5-1a/AP1 and those for the second are Ds5–2a/AP2″. Another flanking sequence from the other end of *Ds* can also be amplified using the primer pair Ds3-1a/AP1 for the first PCR and Ds3–2a/AP2″ for the second (Table 2.2 and Figure 2.4).
2. For the first PCR reaction use 1 μl of the ligation sample in a total volume of 30 μl with 3 μl of 10 × PCR buffer, 3 μl of 10 × dNTPs (2 mM each), 1 μl of 20 μM Ds5-1a (or Ds3-1a) primer, 1 μl of 20 μM AP1 primer, 0.1 μl of *Taq* polymerase, and 20.9 μl of water. (*Option*: Solid pin multiblot replicators for 96-well plates are convenient for transferring 1 μl from each well to a new 96-well PCR plate.)

1. BLASTN search **of *Arabidopsis* genome sequences.**
   → **Prediction of the Insertion Site in each line.**

2. **Getting the** Closest CDSs **to the insertion sites.**
   → **List of Coding Regions.**

3. BLASTP search **of the entire protein database.**
   → **Key Word search.**

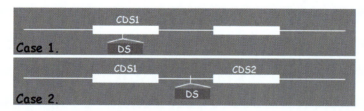

Figure 2.3 Sequence analysis and construction of the database. In scheme, white square boxes represent gene-coding sequences (CDS), white bars represent noncoding regions of the genome, and dark boxes represent Ds elements (*DS*) inserted in the genome

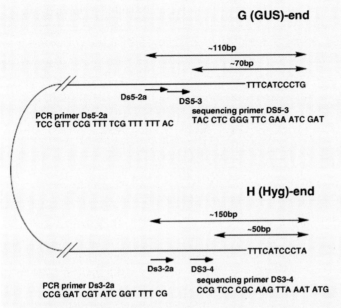

Figure 2.4 Schematic representation of PCR and sequencing primers and the ends of the Ds region. The G-end is on the upstream side of the β-glucuronidase (GUS) reporter gene and the H-end is on the downstream side of the *aph4* gene that confers hygromycin (Hyg) resistance [5]. TTTCATCCCTG and TTTCATCCCTA are the bordering sequences of the *Ds* transposon at the G- and H-ends, respectively. Arrowheads marked as Ds5-2a, Ds5-3, Ds3-2a, and Ds3-4 show the location of hybridized sites in the *Ds* region for each primer. PCR primers Ds5-1a and Ds3-1a described in the text hybridize to a sequence further inside *Ds* than do the sequences Ds5-2a and Ds3-2a, respectively.

3. The first PCR program is as follows: (i) 7 cycles of 94 °C for 25 s and 72 °C for 3 min; and (ii) 32 cycles of 94 °C for 25 s and 67 °C for 3 min.
4. After the PCR reaction, spin down with a swing-out rotor.
5. Use 1 µl of the first PCR sample in a total volume of 30 µl for the second PCR reaction with 3 µl of 10 × PCR buffer, 3 µl of 10 × dNTPs (2 mM each), 1 µl of 20 µM Ds5-2a (or Ds3-2a) primer, 1 µl of 20 µM AP2″ primer, 0.1 µl of *Taq* polymerase, and 20.9 µl of water.
6. The second PCR program is as follows: (i) 14 cycles of 94 °C for 30 s, 63 °C (lowered by 0.5 °C per cycle) for 30 s, and 72 °C for 3 min; (ii) 25 cycles of 94 °C for 30 s, 56 °C for 30 s, and 72 °C for 3 min; and (iii) 72 °C for 10 min.
7. After the PCR reaction, spin down with a swing-out rotor.
8. Mix 20 µl of the second PCR reaction mixture with 20 µl of polyethylene glycol solution (2 M NaCl, 18% PEG8000).
9. Incubate at room temperature for 30 min.
10. Centrifuge at 1970 × g (4000 rpm) for 20 min at room temperature with a swing-out rotor.
11. Remove the supernatant fluid and add 50 µl of 70% EtOH to each well.

12. Centrifuge at 1970 × g (4000 rpm) for 5 min at room temperature with a swing-out rotor.
13. Remove the supernatant fluid by inverting the plate on a paper towel.
14. Incubate at 37 °C for 15 min to dry the DNA samples.
15. Add 10 µl of water and incubate at room temperature for 30 min to dissolve PCR products.
16. One-pass sequencing is carried out using 4 µl of the concentrated and cleaned second PCR product using Ds5-3 (or Ds3-4 primer) in an ABI PRISM® Dye Terminator Cycle Sequencing Ready Reaction Kit (Perkin-Elmer) [24] (Table 2.2 and Figure 2.4).

Sequence Analysis

This section describes the treatment of the flanking sequence data and the construction of a searchable database (Figure 2.3).

1. Native sequence results should include some of the *Ds* elements and adaptor sequences. These sequences are removed when recognized to make sequence data more useful.
2. Delete plant lines which include an untransposed *Ds* element. A small fraction of plants screened using this system have been reported to have an untransposed *Ds* insertion [11].
3. Cleaned sequences are subjected to a BLASTN search of the National Center for Biotechnology Information (www.ncbi.nlm.nih.gov). The position with the highest similarity score is defined as the insertion site.
4. In the case of the insertion in a coding region, this gene is designated as the tagged gene. In the other case (when *Ds* is inserted into an intergenic region between predicted genes), the two genes on either side are obtained with the distances to the predicted initiation codon (minus digit) or stop codon (plus digit) of these genes (Figure 2.3).

2.3
Applications

An adaptor-ligation method was used instead of conventional TAIL-PCR to amplify fragments near the *Ds* transposon in order to determine the insertion sites. TAIL-PCR requires a higher quality of DNA sample and precise manipulation of the cycling conditions to obtain a consistent PCR product. In addition, this method sometimes gives multiple amplified fragments that interfere with the sequence determination [20]. By contrast, the adaptor-ligation method does not demand such delicate experimental procedures and is more suitable for an automated high-throughput system. We succeeded in determining the *Ds* insertion sites of around 18 000 lines by amplifying DNA segments on at least one side of the *Ds* flanking region of each insertion using two sets (one is routine, the other is optional) of restriction enzymes

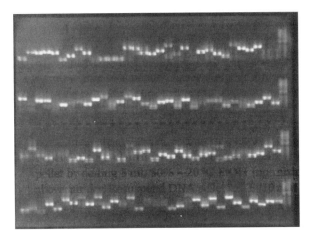

Figure 2.5 Example of PCR products amplified by the adaptor-ligation method. In each lane, 5-μl samples from the second PCR reaction were electrophoresed in an agarose gel. Using high-throughput analysis, DNA extracts from 192 plant lines and 384 samples for both sides of the transposon insertion were treated at one time.

and adaptors (Figure 2.5). We emphasize the importance of the determination of both flanking sequences because it allows identification of experimental errors such as sample contamination or chromosomal rearrangements, including deletion of DNA segments caused by transposon insertions. In addition, the 8-nucleotide tandem repeat of the inserted genome locus, which is a hallmark of a normal *Ds* insertion, can be detected by comparing the sequences of both sides of the transposon insertion sites. Alonso *et al.* [4, 23] used the adaptor-ligation method for analyses of T-DNA tagged lines. As they introduced, one application of this technique is to identify the location of HIV virus integration into the human genome [23, 25].

In our analysis of about 18 000 lines with unique *Ds* insertion positions, the whole genome distribution pattern of the *Ds* insertion sites indicates that about half of the *Ds* elements were transposed within the same chromosomes of the donor sites of the parental lines and the other *Ds* elements were transposed to other chromosomes [13]. Although it has been reported that *Ds* tends to transpose to linked sites, the insertion of the transposon was fairly well distributed in the *Arabidopsis* genome [13]. Additionally, 46% of the transposed *Ds* elements were inserted in known or assumed protein-coding regions including exon and intron, while 30–35% of T-DNA were inserted to these regions [19, 20]. It seems that *Ds*-tagged lines have a higher rate of insertion into coding regions than do T-DNA-tagged lines. Insertional mutant resources are used for both forward and reverse genetic studies. A higher tagged rate of the mutant phenotype, which indicates the rate of correspondence between the insertion mutation and the observed phenotype, is demanded. In a preliminary experiment, the tagged ratio of the mutant phenotype was around 40% in our *Ds*-inserted mutants. It is a little better than or the same as the case of T-DNA resources [26]. These findings suggest that *Ds*-tagged lines would be suitable for phenotypic and gene functional analyses of disrupted genes. This is expected to be one of the most useful resources for functional genomics in plant research.

2.4
Perspectives

As we have shown, transposon insertional mutagenesis has several advantages. Owing to fewer unexpected mutations and a single insertion per line in the various transposon systems, transposon insertion lines appear to be suitable for large-scale or systematic phenotypic analyses, including phenome or metabolome analyses. As a first step in phenome analysis, 4000 homozygous *Arabidopsis Ds* gene disruption lines have already been used [26]. Even relatively weak phenotypes, which cannot be detected by traditional genetic screening, were efficiently characterized using established *Ds* gene disruption lines.

Due to its utility and high efficiency in obtaining insertional mutations, particularly in plant systems with a lower potential for T-DNA insertion mutagenesis due to less efficient agrobacteria-mediated transformation, *Ds* transposon mutagenesis has been used in many plants other than *Arabidopsis*. In tomato, *Brassica*, legumes, soybean, rice, sorghum, barley, aspen, and poplar, *Ac/Ds* transposon systems were used for constructing large insertional mutant lines [27–35]. In addition, other transposon systems such as retroposons have been used for constructing mutant lines in *Arabidopsis*, rice, lettuce, and legumes [36–40]. In plants, there are many transposon systems with a variety of characteristics, including relocation efficiency, target sequence preference, and copy number. Therefore, transposon mutagenesis is a useful method for constructing mutant lines and cultivars in any crop. The method described here will be invaluable for understanding plant genomics and functionality.

References

1 Parinov, S. and Sundaresan, V. (2000) Functional genomics in *Arabidopsis*: large-scale insertional mutagenesis complements the genome sequencing project. *Curr. Opin. Biotechnol.*, **11**, 157–161.
2 Krysan, P.J., Young, J.C., and Sussman, M.R. (1999) T-DNA as an insertional mutagen in *Arabidopsis*. *Plant Cell*, **11**, 2283–2290.
3 Ramachandran, S. and Sundaresan, V. (2001) Transposons as tools for functional genomics. *Plant Physiol. Biochem.*, **39**, 243–252.
4 Alonso, J.M., Stepanova, A.N., Leisse, T.J., Kim, C.J., Chen, H., Shinn, P., Stevenson, D.K., Zimmerman, J., Barajas, P., Cheuk, R., Gadrinab, C., Heller, C., Jeske, A., Koesema, E., Meyers, C.C., Parker, H., Prednis, L., Ansari, Y., Choy, N., Deen, H., Geralt, M., Hazari, N., Hom, E., Karnes, M., Mulholland, C., Ndubaku, R., Schmidt, I., Guzman, P., Aguilar-Henonin, L., Schmid, M., Weigel, D., Carter, D.E., Marchand, T., Risseeuw, E., Brogden, D., Zeko, A., Crosby, W.L., Berry, C.C., and Ecker, J.R. (2003) Genome-wide insertional mutagenesis of *Arabidopsis thaliana*. *Science*, **301**, 653–657.
5 Fedoroff, N.V. and Smith, D.L. (1993) A versatile system for detecting transposition in *Arabidopsis*. *Plant J.*, **3**, 273–289.
6 Wessler, S.R., Baran, G., Varagona, M., and Dellaporta, S.L. (1986) Excision of *Ds*

produces waxy proteins with a range of enzymatic activities. *EMBO J.*, **5**, 2427–2432.

7 Smith, D., Yanai, Y., Liu, Y.G., Ishiguro, S., Okada, K., Shibata, D., Whittier, R.F., and Fedoroff, N.V. (1996) Characterization and mapping of *Ds*-GUS T-DNA lines for targeted insertional mutagenesis. *Plant J.*, **10**, 721–732.

8 Muskett, P.R., Clissold, L. Marocco, A., Springer, P.S., Martienssen, R., and Dean, C. (2003) A resource of mapped dissociation launch pads for targeted insertional mutagenesis in the *Arabidopsis* genome. *Plant Physiol.*, **132**, 506–516.

9 Parinov, S., Sevugan, M., De, Y., Yang, W.C., Kumaran, M., and Sundaresan, V. (1999) Analysis of flanking sequences from *dissociation* insertion lines: a database for reverse genetics in *Arabidopsis*. *Plant Cell*, **11**, 2263–2270.

10 Tissier, A.F., Marillonnet, S., Klimyuk, V., Patel, K., Torres, M.A., Murphy, G., and Jones, J.D. (1999) Multiple independent defective suppressor-mutator transposon insertions in *Arabidopsis*: a tool for functional genomics. *Plant Cell*, **11**, 1841–1852.

11 Raina, S., Mahalingam, R., Chen, F., and Fedoroff, N. (2002) A collection of sequenced and mapped *Ds* transposon insertion sites in *Arabidopsis thaliana*. *Plant Mol. Biol.*, **50**, 93–110.

12 Ito, T., Motohashi, R., Kuromori, T., Mizukado, S., Sakurai, T., Kanahara, H., Seki, M., and Shinozaki, K. (2002) A new resource of locally transposed *Dissociation* elements for screening gene-knockout lines *in silico* on the *Arabidopsis* genome. *Plant Physiol.*, **129**, 1695–1699.

13 Kuromori, T., Hirayama, T., Kiyosue, Y., Takabe, H., Mizukado, S., Sakurai, T., Akiyama, K., Kamiya, A., Ito, T., and Shinozaki, K. (2004) A collection of 11 800 single-copy *Ds* transposon insertion lines in *Arabidopsis*. *Plant J.*, **37**, 897–905.

14 Ito, T., Motohashi, R., Kuromori, T., Noutoshi, Y., Seki, M., Kamiya, A., Mizukado, S., Sakurai, T., and Shinozaki, K. (2005) A resource of 5814 *Dissociation* transposon-tagged and sequence-indexed lines of *Arabidopsis* transposed from start loci on chromosome 5. *Plant Cell Physiol.*, **46**, 1149–1153.

15 Miyao, A., Iwasaki, Y., Kitano, H., Itoh, J., Maekawa, M., Murata, K., Yatou, O., Nagato, Y., and Hirochika, H. (2007) A large-scale collection of phenotypic data describing an insertional mutant population to facilitate functional analysis of rice genes. *Plant Mol. Biol.*, **63**, 625–635.

16 van Enckevort, L.J., Droc, G., Piffanelli, P., Greco, R., Gagneur, C., Weber, C., González, V.M., Cabot, P., Fornara, F., Berri, S., Miro, B., Lan, P., Rafel, M., Capell, T., Puigdomènech, P., Ouwerkerk, P.B., Meijer, A.H., Pe', E., Colombo, L., Christou, P., Guiderdoni, E., and Pereira, A. (2005) EU-OSTID: a collection of transposon insertional mutants for functional genomics in rice. *Plant Mol. Biol.*, **59**, 99–110.

17 Kim, C.M., Piao, H.L., Park, S.J., Chon, N.S., Je, B.I., Sun, B., Park, S.H., Park, J.Y., Lee, E.J., Kim, M.J., Chung, W.S., Lee, K.H., Lee, Y.S., Lee, J.J., Won, Y.J., Yi, G., Nam, M.H., Cha, Y.S., Yun, D.W., Eun, M.Y., and Han, C.D. (2004) Rapid, large-scale generation of *Ds* transposant lines and analysis of the *Ds* insertion sites in rice. *Plant J.*, **39**, 252–263.

18 Liu, Y.G. and Whittier, R.F. (1995) Thermal asymmetric interlaced PCR: automatable amplification and sequencing of insert end fragments from P1 and YAC clones for chromosome walking. *Genomics*, **25**, 674–681.

19 Sessions, A., Burke, E., Presting, G., Aux, G., McElver, J., Patton, D., Dietrich, B., Ho, P., Bacwaden, J., Ko, C., Clarke, J.D., Cotton, D., Bullis, D., Snell, J., Miguel, T., Hutchison, D., Kimmerly, B., Mitzel, T., Katagiri, F., Glazebrook, J., Law, M., and Goff, S.A. (2002) A high-throughput *Arabidopsis* reverse genetics system. *Plant Cell*, **14**, 2985–2994.

20 Szabados, L., Kovacs, I., Oberschall, A., Abrahám, E., Kerekes, I., Zsigmond, L., Nagy, R., Alvarado, M., Krasovskaja, I., Gál, M., Berente, A., Rédei, G.P., Haim, A.B., and Koncz, C. (2002) Distribution of 1000 sequenced T-DNA tags in the *Arabidopsis* genome. *Plant J.*, **32**, 233–242.

21 Yephremov, A. and Saedler, H. (2000) Technical advance: display and isolation of transposon-flanking sequences starting

from genomic DNA or RNA. *Plant J.*, **21**, 495–505.

22 Sakurai, T., Satou, M., Akiyama, K., Iida, K., Seki, M., Kuromori, T., Ito, T., Konagaya, A., Toyoda, T., and Shinozaki, K. (2005) RARGE: a large-scale database of RIKEN *Arabidopsis* resources ranging from transcriptome to phenome. *Nucleic Acids Res.*, **33**, D647–650.

23 O'Malley, R.C., Alonso, J.M., Kim, C.J., Leisse, T.J., and Ecker, J.R. (2007) An adapter ligation-mediated PCR method for high-throughput mapping of T-DNA inserts in the *Arabidopsis* genome. *Nat. Protoc.*, **2**, 2910–2917.

24 Tsugeki, R., Kochieva, E.Z., and Fedoroff, N.V. (1996) A transposon insertion in the *Arabidopsis SSR16* gene causes an embryo defective lethal mutation. *Plant J.*, **10**, 479–489.

25 Schröder, A.R.W., Shinn, P., Chen, H., Berry, C., Ecker, J.R., and Bushman, F. (2002) HIV-1 integration in the human genome favors active genes and local hotspots. *Cell*, **110**, 521–529.

26 Kuromori, T., Wada, T., Kamiya, A., Yuguchi, M., Yokouchi, T., Imura, Y., Takabe, H., Sakurai, T., Akiyama, K., Hirayama, T., and Shinozaki, K. (2006) A trial of phenome analysis using 4000 *Ds*-insertional mutants in gene-coding regions of *Arabidopsis*. *Plant J.*, **47**, 640–651.

27 Meissner, R., Chague, V., Zhu, Q., Emmanuel, E., Elkind, Y., and Levy, A.A. (2000) Technical advance: a high throughput system for transposon tagging and promoter trapping in tomato. *Plant J.*, **22**, 265–274.

28 Palmer, R.G. and Horner, H.T. (2000) Genetics and cytology of a genic male-sterile, female-sterile mutant from a transposon-containing soybean population. *J. Hered.*, **91**, 378–383.

29 McKenzie, N., Wen, L.Y., and Dale, J. (2002) Tissue-culture enhanced transposition of the maize transposable element *Dissociation* in *Brassica oleracea* var. "Italica". *Theor. Appl. Genet.*, **105**, 23–33.

30 Brutnell, T.P. and Conrad, L.J. (2003) Transposon tagging using *Activator* (*Ac*) in maize. *Methods Mol. Biol.*, **236**, 157–176.

31 Kolesnik, T., Szeverenyi, I., Bachmann, D., Kumar, C.S., Jiang, S., Ramamoorthy, R., Cai, M., Ma, Z.G., Sundaresan, V., and Ramachandran, S. (2004) Establishing an efficient *Ac/Ds* tagging system in rice: large-scale analysis of *Ds* flanking sequences. *Plant J.*, **37**, 301–314.

32 Fladung, M., Deutsch, F., Honicka, H., and Kumar, S. (2004) T-DNA and transposon tagging in aspen. *Plant Biol.*, **6**, 5–11.

33 Singh, J., Zhang, S., Chen, C., Cooper, L., Bregitzer, P., Sturbaum, A., Hayes, P.M., and Lemaux, P.G. (2006) High-frequency *Ds* remobilization over multiple generations in barley facilitates gene tagging in large genome cereals. *Plant Mol. Biol.*, **62**, 937–950.

34 Ayliffe, M.A., Pallotta, M., Langridge, P., and Pryor, A.J. (2007) A barley activation tagging system. *Plant Mol. Biol.*, **64**, 329–347.

35 Qu, S., Desai, A., Wing, R., and Sundaresan, V. (2008) A versatile transposon-based activation tag vector system for functional genomics in cereals and other monocot plants. *Plant Physiol.*, **146**, 189–199.

36 Hirochika, H., Sugimoto, K., Otsuki, Y., Tsugawa, H., and Kanda, M. (1996) Retrotransposons of rice involved in mutations induced by tissue culture. *Proc. Natl. Acad. Sci. USA*, **93**, 7783–7788.

37 Mazier, M., Botton, E., Flamain, F., Bouchet, J.P., Courtial, B., Chupeau, M.C., Chupeau, Y., Maisonneuve, B., and Lucas, H. (2007) Successful gene tagging in lettuce using the *Tnt1* retrotransposon from tobacco. *Plant Physiol.*, **144**, 18–31.

38 Tadege, M., Wen, J., He, J., Tu, H., Kwak, Y., Eschstruth, A., Cayrel, A., Endre, G., Zhao, P.X., Chabaud, M., Ratet, P., and Mysore, K.S. (2008) Large-scale insertional mutagenesis using the *Tnt1* retrotransposon in the model legume *Medicago truncatula*. *Plant J.*, **54**, 335–347.

39 d'Erfurth, I., Cosson, V., Eschstruth, A., Lucas, H., Kondorosi, A., and Ratet, P. (2003) Efficient transposition of the *Tnt1* tobacco retrotransposon in the model legume *Medicago truncatula*. *Plant J.*, **34**, 95–106.

40 Okamoto, H. and Hirochika, H. (2000) Efficient insertion mutagenesis of *Arabidopsis* by tissue culture-induced activation of the tobacco retrotransposon *Tto1*. *Plant J.*, **23**, 291–304.

3
Use of Mutants from T-DNA Insertion Populations Generated by High-Throughput Screening

Ralf Stracke, Gunnar Huep, and Bernd Weisshaar

3.1
Introduction

Upon completion of the genome sequence of the model plant *Arabidopsis thaliana* in the year 2000 [1], research emphasis focused on the assignment of biological function to all individual genes. At this time experimental evidence for the respective function was available for only about 10% of the initially predicted 25 500 genes [1]. Determination of gene function for the remaining 90% has developed in to a major task for researchers in the *Arabidopsis* field. Mutant analysis provides a reliable way to assign gene function [2]. Commonly, in a first step, genes activity is perturbed randomly. This approach, which is used in both forward and reverse genetics, may be followed up by different methods. A broad range of physical, chemical, and biological agents like fast neutrons, γ-radiation, carbon ions, ethylmethane sulfonate (EMS), transposons, or T-DNA can be utilized to create alterations in *Arabidopsis thaliana*'s genome. A major problem in using random mutagenesis is that the induced genetic alterations in the genome of individual plants have to be determined. This requires enormous efforts in such mutants generated by physical or chemical methods. In the case of insertional mutagenesis, based on the insertion of foreign DNA into a gene, researchers take advantage of the fact that the sequence of the artificially induced mutagen (DNA) is usually known and acts as a marker for subsequent identification of the mutation. For this reason insertional mutagenesis is widely used in the course of reverse genetics. Large collections of gene-indexed mutations have been created in the process of generating tools for reverse genetics (reviewed in [3]). In *Arabidopsis*, this involves the use of either transposable elements or T-DNA.

The most commonly used way of insertional mutagenesis relies on the natural ability of the soil-borne bacterium *Agrobacterium tumefaciens* to introduce bacterial DNA into the genomes of dicots like *A. thaliana* [4, 5]. The bacterium uses this ability to change the plants metabolism to its own needs. The T-DNA is naturally part of the so-called tumor-inducing Ti-plasmid. During transformation, the T-DNA, which is flanked by 25-bp imperfect repeats (left border (LB) and right border (RB)), is

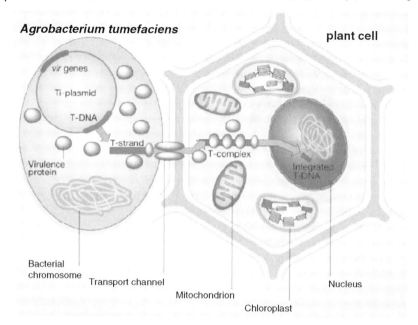

Figure 3.1 Mutagenesis by A. tumefaciens-mediated T-DNA integration. Agrobacterium contains a Ti-plasmid that includes virulence (vir) genes and a T-DNA region. Wounded plant cells produce phenolic defense compounds, which can trigger the expression of the agrobacterium vir genes. The encoded virulence (Vir) proteins process the T-DNA region from the Ti-plasmid, producing a "T-strand." After the bacterium attaches to a plant cell, the T-strand and several types of Vir proteins are transferred to the plant through a transport channel. Inside the plant cell, the Vir proteins interact with the T-strand, forming a T-complex. This complex targets the nucleus, allowing the T-DNA to integrate into the plant genome. (Figure and legend reprinted by permission from Macmillan Publishers Ltd; taken from [9].)

transferred to the plant cell and imported into the nucleus with the help of several virulence proteins (Figure 3.1). Finally, the T-DNA is integrated into the plant genome [6]. Large parts of the Ti-plasmid – the part between LB and RB – can be replaced by genetically engineered sequences without affecting the efficiency of the DNA transfer [7]. By using modified variants of the Ti-plasmid, lacking the parts that change the plants metabolism and containing additional sequences that introduce features of interest into the plant genomes, it is possible to produce transgenic plants with entirely new properties [8]. These properties can be reasoned in the artificially introduced DNA sequence itself and/or on the insertion site of the T-DNA. A disruption of a gene sequence by the T-DNA can lead to a knockout (loss-of-function) or a knockdown of this gene. Due to enhancing elements within the T-DNA, knockon (gain-of-function) of genes close to the insertion site is possible.

Starting in the late 1990s, large transposon and T-DNA mutagenized populations have been built up, allowing the identification of "loss-of-function" mutants of A. thaliana via reverse genetics approaches to unravel gene function [10–17]. These mutagenized populations were analyzed by searching for mutants in DNA pools by polymerase chain reaction (PCR)-based screens using T-DNA border-specific primers in combination with gene-specific primers (GSPs). Genomic DNA from up to 2000 mutant plants was pooled and analyzed by several rounds of PCR [18–20]. Depending on the population size and the pooling scheme, this method is time-consuming and laborious.

A big improvement was the availability of flanking sequence tag (FST)-based T-DNA populations, for which the sequences of PCR-amplified insertion sites from individual mutants are stored in a database. Users can apply BLAST or keyword searches to easily select lines with knockout alleles according to their special interest [17, 21–24]. Also, the availability of data on the location of the insertion site, such as in an intron or exon, allows the prioritization of candidate lines. Based on the assumption that T-DNA insertion into the *A. thaliana* genome is random, it is estimated that in 180 000 independent T-DNA lines a single specific knockout allele of a 2.1-kb gene can be detected with 95% probability [25].

An advantage of using T-DNAs as the insertional mutagen, contrary to transposons [12, 26], is that T-DNA insertions will not transpose subsequent to integration within the genome, and are therefore chemically and physically stable through multiple generations [25]. In the T-DNA insertion process the site of integration within the genome is neither targeted nor sequence-specific. Nevertheless, analyses of the integration sites of thousands of plant lines showed that T-DNA insertions are not randomly distributed in the genome. The integration sites are found preferentially in intergenic regions compared to genic regions, and the integration events seem to be associated with gene density since higher frequencies of insertions were observed in gene-rich regions and lower frequencies around centromeric regions that contain fewer genes [19, 27].

As the generation of large populations of transgenic plants for large-scale insertional mutagenesis approaches is expensive and time-consuming, many consortia have formed to provide the scientific community with such material. There are several *A. thaliana* T-DNA insertion collections publicly available that are easily searchable via the FST database, which offer predominantly loss-of-function mutants to the scientific community. The four biggest publicly accessible *A. thaliana* T-DNA insertion line collections (Table 3.1) are SALK (Salk Institute T-DNA insertion lines [23]), GABI-Kat (Genomanalyse im biologischen System Pflanze-Kölner *Arabidopsis* T-DNA lines [24]), SAIL (Syngenta *Arabidopsis* Insertion Library [17]) and FLAGdb^{++} (FunctionaL Analysis of the *Arabidopsis* Genome Database, also known as INRA/Versaille lines [22]).

Table 3.1 The four most used *A. thaliana* T-DNA insertion collections (in alphabetical order).

Collection	URL
GABI-Kat	www.gabi-kat.de
FLAGdb^{++}	http://urgv.evry.inra.fr/projects/FLAGdb++/HTML/index.shtml
SAIL	http://www.arabidopsis.info/CollectionInfo?id=47
SALK	http://signal.salk.edu/tabout.html

With the exception of FLAGdb^{++} lines, which can only be ordered through the Institute National de la Recherche Agronomique (INRA) directly, T-DNA insertion lines can usually be ordered from seed stock centers, namely the *Arabidopsis* Biological Resource Center (http://www.biosci.ohio-state.edu/~plantbio/Facilities/abrc/index.html) and the Nottingham *Arabidopsis* Stock Centre (arabidopsis.info). Here, researchers can order seeds of unconfirmed and confirmed SALK and SAIL lines as well as from confirmed GABI-Kat lines. Unconfirmed GABI-Kat lines can be ordered at GABI-Kat itself and are sent out after successful confirmation.

The assignment of a T-DNA insertion to a gene of interest in a specific line of one of the available collections is initially based on the comparison of the annotated *A. thaliana* genome and the FST. To confirm the assumed insertion locus, a PCR with one GSP and one T-DNA-specific primer on genomic DNA of plants from the respective line is usually carried out. If the sequence of the resulting PCR product fits the original FST, the line is considered as confirmed.

The next sections focus on the generation and the use of *A. thaliana* mutants from T-DNA insertion populations, and discuss certain aspects in more detail to consider why working with T-DNA insertion mutants is a feasible way to unravel gene functions.

3.2
Methods and Protocols

In this section the GABI-Kat T-DNA insertion collection, which is similar to the other large T-DNA insertion collections mainly used as a resource for knockout lines, serves as a showcase. For simplicity reasons, the description focuses on the part of the population transformed with the binary transformation vector pAC161.

3.2.1
Plant Material and Growth Conditions

A. thaliana plants (accession Columbia-0) are grown for 4 weeks under short-day (8-h light/16-h dark) conditions and then transferred to long-day (16-h light/8-h dark) conditions. Temperature is regulated during the light period to $22 \pm 2\,°C$ and in the dark phase to $20 \pm 2\,°C$ with a relative humidity of 55–60% [28]. Plants are fertilized with liquid nitrogen phosphate potassium.

3.2.2
Plasmid Design

The binary transformation vector pAC161 (Figure 3.2; GenBank accession number AJ537514), originally designed for activation tagging, is used to generate the T-DNA population. The vector contains the *SUL*r open reading frame (ORF) for resistance against the herbicide sulfadiazine (4-amino-*N*-2-pyrimidinylbenzenesulfonamide; Sigma-Aldrich), allowing selection of transformed plants in the greenhouse [29].

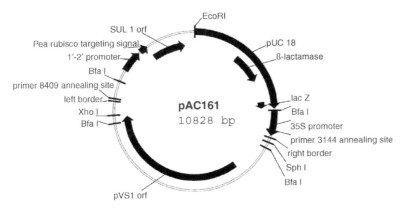

Figure 3.2 Plasmid map of the binary vector pAC161 (accession number AJ537514) containing the 5799-bp T-DNA used in the generation of the majority of GABI-Kat lines. pAC161 contains the *SULr* ORF driven by the 1'-2' promoter. The 35S CaMV promoter located at the RB can act as an activation tagging element after T-DNA integration. Structural parts and relevant restriction sites of the vector are marked. (From [24].)

pAC161 contains a full-length 35S cauliflower mosaic virus (CaMV) promoter located at the T-DNA RB and the 1'-2' promoter [30] driving the expression of the *SULr* ORF [31].

3.2.3
Agrobacterium Culture

For plant infiltration, an *A. tumefaciens* standard strain like GV3101 [32] carrying the pAC161 plasmid is used. *A. tumefaciens* suspension cultures are inoculated with a single colony and grown for 2 days at 28 °C and 220–250 rpm in 5 ml YEB liquid culture medium (1 g/l yeast extract, 5 g/l beef extract, 5 g/l tryptone/peptone, 5 g/l sucrose, 2 mM $MgSO_4$) containing the antibiotics carbenicillin (100 μg/ml), rifampicin (50 μg/ml), and gentamycin (40 μg/ml). The entire preculture is used to inoculate 500 ml fresh medium containing 100 μg/ml carbenicillin and 100 μg/ml rifampicin, which is incubated overnight with rapid shaking.

3.2.4
Plant Transformation and T1 Seed Harvesting

Five to six plants are grown in 8 cm × 8 cm pots and infiltrated according to Clough and Bent [33] when the first flower buds start to open 5–8 days after clipping of the main inflorescence. Prior to dipping, plants were intensely watered to reduce uptake of *A. tumefaciens* suspension into soil. The flowerbuds are submerged for 5–10 s into 500 ml *A. tumefaciens* overnight culture mixed 1 : 1 (v/v) with 500 ml 10% sucrose solution containing 0.02% Silwet L-77 (Lehle Seeds). T1 seeds from these transformed T0 plants are bulk-harvested at maturity. T1 seeds are stored in paper bags at 5 °C and 15% relative humidity.

3.2.5
Sulfadiazine Selection of Transgenic T_1 Plants

Aliquots of 300 mg T1 seed material representing about 15 000 seeds are spread on a (22 cm × 32 cm) prewashed rockwool "grodan mats" (Grodania) for weekly selection of transgenic sulfadiazine-resistant plants. Seeds are mixed with quartz sand 1 : 6 (v/v) and uniformly spread with a salt shaker onto the surface of grodan mats prewetted with water containing 0.15% liquid fertilizer (WUXAL Super) and 7.5 mg/l sulfadiazine. Covered trays are stored for 5 days at 4 °C in a dark room to synchronize seed germination before transfer into the greenhouse. T1 transformants are identified as sulfadiazine-resistant seedlings that produce clearly visible green secondary leaves and develop well, whereas sensitive wild-type plants bleach and die. Dependent on the developmental stage, 14- to 21-day-old plantlets are individually transplanted with forceps on trays in a 96-well grid. Plantlets are covered for another 5–6 days for root regeneration and are finally grown to maturity under long-day (16-h light/8-h dark) conditions.

3.2.6
DNA Preparation from Sulfadiazine-Selected T1 Plants

The term T1 is used for the seeds and plants that grow out of the embryos in these seeds, which are set by the infiltrated plants (T0 plants) (Figure 3.3). T1 plants are hemizygous for the T-DNA insertion and contain the selectable marker.

Preparation of genomic DNA from 300–500 mg leaf material is adapted to the use of 96-well polypropylene 2.2-ml deep-well blocks (Ritter) and an automated 96-channel pipettor (Biomek FX; Beckman). Blocks containing the leaf material of individual T1 plants are stored at −80 °C until DNA preparation. Two stainless steel beads (3.5 mm diameter) per well are added to the frozen leaf material. After cooling on dry ice, leaf material is homogenized by shaking the blocks for 30–60 s at 30 Hz in a homogenizer (Retsch 300 Matrix Mill; Qiagen). To prevent cross-contamination of homogenized leaf powder adhering to the lid, blocks are afterwards centrifuged for 2 min at 6000 × g in a Sigma 4K15 centrifuge precooled to 4 °C. Thawing of the leaf material should be avoided by cooling the blocks on dry ice between the single work steps.

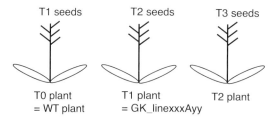

Figure 3.3 Nomenclature for seeds and plants. T0 plants are the plants being infiltrated. The nomenclature is consistent with the one provided by Feldmann [34].

Genomic DNA from rosette leaf material is prepared with a modified cetyltrimethylammonium (CTAB)-DNA preparation protocol [35, 36] where chloroform is replaced by dichloromethane (CH_2Cl_2). An aliquot of 0.6 ml prewarmed (70 °C) CTAB buffer-1 (2% CTAB, 0.1 M Tris–HCl, pH 8.0, 20 mM EDTA, 1.4 M NaCl, 0.5% polyvinylpyrrolidone (PVP); before usage add 1/100 volumes from 1 M dithiothreitol stock solution) is added in each deep well. An incubation for 10 min at 65 °C in a water bath with occasional shaking leads to inactivation of DNase. After recooling to room temperature, 0.3 ml CH_2Cl_2 is added with a handheld multichannel pipettor and the samples are mixed gently. The deep-well blocks are centrifuged for 30–45 min at 6000 × g. An aliquot of 1.2 ml CTAB buffer-2 (1% CTAB, 50 mM Tris–HCl, pH 8.0, 10 mM EDTA, 0.25% PVP) is put in each well of a new deep-well block and the upper aqueous phase (600 µl) from the first block is added. After careful closure of the blocks with sealing mats, DNA is precipitated in 1 h (or overnight) at room temperature. Afterwards the block is centrifuged at room temperature for 45–60 min at 6000 × g. The supernatant is discarded and the pellets are redissolved in 0.3 ml 1 M NaCl. After addition of 0.3 ml isopropanol, the block is gently swung to precipitate the DNA again. DNA is gathered at the bottom of the deep-well block by centrifugation for 45–60 min at 6000 × g. After the supernatant has been discarded, the DNA pellets are washed with 150 µl 70% ethanol. Finally, DNA is dried for about 2–4 h at room temperature and redissolved in 60 µl TE/RNase (10 mM Tris–HCl, pH 8.0, 0.1 mM EDTA, 7.5 µg/ml) in 10 min at room temperature and stored at −20 °C.

3.2.7
FST Production

The generation of FSTs is based on high-throughput methods described by Devic et al. [37], Spertini et al. [38], Balzergue et al. [21], and Steiner-Lange et al. [39]. The workflow is modified and optimized for the use of the GABI-Kat population [40], allowing mainly the amplification of LB insertion site fragments. The procedure of adapter ligation-based FST production requires two nested amplifications and one cycle-sequencing reaction for each border (LB and RB). DNA amplifications are carried out in 96-well Thermo-Fast plates (ABgene).

Based on the pAC161 sequence (Figure 3.4), *Bfa*I (5′-C∧TAG-3′) was selected as the restriction enzyme of choice for the generation of DNA fragments spanning the T-DNA to genomic DNA junctions. About 100 ng of genomic DNA per T1 line is digested with 4–5 U *Bfa*I (New England Biolabs) in 20 µl total volume for approximately 5 h at 37 °C. The reaction mixture contains 5 µg RNase A (Roche).

An aliquot of 5–10 µl from the restriction reaction is used for ligation with 12.5 pmol of the asymmetrical adapter T732/APL16OH, which is generated by the annealing of two oligonucleotides, T732 (5′-GAGTAATACGACTCACTATAGGGAG-GATCCCA-3′) and APL16OH (5′-TATGGGATCACATTAA-ddC-3′), at an equimolar concentration of 25 µM. The adapter is ligated in 20 µl total volume, with approximately 3 Weiss U of T4 DNA ligase (New England Biolabs) in 11 h at 14 °C. Subsequently, the ligase is inactivated by incubation at 65 °C for 20 min.

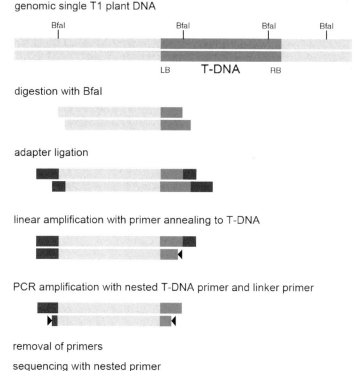

Figure 3.4 Strategy for production of FSTs. The figure gives an overview of how sequences of T-DNA flanking genomic fragments are created. The strategy is shown using for LB adjacent sequences as an example.

DNA extracted from T1 leaves can contain the T-DNA plasmid. To prevent amplification of plasmid-borne *Bfa*I fragments, a postligation restriction enzyme digest (with the 6-bp cutters *Xho*I, 5′-C∧TCGAG-3′ or *Sph*I, 5′-GCATG∧C-3′) is included that splits such fragments (the losses of FSTs containing these enzyme restriction sites is estimated to be moderate). An aliquot of 5 µl of the ligation mixture is digested either with *Xho*I (for the LB) or *Sph*I (for the RB) for 5 h in 15 µl at 37 °C.

The first linear amplification is performed in 55 µl with the whole *Xho*I or *Sph*I digest using primer 8603 (5′-CCCATTTGGACGTGAATGTAGACAC-3′). When postligation digestions are not performed, a 5-µl aliquot of the ligation reaction is directly subjected to the first amplification.

First linear amplification reaction:

10 × PCR buffer, incl. 15 mM MgCl$_2$	5.5 µl
10 mM dNTPs	0.5 µl
10 µM T-DNA LB primer 8603	1 µl
*Xho*I digest reaction	15 µl

(*Continued*)

10 × PCR buffer, incl. 15 mM MgCl₂	5.5 µl
Taq DNA polymerase	0.3 µl
H₂O	32.7 µl
Final volume	55 µl

PCR scheme:

94 °C	2 min
94 °C	30 s
64 °C	1 min
72 °C	1.5 min
30 cycles	
72 °C	3 min
15 °C	∞

To reduce multiband patterns after the second amplification reaction, a special design of adaptor-specific primers is used. The primers adpA, adpC, adpG, and adpT are homologous to the 3′ part of the adapter sequence and extended with the last 3′-end nucleotide being different in all four adapter primers. Two second amplification reactions, either with adpA and adpC or with adpG and adpT, are performed for each line preceding sequencing. An aliquot of 1.5 µl from the first amplification reaction is used as a template for the second PCR amplification in 50 µl volume with a nested T-DNA primer (8474, 5′-ATAATAACGCTGCGGACATCTACATTTT-3′ for LB and 2895, 5′-CAGTCTCAGAAGACCAAAGGGC-3′ for RB) and adapter primers adpA (5′-CTCACTATAGGGAGGATCCCATAGA-3′) and adpC (5′-CTCACTATAGGGAG-GATCCCATAGC-3′) or adpG (5′-CTCACTATAGGGAGGATCCCATAGG-3′) and adpT (5′-CTCACTATAGGGAGGATCCCATAGT-3′) at 0.2 mM each.

Second amplification reaction 1:

10 × PCR buffer, incl. 15 mM MgCl₂	5 µl
10 mM dNTPs	0.5 µl
10 µM T-DNA LB primer 8474	1 µl
10 µM adpA	1 µl
10 µM adpC	1 µl
Template DNA from linear amplification	1.5 µl
Taq DNA polymerase	0.3 µl
H₂O	39.7 µl
Final volume	50 µl

Second amplification reaction 2:

10 × PCR buffer, incl. 15 mM MgCl₂	5 µl
10 mM dNTPs	0.5 µl
10 µM T-DNA LB primer 8474	1 µl

(*Continued*)

10 × PCR buffer, incl. 15 mM MgCl$_2$	5 µl
10 µM adpG	1 µl
10 µM adpT	1 µl
Template DNA from linear amplification	1.5 µl
Taq DNA polymerase	0.3 µl
H$_2$O	39.7 µl
Final volume	50 µl

PCR scheme:

94 °C	2 min
94 °C	30 s
64 °C	30 s
72 °C	1.5 min
30 cycles	
72 °C	2 min
15 °C	∞

PCR products are analyzed by electrophoresis on a 1.5% agarose gel.

3.2.8
Sequencing and Computational Sequence Analysis

The PCR products are purified over a Sephadex G50 (Amersham Biosciences) superfine column. Dry Sephadex is filled into 96-well Multiscreen HV-plates (Millipore) and 0.3 ml water is added per well for swelling. After 3 h, the plates are centrifuged at 910 × g for 5 min, washed with 150 µl water, and again centrifuged for 5 min at 910 × g. An aliquot of 16 µl of the second amplification reaction is applied to the minicolumns and centrifuged. Aliquots of 2.5 µl of the eluate are used for cycle sequencing on a 3700 ABI PRISM® 96-capillary sequencer (Applied Biosystems) using Big Dye Terminator chemistry and primer 8409 (5′-ATATTGACCATCATACT-CATTGC-3′). Unincorporated fluorescent nucleotides are removed by another centrifugation through a separate Sephadex minicolumn.

The generated trace files are fed into a computational data-processing pipeline [41]. Information on the source plant, for which the PCR fragment is analyzed, is written into the respective trace file via the sample description file of the sequencer. Each resulting sequence is assayed for T-DNA vector sequence before alignment against the *A. thaliana* nuclear genome sequence using BLASTN. An FST is called a "genome hit" when the sequence shows significant similarity (BLAST *e*-value lower than 5×10^{-4}) to the *A. thaliana* genomic sequence as described by The Institute for Genomic Research. Since the *A. thaliana* AGI gene code is taken as a basis for searching and identifying insertions, the terms "genome hit" and "gene hit" are used to describe the quality of FSTs. An FST is a gene hit when it is a genome hit and the expected insertion site is located between 300 bp upstream of the ATG and 300 bp downstream of the STOP codon of an annotated gene. An FST qualifies as a "CDSi hit" if the predicted insertion locates between ATG and STOP (coding sequence (CDS) plus intron sequences) of an annotated gene. Resulting data are stored in a SimpleSearch database and FST data are submitted to EMBL/GenBank/DNA Data Bank of Japan.

3.2.9
Genetic Analysis of T-DNA Insertions

Seeds (up to 2500) are surface-sterilized for a maximum of 5 min in 33% v/v DanKlorix (Colgate-Palmolive) containing 0.1% *N*-laurylsarcosine (Merck-Schuchardt), washed 5 times with sterile water and spread on sterile 7-cm round filters (Schleicher & Schuell) placed in Petri dishes. Fifty T2 seeds are transferred under sterile conditions onto MS/agar plates containing MS inorganic salts (Ducheva, M-0221), 0.3% sucrose, 0.8% w/v agar, vitamins (1 mg/ml biotin, 1 mg/ml nicotinic acid, 1 mg/ml pyridoxin, 20 mg/ml) and 7.5 mg/l sulfadiazine (4-amino-*N*-[2-pyrimidinyl]benzene-sulfonamide-Na; Sigma, S-6387) with a toothpick. On each plate, five wild-type seeds are added as control. Seed germination is synchronized by incubation at 4 °C in the dark for 3–5 days prior to the transfer of the plates to the growth chamber (22 °C; 16 h light/8 h dark). Germination rate and ratios of resistant to sensitive seedlings are obtained by evaluation 12–14 days after transfer to light (Figure 3.5). With this method, the generally observed germination rates are above 90%. Seedlings producing clearly visible green secondary leaves and/or showing normal root development are scored as resistant. Sensitive plants containing no active resistance marker gene bleach and die. Assuming that sulfadiazine resistance segregates as an independent dominant locus, a unique integration locus is concluded when 60–85% of the germinated plants survive selection (Figure 3.5a). The existence of two or more loci is concluded when 86–100% of the germinated seedlings are resistant (Figure 3.5b). Lines displaying survival rates lower than 60% are scored as inconclusive.

3.2.10
DNA-Preparation for Confirmation of FST Predicted Insertion Sites

For confirmation analysis of the predicted insertion sites, T2 plantlets from the segregation analysis (see Section 3.2.9) are used. Genomic DNA is prepared according to the modified CTAB method described above (see Section 3.2.6).

About four to six sulfadiazine-resistant T2 plantlets from the segregation analysis (about 1 g plant material) are pooled in a 1.5-ml reaction tube. After addition of 150 µl CTAB buffer-1 and some sea sand, the plantlets are ground with a glass pestle fixed in a RZR 2020 drilling machine (Heidolph). A 5-min incubation at 70 °C and recooling to room temperature are followed by the addition of 300 µl CH_2Cl_2 and gentle mixing. The tubes are centrifuged for 20 min in a table centrifuge at full speed. The upper aqueous phase is transferred to a new 1.5-ml reaction tube and DNA is precipitated in 5 min by the addition of 2 volumes CTAB buffer-2. After centrifugation for 20 min in

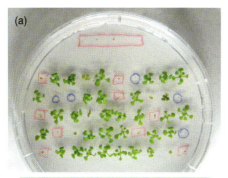
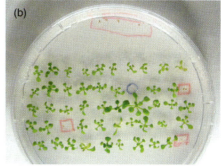
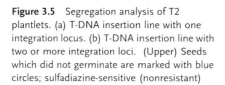

Figure 3.5 Segregation analysis of T2 plantlets. (a) T-DNA insertion line with one integration locus. (b) T-DNA insertion line with two or more integration loci. (Upper) Seeds which did not germinate are marked with blue circles; sulfadiazine-sensitive (nonresistant) plantlets are marked with red boxes; five wild-type seeds, used as sulfadiazine-sensitive control, are seen at the top. (Lower) Screenshots of segregation analysis results display from the GABI-Kat LIMS database.

a table centrifuge at full speed, the supernatant is discarded and the pellet is redissolved in 100 μl 1 M NaCl. DNA is precipitated for a second time by adding 100 μl isopropanol, followed by a 5-min incubation at room temperature and centrifugation for 20 min in a table centrifuge at full speed. After the supernatant has been discarded, the DNA pellet is washed with 200 μl 70% ethanol. The DNA is dried for about 2 h at room temperature and redissolved in 50 μl TE/RNase. T2 mixed DNA is stored at 4 °C.

3.2.11
Confirmation PCR

For each of the individual T-DNA insertions, a GSP (T_m 60 °C) is designed with the program Primer3 (http://www-genome.wi.mit.edu/genome_software/other/

primer3.html) taking into account T-DNA insertion position and the generation of a desired PCR product size of 500–800 bp.

The quality of the template DNA is tested simultaneously on the presence of T-DNA with the *SULr* ORF-specific primers Sul2 (5′-GTCGAACCTTCAAAAGCTGAAGT-3′) and Sul4 (5′-ATTTCACACAGGAAACAGCTATGA-3′). Confirmation and *SULr* ORF PCR are carried out in the same cycler with 1 µl of T2 mixed genomic DNA in a final reaction volume of 50 µl.

Confirmation PCR:

10 × PCR buffer, incl. 15 mM MgCl$_2$	5 µl
10 mM dNTPs	0.5 µl
10 µM T-DNA LB primer 8409	1 µl
10 µM GSP1	1 µl
Template DNA	1 µl
Taq DNA polymerase	0.3 µl
H$_2$O	41.2 µl
Final volume	50 µl

SULr ORF PCR:

10 × PCR buffer, incl. 15 mM MgCl$_2$	5 µl
10 mM dNTPs	0.5 µl
10 µM Sul2 primer	1 µl
10 µM Sul4 primer	1 µl
Template DNA	1 µl
Taq DNA polymerase	0.3 µl
H$_2$O	41.2 µl
Final volume	50 µl

PCR scheme:

94 °C	2 min
94 °C	30 s
59 °C	30 s
72 °C	2 min
37 cycles	
72 °C	2 min
15 °C	∞

PCR products are analyzed by electrophoresis on a 1.5% agarose gel.

When the *SULr* ORF-specific amplification fails, new genomic DNA from the respective line is prepared and the confirmation PCR is repeated. See Figure 3.6.

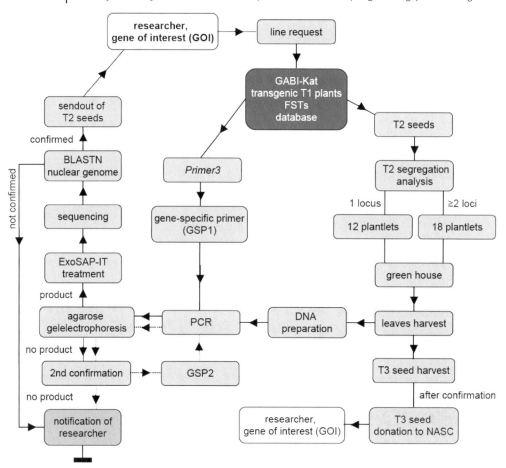

Figure 3.6 Flowchart of workflow in the confirmation process after line request.

3.2.12
Sequencing and Computational Sequence Analysis

Electrophoresis-checked PCR products are purified enzymatically with the ExoSAP-IT kit (USB). The ExoSAP-IT mixture of *Escherichia coli* exonuclease I and shrimp alkaline phosphatase eliminates residual single-stranded PCR primers by exonuclease, and degrades unincorporated dNTPs into nucleosides to avoid an imbalanced dNTP : ddNTP ratio in the sequencing reaction. An aliquot of 1 μl ExoSAP-IT reaction mixture is added to 6 μl from the PCR reaction. An incubation for 15 min at 37 °C is followed by enzyme inactivation at 80 °C for a further 15 min. Prepared products of confirmation PCR are sequenced using BigDye terminator chemistry on a 3700 ABI PRISM 96-capillary sequencer.

The generated trace files are processed as described above (Section 3.2.8). Resulting sequences are assayed for T-DNA vector sequences and aligned against the *A. thaliana* nuclear genome sequence using BLASTN.

3.2.13
Seed Donation

If confirmation is successful, segregating T2 seeds are delivered to those who ordered the lines. T3 seeds of confirmed lines are given to the Nottingham *Arabidopsis* Stock Centre, where they can be ordered as a set of T3 seeds. The T3 seeds are harvested from individual sulfadiazine-resistant T2 plants, which usually stem from the segregation analysis of the respective insertion (see Section 3.2.9). Each line within a set is a sister line of the other lines in that set and as a result, the T3 set has a high probability of containing seeds from at least one homozygous T2 plant for a given insertion line. Since it takes another generation to harvest, process, and prepare T3 seeds, which are finally given to the Nottingham *Arabidopsis* Stock Centre, the researcher gets T2 seeds of the requested line significantly before becomes available from the stock centre. See Figure 3.7.

Figure 3.7 Cataloged GABI-Kat seed collection of T-DNA mutagenized *A. thaliana* plants. Seeds are stored in paper bags, and subsequently preserved in cabinets (Lista) at a temperature of 9 °C and a relative humidity of 20–25%. Picture from the Max-Planck-Institute for Plant Breeding Research, Cologne, Germany.

3.2.14
Identification of Homozygous Mutants

In most cases the researcher wants to identify a homozygous T-DNA insertion plant, especially in the case of recessive genes. Depending on the available material this has to be done in individual T2 plants or in a T3 plant set. The easiest way doing this is by PCR-based genotyping, using the primers from the confirmation PCR (Section 3.2.11) and genomic DNA prepared from rosette leaves by standard methods. Therefore, the researcher needs to check which primers were used to generate the respective FST/confirmation sequence. It should be noted that the insertion position deduced from FSTs is not absolutely reliable, especially when no T-DNA sequence is detected at the beginning of the FST (and also depending on the quality of the respective sequence read). The presence of the wild-type allele is tested by PCR using two GSPs (Figure 3.8, GSP1 and GSP2) flanking the T-DNA insertion on both sites. PCR extension time is selected in such a way that *Taq* DNA polymerase is not able to produce a PCR product including the total T-DNA of about 5.8 kb. Thus, a wild-type allele-PCR resulting in no PCR product indicates the presence of a T-DNA insertion or the failure of PCR (due to many reasons).

Genotyping PCR, wild-type allele:

10 × PCR buffer, incl. 15 mM $MgCl_2$	2.5 µl
10 mM dNTPs	0.5 µl
10 µM GSP1	0.5 µl
10 µM GSP2	0.5 µl
Individual T2 plant DNA	1 µl
Taq DNA polymerase	0.3 µl
H_2O	19.7 µl
Final volume	25 µl

Genotyping PCR, insertion allele:

10 × PCR buffer, incl. 15 mM $MgCl_2$	2.5 µl
10 mM dNTPs	0.5 µl
10 µM T-DNA LB primer 8409	0.5 µl
10 µM GSP1	0.5 µl
Individual T2 plant DNA	1 µl
Taq DNA polymerase	0.3 µl
H_2O	19.7 µl
Final volume	25 µl

(a)
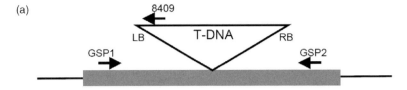

(b)
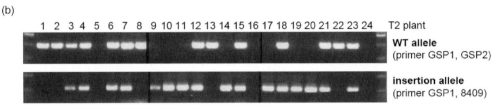

Figure 3.8 PCR-based identification of homozygous T-DNA insertion lines. (a) Strategy for PCR-based genotyping with GSPs and T-DNA-specific primer (8409). (b) Example analysis of PCR products by agarose electrophoresis. Not evaluable: 5, 16, 24; wild-type plants: 1, 2, 8, 13, 22; hemizygous: 3, 4, 6, 7, 12, 15, 18, 21, 23; homozygous: 9, 10, 11, 14, 17, 19, 20.

PCR scheme:

94 °C	3 min
94 °C	20 s
56 °C	20 s
72 °C	1 min
34 cycles	
72 °C	2 min
15 °C	∞

PCR products are analyzed by electrophoresis on a 1% agarose gel. A typical result of genotyping PCRs is shown in Figure 3.8(b).

3.3
Applications and Considerations for Work with T-DNA Insertion Mutants

The practical application of T-DNA insertion line collections, which is described above in detail for GABI-Kat lines, is straightforward for researchers in most cases. However, some general features of T-DNA insertion lines have to be taken into account in order to fully understand certain aspects and to obtain the greatest benefits out of the use of such lines [42].

3.3.1
Unconfirmed T-DNA Insertion Lines

Sometimes, in the case of unconfirmed T-DNA insertion lines (see Section 3.1), the gene of interest is not tagged in the line received. It is possible to receive a line that does not have a tag in the position predicted from the database. However, even if no T-DNA sequences are detected, it is worth trying to modify the position of the GSPs and by that extending the genomic region covered, before assuming that this is the case.

3.3.2
Use of Selectable Marker

Due to transgene silencing at the transcriptional or post-transcriptional level, T-DNA insertion lines may fail to express their selectable marker gene (*SULr* in the case of GABI-Kat lines) after two or more generations [43]. Consequently, the selectable marker and the insertion into the gene of interest are not necessarily genetically linked. For these reasons, one should not rely on the expression of the selectable marker gene for identification of homozygous T-DNA insertion lines. This issue of identification can easily be overcome by the application of PCR-based genotyping (Section 3.2.1.4).

3.3.3
Aberrant T-DNA Insertions

In a significant number of T-DNA transformants the occurrence of aberrant, truncated T-DNA insertions, missing one or both ends, has been reported [44]. Therefore, it is important that primer annealing sites are not designed too close (at least 300 bp away) to the predicted T-DNA insertion site when screening for homozygous T-DNA insertion lines.

As T-DNA insertions often occur in tandem [45], both T-DNA/T-DNA and T-DNA/flanking genomic DNA junctions might correspond to the same T-DNA border for any given line. This can confuse analysis, especially since several complex arrangements of the individual T-DNA stretches are possible. A head-to-tail arrangement would result in a correct LB border on one side of the genomic plant DNA and an RB border at the other side (LB–RB–LB–RB). On the other hand, a tail-to-tail arrangement results in two LB junctions (LB–RB–RB–LB) and consequently a head-to-head arrangement leads to two RB junctions (RB–LB–LB–RB). To cope with these arrangements in the characterization of the T-DNA insertion, the choice of primer combinations for genotyping or confirmation of the insertion has to be adopted in a way that all possible arrangements are covered.

In rare cases the transfer of Ti-plasmid-derived, non-T-DNA fragments has been observed [46]; in that case a junction between T-DNA border and plant genomic DNA can be missing. Recently, it has been shown that even fragments of genomic DNA from *A. tumefaciens* can be integrated into the plant genome at the insertion site [47]. The molecular mechanism for this observation is not understood. Apparently, such rare cases might complicate the characterization of the respective lines massively.

Truncated, tandem, or multiple insertions (see Section 3.3.4) that often occur in complex patterns can result in complicated amplification products during FST generation and confirmation, and are considered as one major reasons for poor-quality sequences. Nevertheless, the majority of T-DNA insertion lines should be analyzable by the standard methods described above.

3.3.4
Multiple T-DNA Insertions

Apart from the problem of tandem T-DNA insertions (see Section 3.3.3), the presence of a second or third T-DNA insertion in the original mutant background might be the reason for a lack of complementation of an observed phenotype. In this case, at least two or three homozygous plants per T-DNA line should be analyzed. Cosegregation of phenotype and mutant can provide information about linkage. For a more formal demonstration, backcrossing to wild-type plants for several generations should be attempted in order to segregate out other T-DNAs if they are inserted in unlinked loci. DNA (Southern) blot analysis can also be performed to determine whether there are other detectable T-DNAs in the background.

3.3.5
T-DNA-Induced Dominant Effects

The integration of T-DNA can sometimes induce dominant effects (e.g., due to effects of regulatory elements located in the T-DNA itself). Dominant effects occur partly due to promoters such as 35S CaMV or $1'$-$2'$ located at or around the T-DNA borders, as well as the presence of leaky terminators such as the 35S terminator, leading to read-through transcription. T-DNA-based dominant effects include (over)expression of full or partial transcripts/proteins (activation lines), and generation of antisense RNA, small RNAs and therefore silencing of other sequence-related genes or flanking genes (antisense lines).

3.3.6
Allelic Series of Mutants

To improve the likelihood of finding a real knockout mutant (in contrast to a *bona fide* knockout or knockdown mutant) it is recommended to find T-DNA insertion alleles from different collections, if possible. Two different alleles from the same collection may generate similar dominant effects from the T-DNA. Ideally, the researcher identifies at least one allele in collections that do not contain 35S in their T-DNA inserts. It has been shown that two or more 35S promoter copies in the same line can lead to trans-inactivation silencing [43].

If only one homozygous T-DNA allele is available and an observed phenotype appears to be recessive, there is an absolute requirement for other supporting evidence in order to correctly attributed function to a particular gene. This could be achieved in three ways. (i) A clear phenotype that reverts back to the wild-type phenotype after complementation of the mutant plants with a wild-type copy of the gene. Thereby the use a genomic fragment comprising 2–3 kb of the promoter, plus the coding sequence including the introns and untranslated regions, and 1–2 kb of the terminator region is recommended. However, the complementation with a large genomic fragment does not absolutely mean that the predicted ORF of the gene is responsible for the phenotype. Hidden ORFs inside the predicted gene or on the flanking sequences could in fact be the ones causing the phenotype, rather than the gene that the researcher predicts. (ii) A noninsertional mutant, such as EMS with a mutation in the same gene displaying the same phenotype as the insertion mutant. (iii) RNA interference/antisense lines suppressing expression of this gene also displaying the same phenotype.

3.3.7
Lethal Knockout Mutants

A complete knockout mutant might be lethal, hence homozygous T-DNA mutants may not survive. A simple silique dissection and examination of aborted seeds may help in identifying such lethal cases. However, lethality may occur after seed setting and during germination. Therefore, seedlings should be examined during the whole growth period. The identification of leaky alleles (e.g., with insertions in the promoter region or introns) could be an alternative attempt.

3.3.8
Search for Knockout Phenotype

It is apparently not always possible to detect phenotypes in homozygous mutant plants. Therefore, it is appropriate to perform detailed phenotypic analysis, to force phenotypes by altering growth conditions or testing the effects of various biotic and abiotic stresses on mutant plants, or to identify molecular phenotypes based on predicted function of the gene. Such a phenotype search includes transcriptomic, proteomic, and metabolic analyses.

3.3.9
Handling of Non-Single-Copy Genes

Sometimes no phenotype can be found even after detailed analyses and the function of the protein might still prove elusive. In such cases, redundant genes and genes of related sequence might need to be mutated. Since it is improbable that two (multiple) independent T-DNA insertions hit two (multiple) genes in a duplicated gene pair, the usefulness of T-DNA insertion seems to be limited on single-copy genes. For that reason other techniques like activation tagging or RNA interference might provide clues to the function of the gene. Alternatively, the generation of double, triple, or multiple mutants can be attempted. This is of course only possible when the redundant genes are not genetically linked too closely. The double, triple, or multiple homozygous mutant might reveal new properties that the single mutants do not have. Efforts are underway to produce a set of double mutants out of the existing T-DNA collections in order to provide the scientific community with a T-DNA insertion solution for duplicated genes.

3.4
Perspectives

Large T-DNA populations have now been in use for more than 10 years and they are still among the most important resources in reverse genetic approaches. In this respect they offer many advantages, but they also have a number of disadvantageous features. The advantages begin with the ability to apply high-throughput techniques for the systematic identification of insertion sites in genome-wide approaches. By the insertion of T-DNA, stable mutations of virtually every gene in *A. thaliana* might be generated. Furthermore, only few unwanted mutations are expected in T-DNA lines, since only a very limited number of insertions – one or two insertion(s) in the majority of cases – is expected in a given line. A further advantage lies in the variability of the method, because it can be adapted for both loss-of-function and gain-of-function studies. Among the limitations is the fact that the principle of T-DNA insertion line generation relies on a (more or less) random mutagenesis, which brings about that a desired mutation in a specific gene might never be found. This covers the problem of generating populations that are large enough to saturate the genome with insertions. Another disadvantage is that T-DNA insertion lines are inappropriate for the in-depth analysis of essential genes, because a given mutation often generates a complete, lethal knockout of the hit gene. Additionally, it is problematic to analyze tandemly repeated genes with T-DNA insertion mutants.

As it stands now, it seems clear that FST-based T-DNA collections will still be – in combination with other resources and methods – an invaluable tool in plant science in the future.

Acknowledgments

The authors greatly appreciate the significant contributions of many individuals to the development and improvement of the portrayed methods and protocols, and apologize that space restrictions do not allow us to list the individual names. The GABI-Kat project is supported by the Bundesministerium für Bildung und Forschung in the context of the German plant genomics program GABI (Förderkennzeichen 0312273).

References

1 The *Arabidopsis* Genome Initiative (2000) Analysis of the genome sequence of the flowering plant *Arabidopsis thaliana*. *Nature*, **408**, 796–815.

2 Meinke, D.W., Cherry, J.M., Dean, C., Rounsley, S.D., and Koornneef, M. (1998) *Arabidopsis thaliana*: a model plant for genome analysis. *Science*, **282**, 662–682.

3 Alonso, J.M. and Ecker, J.R. (2006) Moving forward in reverse: genetic technologies to enable genome-wide phenomic screens in *Arabidopsis*. *Nat. Rev. Genet.*, **7**, 524–536.

4 Van Larebeke, N., Engler, G., Holsters, M., Van den Elsacker, S., Zaenen, I., Schilperoort, R.A., and Schell, J. (1974) Large plasmid in *Agrobacterium tumefaciens* essential for crown gall-inducing ability. *Nature*, **252**, 169–170.

5 Gelvin, S.B. (1998) The introduction and expression of transgenes in plants. *Curr. Opin. Biotechnol.*, **9**, 227–232.

6 Tzfira, T., Li, J., Lacroix, B., and Citovsky, V. (2004) *Agrobacterium* T-DNA integration: molecules and models. *Trends Genet.*, **20**, 375–383.

7 Hernalsteens, J., Van Vliet, F., De Beuckeleer, M., Depicker, A., Engler, G., Lemmers, M., Holsters, M., Van Montagu, M., and Schell, J. (1980) The *Agrobacterium tumefaciens* Ti plasmid as a host vector system for introducing foreign DNA in plant cells. *Nature*, **287**, 654–656.

8 Hoekema, A., Hirsch, P.R., Hooykaas, P.J.J., and Schilperoort, R.A. (1983) A binary plant vector strategy based on separation of *vir-* and T-region of the *Agrobacterium tumefaciens* Ti-plasmid. *Nature*, **303**, 179–180.

9 Gelvin, S.B. (2005) Agricultural biotechnology: gene exchange by design. *Nature*, **433**, 583–584.

10 Azpiroz-Leehan, R. and Feldmann, K.A. (1997) T-DNA insertion mutagenesis in *Arabidopsis*: going back and forth. *Trends Genet.*, **13**, 152–156.

11 Bouchez, D. and Höfte, H. (1998) Functional genomics in plants. *J. Plant Physiol.*, **118**, 725–732.

12 Wisman, E., Hartmann, U., Sagasser, M., Baumann, E., Palme, K., Hahlbrock, K., Saedler, H., and Weisshaar, B. (1998) Knock-out mutants from an En-1 mutagenized *Arabidopsis thaliana* population generate new phenylpropanoid biosynthesis phenotypes. *Proc. Natl. Acad. Sci. USA*, **95**, 12432–12437.

13 Parinov, S., Sevugan, M., Ye, D., Yang, W.C., Kumaran, M., and Sundaresan, V. (1999) Analysis of flanking sequences from *Dissociation* insertion lines: a database for reverse genetics in *Arabidopsis*. *Plant Cell*, **11**, 2263–2270.

14 Speulman, E., Metz, P.L., van Arkel, G., te Lintel Hekkert, B., Stiekema, W.J., and Pereira, A. (1999) A two-component enhancer–inhibitor transposon mutagenesis system for functional analysis of the *Arabidopsis* genome. *Plant Cell*, **11**, 1853–1866.

15 Tissier, A.F., Marillonnet, S., Klimyuk, V., Patel, K., Torres, M.A., Murphy, G., and Jones, J.D. (1999) Multiple independent defective suppressor-mutator transposon insertions in *Arabidopsis*: a tool for functional genomics. *Plant Cell*, **11**, 1841–1852.

16 Rios, G., Lossow, A., Hertel, B., Breuer, F., Schaefer, S., Broich, M., Kleinow, T., Jasik, J., Winter, J., Ferrando, A., Farras, R., Panicot, M., Henriques, R., Mariaux, J.B., Oberschall, A., Molnar, G., Berendzen, K., Shukla, V., Lafos, M., Koncz, Z., Redei, G.P., Schell, J., and Koncz, C. (2002) Rapid identification of *Arabidopsis* insertion mutants by non-radioactive detection of T-DNA tagged genes. *Plant J.*, **32**, 243–253.

17 Sessions, A., Burke, E., Presting, G., Aux, G., McElver, J., Patton, D., Dietrich, B., Ho, P., Bacwaden, J., Ko, C., Clarke, J.D., Cotton, D., Bullis, D., Snell, J., Miguel, T., Hutchison, D., Kimmerly, B., Mitzel, T., Katagiri, F., Glazebrook, J., Law, M., and Goff, S.A. (2002) A high-throughput *Arabidopsis* reverse genetics system. *Plant Cell*, **14**, 2985–2994.

18 McKinney, E.C., Ali, N., Traut, A., Feldmann, K.A., Belostotsky, D.A., McDowell, J.M., and Meagher, R.B. (1995) Sequence-based identification of T-DNA insertion mutations in *Arabidopsis*: actin

mutants *act2-1* and *act4-1*. *Plant J.*, **8**, 613–622.

19. Krysan, P.H., Young, J.C., Tax, F., and Sussman, M.R. (1996) Identification of transferred DNA insertions within *Arabidopsis* genes involved in signal transduction and ion transport. *Proc. Natl. Acad. Sci. USA*, **93**, 8145–8150.

20. Winkler, R.G., Frank, M.R., Galbraith, D.W., Feyereisen, R., and Feldmann, K.A. (1998) Systematic reverse genetics of transfer-DNA-tagged lines of *Arabidopsis*. Isolation of mutations in the cytochrome p450 gene superfamily. *J. Plant. Physiol.*, **118**, 743–750.

21. Balzergue, S., Dubreucq, B., Chauvin, S., Le-Clainche, I., Le Boulaire, F., de Rose, R., Samson, F., Biaudet, V., Lecharny, A., Cruaud, C., Weissenbach, J., Caboche, M., and Lepiniec, L. (2001) Improved PCR-walking for large-scale isolation of plant T-DNA borders. *Biotechniques*, **30**, 496–504.

22. Samson, F., Brunaud, V., Balzergue, S., Dubreucq, B., Lepiniec, L., Pelletier, G., Caboche, M., and Lecharny, A. (2002) FLAGdb/FST: a database of mapped flanking insertion sites (FSTs) of *Arabidopsis thaliana* T-DNA transformants. *Nucleic Acids Res.*, **30**, 94–97.

23. Alonso, J.M., Stepanova, A.N., Leisse, T.J., Kim, C.J., Chen, H., Shinn, P., Stevenson, D.K., Zimmerman, J., Barajas, P., Cheuk, R., Gadrinab, C., Heller, C., Jeske, A., Koesema, E., Meyers, C.C., Parker, H., Prednis, L., Ansari, Y., Choy, N., Deen, H., Geralt, M., Hazari, N., Hom, E., Karnes, M., Mulholland, C., Ndubaku, R., Schmidt, I., Guzman, P., Aguilar-Henonin, L., Schmid, M., Weigel, D., Carter, D.E., Marchand, T., Risseeuw, E., Brogden, D., Zeko, A., Crosby, W.L., Berry, C.C., and Ecker, J.R. (2003) Genome-wide insertional mutagenesis of *Arabidopsis thaliana*. *Science*, **301**, 653–657.

24. Rosso, M.G., Li, Y., Strizhov, N., Reiss, B., Dekker, K., and Weissbaar, B. (2003) An *Arabidopsis thaliana* T-DNA mutagenised population (GABI-Kat) for flanking sequence tag based reverse genetics. *Plant Mol. Biol.*, **53**, 247–259.

25. Krysan, P.J., Young, J.C., and Sussman, M.R. (1999) T-DNA as an insertional mutagen in *Arabidopsis*. *Plant Cell*, **11**, 2283–2290.

26. Martienssen, R.A. (1998) Functional genomics: probing plant gene function and expression with transposons. *Proc. Natl. Acad. Sci. USA*, **95**, 2021–2026.

27. Li, Y., Rosso, M.G., Ulker, B., and Weisshaar, B. (2006) Analysis of T-DNA insertion site distribution patterns in *Arabidopsis thaliana* reveals special features of genes without insertions. *Genomics*, **87**, 645–652.

28. Corbesier, L., Gadisseur, I., Silvestre, G., Jacqmard, A., and Bernier, G. (1996) Design in *Arabidopsis thaliana* of a synchronous system of floral induction by one long day. *Plant J.*, **9**, 947–952.

29. Hadi, M.Z., Kemper, E., Wendeler, E., and Reiss, B. (2002) Simple and versatile selection of *Arabidopsis* transformants. *Plant Cell Rep.*, **21**, 130–135.

30. Velten, J., Velten, L., Hain, R., and Schell, J., (1984) Isolation of a dual plant promoter fragment from the Ti plasmid of *Agrobacterium tumefaciens*. *EMBO J.*, **3**, 2723–2730.

31. Guerineau, F., Brooks, L., Meadows, J., Lucy, A., Robinson, C., and Mullineaux, P. (1990) Sulfonamide resistance gene for plant transformation. *Plant Mol. Biol.*, **15**, 127–136.

32. Hood, E.E., Gelvin, S.B., Melchers, L.S., and Hoekema, A. (1993) New *Agrobacterium* helper plasmids for gene transfer to plants. *Transgenic Res.*, **2**, 208–218.

33. Clough, S.J. and Bent, A.F. (1998) Floral dip: a simplified method for *Agrobacterium*-mediated transformation of *Arabidopsis thaliana*. *Plant J.*, **16**, 735–743.

34. Feldmann, K. (1992) T-DNA insertion mutagenesis in *Arabidopsis*: Seed infection/transformation, in *Methods in Arabidopsis Research* (eds C. Koncz, N.-H. Chua, and J. Schell), World Scientific, Singapore, pp. 275–289.

35. Dellaporta, S.L., Wood, J., and Hicks, J.B., (1983) A plant DNA minipreparation: Version II. *Plant Mol. Biol. Rep.*, **1**, 19–21.

36. Jobes, D.V., Hurley, D.L., and Thien, L.B. (1995) Plant DNA isolation: a method to efficiently remove polyphenolics,

polysaccharides, and RNA. *Taxon*, **44**, 379–386.

37 Devic, M., Albert, S., Delseny, M., and Roscoe, T.J. (1997) Efficient PCR walking on plant genomic DNA. *Plant Physiol. Biochem.*, **35**, 331–339.

38 Spertini, D., Beliveau, C., and Bellemare, G. (1999) Screening of transgenic plants by amplification of unknown genomic DNA flanking T-DNA. *Biotechniques*, **27**, 308–314.

39 Steiner-Lange, S., Gremse, M., Kuckenberg, M., Nissing, E., Schachtele, D., Spenrath, N., Wolff, M., Saedler, H., and Dekker, K. (2001) Efficient identification of *Arabidopsis* knock-out mutants using DNA-arrays of transposon flanking sequences. *Plant Biol.*, **3**, 391–397.

40 Strizhov, N., Li, Y., Rosso, M.G., Viehoever, P., Dekker, K.A., and Weisshaar, B. (2003) High-throughput generation of sequence indexes from T-DNA mutagenized *Arabidopsis thaliana* lines. *Biotechniques*, **35**, 1164–1168.

41 Li, Y., Rosso, M.G., Strizhov, N., Viehoever, P., and Weisshaar, B. (2003) GABI-Kat SimpleSearch: a flanking sequence tag (FST) database for the identification of T-DNA insertion mutants in *Arabidopsis thaliana*. *Bioinformatics*, **19**, 1441–1442.

42 Ulker, B., Li, Y., Rosso, M.G., Logemann, E., Somssich, I.E., and Weisshaar, B. (2008) T-DNA-mediated transfer of *Agrobacterium tumefaciens* chromosomal DNA into plants. *Nat. Biotechnol.*, **26**, 1015–1017.

43 Daxinger, L., Hunter, B., Sheikh, M., Jauvion, V., Gasciolli, V., Vaucheret, H., Matzke, M., and Furner, I. (2008) Unexpected silencing effects from T-DNA tags in *Arabidopsis*. *Trends Plant Sci.*, **13**, 4–6.

44 Castle, L.A., Errampalli, D., Atherton, T.L., Franzmann, L.H., Yoon, E.S., and Meinke, D.W. (1993) Genetic and molecular characterization of embryonic mutants identified following seed transformation in *Arabidopsis*. *Mol. Gen. Genet.*, **241**, 504–514.

45 Heberle-Bors, E., Charvat, B., Thompson, D., Schernthaner, J.P., Barta, A., Matzke, A.J.M., and Matzke, M.A. (1988) Genetic analysis of T-DNA insertions into the tobacco genome. *Plant Cell Rep.*, **7**, 571–574.

46 Kononov, M.E., Bassuner, B., and Gelvin, S.B. (1997) Integration of T-DNA binary vector 'backbone' sequences into the tobacco genome: evidence for multiple complex patterns of integration. *Plant J.*, **11**, 945–957.

47 Ulker, B., Peiter, E., Dixon, D.P., Moffat, C., Capper, R., Bouche, N., Edwards, R., Sanders, D., Knight, H., and Knight, M.R. (2008) Getting the most out of publicly available T-DNA insertion lines. *Plant J.*, **56**, 665–677.

4
Making Mutations is an Active Process: Methods to Examine DNA Polymerase Errors

Kristin A. Eckert and Erin E. Gestl

Abstract

Endogenous metabolic processes and environmental exposures result in the formation of DNA lesions. The conversion of lesions into mutations is controlled by DNA repair efficiency and by the accuracy of DNA polymerases. The failure to complete lesion repair prior to replication may require DNA synthesis to proceed in the presence of DNA damage – a process we will refer to as translesion synthesis (TLS). When DNA polymerase error discrimination is compromised during TLS, mutations are introduced into an organism's genome. In this chapter, we describe genetic and biochemical approaches to examine damage-induced DNA polymerase errors *in vitro*. Genetic assays use circular single-stranded DNA templates containing a reporter gene, and score polymerase-induced errors as mutations after transfection of bacteria. Forward mutation assays are advantageous, in that most types of errors within multiple sequence contexts can be scored. Biochemical methods to study polymerase errors use chemically synthesized, site-specific lesion-containing oligonucleotides as DNA templates. Quantitation of reaction products is used to measure the extent of lesion inhibition and the efficiency of TLS. In this way, distinct steps in the TLS reaction can be analyzed. When both approaches use the same DNA sequence as a template for synthesis, the genetic and biochemical assays provide complementary data regarding DNA polymerase error discrimination mechanisms.

4.1
Introduction

In all kingdoms of life, DNA polymerases can be classified into four distinct families (A, B, X, and Y), based on primary sequence homology [1–3]. The enzymes of different families share the same catalytic mechanism, namely DNA-dependent

The Handbook of Plant Mutation Screening. Edited by Günter Kahl and Khalid Meksem
Copyright © 2010 WILEY-VCH Verlag GmbH & Co. KGaA, Weinheim
ISBN: 978-3-527-32604-4

phosphodiester bond formation [4]. However, the families are postulated to have different functions during DNA metabolic pathways, such as DNA replication, repair, and recombination [5–8]. The number of distinct DNA polymerases present in model organisms ranges from five in *Escherichia coli* [9], to 12 in *Arabidopsis thaliana* [6], and to 15 in *Homo sapiens* [8]. The physical structures of polymerases differ among the families [10, 11] and dictate several important error discrimination parameters of a DNA polymerase, including inherent accuracy [12], the ability to bypass a DNA lesion [9, 13], and the accuracy of translesion synthesis (TLS). In addition, some polymerases have an associated $3' \rightarrow 5'$ exonuclease "proofreading" activity, which selectively removes misincorporated nucleotides. As a result, replicative DNA polymerases are characterized by an intrinsically high accuracy of DNA synthesis [12, 14].

Experimental systems used to examine errors produced by purified DNA polymerases *in vitro* include both genetic and biochemical approaches. We and others have described genetic *in vitro* DNA synthesis assays that utilize circular single-stranded DNA (ssDNA) molecules containing a reporter gene as DNA templates [15, 16]. The *in vitro* forward mutation assay developed in our laboratory can be used to examine DNA polymerase processing of DNA lesions into mutations [16]. Our mutational reporter gene is the herpes simplex virus type 1 thymidine kinase (HSV-*tk*) gene. Importantly, our method does not introduce the DNA template molecule into *E. coli* for selection of mutants, limiting the background mutation frequency of the assay. In all genetic approaches, detectable damage-induced polymerase mutations result from productive polymerase TLS.

Biochemical methods provide a complementary means to study lesion inhibition and mutagenic TLS, and have been used extensively by several laboratories to study DNA polymerase errors [17–23]. In the biochemical approaches, chemically synthesized, linear oligonucleotides containing site-specific lesions are used as templates for synthesis. Quantitative analyses of the DNA reaction products are used to measure the extent of lesion inhibition versus the efficiency of TLS. In this way, distinct steps in the TLS reaction can be analyzed in relation to overall polymerase error discrimination.

4.2
Methods and Protocols

4.2.1
Overview of the Genetic Assay

A schematic of the HSV-*tk* forward mutation assay [16, 24] is shown in Figure 4.1(a), with letter/number combinations referring to the following protocol sections. A realistic timeline for the assay from the point of DNA modification to HSV-*tk* frequency determination is 10 days. The most labor intensive step is preparation of the gapped duplex (GD), which can be performed on a large scale and stored indefinitely at $-20\,°C$.

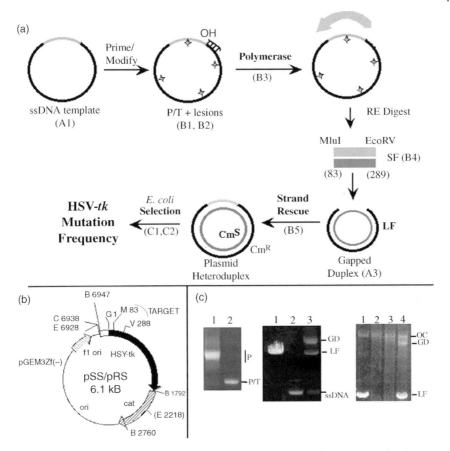

Figure 4.1 Schematic of the HSV-*tk in vitro* assay. (a) Measurement of DNA polymerase errors. Letters and numbers in parentheses refer to Section 2.4.1 for the genetic assay. Circular ssDNA molecules are used as templates for DNA polymerase reactions. The template is primed with an oligonucleotide and may be modified with mutagen to produce random DNA lesion. Purified polymerase is added to the reactions, and after a desired time the DNA reaction products are digested with restriction enzymes *Mlu*I and *Eco*RV. The resulting SF is purified by selective precipitation. To recover and analyze these DNA fragments for the presence of mutations, a GD molecule is used. The GD molecule is formed by hybridization of a linear Cm^R DNA fragment (LF) to a Cm^S ssDNA, thereby forming a molecule whose single-stranded region is complementary to the polymerase-produced DNA fragments. DNA synthesis fragments containing potential mutations within the HSV-*tk* gene are rescued by hybridization to the GD to form heteroduplex plasmid molecules, which are then used to transform *E. coli*. Incubation of the transformed bacteria in the presence of Cm kills bacteria harboring plasmids that are progeny of the Cm^S strand of the hybridized GD; therefore, the only bacteria that survive selection are those that are progeny of the "rescued" SF strand. HSV-*tk* mutant plasmids are selected by plating the bacteria in the presence of FUdR. The resulting HSV-*tk* mutant frequency is a direct measure of the proportion of DNA fragments containing polymerase-induced mutations. (b) Vectors for production of DNA forms. Numbering for all vectors begins with the unique *Bgl*II site within the HSV-*tk* gene. The HSV-*tk* gene and 3′-noncoding sequences (black arrow), driven by the tetracycline promoter (open bar), were cloned into the pGem3Zf(–) vector, such that the HSV-*tk* sense

(*Continued*)

(*Continued*) strand is produced from the f1 origin. The location of the *in vitro* target sequence is indicated. Two versions of each type of construct were produced – one that encodes a functional *cat* gene (pRS) and one that encodes a nonfunctional *cat* gene (pSS). The pSS1 vector is used to isolate ssDNA. The pRS1 plasmid is Cm resistant and used to create LF DNA. The sites of various restriction enzymes are indicated: B, *Bam* HI; C, *Sac*I; E, *Eco*RI; G, *Bgl*II; M, *Mlu*I; V, *Eco*RV. (c) Agarose gel analyzes of DNA forms. (Left panel) Polymerase synthesis products (lane 1) migrate more slowly than the P/T substrate (lane 2) through an 0.8% agarose gel, TAE buffer. (Middle panel) Hybridization of the LF (lane 1) to ssDNA (lane 2) creates the GD molecule (lane 3), which migrates with slower mobility in the 0.8% agarose gel. (Right panel) Successful hybridization of SF to the GD creates plasmid heteroduplex molecules (lanes 2 and 3) that migrate at the same position as an OC standard (lanes 1 and 4). This 0.8% agarose gel was run for around 16 h to resolve the DNA forms.

Preparation of GD Molecules

Purification of ssDNA

The pSS1 vector (Figure 4.1b) is a derivative of the pGem3Zf(−) phagemid vector (Promega), which can be used to produce both single- and double-stranded DNA. See *Molecular Cloning: A Laboratory Manual* [25] for further information regarding phagemid vectors.

- Inoculate 500 ml of 2XYT media [25] with 5 ml of an overnight culture of F′ *E. coli* strain DH5αIQ carrying the pSS1 plasmid. Add antibiotics to select for phagemid (50 μg/ml carbenicillin (Carb)) and for F′ (10 μg/ml kanamycin) plasmids.
- Incubate 37 °C shaker ∼3 h until the turbidity of the culture as measured by OD_{550} is 0.5–0.6 (log phase). Infect bacterial culture with 10 ml of R408 helper phage (Promega). The phage titer should be $>1 \times 10^{11}$ pfu/ml. Incubate 37 °C shaker, 3 h. Pellet bacteria in centrifuge: ∼2000 × g, 30 min, 0 °C.
- Decant 400 ml viral supernatant into a graduated cylinder, taking extra care not to transfer any bacteria. Discard the pellet. Add 100 ml polyethylene glycol (PEG) solution (20% PEG/3.75 M NH_4OAC) to the supernatant and mix well by inversion. Transfer to centrifuge bottles. Ice 30 min.
- Pellet phage in centrifuge: ∼2000 × g, 30 min, 0 °C. Carefully pour off supernatant and thoroughly remove ALL liquid from sides of centrifuge bottles with cotton swabs. (*Note*: Residual PEG will inhibit resuspension of virus.) If desired, store pellet in −20 °C freezer.
- Resuspend viral pellet in 10 ml phenol extraction buffer (PEB) (100 mM Tris, pH 8.0, 300 mM NaCl, 1 mM EDTA, pH 8.0). Transfer to two, 30-ml glass tubes (5 ml per tube).
- Equilibrate phenol with PEB [25]. Extract viral suspension with phenol by adding 10 ml (2 volumes) PEB-equilibrated phenol to each glass tube. Invert several times or swirl tube to mix phases well. Separate aqueous and organic phases by centrifugation, ∼4000 × g, 10 min, 15 °C.

- Carefully remove the top, aqueous phase and transfer to two clean glass tubes. Repeat phenol extraction.
- Remove aqueous phase from each tube and place in a clean tube. Add 5 ml chloroform : isoamyl alcohol (24 : 1, v/v) to each tube; mix by inversion. Let stand ∼5 min to separate phases. Remove top, aqueous phases and combine into one clean, 50-ml centrifuge tube. Measure final aqueous volume.
- Precipitate DNA by adding 10 M ammonium acetate to 2.0 M final concentration and 2 volumes absolute ethanol (EtOH), mixing well between additions. Incubate at −20 °C, 30 min or overnight.
- Pellet DNA in centrifuge: ∼12 000 × g, 40 min, 4 °C. Discard liquid. Wash pellet by adding 5 ml, 80% −20 °C EtOH (no mixing); centrifuge 20 min as above; air dry. Resuspend DNA pellet in TE (10 mM Tris–HCl, pH 8.0, 1 mM EDTA, pH 8.0) to a concentration of 0.5–2.0 mg/ml (generally, ∼200 µl). Store in −20 °C freezer.
- Determine DNA concentration by OD_{260} measurement. Analyze purity of ssDNA by loading 500 ng on a 0.8% agarose gel + ethidium bromide (EtBr, 0.5 µg/ml unless otherwise indicated) using Tris-acetate (TAE) running buffer [25]. (*Note*: Do not use Tris-borate (TBE) buffer for analyses of ssDNA forms). Two bands should be visible: R408 helper phage = 6.4 kb; pSS1 = 6.1 kb. The pSS1 band should be somewhat more prominent than the R408 band. The presence of extraneous bands indicates bacterial carryover and DNA preparations containing >20% extraneous forms should be discarded.

Preparation of Large Fragment (LF)

- Empirically determine reaction conditions for complete digestion of pRS1 plasmid DNA (Figure 4.1b) by each enzyme, *MluI* and *EcoRV*, in individual reactions. Start with 1–2 U of enzyme per µg DNA and a final DNA concentration of 500 ng/µl. (*Note*: High concentration forms of each enzyme (40–50 U/µl) are available from commercial suppliers.) Digest at 37 °C, 1 h.
- Analyze DNA fragments (∼500 and 100 ng) on 0.8% agarose gel containing 2 µg/ml EtBr to resolve uncut (supercoiled), single-cut (open circle (OC)) and fully cut (linear) DNA forms. (*Note*: It is very important at this stage to confirm that each enzyme digests DNA to completion and that no OC form is visible.)
- When conditions for complete digestion have been determined, scale up the digest proportionally to digest ∼1 mg plasmid DNA. Use multiple tubes, not exceeding 200 µl total volume per 1.5-ml microfuge tube.
- Digest first with *EcoRV*, 37 °C, 1 h. Add *MluI* along with appropriate buffer. Incubate 37 °C, 1;h. Analyze ∼1 µg and 200 ng of digested DNA by electrophoresis through a 0.8% agarose gel. A complete digest will yield two fragments of 203 and 5912 bp.
- Perform EtOH precipitation of DNA as follows. Add ammonium acetate to 2.0 M final concentration, followed by 2 volumes EtOH. Incubate −20 °C,

~30 min. Microfuge at 13 000 rpm, 20 min, 4 °C. Discard liquid. Add half volume, cold 80% EtOH (−20 °C). Repeat centrifugation, 5 min. Remove all liquid; briefly air dry pellet. Resuspend DNA in TE (100–200 µl) to a final concentration of 0.5–1 µg/µl.
- The 5.9-kb fragment (LF) is separated from the 203-bp fragment by selective PEG precipitation. In a microfuge tube, combine digested DNA, 5 M NaCl, and 30% PEG (molecular weight 8000) to final concentrations of 0.4 µg/µl DNA, 0.55 M NaCl, and 5% PEG, in a total volume of 200 µl per 1.5-ml microfuge tube. Incubate in a 37 °C water bath overnight.
- Pellet LF in microfuge, 12 000 rpm, 20 °C, 15 min. Carefully remove all supernatant. Resuspend pellets in pure water (not TE) to 1 µg/µl. Combine pellets and EtOH-precipitate DNA as above.
- Resuspend DNA in TE to ~1 µg/µl final concentration, assuming 100% recovery of DNA.
- Analyze success of PEG separation by running DNA samples (1.5 µl) on 1.0% agarose gel + EtBr, with digest standards. If PEG precipitation was successful, the pellet will contain the 5.9-kb fragment with little or no 203-bp fragment. If residual 203-bp fragment can be detected in the LF preparation, repeat PEG precipitation.
- Determine concentration of final LF preparation by OD_{260}. This scale of LF purification typically yields enough DNA for ~8 GD preparations (below). Store indefinitely in −20 °C freezer.

Formation of GD

The proportions of ssDNA and LF needed to form the GD must be determined empirically for each new DNA preparation. Set up small-scale (25 µl) hybridizations using 2 µg LF per reaction, and 0.5, 1, and 2 µg ssDNA, following the procedure below. The goal is to maximize yield of GD while minimizing amount of unhybridized ssDNA circles and LF (Figure 4.1c, left panel). We typically find a 1 : 1 molar ratio of ssDNA to LF (1 µg ssDNA to 2 µg LF) to give the best yield of GD.

- Dilute LF in H_2O (final concentration in hybridization, 250 ng/µl) and incubate at 85 °C for 9 min to denature the duplex. Add ssDNA to the tube and mix well with pipetteman tip; incubate 85 °C, 1 min to disrupt secondary structure.
- Ice 5 min; pulse-spin. Add standard sodium citrate (SSC) [25] to 2 × final concentration. Incubate at 60 °C for 30 min to promote hybridization. Cool to room temperature.
- EtOH precipitate DNA as above. Resuspend DNA in 10 µl of TE.
- Analyze success of hybridization by loading 4 µl DNA on an 0.8% agarose gel + EtBr, using 500 ng ssDNA and LF as markers. The ssDNA + LF should be converted to the GD form, which migrates as a broad band of slower mobility than the LF (Figure 4.1c).

- For large-scale preparation, repeat hybridization using the ssDNA:LF ratio that promotes the best GD formation, using 100 μl total volume per 1.5-ml microfuge tube.
- Gel-purify GD the same day using 0.8% agarose gel + EtBr. We generally load the result of six tubes of hybridization into one 15 cm × 20 cm gel, using a large prep comb. Run the gel in TAE buffer at 120 V for 2–2.5 h, with a circulating pump. Purify the GD form of DNA from the agarose gel using any commercially available silica kit (e.g., GeneClean; MP Biomedicals).
- Typically, we obtain ~4–5 μg of DNA from this scale of purification – enough GD for 20–25 polymerase reactions. GD can be stored indefinitely in –20 °C freezer.

DNA Polymerase Reactions

Preparation of Primer/Template (P/T) DNA Substrates

DNA synthesis templates are created by hybridization of an oligonucleotide complementary to pSS1 ssDNA, at a 1:1 molar ratio. Typically, we use a gel-purified 20mer of the sequence 5'-GGTACGTAGACGATATCGTC-3'. This primer encodes the unique *Eco*RV restriction site at position 288 (Figure 4.1b), and will initiate DNA synthesis at position 282 of the HSV-*tk* gene.

- Combine 20 pmol pSS1 ssDNA (2.0 μg/pmol), 20 pmol oligonucleotide primer, SSC (1 × final concentration), and purified (deionized) H_2O in 200 μl total volume (0.1 pmol/μl final concentration of P/T).
- Place the tube in a beaker of hot water (at least 70 °C) and let cool on bench to ~35 °C. Pulse-spin. Store hybridized DNA at −20 °C. (*Note*: Avoid further heating of the P/T, as elevated temperatures increase the rate of spontaneous base loss and cytosine deamination.)

DNA Modification by Alkylating Agents

Note: Alkylating agents are carcinogenic. Wear gloves, lab coat, and safety glasses. Decontaminate liquids containing alkylating agents with 0.2 N NaOH.

- Dilute P/T DNA to a concentration of 20 fmol/μl (20 nM) in PEN buffer (10 mM sodium phosphate, pH 7.4, 1 mM EDTA, 0.1 mM NaCl). Each polymerase reaction will require 2 pmol P/T DNA.
- Freshly prepare a 1 M *N*-methyl-*N*-nitrosourea (MNU) stock solution in dimethylsulfoxide (DMSO) solvent, and use immediately in the DNA modification reaction. To perform a dose–response experiment, prepare serial dilutions of MNU in solvent.
- Add 1/10 volumes MNU to the P/T DNA; incubate at 37 °C for up to 60 min. Typically, we use a dose range of 0.5–20 mM MNU for polymerase studies.

- Quench the reaction by diluting the DNA in ice-cold PEN and washing with 20–40 volumes of ice-cold PEN using a Centricon-30 (Amicon) filtration device at 4 °C. Concentrate modified DNA sample using a Centricon-30 or Microcon-30 device. Final volume should be equal to starting volume of P/T
- Modified DNA should be kept on ice and used in the polymerase reactions within 1 h to avoid depurination of DNA adducts.

DNA Polymerase Reactions

DNA polymerase reaction buffers are unique for each enzyme [26], but generally contain a divalent cation (Mg^{2+}), dNTP substrates, and a salt (NaCl, KCl) in a Tris or phosphate buffer for optimal pH at the desired reaction temperature. The composition of the reaction buffer can affect DNA polymerase accuracy [27, 28] The molar ratios of template DNA to polymerase also must be optimized for each enzyme, such that synthesis proceeds at least 200 nucleotides to include the *Mlu*I restriction site at HSV-*tk* position 90.(step_list)

- The amount of P/T DNA per standard polymerase reaction is 2 pmol in a 50-μl reaction (40 nM). Preincubate P/T and reaction buffer at desired reaction temperature, 3 min. Add polymerase; mix well. Incubate for desired time. Stop reaction by adding 1.5 μl of 0.5 M EDTA; keep on ice or store in −20 °C freezer.
- Low-resolution analysis of polymerase reaction products. Separate reaction products from starting substrate by electrophoresis of 1/10 reaction volume through a 0.8% agarose + EtBr gel, TAE running buffer, using hybridized ssDNA as a standard. Polymerase reaction products will appear as a broad band or smear of slower migration than the starting ssDNA P/T substrate.
- High resolution (optional). The full extent of DNA synthesis can be determined precisely by performing parallel reactions (0.2 pmol DNA, same molar ratios of enzyme to substrate as above) supplemented with 5 μCi of [α-^{32}P] dCTP (3000 Ci/mmol). Reaction products are separated by electrophoresis through an 8–12% denaturing polyacrylamide gel, TBE running buffer, using a DNA sequencing ladder as a marker.

Isolation of Small Fragment (SF)

- Heat polymerase reaction to 68 °C for 3 min to inactivate the polymerase.
- Wash polymerase reaction products by adding 450 μl TE and concentrate using a Microcon-30 ultrafiltration unit (Amicon). Adjust final volume of sample to 20 μl with TE.
- Add 40 U *Eco*RV and 40 U *Mlu*I in reaction buffer; incubate 37 °C, 60 min. To separate the SF from larger DNA products and P/T DNA, add to the digest: 10 μl, 5 M NaCl; 20 μl, 30% PEG8000; 30 μl H_2O (final concentrations 0.54 M NaCl, 6% PEG8000, and 20 nM DNA). Incubate on ice overnight in cold room.
- Pellet DNA in microfuge: 13 000 rpm, 15 min, 4 °C. Transfer supernatant to clean microfuge tube and discard pellet.

- Precipitate SF from the supernatant by adding 200 µl ethanol and incubating at −20 °C, 15 min.
- Pellet SF in microfuge 13 000 rpm, 15 min, 4 °C. Wash pellet by adding 100 µl, 80% EtOH and repeating centrifugation. Remove all liquid and air-dry pellet.
- Resuspend SF in 200 µl H$_2$O and concentrate to 10 µl using a Microcon-30 device. (*Note*: Do not omit this step and resuspend in 10 µl directly, as salt carryover will inhibit later hybridization.)
- Determine the concentration and purity of SF by agarose gel electrophoresis of 2 µl using DNA low mass ladder standard (Invitrogen). Typically, polymerase reactions that have completed synthesis yield 50–100 ng purified SF.

Formation of Heteroduplex Plasmid

- Combine SF (35 ng–150 ng, 0.25–1.2 pmol) and purified H$_2$O in a total volume of 10.5 µl, in a small (500 µl) microfuge tube. Incubate at 85 °C for 5 min to promote denaturation. Add 2 µl, 5 × SSC (75 mM NaCl, 7.5 mM Na citrate) as a drop to side of tube (do not mix with SF at this time).
- Remove tube and add 2.5 µl GD (200–225 ng; final concentration 15 ng/µl) to the tube. At this time, thoroughly mix GD and SSC with SF DNA with a pipette. Place in 45 °C water bath, 60 min to promote hybridization of SF to GD.
- Remove 2 µl (30 ng) of the hybridization and mix with 6 µl pure H$_2$O for mutational analysis. Freeze −20 °C until needed.
- Analyze the remaining hybridization mixture by electrophoresis through an 0.8% agarose gel (15 cm × 25 cm) + EtBr at 90 V for at least 12 h or overnight, with circulation. Load GD and OC standards. OC standard is prepared by performing a limiting digest of double-stranded DNA with a single-cut restriction enzyme, such as *Mlu*I, followed by EtOH precipitation of the DNA.
- Successful hybridization of SF to GD yields DNA products that migrate coincident with the OC standard with no evidence of products the size of the GD (see Figure 4.1c, right panel).

Bacterial Selection for Plasmids Containing HSV-*tk* Mutations

The selection scheme for mutant HSV-*tk* genes involves inhibition of *de novo* dTMP synthesis in *E. coli* strain FT334 (*hsdS20, supE44, lacY1, proA2, thi1, ara14, galK2, xyl5, mtl1, leuB6, rpsL20, recA13, upp, tdk*) [29]. The pyrimidine analog 5-fluoro-2′-deoxyuridine (FUdR) is cytotoxic upon conversion by wild-type thymidine kinase to FdUMP– an irreversible inhibitor of thymidylate synthetase. In the presence of FUdR, bacteria carrying wild-type HSV-*tk* genes will die from lack of dTTP while bacteria carrying mutant HSV-*tk* genes will survive. Selection for plasmid-encoded HSV-*tk* mutants is performed using VBA minimal media. Media containing yeast extract are avoided, as yeast extract contains high concentrations of thymidine, which will interfere with the FUdR selection.

Preparation of VBA Selective Media

VBA selection medium is $1 \times$ Vogel–Bonner (VB) minimal salts, 0.3 mM each of 19 amino acids (no asparagine); 30 mM glucose; 40 µg/ml thiamine; 40 µg/ml each cytidine, guanosine, and adenosine (rN mix); 500 µg/ml uridine.

- Prepare $5 \times$ VB minimal salts. Per liter: 1.0 g $MgSO_4 \cdot 7H_2O$; 10.0 g citric acid, monohydrate; 50.0 g K_2HPO_4, anhydrous; 17.5 g $NaHNH_4PO_4 \cdot 4H_2O$ (*Note*: Dissolve salts in order.) Store at 4 °C.
- Prepare stock solutions for supplements: $10 \times$ amino acid mix (3 mM each of 19 amino acids); $100 \times$ thiamine; $100 \times$ uridine; and $200 \times$ rN mix. Each supplement should be filter sterilized, and stored at 4 °C. A 2 M glucose stock solution can be prepared, autoclaved, and stored at 4 °C.
- To prepare 1 l of VBA selective agar plates: autoclave in separate flasks: (A) 500 ml H_2O + 15 g agar and (B) 200 ml $5 \times$ VB salts + 160 ml H_2O. (*Note*: VB salts will precipitate from solution if autoclaved in the presence of agar). Temper to 55 °C, then aseptically add supplements to flask B and combine with flask A, for a total of 1 l. For VBAChlor plates, add 50 µg/ml chloramphenicol (Cm); for VBA FChlor plates, add 50 µg/ml Cm and 40 µM FUdR before pouring plates.
- Supplemented VBA top agar is 0.75% agar (autoclaved separately) in $1 \times$ VB minimal salts plus supplements.

Electroporation of Heteroduplex Molecules and Selective Plating

- Prepare electrocompetent FT334 bacteria following standard molecular protocols [25], with the exception that the final bacterial resuspension must be done using 10% glycerol. (*Note*: Use electroporation and not chemical transformation of bacteria for optimal recovery of heteroduplex plasmid DNA.)
- Electroporate 1–2 µl heteroduplex plasmid DNA (from Step B5 above) and 50 µl FT334 competent cells: 2.0 kV, 400 ohms, 25 µFD. Immediately add 1 ml SOC broth [25]. (*Note*: The time constant range for these settings is 8.5–10; if the time constant is below this range, dilute DNA with water and repeat electroporation.)
- Transfer cells to a 15-ml conical tube and add an additional 1 ml SOC (2 ml total volume). Incubate 37 °C shaker, 2 h. (*Note*: A 2-h expression period is needed for full expression of wild-type HSV-*tk* protein.)
- Pellet bacteria in centrifuge, $2000 \times g$, 5 min, 4 °C. Aspirate off SOC media and resuspend cells by vortexing in 1 ml VBA medium. Keep on ice.
- Aliquot 9 ml of supplemented VBA top agar into 20-mm test tubes (one per plate) and temper to 45 °C in a dry bath or water bath. Prepare $10 \times$ drug solutions in sterile water for addition to top agar: 500 µg/ml Cm for VBA Chlor plates and 400 µM FUdR + 500 µg/ml Cm for VBA FChlor plates. Drug supplements can be stored in −20 °C freezer until use.

- Dilute bacteria as needed using sterile water. For each plating, combine 1 ml 10 × drug solution and bacterial dilution in a 9-ml top agar tube; mix well and pour onto appropriate VBA Chlor or VBA FChlor agar plate. For accurate sampling, each dilution should be plated in duplicate.
- Incubate 37 °C, 12–18 h, then room temperature overnight. Colonies in agar appear white and ovoid.
- Count colonies. Calculate the viable cells per milliliter on each type of selection by dividing the average number of colonies per plate by the dilution factor.

Calculation of HSV-*tk* Mutant Frequency (Table 4.1)

- The HSV-*tk* mutant frequency is defined as: (FUdR^{R}Cm^R viable cells/ml)/(Cm^R viable cells/ml).
- The polymerase mutant frequency should be determined in several independent experiments and reported as the mean with standard deviation.

Determination of ssDNA Background Mutation Frequency

- Prepare selective agar plates as above, except replace Cm with 250 μg/ml Carb. We have found Carb to be superior to ampicillin for this selection protocol.
- Electroporate 50 μl FT334 bacteria with 500 ng pSS1 ssDNA as above. After a 2-h expression period, resuspend bacteria in 1 ml VBA supplemented medium. Plate cells the same as above, replacing Cm with Carb (250 μg/ml final concentration) in the top agar.
- Calculate the HSV-*tk* mutant frequency as in Table 4.1. This frequency is variable among ssDNA preparations and should be determined for each new ssDNA purification. Preparations with unusually high mutant frequencies ($>1 \times 10^{-4}$) should be discarded.

Table 4.1 Examples of mutation frequency determination in the HSV-*tk* assay.

Polymerase	Experiment	Viable cells/ml		HSV-*tk* mutant frequency ($\times 10^{-4}$)
		FUdR^{R}Cm^R	Cm^R	
Klenow fragment, Exo$^-$	1	42	42000	10
	2	71	34000	21
	3	110	110000	10
	4	61	48000	13 (mean 14 ± 4.5)
Polymerase β	1	83	48000	18
	2	3200	920000	35 (mean: 26)

Generation of Mutational Spectra

The postelectroporation doubling time of FT334 in SOC media at 37 °C is around 45 min, while full FUdR resistance requires a 2-h recovery period. The following procedure is used to ensure independence of the mutants selected for DNA sequencing.

- Electroporate bacteria as above and place on ice immediately, in 1 ml SOC.
- From the expected number of mutants, calculate the volume of SOC (x) that will give rise to ~20 FUdR^{R}Cm^R colonies, using the equation: $x = 1$ ml/(#FUdR^{R}Cm^R cells/20). Aliquot (x) µl of cells into multiple tubes containing 1 ml VBA media. For example, a DNA sample that yields 1.0×10^3 FUdR^{R}Cm^R mutants would be aliquoted at 20 µl per tube, for up to 50 tubes.
- Incubate 2 h, 37 °C shaker. Plate the entire 1 ml individual culture tube onto one selective VBA FChlor plate, using top agar as above. Incubate 37 °C, 12–18 h.
- Using a sterile toothpick, transfer one FUdR^R mutant from each plate to a master LB agar grid plate containing 50 µg/ml Cm [25]. Isolate plasmid DNA from LB + 50 µg/ml Cm overnight cultures by any standard method [25]. (Note: pSS1 is a high-copy plasmid.)
- Determine the DNA sequence of the HSV-*tk* gene within the *Mlu*I–*Eco*RV region of each mutant by standard dideoxy DNA sequence analysis of plasmid DNA [25]. A typical mutant spectrum is derived from sequence changes observed in 50–100 independent mutants, isolated from at least two independent polymerase reactions.
- To calculate the polymerase error frequency of a particular class or specific type of mutation, first determine the mean mutation frequency for the reactions that were sampled to create the mutational spectra (Table 4.2). Next, determine the proportion of mutants within the desired class (i.e., base substitutions). Calculate the estimated polymerase base substitution error frequency by multiplying the proportion of base substitutions by the observed mean mutant frequency (Table 4.2).

Table 4.2 Examples of mutational specificity determination.

	KF polymerase	Polymerase β
Observed HSV-*tk* frequency[a]	14×10^{-4}	26×10^{-4}
Number sequenced	67	98
no mutation *Mlu*I–*Eco*RV[b]	11	4
base substitutions	37 (0.55)[c]	25 (0.26)
frameshifts	17 (0.46)	65 (0.66)
other	2 (0.03)	4 (0.04)
Base substitutions error frequency	7.7×10^{-4}	6.8×10^{-4}
Frameshift error frequency	3.6×10^{-4}	17×10^{-4}

a) Mean frequency of reactions sampled for mutational specificity.
b) Represents pre-existing mutations in the GD molecule, lying outside the *Mlu*I–*Eco*RV target (background mutant).
c) Proportion of total mutants sequenced.

4.2.2
Overview of the Biochemical Assay for TLS

Error discrimination by a DNA polymerase can be broken down into distinct biochemical reactions [30]. Likewise, the efficiency of complete polymerase TLS (Figure 4.2a) can be described in at least two steps: incorporation of a dNTP opposite a template DNA lesion and extension synthesis from a lesion-containing 3′-terminal

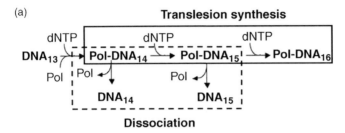

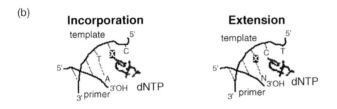

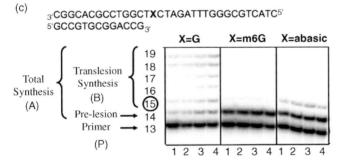

Figure 4.2 Schematic of the biochemical TLS assay. (a) Steps in polymerase error discrimination. (b) Complete TLS requires the independent steps of dNTP incorporation opposite a template lesion (left panel) and extension from a 3′-terminal lesion-containing base pair (right panel). For either step, biochemical assays can be used to measure the kinetics of phosphodiester bond formation between the 3′-OH and the dNTP. (c) Quantitation of the biochemical assays. The sequence at the top is the P/T oligonucleotides. The bottom panel is a representative acrylamide gel showing three reactions using an unmodified P/T (X = G) and either of two P/T containing a DNA lesion (X = m6G or abasic) at position 15 (right), indicated by the circled number. The bands that are included for quantification are shown to the left of the gel: unextended primer (P), percentage total synthesis (A) and percentage TLS (B). Lanes 1–4 represent increasing dNTP concentration.

base pair (Figure 4.2b). Dissociation of the polymerase from the growing strand at the site of a lesion limits the amount of complete TLS (Figure 4.2a).

Complete TLS Assay

General considerations for DNA P/T Design

- The sequence of the template oligonucleotide should have some genetic or biological relevance. For example, the sequence may represent a gene that is known to be mutated by a given chemical in the target organism. We use HSV-*tk* sequences that correspond to mutational hotspots observed in the *in vitro* assay (above).
- Chemically synthesized oligonucleotides of defined sequence can be purchased from several commercial suppliers. The purity of both the template and priming oligonucleotides is of utmost concern for the biochemical assays. We generally use high-performance liquid chromatography-purified template oligonucleotides. Priming oligonucleotides (15–20mers) should be gel-purified to ensure the absence of $N+1$ or $N-1$ sized products.
- The length of the primer and template oligonucleotides is dependent upon the footprint of the DNA polymerase binding to the duplex P/T stem and the ssDNA template. Template oligonucleotides are generally twice as long as the priming oligonucleotides. An example of the 13mer priming/33mer template oligonucleotides we have used for the Klenow fragment of *E. coli* DNA polymerase I (KF polymerase) and T4 polymerase is shown in Figure 4.2(c).
- Template oligonucleotides containing a site-specific DNA lesion can be purchased directly (Midland Certified Reagents) or synthesized by incorporating modified dNTP phosphoramidites (available from Glen Research). The presence of the lesion in the final template preparation should be verified prior to use, as not all adducts are stable under conditions of oligonucleotide synthesis.

DNA P/T Preparation

- Radioactively label the 5'-OH of the priming oligonucleotide with [γ-^{32}P]ATP, using 16 pmol oligonucleotide and T4 polynucleotide kinase, according to the manufacturer's instructions and general protocols [25]. The final DNA concentration should be 0.4 pmol/μl (0.4 μM). This labeling reaction yields enough for four sets of hybridization reactions.
- Remove unincorporated ATP using G-25 Sephadex columns, according to the manufacturer's protocol (Roche Diagnostics).
- Combine equal molar ratios of the labeled primer oligonucleotide (4 pmol) and template oligonucleotides (4 pmol) in 1 × SSC solution (final P/T hybrid concentration, 40.0 fmol/μl or 40 nM). The lesion-containing template should

be compared to a control unmodified template of the same sequence, in side-by-side experiments.
- Heat to >70 °C and cool slowly to room temperature (~1 h) to promote hybridization.
- Store samples at −20 °C until the reactions are performed. The above hybridization reaction yields enough for approximately 20 different insertion reactions.

DNA Polymerase Reactions

To ensure that polymerase molecules do not dissociate from the growing DNA strand and rebind to a previously extended P/T molecule, the TLS reactions should be completed under enzyme-limiting concentrations. These must be determined empirically by varying the DNA: polymerase molar ratio, such that the total synthesis is between 10 and 40% (see [31] for detailed explanation.)

- Combine hybrid DNA in optimal polymerase reaction buffer to a final concentration of 6.7 nM. To assess the overall efficiency of polymerase TLS, the reactions should be performed with a buffer containing all four dNTP substrates present in equal concentration. The range of [dNTP] we have used in our studies has been 20–200 µM. (*Note*: Very high [dNTP] are unlikely to be physiologically relevant and may give experimental artifacts.)
- Add polymerase to start the reaction (time = 0) giving a total reaction volume of 30 µl.
- Remove 5–6 µl aliquots at various times (i.e., $t = 30, 60, 90,$ and 180 s) and place in microfuge tubes containing an equal amount of formamide stop dye (5 mM EDTA, 0.1% xylene cyanol, 0.1% bromophenol blue in formamide). For the last time point, stop dye can be added directly to the reaction tube.
- Store samples at −20 °C until electrophoresis.

Control Reactions

- No polymerase control: shows the purity of the priming oligonucleotide and should contain a single band representing the desired length. Multiple bands present in this control indicate an impure oligonucleotide.
- DNA polymerase in molar excess: should result in complete extension of the available P/T complex. We generally use a commercially available enzyme, such as the exonuclease-deficient KF polymerase. If nonextended primer exists following the reaction incubation, the P/T hybridization procedure was incomplete.

Analysis of Polymerase Reaction Products

- Prepare a denaturing 16% polyacrylamide gel (43 cm in length), TBE buffer. Preheat gel to ~45 °C.

- Denature samples by heating at >85 °C for 10 min and placing immediately on ice.
- Load 3.0–5.0 µl of the samples (10.0–16.7 fmol product) per lane, along with controls. Electrophorese at 40 W for 5 min to allow the products to enter the gel matrix and then at a constant 100 W until the bromophenol blue dye is near the bottom of the gel, ∼1 h. Running time may need to be varied according to size of the priming oligonucleotide.
- Cover gels with plastic wrap and expose to a phosphor screen (Molecular Dynamics or equivalent), 4 h to overnight. Scan the screen. Alternatively, the gels may be dried and exposed to X-ray film, followed by densitometer quantitation of band intensity.

Calculation of TLS

- For each polymerase reaction lane, quantitate the radioactivity present in the primer band (P) and in individual bands representing prelesion or TLS reaction products (A and B) using ImageQuant (or equivalent) software (see Figure 4.2c).
- Determine the gel background by quantitating a box in an area of the polyacrylamide gel image that was not loaded with DNA. The box drawn to represent gel background must be of the same size as that drawn for the reactions. We have observed that this background varies for each gel. Subtract the gel background from the product to yield the corrected amount of reaction product for each length of DNA (corrected A and B).
- Quantitate the radioactivity present at the primer position in the polymerase excess control. This value represents unhybridized, free primer present in the P/T substrate preparation.
- Determine the amount of primer in each polymerase reaction lane. To correct for the extent of substrate that is productively hybridized, subtract the free primer value (from the polymerase excess control) from the primer bands of each of the reaction products to estimate the unextended P/T substrate (corrected P).
- Reactions can be analyzed as follows:
  - % Total Synthesis = [Corrected A/(Corrected A + Corrected P)] × 100. This is a measure of the total amount of substrate utilized in the reaction.
  - % TLS = [Corrected B/(Corrected B + Corrected A)] × 100. This is a measure of the extent of complete lesion bypass in the reaction.

Polymerase Dissociation Assay

The dissociation assay was designed such that the dissociation/reassociation equilibrium of a polymerase during DNA synthesis from a given P/T may be quantitated by both the loss and gain of a signal (Figure 4.3a). The ability of the polymerase to equally bind the two complexes is crucial to this experimental

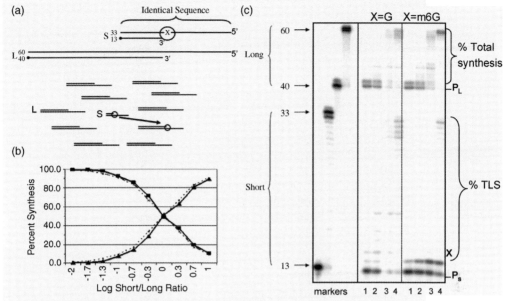

Figure 4.3 Polymerase dissociation assay. (a) Schematic of the assay. The short primer/template and long trap complexes share a region of identical DNA sequence surrounding the 3′-OH. A lesion present in the short complex is indicated by an X. The polymerase (indicated by an oval) is initially bound to a short P/T complex. The reactions are initiated by the simultaneous addition of dNTPs and a 10-fold molar excess of a long trap complex, to act as a competitor for free polymerase. If the polymerase dissociates from the short complex, it will likely reassociate with the long trap complex and begin synthesis. Therefore, the amount of synthesis on the long complex is a measure of the dissociation equilibrium from the short complex. (b) Mixing experiment of short: trap complexes. Ratios of short: trap complex range from 10 : 1 to 1 : 100. Solid lines represent experimental values, Dashed lines represent theoretical values. ▲, Short complex. ■, Trap complex. (Note: The ratio of total synthesis of short complex: total synthesis of trap complex was used to determine equality of polymerase recognition.) (c) Representative gel of the dissociation assay. Oligonucleotide length is noted on the right with 33 and 60 being full-length products of the short and trap complexes, respectively. The products of two polymerase reactions are shown – one in which the short complex was unmodified (X = G) and one in which the short complex contained an m⁶G lesion. Lanes 1–4 represent increasing reaction time (1–30 min).

design, and needs to be experimentally tested by mixing defined ratios of each complex with unbound polymerase and quantitating the amount of product formed (Figure 4.3b).

DNA P/T Preparation

- Two types of P/T hybrids are used in this assay. The first, referred to as the short complex, can be the same substrates used in the TLS experiments. The second, referred to as the long complex, consists of priming and template oligonucleotides of greater length than the short complexes. The priming oligonucleotide

for the long complex should be at least 2 nucleotides greater than the short template, so that the products from each primer can be resolved on a polyacrylamide gel (see Figure 4.3c). In our experiments, the long complex consisted of a 60mer template and a 40mer primer. The sequence surrounding the 3'-OH should be identical for both complexes to ensure equal substrate recognition and initial binding.
- 5' end-label and purify the short and long primer oligonucleotides as described above.
- Prepare three different P/T hybrids:
  - Hybrid 1: ^{32}P-labeled short primer (6 pmol) hybridized at 1 : 1 molar ratio to short template DNA (0.1 μM final).
  - Hybrid 2: ^{32}P-labeled long primer (6 pmol) hybridized at 1 : 1 molar ratio to long template DNA (0.1 μM final).
  - Hybrid 3: unlabeled long primer (120 pmol) hybridized to long template oligonucleotide (120 pmol; 0.4 μM final).

Dissociation Experiments

In the dissociation assay, the long trap is present in a 10 molar excess over the short complex; however, only 10% of the long complex is radioactively labeled to equalize the intensity of the bands of the long and short complexes after gel electrophoresis. The DNA: polymerase ratios and controls (for both short and long complexes) are the same as those used in the TLS experiments.
- Prepare the trap solution (22.5 μl) containing 3.00 μl Hybrid 2 (13.3 nM final concentration) and 6.75 μl Hybrid 3 (120 nM final concentration), in 2 × reaction buffer (4 × dNTPs).
- Prebind the polymerase to the short hybrid complex by incubating at room temperature for 3 min. Reactions (22.5 μl) should contain 3.0 μl Hybrid 1 (13.33 nM final concentration) and 2 × reaction buffer without dNTPs.
- Add the trap solution (containing dNTPs) to the prebound polymerase solution (equal volumes) and begin incubation at desired reaction temperature (time 0). Final DNA concentrations are: 6.67 nM Hybrid 1; 66.67 nM Hybrids 2 + 3.
- Remove 5-μl aliquots at various times and place in microfuge tubes containing an equal amount of formamide stop dye.
- Store at −20 °C until electrophoresis.

Analysis and Calculation of Polymerase Dissociation

- Denature samples by heating at >85 °C for 10 min and placing immediately on ice.
- Load 3–5 μl of the samples per lane on a preheated, denaturing 16% polyacrylamide gel. Each lane contains ∼300 fmol (40–67 fmol labeled) of product.
- Electrophoresis, exposure, and quantitation of gels are the same as stated above.

- Calculate the percentage TLS for the short complex and the percentage total synthesis for the long complex as above.
- Calculate the dissociation ratio as percentage TLS short complex/percentage total synthesis long complex. The data may be interpreted by graphing the dissociation ratio versus time. If the ratio increases with time, the polymerase is remaining bound to the short complex and completing TLS more often than dissociating. If the ratio decreases with time, the polymerase is dissociating from the short complex, and binding to and extending the long complex.

4.3
Applications

4.3.1
General Features of the *In Vitro* Genetic Assay

The major strength of using forward mutation assays to study DNA polymerase errors is that base substitution errors, frameshift mutations, deletions, and more complex errors can be scored within multiple sequence contexts. DNA polymerase error rates are affected by sequence context and the efficiency of chemical modification of DNA can be variable depending on the DNA sequence [32, 33]. Therefore, it is important to analyze mutations arising within a range of sequences in order to accurately determine polymerase-induced mutations. Random template modification allows for determination of the base lesions and sequence contexts that are most prone to errors by a given polymerase. The experimental design of the HSV-*tk* assay is versatile, and can be applied to the study of any type of DNA-damaging lesion and DNA polymerase combination.

The *Mlu*I–*Eco*RV target region of the HSV-*tk* gene encodes the enzyme ATP-binding site and is sensitive to amino acid changes. To date, we have detected within this region all 12 possible base substitution mutations arising within 57 distinct sequence contexts. The target sequence for the *in vitro* assay is operationally defined by the restriction enzymes used to construct the GD molecule. Therefore, the mutagenic target can be easily altered to encompass different sequence contexts within the 1150-bp HSV-*tk* coding sequence (GenBank accession number V00470) by using a different pair of unique restriction enzymes. The experimental design (Cm^R/Cm^S GD molecule) ensures exclusive analysis of mutations derived from the DNA strand produced during *in vitro* DNA synthesis. We measured full recovery of input mutant DNA molecules in reconstitution experiments, where the observed mutant frequency was equal to the known mutant SF fraction in the sampled DNA population [16].

The background mutation frequency for the HSV-*tk* assay is determined by spontaneous mutations arising during plasmid replication in *E. coli* and by DNA damage introduced during experimental manipulations. The spontaneous HSV-*tk* mutation frequency of the pRS1 plasmid is around 6×10^{-5}. This value represents mutations occurring anywhere within the HSV-*tk* gene. Therefore, some GD molecules may harbor spontaneous mutations occurring within the gene, but outside of the *Mlu*I–*Eco*RV target sequence. In addition, base damage can occur during manipulations to form the GD. Such molecules, when converted to plasmid heteroduplexes by hybridization of a wild-type SF, will be scored as a FUdR^{R}Cm^R mutant (see Table 4.2). We have estimated the frequency of this background to be around 1×10^{-4} [24]. Spontaneous mutations arising in the ssDNA that is used as a template for the DNA polymerase reaction is a second source of background errors. Such mutations, when accurately copied by the DNA polymerase, will result in mutant SF that are captured by hybridization to the GD. The mutant frequency for pSS1 ssDNA is around 6×10^{-5}. Overall, the combined background mutant frequency for the HSV-*tk in vitro* assay is relatively low, around 2×10^{-4}, comparable to other *in vitro* genetic assays [34, 35].

4.3.2
Polymerase Accuracy in the Absence of DNA Damage

Replicative DNA polymerases in prokaryotic and eukaryotic cells are generally associated with a $3' \rightarrow 5'$ proofreading exonuclease that acts in conjunction with the polymerase function to remove errors from the nascent DNA strand. KF polymerase is a member of the A family of polymerases and contains a $3' \rightarrow 5'$ proofreading exonuclease domain. The strength of the KF polymerase proofreading function is dependent upon the dNTP substrate concentration [36, 37]. The HSV-*tk* mutant frequency of KF polymerase at 50 µM dNTPs is around 4×10^{-4}, while the frequency using 1 mM dNTPs is around 9×10^{-4}. Genetic inactivation of the KF polymerase proofreading function by amino acid changes within the exonuclease domain results in a mutant frequency of around 15×10^{-4} (1 mM dNTPs, Table 4.1). The bacteriophage T4 DNA polymerase is a member of the B family and is often used as a model replicative polymerase [38]. T4 polymerase is a highly accurate enzyme, as measured by an HSV-*tk* mutant frequency of around 3×10^{-4} (a value that is not significantly greater than the background frequency). Mammalian DNA polymerase β is a member of the X family of polymerases and does not contain an associated proofreading exonuclease activity. This enzyme produces HSV-*tk* mutants at a frequency of around 24×10^{-4} (Table 4.1) (about 10-fold above background frequency).

In addition to quantitative differences in error frequency, each DNA polymerase also differs qualitatively in the types of errors produced. This can be readily visualized by creating a mutational spectrum, such as that shown in Figure 4.4(a). The polymerase β HSV-*tk* error spectrum is dominated by one and two base frameshift errors within repeated sequences [16, 24], similar to the *lacZ* reporter gene [34, 35]. Polymerase errors are often clustered at specific sites ("hotspots"), such as the CCC sequence at positions 147–149 and the TATATA sequence at positions 214–219 in the polymerase β spectrum (Figure 4.4a). The absolute frequency for a specific type of polymerase error can be calculated using the mutational spectrum (Table 4.2).

4.3.3
Mutational Processing of Alkylation Damage by DNA Polymerases

The mutational conversion of DNA lesions is determined by the chemical structure of the DNA adduct [39] as well as by the discrimination parameters of the DNA polymerase. To demonstrate use of the HSV-*tk* assay for the study of damage-induced polymerase errors, we present data for the monofunctional alkylating agent, MNU. We have measured linear MNU mutational dose–response curves for several DNA polymerases using our assay [36, 40, 41]. A dose–response relationship is important for the interpretation of the observed polymerase errors as being caused by modification of the DNA template. The magnitude of the polymerase mutational response with treatment, relative to the solvent control, is an indicator of the proportion of mutants within a spectrum that are likely to be caused by DNA damage, rather than by an inherent polymerase error. For example, within a mutational spectrum of 100 independent mutants, a 10-fold increase in mutant frequency with DNA treatment would correspond to an average of 90/100 damage-induced mutants and 10/100 inherent polymerase errors. Thus, the size of the mutational spectrum that needs to be created depends upon the magnitude of the dose-response. In addition, mutational spectra should be generated from reactions utilizing both solvent control and treated templates. For example, as expected due to the production of premutagenic O^6-methylguanine (m^6G) lesions, MNU treatment increased polymerase β-produced G $\rightarrow$ A transitions – an error rarely observed in the control spectrum (Figure 4.4b). The absolutely frequency of this base substitution with MNU treatment was 160×10^{-4} (20-fold higher than the DMSO control).

To determine whether the 3' $\rightarrow$ 5' exonuclease activity can remove premutagenic lesion-containing mispairs, we compared exonuclease-proficient and -deficient forms of T4 polymerase [41]. The MNU mutation rate, defined by the slope of the dose–response curves, was 7-fold greater for the exonuclease-deficient form, relative to the proficient form, demonstrating that the exonuclease of T4 polymerase is able to remove around 85% of total alkylation mispairs. Mutational spectra were derived for each polymerase to determine which mispairs are removed by the exonuclease (Figure 4.4c). Mutational spectra derived from the exonuclease-deficient form contained G $\rightarrow$ A, C $\rightarrow$ A, C $\rightarrow$ T, and A $\rightarrow$ T base substitutions, whereas that of the exonuclease-proficient form contained exclusively G $\rightarrow$ A transitions. Therefore, the exonuclease efficiently removes mispairs involving methylated C and A template lesions, but inefficiently removes the m^6G/T mispair.

4.3.4
DNA Lesion Discrimination Mechanisms

The biochemical assays are a powerful method to elucidate molecular interactions that are important for DNA polymerase discrimination against lesion-induced errors. We have used the approaches described above to analyze the structure–function

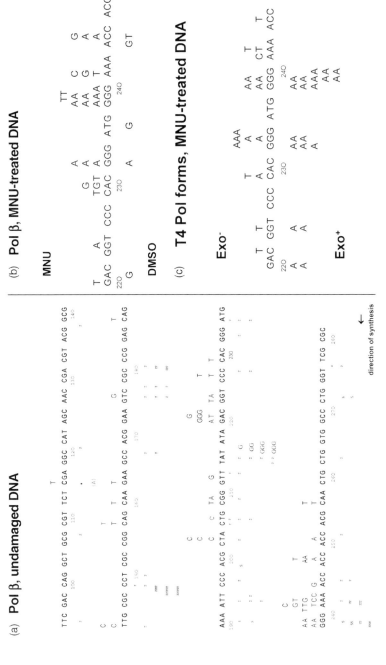

Figure 4.4 Mutational spectra for polymerase errors within the HSV-*tk* target. (a) Mutations produced by polymerase β using an undamaged pSS1 DNA template. The DNA template sequence shown (middle line) is the 203-bp HSV-*tk* mutational target. DNA synthesis is initiated at position 282 and proceeds right to left through the indicated sequence. Base substitution mutations are indicated above the sequence and frameshift errors are indicated below the sequence. Each symbol represents one independent mutant: Δ, one base deletion; ◊, two base deletion; ▲, one base addition. (b) Base substitution errors produced by polymerase β using MNU-treated template. The HSV-*tk* target is shown in the middle line (for clarity, only a partial sequence is shown). Base substitutions observed using MNU templates are indicated above the line and those observed on DMSO solvent templates are shown below the line. (c) Errors produced by T4 polymerase using MNU-treated templates. Errors produced by the exonuclease-deficient form are shown above the line, while those produced by the exonuclease-proficient form in side-by-side experiments are shown below the line. For clarity, only a partial sequence is shown.

relationships between KF polymerase and DNA during TLS. Exonuclease-deficient KF polymerase variants containing single amino acid substitutions (R668A, Y766A, and Q849A) were compared to an exonuclease-deficient "wild-type" counterpart (D424A) [42]. The ability of the polymerase variants to perform TLS using templates containing either an m^{6}G or an abasic lesion was determined, relative to an unmodified control template (Figure 4.5a). The percentage TLS by the wild-type (D424A) polymerase decreased from 73% on the unmodified template, to 10% on the m6G template, and 17% on the abasic template. The Y766A variant, which lacks dNTP base-stacking interactions, has a similar TLS efficiency to that of its

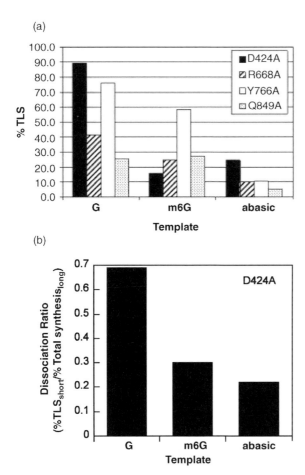

Figure 4.5 KF polymerase TLS and dissociation in the presence of DNA lesions. (a) Polymerase TLS using G, m^{6}G, or abasic templates. Filled, D424A; hatched, R668A; open, Y766A; stippled, Q849A. Data were collected at time = 30 s and are the mean of two or three independent experiments. (b) KF polymerase dissociation ratios on G, m6G, or abasic short templates. Data are the average ratios of time points 0.5–3 s in three independent experiments.

wild-type counterpart regardless of template. In contrast, the R668A and Q849A variants, which lack primer and template strand interactions to the DNA minor groove, respectively, show significantly reduced TLS on all templates. Qualitatively, the Q849A variant could not incorporate any dNTP substrate opposite either lesion.

Dissociation of a DNA polymerase from a lesion-containing template is a critical step in limiting the production of errors. Moreover, complete TLS may involve several polymerases – the replicative polymerase, often inhibited at a lesion, and specialized polymerases for insertion opposite the lesion and for extension from the lesion base pair [43–45]. This situation necessitates polymerase dissociation from the lesion template prior to transfer of the nascent DNA strand to another enzyme. We measured the dissociation equilibrium of exonuclease-deficient KF polymerase in the presence of lesions [42]. On undamaged templates, we observed a dissociation ratio of around 0.7, indicating limited KF polymerase dissociation from the short template and rebinding to the long complex over the 3-min time course (Figure 4.5b). In contrast, the dissociation ratio decreased to 0.3 for the m^{6}G template and 0.22 for the abasic template, indicating that KF polymerase dissociated from lesion-containing complexes at an increased rate. KF polymerase variants lacking DNA minor groove interactions displayed increased dissociation from DNA templates, but did not display an increased level of lesion-induced polymerase dissociation.

4.4
Perspectives

We [46, 47] and others [14, 48, 49] have used *in vitro* genetic assays extensively to study DNA polymerase errors in the absence of DNA damage. For example, our laboratory has analyzed the effects of tumor-associated polymerase β variants on polymerase accuracy [50, 51]. The strength of the forward assay was apparent in these studies, as we were able to uncover increased production of specific types of errors by each variant, often within specific sequence contexts. For other polymerases, such as the replicative polymerase δ, we have performed gap-filling reactions using the GD directly as a DNA substrate. This approach is advantageous when purified polymerase amounts are limiting, as less DNA substrate is required in the polymerase reaction. However, extra biochemical steps must be taken to ensure that the gap-filling reaction is complete. Moreover, the gap-filling approach cannot be used to study the effects of DNA damage, as the damaged template will be introduced into and processed into mutations by *E. coli*.

A potential new adaptation of the genetic assay is to create long (around 100 bases) oligonucleotides containing site-specific lesions as DNA templates for the polymerase reaction, in place of randomly modified ssDNA templates. The resulting SF products can be rescued for mutational analyses using the GD method, and will be a sensitive measure of the accuracy of TLS using a defined combination of DNA lesion, DNA sequence context, and polymerase.

More recently, we have modified the HSV-*tk* mutational target to include artificial microsatellite sequences [52]. We have used these new substrates to analyze DNA polymerase errors occurring at mono-, di-, and tetranucleotide repeats of varying length and sequence composition [24, 53, 54]. Another advance that we have made is to create pairs of vectors that produce ssDNA forms of the cDNA sequence. This allows us to model errors produced on both strands of the replication fork. Using vector pairs, we have observed significant differences in polymerase error rates and mutational hotspots on the complementary HSV-*tk* strands [24, 53].

While the genetic assays provide an extensive amount of data regarding the frequencies with which various polymerases produce specific mutations, the precise rates of nucleotide incorporation versus extension (Figure 4.2b) cannot be ascertained. Numerous laboratories have performed detailed examinations of the individual steps in DNA polymerase error discrimination, using biochemical methods. The complete TLS reactions described above are performed with all four dNTP substrates present in equal concentration, so that the polymerase has a choice of which base to incorporate into the nascent strand. A widely applied variation of this technique is used to determine the efficiency of polymerase incorporation of one specific nucleotide substrate [55]. By varying the length of the priming oligonucleotide, the individual steps of incorporation and extension can be separately monitored. This approach can be used for both undamaged [56] and lesion-containing [20] templates, and allows the experimental derivation of steady-state [17, 18] and pre-steady-state [23, 57, 58] kinetic parameters. We also have adapted the biochemical approach to examine T4 polymerase $3' \rightarrow 5'$ exonuclease activity at m^6G lesion base pairs [41]. When the same target sequence is used as a template in both the genetic and biochemical assays of DNA polymerase accuracy, the biochemical studies provided a direct mechanistic explanation of the genetic mutational spectrum [28, 59].

Acknowledgments

Research to develop and use the HSV-*tk* assay was supported by the American Cancer Society grants CN-144 and RPG-95-075, and by the National Institutes of Health grant R01 CA73649 to K.A.E. Special thanks to Kim Duncan for her support and suggestions during the preparation of this chapter.

References

1 Delarue, M., Poch, O., Tordo, N., Moras, D., and Argos, P. (1990) An attempt to unify the structure of polymerases. *Protein Eng.*, **3**, 461–467.

2 Braithwaite, D.K. and Ito, J. (1993) Compilation, alignment and phylogenetic relationships of DNA polymerases. *Nucleic Acids Res.*, **21**, 787–802.

3 Ohmori, H., Friedberg, E.C., Fuchs, R.P., Goodman, M.F., Hanaoka, F., Hinkle, D., Kunkel, T.A., Lawrence, C.W., Livneh, Z., Nohmi, T., Prakash, L., Prakash, S., Todo, T., Walker, G.C., Wang, Z., and Woodgate, R. (2001) The Y-family of DNA polymerases. *Mol. Cell*, **8**, 7–8.

4 Steitz, T.A. (1999) DNA polymerases: structural diversity and common mechanisms. *J. Biol. Chem.*, **274**, 17395–17398.

5 Bebenek, K. and Kunkel, T.A. (2004) Functions of DNA polymerases. *Adv. Protein Chem.*, **69**, 137–165.

6 Garcia-Diaz, M. and Bebenek, K. (2007) Multiple functions of DNA polymerases. *CRC Crit. Rev. Plant Sci.*, **26**, 105–122.

7 Hubscher, U., Maga, G., and Spadari, S. (2002) Eukaryotic DNA polymerases. *Annu. Rev. Biochem.*, **71**, 133–163.

8 Sweasy, J.B., Lauper, J.M., and Eckert, K.A. (2006) DNA polymerases and human diseases. *Radiat Res.*, **166**, 693–714.

9 Goodman, M.F. (2002) Error-prone repair DNA polymerases in prokaryotes and eukaryotes. *Annu. Rev. Biochem.*, **71**, 17–50.

10 Wang, J., Sattar, A.K.M.A., Wang, C.C., Karam, J.D., Konigsberg, W.H., and Steitz, T.A. (1997) Crystal structure of a pol α family replication DNA polymerase from bacteriophage RB69. *Cell*, **89**, 1087–1099.

11 Yang, W. (2003) Damage repair DNA polymerases Y. *Curr. Opin. Struct. Biol.*, **13**, 23–30.

12 Kunkel, T.A. (2004) DNA replication fidelity. *J. Biol. Chem.*, **279**, 16895–16898.

13 Prakash, S., Johnson, R.E., and Prakash, L. (2005) Eukaryotic translesion synthesis DNA polymerases: specificity of structure and function. *Annu. Rev. Biochem.*, **74**, 317–353.

14 Thomas, D.C., Roberts, J.D., Sabatino, R.D., Myers, T.W., Tan, C.-K., Downey, K.M., So, A.G., Bambara, R.A., and Kunkel, T.A. (1991) Fidelity of mammalian DNA replication and replicative DNA polymerases. *Biochemistry*, **30**, 11751–11759.

15 Bebenek, K. and Kunkel, T.A. (1995) Analyzing the fidelity of DNA polymerases, in *Methods in Enzymology* (ed. J.L. Campbell), Academic Press, New York, pp. 217–232.

16 Eckert, K.A., Hile, S.E., and Vargo, P.L. (1997) Development and use of an *in vitro* HSV-tk forward mutation assay to study eukaryotic DNA polymerase processing of DNA alkyl lesions. *Nucleic Acids Res.*, **25**, 1450–1457.

17 Dosanjh, M.K., Essigmann, J.M., Goodman, M.F., and Singer, B. (1990) Comparative efficiency of forming m^{4}T · G versus m^{4}T · A base pairs at a unique site by use of *Escherichia coli* DNA polymerase I (Klenow fragment) and *Drosophila melanogaster* polymerase α-primase complex. *Biochemistry*, **29**, 4698–4703.

18 Dosanjh, M.K., Galeros, G., Goodman, M.F., and Singer, B. (1991) Kinetics of extension of O^6-methylguanine paired with cytosine or thymine in defined oligonucleotide sequences. *Biochemistry*, **30**, 11595–11599.

19 Paz-Elizur, T., Takeshita, M., Goodman, M., O'Donnell, M., and Livneh, Z. (1996) Mechanism of translesion DNA synthesis by DNA polymerase II: comparison to DNA polymerases I and III core. *J. Biol. Chem.*, **271**, 24662–24669.

20 Randall, S.K., Eritja, R., Kaplan, B.E., Petruska, J., and Goodman, M.F. (1987) Nucleotide insertion kinetics opposite abasic lesions in DNA. *J. Biol. Chem.*, **262**, 6864–6870.

21 Shibutani, S. and Grollman, A. (1993) On the mechanism of frameshift (deletion) mutagenesis *in vitro*. *J. Biol. Chem.*, **268**, 11703–11710.

22 Shibutani, S., Takeshita, M., and Grollman, A.P. (1991) Insertion of specific bases during DNA synthesis past the oxidation-damaged base 8-oxodG. *Nature*, **349**, 431–434.

23 Gurge, L.L. and Guengrich, F.P. (1998) Pre-steady-state kinetics of nucleotide insertion following 8-oxo-7,8-dihydroguanine base pair mismatches by bacteriophage T7 DNA polymerase exo⁻. *Biochemistry*, **37**, 3567–3574.

24 Eckert, K.A., Mowery, A., and Hile, S.E. (2002) Misalignment-mediated DNA polymerase beta mutations: comparison of microsatellite and frame-shift error rates using a forward mutation assay. *Biochemistry*, **41**, 10490–10498.

25 Sambrook, J. and Russell, D.W. (2001) *Molecular Cloning: A Laboratory Manual*, 3rd edn, Cold Spring Harbor Laboratory Press, Cold Spring Harbor, NY.

26 Campbell, J.L. (ed.) (1995) *DNA Replication, Methods in Enzymology*, vol. 262, Academic Press, San Diego, CA.

27 Eckert, K.A. and Kunkel, T.A. (1990) High fidelity DNA synthesis by the *Thermus aquaticus* DNA polymerase. *Nucleic Acids Res.*, **18**, 3739–3744.

28 Eckert, K.A. and Kunkel, T.A. (1993) Effect of reaction pH on the fidelity and processivity of exonuclease-deficient Klenow polymerase. *J. Biol. Chem.*, **268**, 13462–13471.

29 Eckert, K.A. and Drinkwater, N.R. (1987) recA-dependent and recA-independent *N*-ethyl-*N*-nitrosourea mutagenesis at a plasmid-encoded herpes simplex virus thymidine kinase gene in *Escherichia coli*. *Mutat. Res.*, **178**, 1–10.

30 Goodman, M.F., Creighton, S., Bloom, L.B., and Petruska, J. (1993) Biochemical basis of DNA replication fidelity. *Crit. Rev. Biochem. Mol. Biol.*, **28**, 83–126.

31 Fygenson, D.K. and Goodman, M.F. (1997) Gel kinetic analysis of polymerase fidelity in the presence of multiple enzyme DNA encounters. *J. Biol. Chem.*, **272**, 27931–27935.

32 Kunkel, T.A. and Bebenek, K. (1988) Recent studies of the fidelity of DNA synthesis. *Biochem. Biophys. Acta*, **951**, 1–15.

33 Singer, B. and Essigmann, J.M. (1991) Site-specific mutagenesis: retrospective and prospective. *Carcinogenesis*, **12**, 949–955.

34 Kunkel, T.A. (1985) The mutational specificity of DNA polymerase β during *in vitro* DNA synthesis. *J. Biol. Chem.*, **260**, 5787–5796.

35 Kunkel, T.A. (1986) Frameshift mutagenesis by eucaryotic DNA polymerases *in vitro*. *J. Biol. Chem.*, **261**, 13581–13587.

36 Eckert, K.A. and Opresko, P.L. (1999) DNA polymerase mutagenic bypass and proofreading of endogenous DNA lesions. *Mutat. Res.*, **424**, 221–236.

37 Kunkel, T.A., Schaaper, R.M., Beckman, R.A., and Loeb, L.A. (1981) On the fidelity of DNA replication. Effect of the next nucleotide on proofreading. *J. Biol. Chem.*, **256**, 9883–9889.

38 Elisseeva, E., Mandal, S., and Reha-Krantz, L.J. (1999) Mutational and pH studies of the 3′ to 5′ exonuclease activity of bacteriophage T4 DNA polymerase. *J. Biol. Chem.*, **274**, 25151–25158.

39 Dipple, A. (1995) DNA adducts of chemical carcinogens. *Carcinogenesis*, **16**, 437–441.

40 Hamid, S. and Eckert, K.A. (2005) Effect of DNA polymerase beta loop variants on discrimination of O^6-methyldeoxy-guanosine modification present in the nucleotide versus template substrate. *Biochemistry*, **44**, 10378–10387.

41 Khare, V. and Eckert, K.A. (2001) The 3′ → 5′ exonuclease of T4 DNA polymerase removes premutagenic alkyl mispairs and contributes to futile cycling at O^6-methylguanine lesions. *J. Biol. Chem.*, **276**, 24286–24292.

42 Gestl, E.E. and Eckert, K.A. (2005) Loss of DNA minor groove interactions by exonuclease-deficient Klenow polymerase inhibits O^6-methylguanine and abasic site translesion synthesis. *Biochemistry*, **44**, 7059–7068.

43 Friedberg, E.C., Lehmann, A.R., and Fuchs, R.P. (2005) Trading places: how do DNA polymerases switch during translesion DNA synthesis? *Mol. Cell*, **18**, 499–505.

44 Johnson, R.E., Washington, M.T., Haracska, L., Prakash, S., and Prakash, L. (2000) Eukaryotic polymerases iota and zeta act sequentially to bypass DNA lesions. *Nature*, **406**, 1015–1019.

45 Prakash, S. and Prakash, L. (2002) Translesion DNA synthesis in eukaryotes: a one- or two-polymerase affair. *Genes Dev.*, **16**, 1872–1883.

46 Opresko, P.L., Shiman, R., and Eckert, K.A. (2000) Hydrophobic interactions in the hinge domain of DNA polymerase beta are important but not sufficient for maintaining fidelity of DNA synthesis. *Biochemistry*, **39**, 11399–11407.

47 Opresko, P.L., Sweasy, J.B., and Eckert, K.A. (1998) The mutator form of polymerase beta with amino acid substitution at tyrosine 265 in the hinge region displays an increase in both base substitution and frame shift errors. *Biochemistry*, **37**, 2111–2119.

48 Bell, J.B., Eckert, K.A., Joyce, C.M., and Kunkel, T.A. (1997) Base miscoding and strand miscoding errors by mutator Klenow polymerases with amino acid substitutions at tyrosine 766 in the O helix

of the fingers subdomain. *J. Biol. Chem.*, **272**, 7345–7351.

49 Minnick, D.T., Bebenek, K., Osheroff, W.P., Turner, R.M. Jr., Astatke, M., Liu, L., Kunkel, T.A., and Joyce, C.M. (1999) Side chains that influence fidelity at the polymerase active site of *Escherichia coli* DNA polymerase I (Klenow fragment). *J. Biol. Chem.*, **274**, 3067–3075.

50 Dalal, S., Hile, S., Eckert, K.A., Sun, K.W., Starcevic, D., and Sweasy, J.B. (2005) Prostate-cancer-associated I260M variant of DNA polymerase beta is a sequence-specific mutator. *Biochemistry*, **44**, 15664–15673.

51 Maitra, M., Gudzelak, A. Jr., Li, S.X., Matsumoto, Y., Eckert, K.A., Jager, J., and Sweasy, J.B. (2002) Threonine 79 is a hinge residue that governs the fidelity of DNA polymerase beta by helping to position the DNA within the active site. *J. Biol. Chem.*, **277**, 35550–35560.

52 Hile, S.E., Yan, G., and Eckert, K.A. (2000) Somatic mutation rates and specificities at TC/AG and GT/CA microsatellite sequences in nontumorigenic human lymphoblastoid cells. *Cancer Res.*, **60**, 1698–1703.

53 Hile, S.E. and Eckert, K.A. (2004) Positive correlation between DNA polymerase alpha-primase pausing and mutagenesis within polypyrimidine/polypurine microsatellite sequences. *J. Mol. Biol.*, **335**, 745–759.

54 Hile, S.E. and Eckert, K.A. (2008) DNA polymerase kappa produces interrupted mutations and displays polar pausing within mononucleotide microsatellite sequences. *Nucleic Acids Res.*, **36**, 688–696.

55 Boosalis, M.S., Petruska, J., and Goodman, M.F. (1987) DNA polymerase insertion fidelity. Gel assay for site-specific kinetics. *J. Biol. Chem.*, **262**, 14689–14696.

56 Mendelman, L.V., Boosalis, M.S., Petruska, J., and Goodman, M.F. (1989) Nearest neighbor influences on DNA polymerase insertion fidelity. *J. Biol. Chem.*, **264**, 14415–14423.

57 Ahn, J., Werneburg, B.G., and Tsai, M.-D., (1997) DNA polymerase □: structure–fidelity relationship form pre-steady-state kinetic analyses of all possible correct and incorrect base pairs for wild type and R283A mutant. *Biochemistry*, **36**, 1100–1107.

58 Dalal, S., Kosa, J.L., and Sweasy, J.B. (2004) The D246V mutant of DNA polymerase beta misincorporates nucleotides: evidence for a role for the flexible loop in DNA positioning within the active site. *J. Biol. Chem.*, **279**, 577–584.

59 Minnick, D.T., Liu, L., Grindley, N.D., Kunkel, T.A., and Joyce, C.M. (2002) Discrimination against purine–pyrimidine mispairs in the polymerase active site of DNA polymerase I: a structural explanation. *Proc. Natl. Acad. Sci. USA*, **99**, 1194–1199.

5
Tnt1 Induced Mutations in *Medicago*: Characterization and Applications

Pascal Ratet, Jiangqi Wen, Viviane Cosson, Million Tadege, and Kirankumar S. Mysore

Abstract

The demonstration of the transposition of the tobacco retroelement *Tnt1* during *in vitro* transformation of *Medicago truncatula* as allowed the construction of a large insertion knockout mutant collection in this model legume. The availability of these mutant collections should boost legume research in the same way that the T-DNA collections have allowed the development of modern molecular genetics in *Arabidopsis*. The currently available 10 000 lines represent over 250 000 *Tnt1* inserts, most of which are independently distributed in the gene-rich regions of the *Medicago* genome. In this chapter, we describe the protocols developed in our laboratories that allow the easy identification of tagged mutants presents in our collections as well as the multiple ways to use this technology to advance legume and plant research.

5.1
Introduction

Insertion mutagenesis is a powerful tool to discover and understand gene function in plants. T-DNA was successfully used as a mutagen in *Arabidopsis thaliana*, but may not be feasible in other plants like legumes. Retrotransposons (class I transposable elements) were also used successfully as mutagens in plants. Members of class I transposable elements multiply via a copy/paste mechanism that can result in the invasion of their host genomes. In the plant kingdom, for example, it is estimated that over 70% of the maize genome is composed of long terminal repeat (LTR) retrotransposons [1] and the Ogre Ty3/Gypsy-like retroelement can represent up to 38% of the *Vicia* genome [2]. Transposition of some of these elements is activated by various stresses and, interestingly, during tissue culture, allowing their use for large-scale insertion mutagenesis in model plants. For example, *Tos17*, an endogenous retrotransposon of rice, and *Tnt1* and *Tto1*, retrotransposons of tobacco, were used for

The Handbook of Plant Mutation Screening. Edited by Günter Kahl and Khalid Meksem
Copyright © 2010 WILEY-VCH Verlag GmbH & Co. KGaA, Weinheim
ISBN: 978-3-527-32604-4

gene tagging in *A. thaliana* and rice, respectively [3–5]. These elements are also good candidates for gene tagging in leguminous plants because they efficiently transpose into genes of their heterologous hosts and because their target site sequences exhibit moderate or no consensus. In addition, they do not transpose in the vicinity of their original location (like DNA transposons), but are rather dispersed in the genome as a result of their mode of transposition.

The *Tnt1* retroelement was isolated following its transposition into the nitrate reductase (*NiaD*) gene of tobacco [6]. We have previously demonstrated that *Tnt1* transposes actively during *in vitro* transformation of *Medicago truncatula* R108 and Jemalong lines [7, 8]. The efficiency of *Tnt1* transposition during the regeneration process results in *M. truncatula* lines carrying multiple *Tnt1* inserts (from four to up to 50 insertions per regenerated plant) and, by consequence, possibly multiple mutations. These insertions are stable during the lifecycle of *M. truncatula* and most of them are genetically independent and can be separated by recombination. In addition, in *Medicago Tnt1* seems to transpose preferentially into genes [8] and its multiplication by transposition can be reinduced by tissue culture. Among the already generated *Tnt1* mutants, several developmental as well as symbiotic *Tnt1*-tagged mutants have already been identified and characterized [7–11].

An important feature of the *Tnt1*-mutated *Medicago* collection is that it can be used for forward as well as reverse genetics studies which can be conducted by polymerase chain reaction (PCR) screening on DNA pools or by sequencing *Tnt1* insertion sites in the transgenic lines. In this chapter, we describe three protocols that allow the characterization of *Tnt1* insertions sites in the mutant *Medicago* plants, as well as the protocol developed for reverse screening on DNA pools of the collection. Applications of these technologies are described in Section 5.3.

5.2
Methods and Protocols

5.2.1
Identification of *Tnt1* Insertion Sites

Three PCR-based protocols (transposon display (TD)-PCR, thermal asymmetric interlaced (TAIL)-PCR, and inverse (I)-PCR; Figure 5.1) are described below for *Tnt1* border characterization. These borders or insertion sites of the element are also called flanking sequence tags (FSTs). All these protocols give good results in our laboratories. Among the three protocols, I-PCR is less efficient compared to other two protocols. Note that these protocols are complementary to some extent in the sense that some FSTs can be more easily isolated with one protocol.

Alternatively, another slightly different transposon display technology [12] can be applied on a *Medicago* segregating population in order to find the *Tnt1* border linked to the phenotype. This technique might be more time-consuming, but also more powerful for the characterization of the tagged locus. This protocol is not described in this chapter.

5.2 Methods and Protocols

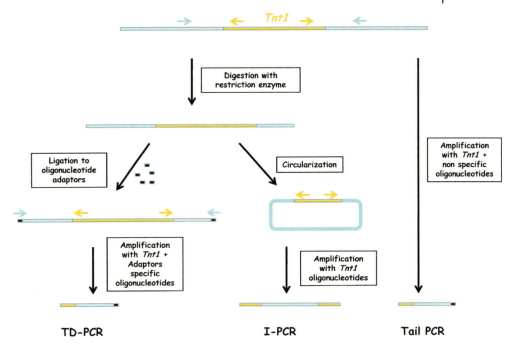

Figure 5.1 Schematic representation of the three PCR techniques used to characterize the Tnt1 borders. Tnt1 sequences are indicated in orange, genomic sequences in grey, oligonucleotides are represented by arrows, and adaptor sequences by black boxes. Only one oligonucleotide couple is represented for each nested PCR.

Protocol 1: TD-PCR

This protocol allows the amplification of the *Tnt1* borders using an oligonucleotide adaptor. We have tested it for both sides of the element, but it seems to work better with the 3′ side of the retrotransposon. This is probably related to oligonucleotides combinations.

Several restriction enzymes or enzyme combinations can be used for the described protocol. We used *Eco*RI (G^AATTC), *Mfe*I (C^AATTG), or a combination of both; *Ase*I (AT^TAAT), *Nde*I (CA^TATG), or a combination of both; and finally *Nde*II (^GATC). These enzymes were chosen because they cut the genomic DNA, alone or in combination, generating 0.5- to 1-kb DNA fragments that are optimal for this technique. For *Nde*II (4-bp cutter) the generated fragments are smaller, but the probability to get each insertion site is increased. However, an internal fragment is generated using the 4-bp cutter that will be the same for all *Tnt1* copies of the genome and will compete with the amplification of the genomic insertion sites. The *Nde*II enzyme was preferred to other isoschizomers recognizing the GATC sequence because it is not sensitive to CG and CNG methylation.

EcoRI and MfeI have the same cohesive ends compatible with the Eco-adaptor. This enzyme combination gives us good results when the *Tnt1* copy number is not very high. Note that these enzymes can be used separately to reduce the PCR complexity if needed. We also use AseI alone but it can be used in combination with NdeI for the Ase-adaptor because the cohesive ends are the same.

Briefly, for this experiment the DNA is digested with one enzyme (or a combination), ligated to an oligonucleotide adaptor and the borders are amplified through a nested (two-step) PCR. The use of the high-quality enzyme (TaKaRa Ex Taq™, pGEM®-T enzyme; www.lonza.com) seems to be very important for the success of this experiment. The PCR products can be visualized on an agarose gel (see Figure 5.2) and also cloned in pGEM-T vector (Promega), for example, before sequencing for insertion site characterization.

Adaptor and Oligonucleotides for the Nested PCRs

The structure of the double-stranded oligonucleotide adaptor is:

5'-CCC CTC GTA GAC TGC GTA CC-3'   (Adaptor 1)
3'-AGC ATC TGA CGC ATG G(XX)$_n$-5'   (Adaptor 2)

Where $(XX)_n$ represent the cohesive end for the restriction used to digest the genomic DNA. The Adaptor 2 complete sequence is given below with the corresponding oligonucleotides (sp) necessary for the two-step PCR.

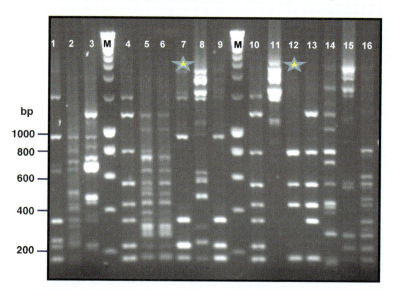

Figure 5.2 Transposon display analysis of *Tnt1*-containing plants. Each PCR fragment represents a *Tnt1* insertion site. Lanes with a star represent parent lines used to generate by *in vitro* culture lines with more inserts. Line 7 is a parent line for samples 1–10 and 15–16. Line 12 is a parent line for sample 13 and 14. M: molecular weight marker.

To construct the 50 mM (10×) double-strand Adaptor, mix equal volumes of the 100 mM Adaptor 1 (5'-CCC CTC GTA GAC TGC GTA CC-3') and Adaptor 2 (specific for each restriction enzyme) solutions, heat the mix at 95 °C for 10 min, and then let the mix cool at room temperature.

Specific Oligonucleotides (Adaptor 2, sp1 and sp2)

Eco-adaptor 2	5'-AAT TGG TAC GCA GTC TAC G-3'
Eco-1 (sp1)	5'-CTC GTA GAC TGC GTA CCA A-3'
Eco-2 (sp2)	5'-CGT AGA CTG CGT ACC AAT T-3'

Ase-adaptor 2	5'-TAG GTA CGC AGT CTA CGA-3'
Ase-1 (sp1)	5'-CTC GTA GAC TGC GTA CCT A-3'
Ase-2 (sp2)	5'-CGT AGA CTG CGT ACC TAA T-3'

NdeII-adaptor 2	5'-GAT CGG TAC GCA GTC TAC GA-3'
NdeII-1 (sp1)	5'-CTC GTA GAC TGC GTA CCG A-3'
NdeII-2 (sp2)	5'-CGT AGA CTG CGT ACC GAT C-3'

The two oligonucleotides used to amplify the 3' region of the *Tnt1* retroelement are:

LTR31	5'-GCT CCT CTC GGG GTC GTG G-3'
LTR4	5'-TAC CGT ATC TCG GTG CTA CA-3'

The two oligonucleotides used to amplify the 5' region of the *Tnt1* retroelement are:

LTR51	5'-CAA AGC TTC ACC CTC TAA AGC C-3'
LTR6	5'-GCT ACC AAC CAA ACC AAG TCA A-3'

Genomic DNA Digestion

Digest 2–3 μg genomic DNA with 10 U of the chosen restriction enzyme in 50 μl final volume for 3 h at 37 °C. Inactivate the enzyme for 20 min at 65 °C.

Ligation

Remember that the adaptor used at this step is specific for the enzyme used to digest the DNA.

Digested DNA	5 µl
Double-stranded adaptor (1×, 5 mM)	1 µl
10 × Ligation buffer	2 µl
Ligase (Biolabs M02025, 400 U/µl)	0.1 µl
H$_2$O	11.9 µl
Total	20.0 µl

The ligation is performed overnight at room temp or in a 16 °C bath. We have also done it 5 h at 16 °C followed by 2 h incubation at 37 °C.

Border Amplification

This is a nested (two-step) PCR.

TD-PCR1

Ligated DNA	2 µl
10 × *Taq* buffer	2 µl
dNTP mix	1.6 µl
10 µM LTR31	0.5 µl
10 µM oligo sp1	0.5 µl
TaKaRa Ex Taq	0.08 µl
H$_2$O	13.32 µl
Total	20.00 µl

TD-PCR1 Program
- 94 °C 2 min *1 time*
- 94 °C 20 s, 60 °C 20 s, 72 °C 2 min *5 times*
- 94 °C 20 s, 58 °C 20 s, 72 °C 2 min *5 times*
- 94 °C 20 s, 56 °C 20 s, 72 °C 2 min *20 times*

TD-PCR2
!!!!Dilute PCR1 100 times!!!!

Diluted DNA from PCR1	2 µl
10 × *Taq* buffer	2 µl
dNTP mix	1.6 µl
10 µM LTR4	0.5 µl
10 µM oligo sp2	0.5 µl
TaKaRa Ex Taq	~8 µl
H$_2$O	13.32 µl
Total	20.00 µl

TD-PCR2 Program
- 94 °C 2 min *1 time*
- 94 °C 20 s, 55 °C 20 s, 72 °C 2 min *10 times*
- 94 °C 20 s, 52 °C 20 s, 72 °C 2 min *25 times*

Fragment Characterization

- Use 5 μl of the PCR2 reaction for gel electrophoresis. Use 1.5% agarose gel in order to have good separation of the fragments (see Figure 5.2).
- For vector cloning of the PCR fragments, the PCR reaction is first purified on a commercial PCR purification column. If using the pGEM-T vector, the reaction mix for cloning in the vector is:

Column purified PCR2	5 μl
10 × Ligation buffer	2 μl
pGEM-T	0.5 μl
Ligase (3 U/μl)	1 μl
H$_2$O	11.5 μl
Total	20.0 μl

The ligation is done overnight at 16 °C and half of the ligation product is used for transformation.

Protocol 2: TAIL PCR

The TAIL-PCR approach enables the amplification of the *Tnt1* borders without adaptor ligation or restriction enzyme digestion [13, 14]. Two rounds of PCR amplification will result in the recovery of *Tnt1* FSTs. This protocol does not require very high quality genomic DNA.

Briefly, for this experiment approximately 50 ng of genomic DNA will be used as template. *Tnt1*-specific primer Tnt1-F and arbitrary degenerate primers (AD1, AD2, AD3, AD5, and AD6) should be used for primary PCR amplification. After the primary round of PCR, a subsequent second round (nested) of PCR amplification is carried out using diluted first-round PCR products as template. Tnt1-F1 and the corresponding degenerate primers will be used for the second-round (nested) PCR. After the second round of PCR amplification, PCR products amplified from the same genomic DNA by various arbitrary primers will be combined and purified using a PCR Purification Kit (Qiagen) and cloned into pGEM-T easy vector. Between 48 and 72 white colonies can be picked for each line and the inserts should be sequenced using Tnt1-F2 primer. Fragments longer than 2 kb could be recovered using this method, but there is no need to sequence the entire length. Most commonly, FSTs 200–600 bp long will be recovered. The use of the high quality enzyme (TaKaRa Ex Taq enzyme) seems to be very important for the success of this experiment.

Primers used for TAIL-PCR

The primers used for TAIL-PCR are:

AD1	5′-NTCGA(G/C)T(A/T)T(G/C)G(A/T)GTT-3′
AD2	5′-NGTCGA(G/C)(A/T)GANA(A/T)GAA-3′
AD3	5′-(A/T)GTGNAG(A/T)ANCANAGA-3′
AD5	5′-(G/C)(G/C)TGG(G/C)STANAT(A/T)AT(A/T)CT-3′
AD6	5′-CG(G/C)AT(G/C)TC(G/C)AANAA(A/T)AT-3′
Tnt1-F	5′-ACAGTGCTACCTCCTCTGGATG-3′
Tnt1-F1	5′-TCCTTGTTGGATTGGTAGCCAACTTTGTTG-3′
Tnt1-F2	5′-TCTTGTTAATTACCGTATCTCGGTGCTACA-3′

TAIL-PCR1

Genomic DNA	2 μl
10 × *Taq* buffer	4 μl
2.5 mM dNTP mix	3.2 μl
20 μM Tnt1-F	0.6 μl
100 μM AD1 (or AD2, 3, 5, 6)	1.2 μl
TaKaRa Ex Taq	0.3 μl
H$_2$O	28.7 μl
Total	40.0 μl

TAIL-PCR1 Program

- 93 °C 1 min, *1 time*
- 95 °C 1 min, *1 time*
- 94 °C 45 s, 62 °C 1 min, 72 °C 2.5 min, *4 times*
- 94 °C 45 s, 25 °C 3 min, Ramp to 72 °C in 3 min, 72 °C 2.5 min, *1 time*
- 94 °C 20 s, 68 °C 1 min, 72 °C 2.5 min, 94 °C 20 s, 68 °C 1min, 72 °C 2.5 min, 94 °C 20 s, 44 °C 1 min, 72 °C 2.5 min, *14 times*
- 72 °C 5 min, *1 time*

TAIL-PCR2

!!!!Dilute PCR1 50 times!!!!

Diluted PCR1	2 μl
10 × *Taq* buffer	4 μl
2.5 mM dNTP mix	3.2 μl
20 μM Tnt1-F1	0.6 μl
100 μM AD1(or AD2, 3, 5, 6)	1.2 μl
TaKaRa Ex Taq	0.3 μl
H$_2$O	28.7 μl
Total	40.0 μl

TAIL-PCR2 Program

- 94 °C 20 s, 64 °C 1min, 72 °C 2.5min, 94 °C 20 s, 64 °C 1min, 72 °C 2.5min, 94 °C 20 s, 44 °C 1min, 72 °C 2.5min, *12 times*
- 72 °C 5min, *1 time*

Fragment Characterization

- Use 5 µl of the combined TAIL-PCR2 products for gel electrophoresis. Use 1.5% agarose gel in order to have a good separation of the fragments. The result of a typical experiment is shown Figure 5.3.
- For cloning of the PCR fragments, the PCR reaction is first purified on a commercial PCR purification column. If using the pGEM-T vector, the reaction mix for cloning in the vector is:

Column purified PCR2	3.5 µl
2 × Ligation buffer	5 µl
pGEM-T	0.5 µl
Ligase (3 U/µl)	1 µl
Total	10.0 µl

The ligation is carried out overnight at 16 °C and half of the ligation product is used for transformation.

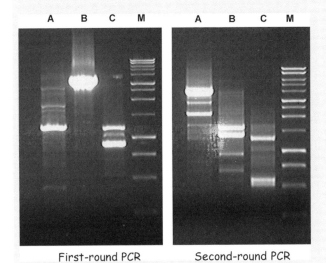

First-round PCR   Second-round PCR

Figure 5.3 A typical TAIL-PCR result. A, B and C are three different lines. M: 1-kb DNA ladder from Promega.

Protocol 3: I-PCR

This protocol allows the amplification of the *Tnt1* borders without using adaptors. Several restriction enzymes or enzyme combinations can be used for the described protocol if these enzymes do not cut inside the retroelement. We used *Eco*RI (G∧AATTC), *Mfe*I (C∧AATTG), or a combination of both (same cohesive ends); and *Ase*I (AT∧TAAT), *Nde*I (CA∧TATG), or a combination of both (same cohesive ends). This enzyme combination gives us good results when the *Tnt1* copy number is not very high. (*Note:* These enzymes can be used separately to reduce the PCR complexity if needed.) We also use *Ase*I alone, but it can be used in combination with *Nde*I because the cohesive ends are the same.

Briefly, for this experiment the DNA is digested with one enzyme (or a combination), ligated in a large volume to favor autoligation and the borders are amplified through a nested (two steps) PCR. The use of the high quality enzyme (TaKaRa Ex Taq enzyme) seems to be very important for the success of this experiment. The PCR products corresponding to the insertion sites can be visualized on an agarose gel and also cloned in pGEM-T vector, for example, before sequencing.

Oligonucleotides used for the Nested PCRs

The four oligonucleotides used in the nested PCR to amplify the *Tnt1* retroelement are:

LTR3	5′-AGT TGC TCC TCT CGG GGT CGT GCT T-3′
LTR4	5′-TAC CGT ATC TCG GTG CTA CA-3′
LTR5	5′-GCC AAA GCT TCA CCC TCT AAA GCC T-3′
LTR6	5′-GCT ACC AAC CAA ACC AAG TCA A-3′

Genomic DNA Digestion

Digest 2–3 µg genomic DNA with 10 U of the chosen restriction enzyme in 50 µl final volume for 3 h at 37 °C. Inactivate the enzyme for 20 min at 65 °C.

Ligation

The ligation is performed in 200 µl. The large volume favors the ligation of the fragments on themselves. This is important for this experiment.

Digested DNA	50 µl
10 × Ligation buffer	20 µl
Ligase (3 U/µl)	2 µl
H$_2$O	128 µl
Total	200.0 µl

The ligation is performed overnight at room temperature or in a 16 °C bath. We have also performed it for 5 h at 16 °C followed by 2 h incubation at 37 °C.

Border Amplification

This protocol is based on a nested PCR using the TaKaRa Ex Taq kit. It can be adapted to other commercial high quality enzymes. The oligonucleotides concentration in the two reactions should be respected. The two PCRs are as follows:

I-PCR 1

Ligation reaction	10 μl
10 × *Taq* buffer	2 μl
dNTP	1,6 μl
10 μM LTR3	0.5 μl
10 μM LTR5	0.5 μl
TaKaRa Ex Taq	0.1 μl
H$_2$O	5.3 μl
Total	20.0 μl

I-PCR1 Program
- 94 °C 2 min, *1 time*
- 94 °C 20 sec, 72 °C 3 min, *30 times*
- 72 °C 5 min, *1 time*

I-PCR2
!!!!Dilute PCR1 100 times!!!!

Diluted I-PCR1	2 μl
10 × *Taq* buffer	2 μl
dNTP mix	1.6 μl
10 μM LTR4	0.5 μl
10 μM LTR6	0.5 μl
TaKaRa Ex Taq	0.1 μl
H$_2$O	13.3 μl
Total	20.0 μl

I-PCR2 Program
- 94 °C 2 min, *1 time*
- 94 °C 20 sec, 60 °C 20 sec, 72 °C 3 min, *30 times*
- 72 °C 5 min, *1 time*

Fragment Characterization

- Use 5 μl of the PCR2 reaction for result characterization. Use 1.5% agarose gel in order to have a good separation of the fragments.

- For vector cloning of the PCR fragments, the PCR reaction is first purified on a commercial PCR purification column. If using the pGEM-T vector (www.promega.com), the reaction mix for cloning in the vector is:

Column purified PCR2	5 µl
10 × Ligation buffer	2 µl
pGEM-T	0.5 µl
Ligase (3 U/µl)	1 µl
H$_2$O	11.5 µl
Total	20.0 µl

The ligation is performed overnight at 16 °C and half of the ligation product is used for transformation.

5.2.2
Reverse Genetic Approach

Reverse genetics is a systematic way of using gene sequence information to look for a phenotype in an effort to integrate this into biological function. Once the sequence of the gene of interest is known, one can identify a *Tnt1*-tagged mutant line in one or both of the following two ways – FST sequencing and screening DNA pools.

5.2.2.1 FST Sequencing
As already described, the plant genomic region that borders the *Tnt1* insert (FST) can be identified by using TD-PCR, TAIL-PCR, I-PCR or a combination thereof. The FSTs are then deposited in a database for public use. At the Samuel Roberts Noble Foundation, we created such a database (http://bioinfo4.noble.org/mutant) to house all the FSTs sequenced by various partners. So far, this database contains over 9000 nonredundant FSTs, but it may be expanded to over 100 000 in the next few years. The first step in a reverse genetic approach is, thus, to look into this database and see if the favorite gene of interest is tagged. If no FST that matches the gene of interest is found, one can resort to a PCR based strategy to screen DNA pools that encompass the whole population.

5.2.2.2 Screening DNA Pools
This strategy (schematized in Figure 5.4) uses a combination of one gene-specific primer and one *Tnt1*-specific primer to selectively amplify the tagged gene of interest from large DNA pools of *Tnt1* lines. Genomic DNA is extracted from individual lines and pooled together in a systematic manner such that fewer subsequent PCR reactions are required to screen the entire population for a particular gene of interest. To screen a very large population of 100 000 or more lines, usually a three-dimensional pooling strategy is adopted as seen, for example, in the T-DNA lines of *A. thaliana*. Since we currently have approximately 10 000 lines and we anticipate saturating the *M. truncatula* genome with less than 20 000 *Tnt1* lines, we resorted

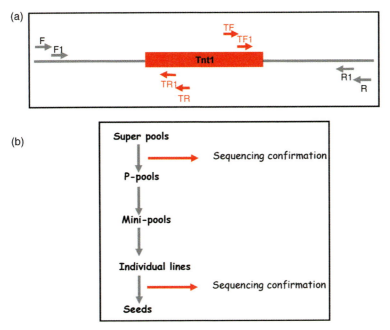

Figure 5.4 Schematic illustration of the reverse genetic screen: (a) illustration of primer positions and (b) reverse screening flowchart.

to a simpler and more efficient one-dimensional pooling strategy. Our superpool contains DNA pooled together from 500 independent lines. Each superpool is then divided into pools and subpools to make final identification of individual lines easier. This technique allows screening of 10 000 lines in 20 superpools with 40 PCR reactions. To increase the specificity, the first-round PCR products are diluted and amplified with a second primer pair that is nested to the first primers. The gene-specific primers should cover the entire length of the gene of interest. For this reason, we usually design two gene-specific primers – one from the 3′ end and the other from the 5′ end of the gene in opposite directions. Similarly, we use two *Tnt1*-specific primers pointing outside from both ends of the *Tnt1* element corresponding to the gene-specific primers. In this way 80 PCR reactions identify insert in any part of the gene in the 10 000 lines and the second round of 40 PCR reactions with nested primers confirm the presence of the insert. This turns out to be quite a powerful tool with astounding efficiency of over 90% for averagely sized gene. Chances are that one could identify multiple alleles for some genes.

5.3
Applications

The *M. truncatula Tnt1*-tagged population is a very useful resource for discovering gene function in legumes. Although the presence of multiple *Tnt1* inserts in each line poses some challenges to clean-up the background, the high probability of finding multiple alleles and the ability to separate most of the inserts by segregation offsets the challenge to make *Tnt1* a uniquely suited and fascinating tool for legume functional genomics. When attempting to use the *Tnt1*-tagged *M. truncatula* resource, one will encounter one or more of the following scenarios.

- Line with a mutant phenotype – no FSTs identified.
- Line with a mutant phenotype and FSTs already identified.
- FST sequence in the *Tnt1* database matches a gene of interest – no mutant phenotype is described in that line.
- Have a gene to work with – no FST or mutant phenotype.

5.3.1
Line with a Mutant Phenotype – No FSTs Identified

Once a year we (www.noble.org) undertake a *Tnt1* mutant phenotype screening event and record obvious visible phenotypes for 1000–2000 lines under one set of environmental conditions whether or not we have initiated sequencing of FSTs from that line. It is therefore possible that you may find a line(s) that shows an interesting phenotype (e.g., altered organ development (AOD), but the *Tnt1* insertion sites have not been isolated or identified yet in that line. One straight forward possibility is, using one of the PCR techniques described above, to isolate most of the insertion sites (FSTs) present in this line. Once these loci are sequenced, the genetic link between the mutant loci and the AOD phenotype can be analyzed by PCR in a segregating population (see Section 5.3.2 for further analysis). At this stage it might be interesting to backcross the line in order to reduce the *Tnt1* copy number. This approach, starting from a line with a *nod*-minus phenotype was applied in order to characterize the first *Tnt1*-tagged symbiotic line [10] carrying an insertion in the symbiotic *NIN* gene.

5.3.2
Line with a Mutant Phenotype and FSTs Already Identified

An interesting phenotype was described in one line and in this line several FSTs were already identified. The original line should be backcrossed to wild-type plants and the progeny analyzed for cosegregation of a putative phenotype with the *Tnt1* insertion sites. The disrupted region can be easily followed in the progeny by PCR, using one oligonucleotide corresponding to one *Tnt1* end and one oligonucleotide in the sequence of the FST locus. Using one oligonucleotide on each side of the insertion site will in addition indicate for each plant if the FST locus is wild-type, heterozygous, or homozygous for the insertion. The genetic link between the phenotype and the homozygous mutant loci can then be confirmed in a segregating population. An insertion site will be considered as a candidate if all mutant plants are homozygous mutant for the tagged locus and if all plants with a wild-type phenotype are wild-type or heterozygous for the tagged locus.

Note that the genetic link does not certify that the tagged locus is responsible for the mutation. This genetic link between one FST locus and the mutation simply indicates that the mutated gene is close to but might be different from the *Tnt1*-tagged region. In order to demonstrate that the tagged locus is really responsible for the mutation, more alleles can be isolated using the reverse genetic approach or by searching the FST database (see Section 5.3.3). If no other allele can be described complementation of the mutation will be needed. The segregation analysis of several alleles was applied to characterize the *SGL1* gene necessary for proper leaf development [11].

5.3.3
FST Sequence in the *Tnt1* Database Matches a Gene of Interest – No Mutant Phenotype is Described in that Line

As mentioned above, the phenotype screen is based on one set of conditions, mainly low nitrogen and low phosphorus symbiotic screen, and phenotyping was incomplete in the sense that only major visible phenotypes were recorded. So, if you find your gene of interest tagged, the first thing you need to do is grow the seeds under your own set of conditions expected to yield a phenotype and genotype your segregating population by PCR. As described in Section 5.3.2, using two gene-specific primers on either side of the insertion site, you can genotype each plant of the segregating population as heterozygote, homozygote or wild-type for the disruption. You will need to backcross the homozygous mutant to wild-type and continue as above.

5.3.4
Have a Gene to Work With – No FST or Mutant Phenotype

This is where the reverse screening of DNA pools comes into play. Design two gene-specific primers, one on either end of your gene, and use these in combination with *Tnt1*-specific primers and screen the entire superpools of DNA. Those that show positive signals need to be repeated with nested primers for confirmation. We also usually sequence the PCR product directly to confirm identity. In those superpools that show confirmed positive signals, the same primer pairs are used in the subsequent screening of pools and subpools. Finally, individual lines of all the positive subpools are tested by the same primer combination one-by-one to identify an individual line (or lines in the case of multiple alleles) that harbor(s) the insertion. Once a tagged line is identified, the same procedure is followed as above to further characterize the mutant including growing under permissive conditions and genotyping the segregating population for a possible phenotype. The homozygous mutant should be backcrossed to wild-type plants and the progeny analyzed for cosegregation of a putative phenotype with the insertion. For efficient reduction of the *Tnt1* copy number, the backcrosses to wild-type plants could be done at each generation (i.e., on the F1 plants). By doing this, we can get rid of half of the inserts at each backcross and the lines with reduced number of *Tnt1* inserts will be more rapidly obtained. Ideally several alleles should be studied in parallel which attributes the observed phenotype to the disrupted gene. If no alleles are found, it is necessary to transform the homozygous mutant with the wild-type genomic fragment to rescue the mutant phenotype. The *mtpim* mutant described by Benlloch *et al.* [9] was isolated using this reverse genetic approach on the small collection described by d'Erfurth *et al.* [7].

5.4
Perspectives

The *Tnt1*-tagged *M. truncatula* collection will boost legume research in the same way as the T-DNA collections have allowed the development of modern molecular genetics in *Arabidopsis*. The currently available 10 000 lines represent over 250 000 *Tnt1* inserts, most of which are independently distributed in the gene-rich regions of the genome. When large-scale FST sequencing is performed on these collections, it will be possible to map most of the *Tnt1* inserts in the *Medicago* genome (www.medicago.org). Finding a mutant in your favorite gene will then be a matter of checking the web site (http://bioinfo4.noble.org/mutant) and ordering the seeds from stock centers analogous to the Salk Institute T-DNA lines. The development of this FST data base corresponding to the majority of the *Tnt1* inserts in the population will, thus, represent a very valuable tool for the scientific research community.

Acknowledgments

This work was supported by the Samuel Roberts Noble Foundation, in part by National Science Foundation Plant Genome grant (DBI 070 3285), and by the European Union (EU FP6-GLIP project FOOD-CT-2004-506 223).

References

1 Benetzen, J.L. (2000) Transposable element contributions to plant gene and genome evolution. *Plant Mol. Biol.*, **42**, 251–269.
2 Neumann, P., Koblízková, A., Navrátilová, A., and Macas, J. (2006) Significant expansion of *Vicia pannonica* genome size mediated by amplification of a single type of giant retroelement. *Genetics*, **173**, 1047–1056.
3 Okamoto, H. and Hirochika, H. (2000) Efficient insertion mutagenesis of *Arabidopsis* by tissue culture-induced activation of the tobacco retrotransposon *Tto1*. *Plant J.*, **23**, 291–304.
4 Courtial, B., Feuerbach, F., Eberhard, S., Rohmer, L., Chiapello, H., Camilleri, C., and Lucas, H. (2001) *Tnt1* transposition events are induced by *in vitro* transformation of *Arabidopsis thaliana*, and transposed copies integrate into genes. *Mol. Genet. Genomics*, **265**, 32–42.
5 Yamazaki, M., Tsugawa, H., Miyao, A., Yano, M., Wu, J., Yamamoto, S., Matsumoto, T., Sasaki, T., and Hirochika, H. (2001) The rice retrotransposon *Tos17* prefers low-copy-number sequences as integration targets. *Mol. Genet. Genomics*, **265**, 336–344.
6 Grandbastien, M.A., Spielmann, A., and Caboche, M. (1989) *Tnt1*, a mobile retroviral-like transposable element of tobacco isolated by plant cell genetics. *Nature*, **337**, 376–380.

7 d'Erfurth, I., Cosson, V., Eschstruth, A., Lucas, H., Kondorosi, A., and Ratet, P. (2003) Efficient transposition of the *Tnt1* tobacco retrotransposon in the model legume *Medicago truncatula*. Plant J., **34**, 95–106.

8 Tadege, M., Wen, J., He, J., Tu, H., Kwak, Y., Eschstruth, A., Cayrel, A., Endre, G., Zhao, P.X., Chabaud, M., Ratet, P., and Mysore, K.S. (2008) Large-scale insertional mutagenesis using the *Tnt1* retrotransposon in the model legume *Medicago truncatula*. Plant J., **54**, 335–347.

9 Benlloch, R., d'Erfurth, I., Ferrandiz, C., Cosson, V., Beltran, J.P., Canas, L.A., Kondorosi, A., Madueno, F., and Ratet, P. (2006) Isolation of *mtpim* proves *Tnt1* a useful reverse genetics tool in *Medicago truncatula* and uncovers new aspects of AP1-like functions in legumes. Plant Physiol., **142**, 972–983.

10 Marsh, J.F., Rakocevic, A., Mitra, R.M., Brocard, L., Sun, J., Eschstruth, A., Long, S.R., Schultze, M., Ratet, P., and Oldroyd, G.E.D. (2007) *Medicago truncatula* NIN is essential for rhizobial-independent nodule organogenesis induced by autoactive calcium/calmodulin-dependent protein kinase. Plant Physiol., **144**, 324–335.

11 Wang, H., Chen, J., Wen, J., Tadege, M., Li, G., Liu, Y., Mysore, K.S., Ratet, P., and Chen, R. (2008) Control of compound leaf development by FLO/LFY ortholog single leaflet1 (SGL1) in *Medicago truncatula*. Plant Physiol., **146**, 1759–1772.

12 Melayah, D., Bonnivard, E., Chalhoub, B., Audeon, C., and Grandbastien, M.A. (2001) The mobility of the tobacco *Tnt1* retrotransposon correlates with its transcriptional activation by fungal factors. Plant J., **28**, 159–168.

13 Liu, Y.G., Mitsukawa, N., Oosumi, T., and Whittier, R.F. (1995) Efficient isolation and mapping of *Arabidopsis thaliana* T-DNA insert junctions by thermal asymmetric interlaced PCR. Plant J., **8**, 457–463.

14 Liu, Y.G., Chen, Y., and Zhang, Q. (2005) Amplification of genomic sequences flanking T-DNA insertions by thermal asymmetric interlaced polymerase chain reaction. Methods Mol. Biol., **286**, 341–348.

Part II
Mutation Discovery

6
Mutation Discovery with the Illumina® Genome Analyzer

Abizar Lakdawalla and Gary P. Schroth

Abstract

The introduction of the Illumina® Genome Analyzer has transformed the discovery of mutations and genome variations. The sequencing-by-synthesis chemistry used with the Genome Analyzer generates more than 35 Gb of high-quality sequence data in 1 week from more than 150 million templates. The increasing read-lengths (currently longer than 100 bp), coupled with short- and long-insert paired-end sequencing capability, and the large number of sequencing templates result in a highly efficient platform for discovering micro- and macro-lesions in complex genomes. The massive data throughput combined with the easiest workflow of any next-generation sequencing technology provides a unique method to discover and validate all genomic variations. Single or multiple nucleotide variations, small and large insertions and deletions, inversions, copy number variations, translocations and complex genomic re-arrangements can be enumerated simultaneously and with single nucleotide resolution. In this chapter, we provide an overview of several effective strategies for maximizing single nucleotide polymorphism (SNP) discovery with genomes of different complexity and size. A step-by-step Genome Analyzer protocol is provided to efficiently discover and validate single nucleotide variations, SNPs, and other genome variations.

6.1
Introduction

6.1.1
Overview of the Illumina Genome Analyzer Sequencing Process

The Illumina® Genome Analyzer system performs sequencing on more than 150 million templates (100–500 bp) in parallel to generate about 35 billion bases of data from one run.

The Handbook of Plant Mutation Screening. Edited by Günter Kahl and Khalid Meksem
Copyright © 2010 WILEY-VCH Verlag GmbH & Co. KGaA, Weinheim
ISBN: 978-3-527-32604-4

FIGURE LEGENDS

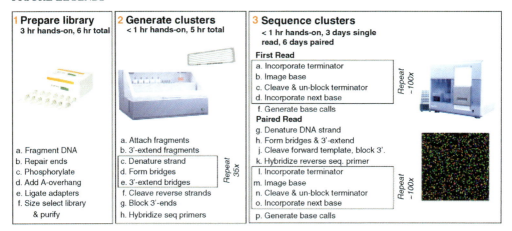

Figure 6.1 Overview of the Genome Analyzer sequencing process. (1) Genomic DNA fragments are blunt-ended, ligated to the Illumina sequencing adaptors, PCR-selected, and purified. (2) The ligated fragments are introduced into an eight-channel flow cell and bridge-amplified *in situ* on the Cluster Station to produce more than 100 million clusters. (3) The flow cell is transferred to the Genome Analyzer where the more than 150 million clusters are sequenced in parallel for more than 100 bases. Reverse templates are resynthesized *in situ* and the other end of the template sequenced.

The sequencing process consists of three stages (Figure 6.1):

- Preparation of libraries from the sample DNA or cDNA.
- Amplification of the single library fragments by a process of bridge amplification in a flow cell.
- Sequencing of the amplified library templates in the flow cell to produce approximately 100 bp of sequence information from each end of the template.

6.1.2
Resequencing Strategies

Efficient discovery of natural or induced mutations associated with plant phenotypes requires the genomic analysis of a large number of diverse individuals at relatively high resolution to separate causal from noncausal variations [1]. The large number of reads produced by the Genome Analyzer can provide a sequencing depth of 10–50 times or greater; that is, each nucleotide position would be sampled more than 10–50 times providing a higher level of statistical confidence for single nucleotide variation validation and the capability to discover low-frequency single nucleotide polymorphisms (SNPs) in a pooled population. Based on the size of the genomic target, one of three primary strategies may be employed; resequencing pools of whole genomes (applicable to smaller genomes), sequencing targeted regions of the genome, and sequencing transcriptomes.

6.1.2.1 Resequencing Whole Genomes (Figure 6.2A)

Sequencing the whole genome provides a comprehensive view of all genetic variants – single nucleotide changes, insertions/deletions, translocations, inversions, large insertions and deletions, and copy number variations [2]. In the case of small genomes, such as the *Arabidopsis* spp. (around 100 Mb genomes), multiple individuals can be pooled together and sequenced in a single run of the Genome Analyzer [3]. At a throughput of around 35 Gb per run, about 5–10 *Arabidopsis* individuals can be pooled, for an average of 10–20 times sequencing depth for each individual or 100–200 times for the population. *Vitis* and *Cucumis* genomes (around 500 Mb size) can also be pooled and sequenced in a limited number of runs on the Genome Analyzer.

Variations in chloroplast DNA are important for evolutionary studies and for screening for induced mutations. The population impact of chloroplast genomes is estimated at about one-fourth of a nuclear locus (haploid state, uniparental transmission, and an average size of 150 kb) [4]. About 1000 pooled chloroplast DNAs can be sequenced in a run of the Genome Analyzer at approximately 100 times average sequencing depth for the pool. Individual samples can be tagged with a specific index to assign variation to a specific individual or groups of individuals. Cronn *et al.* [5] have described a comprehensive method based on polymerase chain reaction (PCR) amplification of chloroplast DNA from *Pinus* spp., followed by the introduction of a barcode index and sequencing on the Genome Analyzer.

6.1.2.2 Targeted Genome Selection

Four approaches, primarily dependent on the size of the region to be sampled, are commonly utilized for resequencing a fraction of the genome.

PCR/Long PCR (Figure 6.2B) For sequencing regions less than a few 100 kb, a series of overlapping PCR (0.1–2 kb) or long-PCR (2–10 kb) products from a sample are pooled together, fragmented, ligated to adaptors, amplified and sequenced on the Genome Analyzer [5, 6]. At 100 times average depth of sequencing, a few hundred samples (100 kb) can be sequenced in one lane of the eight-lane flow cell for a total of about a 1000 samples per run. Products from amplified fragment length polymorphism, Targeting induced local lesions in genomes (TILLING) and similar mutation screening protocols can also be pooled and sequenced to confirm mutations.

Bacterial artificial chromosomes/yeast artificial chromosomes (Figure 6.2C) Bacterial artificial chromosome (BAC)/yeast artificial chromosome (YAC) clones (100 kb–1 Mb) can be fragmented, and the fragments attached to a clone-specific index sequence, pooled, and sequenced on the Genome Analyzer.

Hybridization selection (Figure 6.2D) Sequencing of contiguous or noncontiguous regions larger than a few megabases, such as all known exons (around 25 Mb), can be performed by capture of target regions with a library of oligonucleotides [7]. Two general oligonucleotides capture approaches are used – a solid-phase method that captures and releases complementary DNA directly from oligonucleotides arrays, and a solution-based method in which oligonucleotides from arrays are excised and

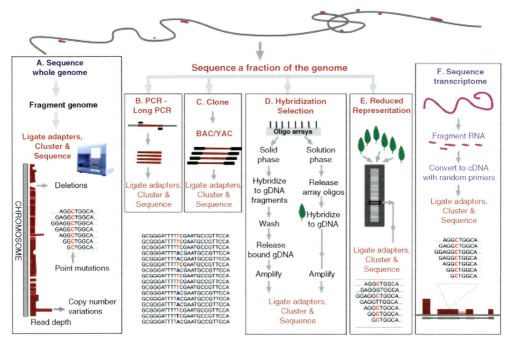

Figure 6.2 Resequencing strategies. Discovering genome variations by sequencing whole genomes (A), targeting regions of the genomes by PCR (B), sequencing of BAC/YAC clones (C), solid-phase hybridization selection or solution-based hybridization selection (D), or by creating reduced representation libraries (E). Discovery of cSNPs by sequencing RNA and alignment of expressed tags to reference genome (F).

then used to selectively amplify the target pool in solution [8]. In solid-phase capture, fragmented DNA is hybridized to customized arrays containing 100 000s of oligonucleotides complementary to the regions of interest, unbound DNA fragments are washed off, and bound DNA is released and processed for sequencing [9, 10]. In liquid-phase hybridization selection, oligonucleotides targeting the region of interest are synthesized on arrays with a cleavable base or sequence, the oligonucleotides are then cleaved off the array and these oligo pools are used to selectively amplify the target regions in a liquid phase [11].

Genome selection by restriction enzymes (Figure 6.2E) Sequencing of reduced representation libraries from restriction enzyme-digested genomic DNAs are useful to simultaneously discover mutations and mutation frequencies. Genomic DNA from 10–100s of different individuals is digested with a restriction enzyme and separated on a gel, a band is excised from this gel, and the extracted DNA is subjected to sequencing-by-synthesis (SBS) on the Genome Analyzer [12]. Sequencing of these reduced representation libraries results in the sampling of the same 1–10% fraction of the genome from multiple individuals. Each individual would be sequenced multiple times giving frequency estimates for variations in the sampled regions. Selection of a specific restriction enzyme requires the evaluation of fragment size distributions and digestion frequencies. Suitability of a restriction enzyme and the appropriate band size to be excised can be indicated by Southern blots probed with labeled genomic DNA to differentiate repetitive from unique regions. Based on the primary focus, sequencing libraries can be created from the repetitive regions or the unique regions of the genome. The method has been successfully in producing SNP arrays for multiple species (Illumina Bovine, Ovine, and Porcine SNP arrays), thus facilitating association studies.

6.1.2.3 Sequencing Transcriptomes (Figure 6.2F)

Sequencing mRNA provides a cost-effective method to generate protein coding region SNP data as sampling the transcriptome reduces the sequencing throughput requirements by two orders of magnitudes in large genomes. Transcriptome sequencing also provides valuable quantitative gene expression information and differences in alternatively spliced RNA isoforms without *a priori* genome annotation information. PolyA RNA is extracted from the same tissue(s) from multiple individuals, fragmented with zinc or magnesium salts, and the RNA fragments converted to cDNA by random priming. These cDNA libraries are then sequenced on the Illumina Genome Analyzer to discover mutations located in the expressed regions of the genome, quantitative gene expression data, and alternative splicing profiles [13–15].

6.2 Methods and Protocols

Steps required to complete paired-end sequencing of genomic DNA are described below [16]. The overall process takes about 6 h for preparing short-insert paired-end libraries, 5 h for cluster amplification, and about 5 days for the generation of 50 bp × 2 paired-end sequence data. Sequencing is enabled for more than 100 bp × 2 paired-end reads. Please refer to the most current product documentation for up-to-date protocols, instrument operation guidelines, and software analyses.

Library Preparation

DNA fragments produced by mechanical fragmentation, enzymatic digestion, cDNA synthesis, from immunoprecipitates, or other means are blunted and phosphorylated (Figure 6.3a). An A-base overhang is added and the Illumina sequencing adaptors are added by ligation (Figure 6.3a). Ligated products are selected by PCR during which process additional sequences are incorporated with tailed primers and the amplified products are separated by gel fractionation followed by purification of a specific size fragments from the gel (Figure 6.3b).

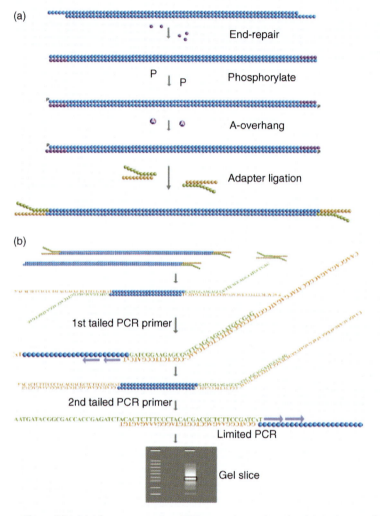

Figure 6.3 (a) Library preparation. DNA fragments are end-repaired to produce blunt ends with a combination of polymerase and exonuclease activity. Blunt-ended fragments are phosphorylated and an A-base overhang is added. The Illumina sequencing adaptors are then added by ligation. These adaptors contain sequences that are complementary to the sequencing primers used for paired-end sequencing. The Y-shaped adaptors prevent concatenation of the adaptors and the overhanging A base insures that ligation is asymmetric such that all fragments will contain a copy of the correct adaptor at each end. (b) Ligated fragments are purified from the pool by limited PCR. PCR is performed with tailed primers that introduce the sequence required for the DNA to bind to the oligos on the surface of the Illumina flow cell. The final PCR product is gel-purified and size-selected depending upon the exact application.

DNA Fragmentation

(Only required if intact genomic DNA is being used.)

1. Place 0.5–5 µg of DNA in 50 µl TE and 700 µl of nebulization buffer in a disposable nebulizer. Mix and cool on ice. Connect to a compressed air source and pressurize the nebulizer at 32–35 psi for 6 min on ice.
2. Recover the DNA (400–600 µl) with a pipette, rinse the nebulizer with 2 ml of the binding buffer from the QIAquick™ PCR Purification Kit.
3. Add the DNA and rinse to the QIAquick spin column, centrifuge at 10 000 × g for 30–60 s. Discard flow-through. Add 0.75 ml of QIAquick PE Buffer. Spin and discard the flow-through again. Spin 1 min to dry the column. Place column in a clean microcentrifuge tube, add 30 µl of QIAquick Buffer EB, incubate 1 min, and elute DNA by spinning for 1 min.

DNA End Repair

1. To 30 µl of fragmented DNA, add 45 µl of water, 10 µl of ligase buffer, 4 µl of dNTPs, 5 µl of T4 DNA polymerase, 1 µl of Klenow, and 5 µl of polynucleotide kinase. Incubate at 20 °C for 30 min.
2. Purify the DNA with the QIAquick column (as in Step 3 in "DNA Fragmentation") eluting with 32 µl of Buffer EB.

A-Addition

1. To purified DNA (32 µl) add 5 µl of Klenow buffer, 10 µl of ATP, and 3 µl of exo-Klenow. Incubate 30 min at 37 °C.
2. Purify with the QIAquick MinElute Kit, eluting with 10 µl of Buffer EB.

Adaptor Ligation

1. To purified DNA (10 µl) add 25 µl of DNA ligase buffer, 10 µl of PE Adaptor mix, and 5 µl of DNA ligase. Incubate for 15 min at room temperature.
2. Purify with the QIAquick Kit eluting in 30 µl of Buffer EB (see Step 3 in "DNA Fragmentation").

Size Selection and Gel Purification

1. To purified DNA (30 µl) add 10 µl of loading buffer and electrophorese on a 2% 1 × TAE agarose gel with ethidium bromide (400 ng/ml) at 120 V for 120 min. Load with an appropriate size standard.
2. Collect DNA from the gel by excising a 2-mm slice of the desired size using the DNA size standard as a guide. Optimal size range is 100–500 bp.
3. Purify the DNA from the gel slice with a QIAquick Gel Extraction Kit followed by purification on a QIAquick column, eluting in 30 µl of EB (see Step 3 in "DNA Fragmentation").

Enrichment with PCR

1. Add the purified DNA from Step 3 of "Size Selection and Gel Purification" to water to a final volume of 23 μl. Use 10 μl of purified DNA if starting with 0.5 μg of genomic DNA or 1 μl if starting with 5 μg of genomic DNA. Add 25 μl of Phusion DNA polymerase, 1 μl of primer PE 1.0, and 1 μl of PE 2.0. Preheat to 98 °C for 30 s and amplify with 10–12 cycles of 98 °C/10 s, 65 °C/30 s, 72 °C/30 s followed by a final extension at 72 °C for 5 min. Hold at 4 °C.
2. Purify with the QIAquick Kit eluting in 50 μl of Buffer EB.

Size Selection and Gel Purification

1. The PCR product from the previous step can be gel purified by loading the PCR reaction products on a 2% 1 × TAE agarose gel, electrophoresing, excising the band, and purifying the DNA as described in Step 1 of "Size Selection and Gel Purification".
2. Measure 260/280 nm absorbance ratio (should be 1.8) and check library on an Agilent BioAnalyzer or equivalent.

Preparing DNA Clusters

Single molecule fragments are captured by oligonucleotides attached to the surface of an eight-channel flow cell. The flow cell oligonucleotides are complementary to the adaptors attached to the DNA fragments during the library preparation step. The oligonucleotides are of two populations, each with a different cleavage specificity (5'-PS-TTTTTTTTTTCAAGCAGAAGACGGCA-TACGAGoxoAT-3' and 5'-PS-TTTTTTTTTTAATGATACGGCGACCACCGA-GAUCTACAC-3'). Library templates hybridized to the lawn of oligonucleotides are extended by DNA polymerase (Figure 6.4a). The original templates are denatured and washed off leaving an attached copy of the template (Figure 6.4b). Newly synthesized copies of the template are amplified by isothermal bridge amplification. Free ends of the bound single-stranded DNA (ssDNA) molecules loop over and hybridize to adjacent lawn oligonucleotides (Figure 6.4c). The hybridized lawn oligonucleotides now serve as a priming site to copy the single-stranded loop creating a double-stranded loop (Figure 6.4d). The newly synthesized double-stranded DNA (dsDNA) bridge is denatured resulting in two ssDNA strands each attached through the lawn oligonucleotides to the glass surface (Figure 6.4e). These ssDNA again loop over to form two new bridges (Figure 6.4f) and this process continues till you have a tight cluster of dsDNA molecules (Figure 6.4g). Every original location where a single DNA template had bound will now contain a cluster of forward and reverse copies of the original template. The reverse strands are now removed by excising at a single base position on one of the two oligonucleotide populations (Figure 6.4h). The 3' ends are blocked and a sequencing primer is hybridized (Figure 6.4i). The whole

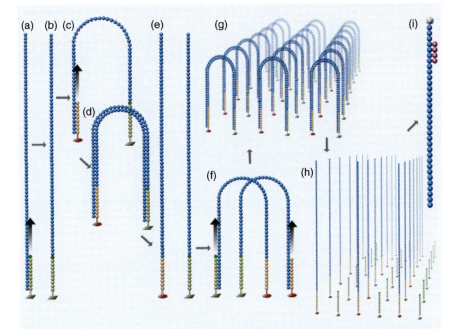

Figure 6.4 Cluster generation. Note that all of the steps described here are automatically performed on the Illumina Cluster Station instrument that is part of the Genome Analyzer system. DNA templates are hybridized to a lawn of oligonucleotide in the flow cell. Single-stranded, single-molecule DNAs are hybridized to the flow cell, extended, and then the original template is denatured and discarded (a and b). The bound DNA is then copied by solid-phase bridge amplification, which involves looping over of the ssDNA to hybridize to an adjacent lawn oligonucleotide (c), extension from the oligonucleotide to produce a dsDNA (d) followed by additional cycles of denaturation (e), bridge formation (f), and extension to form a discrete clonally amplified cluster containing both forward and reverse strands (g). Reverse strands are cleaved at one of the two lawn oligonucleotide families, and then the hybridized strand is washed out of the flow cell (h). Finally, 3′ ends of the templates are blocked and a universal sequencing primer is hybridized to each cluster of templates in the flow cell (i).

process occurs in parallel to generate more than 100 million clusters that are now ready for sequencing.

The cluster amplification process is performed on the Cluster Station as in the following protocol.

Template Hybridization and Amplification

1. Thaw frozen reagents at room temperature and place on ice.
2. Dilute DNA to a final concentration of 0.5 nM in 18 μl of EB solution, add 1 μl of 2 N NaOH. Vortex and incubate for 5 min at room temperature.
3. Further dilute the DNA to a final concentration of 1–8 pM to a final volume of 1 ml of ice-cold hybridization buffer. Place 120-μl aliquots of the DNA in an

eight-well strip tube. Between one and eight libraries can be run simultaneously.
4. Start the Cluster Station software, and select the recipe *Amplification_Linearization_Blocking_v....* Click *Start*, the software will walk you through all the steps below.
   - Preparing and loading Cluster Station reagents (15 ml amplification mix, 10 ml of wash buffer, 8 ml formamide, 10 ml amplification premix, 15 ml hybridization buffer).
   - Prewashing the Cluster Station fluid lines.
   - Loading of the flow cell, hybridization manifold, and reagents.
   - Pre-rinsing the flow cell with hybridization buffer (in eight-well strip tube containing 140 µl in each well).
   - Loading of DNA templates into the flow cell (instrument will hybridize DNA over 25 min).
   - Washing of the flow cell to remove unbound DNA with wash buffer (in an eight-well strip tube containing 100 µl in each well).
   - Initial 3' extension of templates hybridized to flow cell oligos (extension mix in eight-well strip tube containing 120 µl in each well).
   - Loading of the amplification manifold and amplification of individual templates to create clonal clusters (2.5 h). Blocking of amplified DNA.
   - Hybridization of sequencing primers.
   - Removal of flow cell.

Sequencing Forward Strand

Sequencing is performed one base at a time with reversible fluorescently labeled terminators (Figure 6.5). All four bases are present during an incorporation event resulting in increased fidelity due to the natural competition of the four bases for the polymerase. After an incorporation cycle the fluorescent color associated with a cluster is imaged. The fluorescent dye and the reversible terminator are now removed allowing the incorporation of the next base. This process is repeated for more than 75 cycles to simultaneously sequence all of the more than 100 million clusters. The reversible terminators ensure that only one base is added at every cycle preventing homopolymeric sequencing errors. Direct color encoding of sequence reads (one base = one color) simplifies data analysis for *de novo* sequencing, sequencing of bisulfite converted genomes, and so on. The following protocol describes the operation of the Genome Analyzer.

Prerun

1. Restart the Genome Analyzer instrument control computer, and log in with the provided user name and password. Ensure that there is enough space on

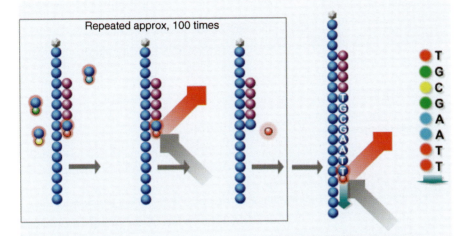

Figure 6.5 Sequencing forward strand. Incorporation takes place in the presence of all four reversibly terminated and fluorescently labeled nucleotides. The clusters are imaged, and the fluorescent dye and the reversible terminator are cleaved off readying the clusters for the next incorporation cycle. These cycles are repeated to derive the sequence of bases.

the data drive (delete or reformat if required). Double click icon to start the Genome Analyzer software.
2. Prewash: load a waste flow cell, and load the wash solution and water. In the software, open the wash recipe file and click on start. The fluidics will be washed for 20 min.
3. Prepare reagents for the sequencing run:

Incorporation mix	Mix 5 ml of 10 × incorporation buffer, 300 µl of Mg^{2+}, 43 ml of dNTP diluent. Add the ff-dNTP. Mix and filter. Add the SBS polymerase.
Cleavage mix	Add 5 ml of 10 × cleavage buffer and 5 ml of the cleavage reagent to the tube containing 40 ml of cleavage mix diluent. Mix. Add the provided cleavage mix additive to the mixture. Filter. Place on ice.
1 × Cleavage buffer	Add 7 ml of the 10× cleavage buffer to 250 ml of cleavage buffer diluent. Mix and store on ice.

4. Load the reagents onto the Genome Analyzer.
5. Prime the system by running the Prime recipe.
6. Clean and install the flow cell, the prism, and check for leaks on the flow cell. Add oil to the prism–flow cell interface.

First Base Incorporation

1. Select *File → Open Recipe* and double-click on *FirstBase.xml*. From the *Run* menu, select *Start*. Dismiss the *Autofocus Calibration* dialog box by clicking on the *No* button.
2. The Genome Analyzer will take approximately 35 min to add the first base and then pause.

Adjusting Focus

1. Align the X-axis by clicking on the *Manual Control/Setup* tab and selecting 0 values for X, Y, and Z. Press *Enter*, this will move the stage to the bottom of the left edge of lane 1 of the flow cell.
   - Set imaging parameters to: *Laser* – Green, *Filter* – None, *Exposure* – 4 ms. Click *Take Picture*. The edge of lane 1 will appear at the center of the screen, aligned with the cross-hair.
   - If the cross-hair is not aligned, type in new values for X (and *Enter*) till the cross-hair is aligned.
   - Select *Instrument → Set Coordinate System → Set Current X as Origin → OK*.
2. Calibrate the Z-axis.
   - In the Tile area, set the coordinates to Lane 4, Column 1, Row 50 (of the flow cell). Set imaging parameters to: *Laser* – Green, *Filter* – None, *Exposure* – 4 msec. Click *Take Picture*. Adjust the Z-value till the clusters are in sharp focus.
   - Select *Instrument → Set Coordinate System → Set Current Z as Origin → OK*.
   - Take images in lane 1 and 8 and note the Z value for being in focus. The differences in Z-value (Tilt) should be not more than 15 000.
3. Autofocus calibration. Click on the *Run* tab, highlight the *UserWait* step right before the "Incorporation" line in Cycle 1, if it is not already selected. Click *Resume*. Click *Yes*. The software automatically performs an autocalibration and shows a table of values. If the values are within the specified range, click *Accept* and then click *OK*.

Sequencing Run, First Read (Forward Strand)

1. Select *File → Open Recipe*. Open the recipe for a 76 cycle paired-end run and click *Start*. Click *OK* to accept the name of the run folder. When the *Autofocus Calibration* dialog box appears, click *No* (you have already calibrated) and the Genome Analyzer starts sequencing.
2. Sequencing run monitoring.
   - Use the included Goldcrest software to review run parameters (cluster intensities, cluster background, number and size of clusters, X–Y coordinates of each cluster, base call confidence levels, etc.).
   - The Run Browser software provides a graphical view of run metrics after a run. The *Flow Cell* window shows a graphical overview of the flow cell with data for every cycle and every tile. The *Report* window enables you to create textual reports on the same data in Crystal Reports (.rpt), .pdf, .xls, .doc, or.

rtf. The *Metric Deviation Report* window summarizes significant cycle-to-cycle deviations of key quality control values, so that problematic cycles in the run can be identified. The *ImageViewer* displays the image for a selected tile. *Chart Windows* allow you to monitor run quality for selected tiles.
- At the end of the first read, ensure that the robocopy script has completed data transfer of all files to the specified network storage.

Sequencing Reverse Strand

After completion of sequencing of the forward strand, the newly formed strand is denatured and washed off, and the reverse strand is resynthesized by allowing the forward template to form a loop with adjacent lawn oligonucleotides followed by 3′ extension from the hybridized lawn oligonucleotides to make a copy of the forward strand. Only the original forward strand is then excised at a cleavable base located at the base of the forward strand. The reverse strand is then sequenced just as before. The reverse strand sequencing process on the Genome Analyzer is described in the following protocol.

1. Delete the run folder from the instrument data drive
2. Prepare reagents for the paired-end module and place in the numbered location in the paired-end module:

Reagent 9	Mix ultra pure water (1600 µl) and 5 × deprotection buffer (400 µl). Place on ice till ready to load.
Reagent 10	Mix ultra pure water (1560 µl) and 5 × deprotection buffer (400 µl), add deprotection enzyme (40 µl). Place on ice till ready to load.
Reagent 11	Mix water (1680 µl), 10 × linearization 2 buffer (200 µl), bovine serum albumin (20 µl), linearization 2 enzyme (100 µl).
Reagent 12	Mix 1 × blocking buffer (1820 µl, Reagent 18), 2.5 mM ddNTP (80 µl), blocking enzyme A (24 µl), blocking enzyme B (100 µl).
Reagent 13	Mix cluster premix (10 ml), 10 mM dNTP Mix (200 µl), *Bst* DNA polymerase (100 µl). Place on ice till ready to load.
Reagent 14	Prepare the cluster premix in a 50 ml tube by mixing water (15 ml), cluster buffer (3 ml), 5 M betaine (12 ml). Filter. Transfer 10 ml of the filtered cluster premix into a 15-ml Falcon tube. Remaining 10 ml will be used to prepare Reagent 13.
Reagent 15	Transfer 8 ml of formamide into a 15-ml Falcon tube.
Reagent 16	Mix hybridization buffer (1492.5 µl) and Rd 2 PE Seq Primer (7.5 µl).
Reagent 17	Mix water (1800 µl) and 10 × linearization 2 buffer (200 µl).

Reagent 9	Mix ultra pure water (1600 µl) and 5 × deprotection buffer (400 µl). Place on ice till ready to load.
Reagent 18	Mix water (4500 µl) with 10 × blocking buffer (500 µl).
Reagent 19	Transfer 4 ml of 0.1 N NaOH into a 15-ml Falcon tube and place.
Reagent 20	Transfer the TE solution into a 15-ml Falcon tube.
Reagent 21	Transfer 10 ml of wash buffer into a 15-ml Falcon tube and place in PE module.

3. Resynthesis of the reverse strand (The reverse strand is generated from the forward strand by discarding the sequencing strand followed by allowing the forward DNA template to loop over and hybridize to flow cell oligos. The single stranded loop is extended from the oligo to produce a dsDNA loop. The loops are denatured and the original forward strand is selectively cleaved leaving the reverse strand to be sequenced).
   - Open the paired-end module recipe xml file and click start. Priming and resynthesis will be automatically performed. Load second-read SBS reagents when prompted and click *OK* to resume the run. Press *OK* after the first base incorporation is complete and looks appropriate. The second read will now be completed automatically.
   - Postrun wash:
     a. Replace the sequencing reagents with wash solutions and water.
     b. Select *Run* → *Open Recipe* → *PostWash.xml* → *Start*.
     c. The wash cycle takes approximately 45 min.
   - Data analysis. For imaging each flow cell is divided into eight lanes, with two columns of imaging per lane (Figure 6.6). Each column is composed of 55 tiles with four monochrome images being generated per tile per cycle for a total of about 600 000 images from a 100 bp × 2 sequencing run. These images are automatically analyzed to extract cluster coordinates and the associated fluorescence intensities for every cycle for every cluster. The resulting intensity table is converted to base calls with quality values to produce a table of reads. These reads are then passed on to alignment or assembly software for resequencing or *de novo* sequencing applications. Multiple data sets can be compared by the Illumina variation detection software, CASAVA. Data output from CASAVA or from the raw reads can be exported into GenomeStudio™ – a sequence viewer that integrates array-based data with sequencing-based reads.

(*Continued*)

6.3 Applications

The Genome Analyzer provides an exceptional platform for resequencing and *de novo* assembly due to its ability to sequence both ends of DNA fragments that are a few

(Continued)

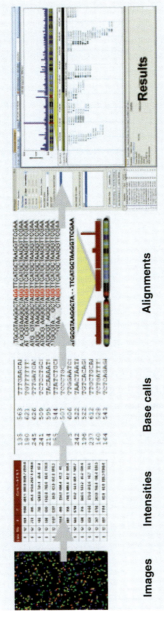

Figure 6.6 Data analysis. Generated images are analyzed to locate clusters and cluster intensity values are recorded for the four bases. The intensity tables are converted to base calls. Sequence reads for each cluster are then aligned against a reference. Alignments from multiple experiments are compared to generate a list of variants. The reads are viewed with a genome browser such as GenomeStudio™.

hundred base pairs to a few kilobases apart. Due to the large number of templates and increasing read-lengths the Genome Analyzer can replace many traditional methodologies based on hybridization such as the chromatin immunoprecipitation ChIP-chip, gene expression, BAC arrays, etc. In addition to genome variation discovery as described here, the Genome Analyzer has been used for the characterization of plant pathogens [17], coevolution of commensals [18], genome-scale discovery and profiling of DNA methylation changes in plants [19, 20], histone modifications and occupancy [21, 22], inter- and intrachromatin interactions, binding of transcription factors [23, 24], RNA polymerase run-on assays [25], profiling of coding [13], noncoding and small RNAs [26, 27], mapping RNA–RNA interactions [28], and studying RNA–protein interactions [29].

6.4
Perspectives

Next-generation sequencing platforms have revolutionized the functional analyses of genomes and genome variation. Offering a data throughput 100–1000 times greater than capillary electrophoresis-based sequencers it now is possible to sequence a broader range of genomes. The Illumina Genome Analyzer permits individual scientists to duplicate the data throughput of genome centers with surprising ease and at extremely low cost as reflected by the high publication rate (more than 450 peer-reviewed publications as of January 2010). With the highest daily throughput, the Genome Analyzer has facilitated ground-breaking discoveries in the biosciences, and has transformed our understanding of the genome, epigenome, transcriptome, and nucleic acid interactome. With continued improvements in the SBS chemistry and optimization of the platform, the Illumina Genome Analyzer will continue providing highly efficient large-scale detection of mutation and genome variations in complex genomes.

References

1. Zhu, C., Gore, M., Buckler, E.S., and Yu, J. (2008) Status and prospects of association mapping in plants. *Plant Genome*, **1**, 5–20.
2. Sturre, M.J.G., Shirzadian-Khorramabad, R., Schippers, J.H.M., Chin-A-Woeng, T.F.C., Hille, J., and Dijkwel, P.P. (2008) Method for the identification of single mutations in large genomic regions using massive parallel sequencing. *Mol Breeding*, **23**, 51–59.
3. Ossowski, S., Schneeberger, K., Clark, R.M., Lanz, C., Warthmann, N., and Weigel, D. (2008) Sequencing of natural strains of *Arabidopsis thaliana* with short reads. *Genome Res.*, **18**, 2024–2033.
4. Birky, C.W. (1978) Transmission genetics of mitochondria and chloroplasts. *Annu. Rev. Genet.*, **12**, 471–512.
5. Cronn, R., Liston, A., Parks, M., Gernandt, D.S., Shen, R., and Mockler, T. (2008) Multiplex sequencing of plant chloroplast genomes using Solexa sequencing-by-synthesis technology. *Nucleic Acids Res.*, **6**, e112.
6. Montgomery, K.T., Iartchouck, O., Li, L., Perera, A., Yassin, Y., Tamburino, A., Loomis, S., and Kucherlapati, R. (2008)

Mutation detection using automated fluorescence-based sequencing. *Curr. Protoc. Hum. Genet.*, **57**, 7.9.1–7.9.31.

7 Hardiman, G. (2008) Ultra-high-throughput sequencing, microarray-based genomic selection and pharmacogenomics. *Future Med.*, **9**, 5–9.

8 Garber, K. (2008) Fixing the front end. *Nat. Biotechnol.*, **26**, 1101–1104.

9 Hodges, E., Xuan, Z., Balija, V., Kramer, M., Molla, M.N., Smith, S.W., Middle, C.M., Rodesch, M.J., Albert, T.J., Hannon, G.J., and McCombie, W.R. (2007) Genome-wide *in situ* exon capture for selective resequencing. *Nat. Genet.*, **39**, 1522–1527.

10 Olson, M. (2007) Enrichment of super-sized resequencing targets from the human genome. *Nat. Methods*, **4**, 891–892.

11 Porreca, G.J., Zhang, K., Li, J.B., Xie, B., Austin, D., Vassallo, S.L., Leproust, E.M., Peck, B.J., Emig, C.J., Dahl, F., Gao, Y., Church, G.M., and Shendure, J. (2007) Multiplex amplification of large sets of human exons. *Nat. Methods*, **4**, 931–936.

12 Van Tassell, C.P., Smith, T.P.L., Matukumalli, L.K., Taylor, J.F., Schnabel, R.D., Lawley, C.T., Haudenschild, C.D., Moore, S.S., Warren, W.C., and Sonstegard, T.S. (2008) SNP discovery and allele frequency estimation by deep sequencing of reduced representation libraries. *Nat. Methods*, **5**, 247–252.

13 Huang, M.-D., Wei, F.-J., Wu, C.-C., Hsing, Y.-I.C., and Huang, A.H.C. (2009) Analyses of advanced rice anther transcriptomes reveal global tapetum secretory functions and potential proteins for lipid exine formation. *Plant Physiol.*, **149**, 694–707.

14 Simon, S.A., Zhai, J., Nandety, R.S., McCormick, K.P., Zeng, J., Mejia, D., and Meyers, B.C. (2009) Short-read sequencing technologies for transcriptional analyses. *Annu. Rev. Plant Biol.*, **60**, 305–333.

15 Wold, B. and Myers, R.M. (2008) Sequence census methods for functional genomics. *Nat. Methods*, **5**, 19–21.

16 Bentley, D.R., Balasubramanian, S., and Swerdlow, H.P. *et al.* (2008) Accurate whole human genome sequencing using reversible terminator chemistry. *Nature*, **456**, 53–59.

17 Boonham, N., Glover, R., Tomlinson, J., and Mumford, R. (2008) Exploiting generic platform technologies for the detection and identification of plant pathogens. *Eur. J. Plant Pathol.*, **121**, 355–363.

18 McCutcheon, J.P. and Moran, N.A. (2007) Parallel genomic evolution and metabolic interdependence in an ancient symbiosis. *Proc. Natl. Acad. Sci. USA*, **104**, 19392–19397.

19 Cokus, S.J., Feng, S., Zhang, X., Chen, Z., Merriman, B., Haudenschild, C.D., Pradhan, S., Nelson, S.F., Pellegrini, M., and Jacobsen, S.E. (2008) Shotgun bisulphate sequencing of the *Arabidopsis* genome reveals DNA methylation patterning. *Nature*, **452**, 215–219.

20 Lister, R., O'Malley, R.C., Tonti-Filippini, J., Gregory, B.D., Berry, C.C., Millar, A.H., and Ecker, J.R. (2008) Highly integrated single-base resolution maps of the epigenome in *Arabidopsis*. *Cell*, **133**, 523–536.

21 Boyle, A.P., Davis, S., Shulha, H.P., Meltzer, P., Margulies, E.H., Weng, Z., Furey, T.S., and Crawford, G.E. (2008) High-resolution mapping and characterization of open chromatin across the genome. *Cell*, **132**, 311–322.

22 Schones, D.E. and Zhao, K. (2008) Genome-wide approaches to studying chromatin modifications. *Nat. Genet.*, **9**, 179–191.

23 Johnson, D.S., Mortazavi, A., Myers, R.M., and Wold, B. (2007) Genome-wide mapping of *in vivo* protein DNA interactions. *Science*, **316**, 1497–1502.

24 Wold, B. and Myers, R.M. (2008) Sequence census methods for functional genomics. *Nat. Methods*, **5**, 19–21.

25 Core, L.J., Waterfall, J.J., and Lis, J.T. (2008) Nascent RNA sequencing reveals widespread pausing and divergent initiation at human promoters. *Science*, **322**, 1845–1848.

26 Gregory, B.D., O'Malley, R.C., Lister, R., Urich, M.A., Tonti-Filippini, J., Chen, H., Millar, A.H., and Ecker, J.R. (2008) A link between RNA metabolism and silencing affecting *Arabidopsis* development. *Dev. Cell*, **14**, 854–866.

27 Stacey, A., Simon, S.A., Zhai, J., Zeng, J., and Meyers, B.C. (2008) The cornucopia of small RNAs in plant genomes. *Rice*, **1**, 52–62.

28 German, M.A., Pillay, M., Jeong, D.-H., Hetawal, A., Luo, S., Janardhanan, P., Kannan, V., Rymarquis, L.A., Nobuta, K., German, R., De Paoli, E., Lu, C., Schroth, G., Meyers, B.C., and Green, P.J. (2008) Global identification of microRNA–target RNA pairs by parallel analysis of RNA ends. *Nat. Biotechnol.*, **26**, 941–946.

29 Montgomery, T.A., Yoo, S.J., Fahlgren, N., Gilbert, S.D., Howell, M.D., Sullivan, C.M., Alexander, A., Nguyen, G., Allen, E., Ahn, J.H., and Carrington, J.C. (2008) AGO1–miR173 complex initiates phased siRNA formation in plants. *Proc. Natl. Acad. Sci. USA*, **105**, 20055–20062.

7
Chemical Methods for Mutation Detection: The Chemical Cleavage of Mismatch Method

Tania Tabone, Georgina Sallmann, and Richard G.H. Cotton

Abstract

The identification of sequence variants is fundamental to plant breeding and research, and the detection and diagnosis of human disorders. The chemical cleavage of mismatch (CCM) is a method to detect point mutations in heteroduplex DNA, by cleaving the chemically modified bases at the site of mismatch. CCM accurately detects, localizes, and reveals the identify of a mutation in polymerase chain reaction (PCR) fragments up to 2 kb in length, allowing the rapid screening of large genomic regions. All four classes of mismatches can be detected and localized using this procedure by incubating PCR heteroduplexes with two mismatch-specific reagents. Hydroxylamine modifies unpaired cytosine residues and potassium permanganate modifies unpaired thymine residues. The DNA duplexes are further incubated with piperidine, which selectivity cleaves the chemically modified DNA bases. Cleavage products are size-separated by electrophoresis, revealing the identity and location of the DNA mutation. The CCM method can efficiently detect point mutations as well as insertions and deletions. The CCM protocol for detecting mutations using liquid-phase and solid-support phase analysis are described in this chapter.

7.1
Introduction

The detection of sequence variants is an important tool for plant breeding and research, as well as the field of medical genetics. While Sanger sequencing represents the "gold standard" in mutation detection, it is expensive to perform for the routine discovery of unknown mutations and may not readily detect heterozygous alleles. Therefore, alternative methods for detecting mutations are implemented to reduce the time and cost of directly sequencing long lengths of DNA that may or may not contain a sequence variant.

The Handbook of Plant Mutation Screening. Edited by Günter Kahl and Khalid Meksem
Copyright © 2010 WILEY-VCH Verlag GmbH & Co. KGaA, Weinheim
ISBN: 978-3-527-32604-4

Methods for mutation detection utilize three main approaches: changes to the physical property of mutant DNA, such as its melting profile or mobility through a gradient (e.g., high-resolution melt curve analysis, single-stranded conformational polymorphism analysis, denaturing high-performance liquid chromatography, and denaturing gradient gel electrophoresis [1–4]), the chemical modification of heteroduplex DNA (e.g., chemical cleavage of mismatch [CCM] [5]), and the enzymatic modification of heteroduplex DNA (e.g., cleavage by T4 endonuclease or the plant endonuclease, CEL1 [6, 7]).

Many of these mutation detection methods exploit the use of heteroduplex DNA, which is formed when polymerase chain reaction (PCR)-amplified DNA from a wild-type and a mutant allele are annealed together by heating followed by slow cooling. The two alleles result from a heterozygous individual or by combining an unknown sample with a wild-type reference sample. However, unlike many of these methods, CCM is not dependent on highly specific melting temperatures, in which a difference of just 1 °C can determine whether a mutation is detected [1–3]. In addition, these methods are commonly limited to the detection of mutations in PCR amplicons that are less than 400 bp in length, whereas CCM is routinely applied to sensitive mutation detection in PCR amplicons that are 2 kb in length. Furthermore, CCM does not require enzymatic modifications, which are often expensive and require optimization to obtain the ideal incubation time or temperature for sensitive and reproducible mutation detection.

Mutation discovery by CCM analysis relies on the modification and cleavage of base mismatches in heteroduplex DNA. Two commonly available reagents are used that specifically detect and modify mismatched pyrimidine bases. Hydroxylamine (NH_2OH) modifies mismatched cytosine (C) residues, while potassium permanganate ($KMnO_4$) modifies mismatched thymine (T) residues [5]. These two modification reactions expose the pyrimidine bases to piperidine-induced β-elimination, resulting in the cleavage of the DNA sugar-phosphate backbone at the site of the mismatched base. Size separation of the cleavage products by electrophoresis confirms both the presence and location of the mutation. In addition, the mismatch specificity of each of the chemical reactions allows the sequence of the mismatch to be inferred. This reduces the need for time-consuming and expensive sequencing reactions by limiting sequencing only to samples that appear to have a novel mutation to confirm the nature and location of the mismatch (Figure 7.1).

The four classes of DNA mismatches (C–T, A–G; G–T, A–C; T–T, A–A; and C–C, G–G), as well as deletion and insertion mutations that give rise to unmatched bases in

Figure 7.1 Schematic representation of the CCM method. Heteroduplexes are formed by denaturing (heating) and reannealing (slow cooling) wild-type and mutant alleles that have been amplified with a HEX-labeled forward primer (yellow) and 6-FAM-labeled reverse primer (blue). Treatment with hydroxylamine modifies mismatched C residues and treatment with potassium permanganate modifies mismatched T residues. The DNA backbone is selectively cleaved at the site of the modified bases after incubation with piperidine. Cleavage

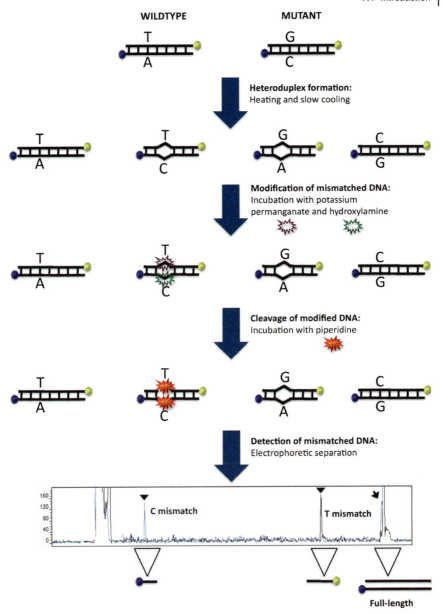

Figure 7.1 (*Continued*) fragments, and hence the presence and location of a mutation, are identified by electrophoretic separation on a polyacrylamide gel or capillary array. Using strand-specific fluorescent primers with hydroxylamine and potassium permanganate reactions electrophoresed as separate samples allows the assignment of the mismatched allele to the correct DNA strand, and permits the genotype to be inferred. In this schematic, the T allele is detected by the cleavage peak on the HEX-labeled sense strand and the C allele is detected by the cleavage peak on the 6-FAM-labeled antisense strand, suggesting that G and A alleles are present on the alternative strands to produce a C–T, A–G mismatch.

heteroduplex DNA, can all be detected using this procedure, as the unpaired bases are readily modified with the mismatch-specific reagents.

The CCM method is highly sensitive and as such is especially suited to the detection of mutations in diluted samples, particularly polyploid organisms (e.g., wheat, potatoes, cotton, and tobacco) and DNA pools (e.g., bulk segregant analysis) in which the presence of a mutation can be obscured by an excess of wild-type alleles. The sensitivity of CCM for mutation detection in diluted DNA samples has been extensively demonstrated in cancer research for the analysis of tumor samples [8–10].

The CCM method is accurate, reliable, and robust for the detection and genotyping of genetic mutations. In addition, the genotyping of known mutations requires only the application of the chemical modifying the mismatch of interest, thereby eliminating the need for two separate chemical incubations. For example, genotyping an A–C mismatch would only require the DNA sample to be incubated with hydroxylamine in order to detect the mismatched C residue. Alternatively, the reaction can be performed with potassium permanganate to detect the G–T mismatch on the antisense DNA strand.

Several modifications to the CCM protocol have been implemented since the method was first described, to improve the simplicity and specificity of the technique. In particular, CCM has been improved by the use of multiple, strand-specific 5′-labeled fluorophores [10, 11], which has increased the sensitivity of mutation detection, enables multiplexing to improve assay throughput and reduces the number of separation reactions that is required. The addition of different 5′-labeled fluorophores, such as hexachlorocarboxyfluorescein (HEX) to the forward primer and 6-carboxyfluorescein (FAM) to the reverse primer, allows the detection and assignment of the mismatched allele to the sense or antisense strand, permitting the genotype to be inferred (Figure 7.2). The introduction of binding DNA to a solid support allows the modifying reagents to be washed away for treatment with piperidine without an ethanol-precipitation step – this permits the adaptation of the technique to robotics for high-throughput automation, and allows easier handling and faster processing [9].

The CCM protocols for detecting mutations using liquid-phase and solid-support phase analysis are described in this chapter, as the two approaches are commonly used and have different advantages according to individual requirements. Liquid-phase analysis requires an ethanol-precipitation step; however, it is cheaper to perform and has a greater sensitivity for detecting mutations in fragments 2 kb or less in length compared to the solid-support phase. The binding of DNA to beads in solid-support phase analysis eliminates the need for an ethanol-precipitation step, allowing faster processing time or adaptation to robotics. However, this approach requires the addition of silica beads, which might not be desirable due to the increased expense. However this approach is ideal for the detection of mutations in PCR fragments 500 bp or less in length, especially for mutations close to the primer site as primer–dimers are removed during the washing steps.

Potassium permanganate modification for the detection of mismatch thymine residues

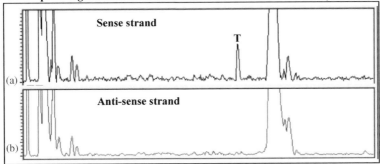

Hydroxylamine modification for the detection of mismatch cytosine residues

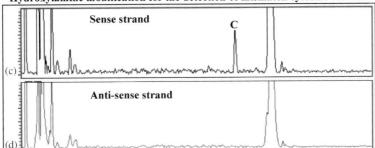

Figure 7.2 CCM analysis of a 650-bp PCR product with a T–G mismatch. (a) Electrophoretic scan of the PCR amplicon following potassium permanganate modification and piperidine cleavage of the HEX-labeled sense (forward) strand and (b) 6-FAM-labeled antisense (reverse) strand. (c) Electrophoretic scan of the PCR amplicon following hydroxylamine modification and piperidine cleavage of the HEX-labeled sense (forward) strand and (d) 6-FAM-labeled antisense (reverse) strand. The presence of a peak in (a) and the absence of a peak in (b) suggests the detection of a mismatched T base on the sense strand. The presence of a peak in (c) and the absence of a peak in (d) suggests the detection of a mismatched C base on the sense strand. This suggests that A and G alleles are present on the antisense strand and the nature of the mutation can therefore be inferred without sequencing as a T–G, C–A mismatch.

7.2 Methods and Protocols

Amplification of wild-Type and Mutant DNA

- Amplify DNA samples, including a known wild-type control, using fluorescent-labeled primers in a 25-μl reaction volume.

Duplex Formation

- Mix equimolar ratios of crude wild-type and mutant DNA amplicons (1–2.5 µg each) in a new PCR tube.
- Anneal heteroduplex DNA in a thermal cycler, using the following conditions: 96 °C for 10 min, 65 °C for 60 min, and 25 °C for 60 min.

Liquid-Phase CCM Analysis

Modification of Mismatched Cytosine Residues
- Add 6 µl (~600 ng) of duplex DNA to 20 µl of 4.2 M NH_2OH and incubate at 37 °C for 40 min (50 min for fragments >1 kb).

Modification of Mismatched Thymine Residues
- Desiccate 6 µl (~600 ng) of duplex DNA by vacuum centrifugation or ethanol precipitation.
- To the dried-down pellet, add 0.2 µl of 100 mM $KMnO_4$ and 19.8 µl of 3 M TEAC, vortex briefly to disrupt the pellet, and incubate at 25 °C for 5 min (10 min for fragments >1 kb).

Cleavage of the Mismatched Residues
- Following incubation of the two duplex reactions, add 200 µl of stop buffer (0.3 M NaAc, pH 5.2, 0.1 mM EDTA, 25 mg/ml tRNA) and 750 µl of ice-cold 100% ethanol to each tube, vortex briefly, and incubate at −20 °C for 30 min.
- Pellet the DNA by centrifugation at 13 000 rpm for 20 min and decant the supernatant.
- Wash the pellet with 400 µl of ice-cold 70% ethanol, centrifuge for 5 min at 13 000 rpm, and decant the supernatant. Air-dry the pellet at room temperature for 15 min.
- Add 8 µl of 100% formamide and 2 µl of 10% piperidine to each pellet, vortex vigorously, and incubate at 90 °C for 30 min to chemically cleave any modified bases and plunge onto ice. The addition of piperidine should be performed in a fumehood and care taken when opening tubes after heating to avoid splashing of the liquid.
- Resolve the cleavage fragments by loading 2–3 µl of reaction mixture onto a 5% denaturing polyacrylamide gel or inject into capillaries.

Solid-Phase CCM Analysis

Binding DNA to a Solid Support
- Bind duplex DNA (100 ng) to a solid support (e.g., UltraClean® Ultra Bind Beads; MO Bio Laboratories), according to the manufacturer's instructions.

- Briefly, incubate 5 μl of Ultra Bind Beads with 100 ng duplex DNA at room temperature for 60 min, with occasional vortexing to keep beads in suspension.
- Pellet DNA-bound silica beads by centrifugation for 5 min at 13 000 rpm and decant the supernatant.
- Wash the DNA-bound silica beads twice, initially with 500 μl and then with 200 μl of Ultra Wash solution.
- Aliquot 100 μl into two separate reaction tubes, centrifuge for 5 min at 13 000 rpm, and decant the supernatant.

Modification of the Mismatched Residues
- To the first aliquot of DNA-bound silica beads add 15 μl of 4.2 M NH_2OH and 15 μl of 3 M TEAC, and incubate for 40 min at 37 °C.
- To the second aliquot of DNA-bound silica beads add 0.6 μl of 100 mM $KMnO_4$ and 29.4 μl of 3 M TEAC and incubate for 5 min at 25 °C.

Cleavage of the Mismatched Residues
- Following incubation with the modifying chemicals, centrifuge the two reactions for 5 min at 13 000 rpm and decant the supernatants. Air-dry the pellets at room temperature for ~15 min.
- Resuspend the two reactions in 8 μl of 100% formamide and 2 μl of 10% piperidine, and incubate at 90 °C for 30 min, releasing the DNA from the silica beads. The addition of piperidine should be performed in a fumehood and care taken when opening tubes after heating to avoid splashing of the liquid.
- To pellet the beads, centrifuge for 5 min at 13 000 rpm and load 2–3 μl of the supernatant onto a 5% denaturing polyacrylamide gel or inject into capillaries for electrophoretic separation of the DNA fragments.

7.3
Applications

CCM is routinely used in our laboratory for mutation screening of patient DNA for schizophrenia, mitochondrial disorders, breast cancer, and congenital nystagmus (Figure 7.2). CCM has been successfully used for the detection and identification of causative mutations in genes implicated in many human diseases, including hemophilia A and B, hereditary angioedema, phenylketonuria, Alzheimer's disease, atherosclerosis, and colorectal cancer [8–10, 12–20].

7.4
Perspectives

A disadvantage of CCM is that multiple operator steps are required, including several incubation steps and a separation step, and the need for two toxic chemicals,

Table 7.1 Advantages and disadvantages of the CCM.

Advantages	Disadvantages
Accurate localization of the site and identify of the mutation	multiple operator steps
Analyze PCR fragments up to 2 kb in length	toxic chemicals
High sensitivity of mutation detection (100%)	
No complex equipment	
Low cost	

hydroxylamine and piperidine (Table 7.1). The observation that potassium permanganate is able to oxidize both thymine as well as cytosine residues that results in a color change from pink to yellow has allowed the development of a simple plate reader assay that detects the development of the color change and hence the presence of a mutation [21]. The oxidation rate of the mismatched bases by potassium permanganate and therefore the presence of a sequence variant is detected by measuring the formation of the oxidization products, which absorbs at visible wavelength of 420 nm [22] using a standard ultraviolet/visible microplate reader. This approach is not based on a cleavage reaction to detect mismatched DNA, thereby eliminating the need for toxic chemicals and time-consuming incubations as well as the separation step that is required by many other techniques. However, unlike CCM, this approach is only suited to the rapid discovery of mutations in PCR fragments less than 500 bp and does not reveal the location or identity of the mutation.

Acknowledgments

The authors thank Elsbeth Richardson and Vincent Sesto for technical assistance, and the National Health and Medical Research Council for financial assistance.

References

1 Wittwer, C.T., Reed, G.H., Gundry, C.N., Vandersteen, J.G., and Pryor, R.J. (2003) High-resolution genotyping by amplicon melting analysis using LCGreen. *Clin. Chem.*, **49**, 853–860.

2 Orita, M., Iwahana, H., Kanazawa, H., Hayashi, K., and Sekiya, T. (1989) Detection of polymorphisms of human DNA by gel electrophoresis as single-strand conformation polymorphisms. *Proc. Natl. Acad. Sci. USA*, **86**, 2766–2770.

3 Oefner, P.J. and Underhill, P.A. (1995) Comparative DNA sequencing by denaturing high-performance liquid chromatography (DHPLC). *Am. J. Hum. Genet.*, **57**, A266.

4 Myers, R.M., Maniatis, T., and Lerman, L.S. (1987) Detection and localization of single base changes by denaturing gradient gel electrophoresis. *Methods Enzymol.*, **155**, 501–527.

5 Orita, M., Iwahana, H., Kanazawa, H., Hayashi, K., and Sekiya, T. (1989) Detection of polymorphisms of human DNA by gel electrophoresis as single-strand conformation polymorphisms. *Proc. Natl. Acad. Sci. USA*, **86**, 2766–2770.

6 Cotton, R.G., Rodrigues, N.R., and Campbell, R.D. (1988) Reactivity of cytosine and thymine in single-base-pair mismatches with hydroxylamine and osmium tetroxide and its application to the study of mutations. *Proc. Natl. Acad. Sci. USA*, **85**, 4397–4401.

7 Youil, R., Kemper, B.W., and Cotton, R.G.H. (1995) Screening for mutations by enzyme mismatch cleavage using T4 endonuclease VII. *Proc. Natl. Acad. Sci. USA*, **92**, 87–91.

8 Oleykowski, C.A., Bronson Mullins, C.R., Godwin, A.K., and Yeung, A.T. (1998) Mutation detection using a novel plant endonuclease. *Nucleic Acids Res.*, **26**, 4597–4602.

9 Lambrinakos, A., Yakubovskaya, M., Babon, J.J., Neschastnova, A.A., Vishnevskaya, Y.V., Belitsky, G.A., D'Cunha, G., Horaitis, O., and Cotton, R.G. (2004) Novel *TP53* gene mutations in tumors of Russian patients with breast cancer detected using a new solid phase chemical cleavage of mismatch method and identified by sequencing. *Hum. Mut.*, **23**, 186–192.

10 Izatt, L., Vessey, C., Hodgson, S.V., and Solomon, E. (1999) Rapid and efficient ATM mutation detection by fluorescent chemical cleavage of mismatch: identification of four novel mutations. *Eur. J. Hum. Genet.*, **7**, 310–320.

11 Tessitore, A., Di Rocco, Z.C., Cannita, K., Ricevuto, E., Toniato, E., Tosi, M., Ficorella, C., Frati, L., Gulino, A., Marchetti, P., and Martinotti, S. (2002) High sensitivity of detection of TP53 somatic mutations by fluorescence-assisted mismatch analysis. *Genes Chromosomes Cancer*, **35**, 86–91.

12 Verpy, E., Biasotto, M., Meo, T., and Tosi, M. (1994) Efficient detection of point mutations on color-coded strands of target DNA. *Proc. Natl. Acad. Sci. USA*, **91**, 1873–1877.

13 Sallmann, G.B., Bray, P.J., Rogers, S., Quince, A., Cotton, R.G., and Carden, S.M. (2006) Scanning the ocular albinism 1 (*OA1*) gene for polymorphisms in congenital nystagmus by DHPLC. *Ophthal. Genet.*, **27**, 43–49.

14 Bidichandani, S.I., Lanyon, W.G., Shiach, C.R., Lowe, G.D., and Connor, J.M. (1995) Detection of mutations in ectopic factor VIII transcripts from nine haemophilia A patients and the correlation with phenotype. *Hum. Genet.*, **95**, 531–538.

15 Rudzki, Z., Duncan, E.M., Casey, G.J., Neumann, M., Favaloro, E.J., and Lloyd, J.V. (1996) Mutations in a subgroup of patients with mild haemophilia A and a familial discrepancy between the one-stage and two-stage factor VIII:C methods. *Br. J. Haematol.*, **94**, 400–406.

16 Montandon, A.J., Green, P.M., Giannelli, F., and Bentley, D.R. (1989) Direct detection of point mutations by mismatch analysis: application to haemophilia B. *Nucleic Acids Res.*, **11**, 3347–3358.

17 Verpy, E., Biasotto, M., Brai, M., Misiano, G., Meo, T., and Tosi, M. (1996) Exhaustive mutation scanning by fluorescence-assisted mismatch analysis discloses new genotype-phenotype correlations in angioedema. *Am. J. Hum. Genet.*, **59**, 308–319.

18 Forrest, S.M., Dahl, H.H., Howells, D.W., Dianzani, I., and Cotton, R.G. (1991) Mutation detection in phenylketonuria by using chemical cleavage of mismatch: importance of using probes from both normal and patient samples. *Am. J. Hum. Genet.*, **49**, 175–183.

19 Liddell, M.B., Bayer, A.J., and Owen, M.J. (1995) No evidence that common allelic variation in the Amyloid Precursor Protein (*APP*) gene confers susceptibility to Alzheimer's disease. *Hum. Mol. Genet.*, **4**, 853–858.

20 Cotton, R.G. and Bray, P.J. (2001) Using CCM and DHPLC to detect mutations in the glucocorticoid receptor in atherosclerosis: a comparison. *J. Biochem. Biophys. Methods*, **47**, 91–100.

21 De Galitiis, F., Cannita, K., Tessitora, A., Martella, F., Di Rocco, Z.C., Russo, A., Adamo, V., Iacobelli, C., Martinotti, S., Marchetti, P., Ficorella, C., Ricevuto, E., and Consorzio Interuniversitario Nazionale Bio-Oncolgia (2006) Novel P53 mutations detected by FAMA in colorectal cancers. *Ann. Oncol.*, **17**, vii78–vii83.

22 Tabone, T., Sallmann, G., Webb, E., and Cotton, R.G.H. (2006) Detection of 100% of mutations in 124 individuals using a standard UV/Vis microplate reader: a novel concept for mutation scanning. *Nucleic Acids Res.*, **34**, e45.

23 Bui, C.T., Sam, L.A., and Cotton, R.G. (2003) UV-visible spectral identification of the solution-phase and solid-phase permanganate oxidation reactions of thymine acetic acid. *Bioorg. Med. Chem. Lett.*, **14**, 1313–1315.

8
Mutation Detection in Plants by Enzymatic Mismatch Cleavage

Bradley J. Till

Abstract

Scientific endeavors in the discovery and genotyping of nucleotide polymorphisms have increased dramatically in recent decades. This is due in part to the large amount of genomic sequence information now available to researchers. It is also driven by the availability of accurate methods of polymorphism discovery that can be easily carried out in many laboratories. In this chapter, I provide a protocol for enzymatic mismatch cleavage using single-strand specific nucleases. A crude enzyme extract from celery is used to cleave single and multiple base-pair mismatches in heteroduplexed DNA templates created by denaturing and annealing polymorphic polymerase chain reaction (PCR) amplicons. Products are size fractionated by denaturing polyacrylamide gel electrophoresis and visualized by fluorescence detection. Cleaved fragments migrating faster than intact PCR amplicons indicate the presence of a nucleotide polymorphism. The molecular weights of cleaved fragments provide the approximate locations of polymorphisms. The method is robust, accurate, low-cost, and scalable, allowing its use in small-scale and high-throughput applications. The protocol utilizes standard methods and reagents, and is flexible in that different readout platforms can be used to identify cleaved products. This has led to its adoption by many labs engaged in the discovery or characterization of mutations and natural polymorphisms in plants and animals.

8.1
Introduction

Enzymes are a fundamental tool of the research biologist and are key components of many experimental approaches. As early as the 1970s, studies showed that a class of enzymes known as single-strand specific nucleases, including S1 nuclease, could preferentially cleave single-stranded, mismatched, regions of otherwise double-stranded DNA duplexes [1–3]. The use of nucleases for the discovery or genotyping of natural polymorphisms and mutations is often referred to as enzymatic mismatch cleavage. In the late 1990s, an enzyme related to S1, called CEL1, was isolated from celery (*Apium graveolens*) that could be optimized to cleave all combinations of single base pair mismatches in short heteroduplexed DNAs [4]. This enzyme was used to develop a sensitive fluorescence-based method for enzymatic mismatch cleavage [5]. Using this method, it was demonstrated that CEL1, mung bean, and S1 nuclease can cleave single base-pair mismatches and small insertions/deletions (Indels) in heteroduplexed amplicons with no apparent sequence bias [6]. Furthermore, similar enzymatic activity was obtained when using crude plant extracts from celery or mung bean sprouts.

Enzymatic mismatch cleavage and fluorescence detection consists of four basic steps (Figure 8.1a). A target genomic region of around 1–1.5 kb is first amplified by polymerase chain reaction (PCR) using gene-specific primers that are fluorescently end-labeled. The forward primer is labeled with infrared dyeIRD700 infrared dye and the reverse primer with IRD800. After amplification, PCR products are denatured and annealed to create heteroduplexed molecules between polymorphic amplicons. This creates single-stranded bulges in DNA duplexes that are the substrate for cleavage by single-strand specific nucleases. Nucleases can recognize and cleave single and multiple base-pair mismatches, making the enzymatic mismatch cleavage suitable for single nucleotide polymorphism (SNP) and Indel discovery. Following cleavage of unpurified PCR products using a crude celery juice extract (CJE), samples are purified and size-fractionated by denaturing polyacrylamide gel electrophoresis (PAGE). DNA is visualized by fluorescence detection using the LI-COR DNA Analyzer®. Two .tiff images are produced per gel run: one for IRD700 fluorescence, representing DNA labeled on the forward end, and one for IRD800-labeled DNA on the reverse end (Figure 8.1b). True nucleotide polymorphisms produce a fragment in both the IRD700 and IRD800 images. The sum of the molecular weights of the fragments equals the size of the full-length PCR product. The molecular weights of the fragments allow the estimation of the position of the polymorphism on the PCR amplicon within approximately 10 bp [7]. This dual-end-labeling strategy increases accuracy by reducing false-positive errors. The entire process takes approximately 10 h.

Any readout platform capable of separating and detecting cleaved DNA fragments can in theory be used for enzymatic mismatch detection, thus providing alternatives to the protocol in this chapter. Standard agarose gel electrophoresis is appealing due to its low run cost and ease of use. For standard native agarose gel electrophoresis, double-strand breaks (DSBs) must be formed at the mismatch. The use of single-strand-specific nucleases to induce DSBs has been described [8]. However, upon template breathing, enzymatic degradation of terminal duplex residues is also expected [9]. This can result in a high degree of background when visualizing nucleic acids by methods that stain the entire duplex rather than by end-labeling. The challenge in optimizing cleavage of small mismatches while limiting the end degradation activity may explain difficulties reported for using

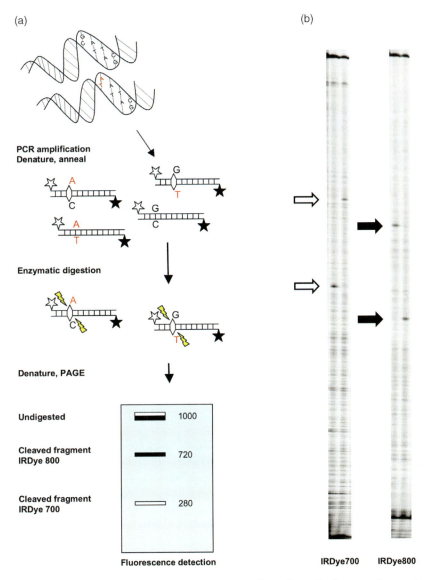

Figure 8.1 Enzymatic mismatch cleavage and fluorescence detection for mutation discovery. PCR is performed using genomic DNA and gene-specific primers (a). The forward primer is labeled with IRD700 and the reverse with IRD800. After amplification, products are denatured and annealed to create heteroduplexed molecules with mismatches in polymorphic regions. Single-stranded mismatches are cleaved by incubating PCR products with a single-strand specific nuclease. Products are denatured and size fractionated by denaturing PAGE. DNA is visualized by fluorescence detection using a LI-COR DNA analyzer. Two images are produced per gel run, one for each dye used (b). Two lanes are shown, each with a unique SNP. True nucleotide polymorphisms produce bands in each gel image whose molecular weights add up to the size of the full-length PCR product (marked by arrows).

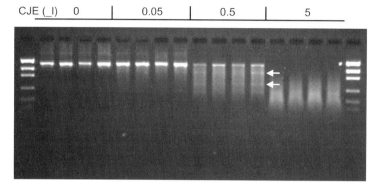

Figure 8.2 Agarose gel image of an enzyme activity test for crude CJE. An approximately 1.5-kb gene fragment was amplified from four rice genotypes. PCR products were denatured and annealed, and 100 ng of PCR product was incubated with 0, 0.05, 0.5, or 5 μl of CJE for 30 min at 45 °C. Samples were electrophoresed through a 1.5% agarose gel and stained with ethidium bromide. Complete degradation of the PCR products is observed upon incubation with 5 μl of enzyme extract, indicating sufficient activity has been obtained from the enzyme extraction procedure. Incubation with 0.5 μl of enzyme extract results in the production of bands consistent with nucleotide polymorphisms identified previously in the four rice genotypes (unpublished, marked by arrows). This exemplifies the ability of single-strand-specific nucleases to generate DSBs at mismatches.

single-strand-specific nucleases to digest single base mismatches [10, 11]. Limiting end degradation while cleaving internal mismatches, however, can be accomplished by optimizing reaction conditions such as pH, temperature, and the concentration of divalent cations, and applications combining CEL1, CEL1-containing Surveyor™ nuclease, or CJE and agarose gels have been described [12–15] (Figure 8.2).

A variety of other readout platforms have been used for enzymatic mismatch cleavage that require only a single-strand nick to be formed at the site of mismatch. This potentially limits the opportunity for end degradation as single-strand nicking is more favorable than inducing DSBs, thus reducing assay noise. Readout platforms include denaturing high-performance chromatography, capillary gel electrophoresis, and denaturing polyacrylamide slab-gel electrophoresis [5, 16–19]. These methods are advantageous over agarose gel electrophoresis because they provide increased sensitivity and dynamic range, allowing greater accuracy and higher throughput via sample pooling and automation. In addition, assay noise due to end degradation can be eliminated entirely by using end-labeled DNA as a substrate. DNA templates that have been subject to end degradation will lose the label and not be part of the observable assay signal.

Slab gel electrophoresis and fluorescence detection using the LI-COR DNA analyzer has been a method of choice for groups engaged in mutation discovery in plants. This may be because the system combines low run costs and the familiarity of standard PAGE with a high sensitivity, dynamic range, and the production of digital gel images suitable for automated band discovery [20–22]. The largest application of enzymatic mismatch cleavage followed by fluorescence detection has been for the reverse genetic technique known as targeting induced local lesions in genomes (TILLING; see Chapter 11 by Rakshit *et al.*) [5, 23]. In TILLING, large mutant populations are screened for rare induced mutations in specific genes. The sensitivity of the method has allowed pooling of samples prior to screening to increase assay throughput. Analysis of around 2000 *Arabidopsis* mutations identified in samples pooled 8-fold showed the expected 2:1 ratio of heterozygous to homozygous mutations, indicating robust mutation discovery and an equal ability to discover mutations represented at a concentration of 1/8th or 1/16th the total sample [7]. The strategy has been widely adopted and TILLING projects, and public screening services in both plants and animals have been developed. Thousands of mutations have been discovered in a variety of organisms including *Arabidopsis*, barley, *Caenorhabditis elegans*, *Drosophila*, maize, rice, soybean, wheat, and zebrafish [17, 24–33]. Furthermore, the approach is advantageous because it offers a low informatics load in comparison to sequencing-based applications. For example, in a highly mutagenized population of *Arabidopsis*, approximately 14 mutations in a 1.5-kb amplicon are identified in a screen of 3000 individuals (http://tilling.fhcrc.org/arab/status.html) [24]. Only 14 nucleotide changes need to be tracked and stored out of the 4.5 Mb screened. The method has also been adopted to identify natural nucleotide polymorphisms in populations, a process sometimes called EcoTILLING [34]. EcoTILLING has been described for haplotype discovery, nucleotide diversity studies, and candidate gene screening in a variety of species, including *Arabidopsis*, human, maize, melon, *Populus trichocarpa*, and rice (http://genome.purdue.edu/maizetilling/EcoTILLING.htm), [15, 22, 35–37].

Studies in wheat and soybean have indicated that the coamplification of homologs or duplicated gene targets prior to mutation screening reduces the accuracy of the enzymatic mismatch cleavage method [29, 32]. The necessity of amplifying a single gene target is therefore a limitation of the method. That the LI-COR DNA analyzer provides only two wavelengths of fluorescence detection also limits multiplexing strategies. Sensitivity of the method may also be limited. Robust detection of heterozygous mutations in pools of eight samples has been described, but similar studies with larger pool sizes have not been published [7]. In spite of these potential limitations, the method has been shown to be accurate for polymorphism detection in pooled samples at a relatively low cost [22]. For example, the *Arabidopsis* TILLING Project offers a mutation discovery screening service that at the time of writing charges a cost-recovery user fee of $1500 to screen a minimum of 3000 mutant individuals (http://tilling.fhcrc.org/files/user_fees.html). The creation of screening services based on the method described in this chapter also highlights its scalability. A screening throughput of around 1500 pooled individuals per day can be achieved by a single technician using a single LI-COR. By adding more LI-COR analyzers, 384 liquid handling and robotic sample loading, a screening throughput of around 12 000 pooled individuals per day can be achieved by two technicians.

8.2
Methods and Protocols

Preparation of CJE for Enzymatic Mismatch Cleavage

Perform all steps at 4 °C.

1. Clean ~0.5 kg of celery stalks (*A. graveolens*) in water and remove leaves.
2. Grind celery using a standard vegetable juicer (optimally) or blender to produce ~400 ml of celery juice. Decant juice to a large beaker and measure the volume of juice.
3. Add 1 M Tris–HCl, pH 7.7 and 0.1 M phenylmethylsulfonate (PMSF) (in isopropanol) to the juice to a final concentration of 0.1 M Tris–HCl and 100 µM PMSF.
4. Centrifuge the solution at $2600 \times g$ for 20 min to pellet solids.
5. Transfer liquid to a larger beaker and bring to 25% saturation of $(NH_4)_2SO_4$ by adding 144 g/l $(NH_4)_2SO_4$.
6. Mix gently for 30 min using a magnetic stir bar and plate.
7. Centrifuge mixture for $15\,000 \times g$ for 40 min.
8. Carefully transfer supernatant to a new large beaker and measure the volume. Bring the solution to 80% saturation in $(NH_4)_2SO_4$ by adding 390 g/l.
9. Mix gently for 30 min using a magnetic stir bar and plate.
10. Centrifuge solution at $15\,000 \times g$ for 1.5 h.
11. Discard supernatant. Suspend pellet in a solution of 0.1 M Tris–HCl, 0.5 M KCl, and 100 µM PMSF (added fresh from 0.1 M stock in isopropanol). Use ~1/10th the volume of starting juice. For example, if starting with 400 ml of juice, suspend pellet in 40 ml of buffer. Make sure the pellet is thoroughly suspended.
12. Transfer suspension into 10 000-Da molecular weight cut-off dialysis tubing (e.g., Spectra/Por 7). Place dialysis tube into a large beaker containing 0.1 M Tris–HCl, 0.5 M KCl, and 100 µM PMSF and gently mix for 1 h. Use 4 l of buffer for every 20 ml of suspension. It may be necessary to split suspension into several dialysis tubes.
13. Discard buffer and replace with fresh 0.1 M Tris–HCl, 0.5 M KCl, and 100 µM PMSF. Mix for 1 h.
14. Repeat buffer exchanges for a total minimum of 4 h of dialysis and 32 l of buffer per 0.5 kg celery. Dialysis steps can be for longer durations with no apparent loss of enzyme activity. It is often convenient to allow the third or fourth step to proceed overnight.
15. Transfer contents of dialysis tubing into a clean tube. Centrifuge at $10\,000 \times g$ for 30 min to clear solution. Aliquot supernatant and store at -80 °C. It is not necessary to store in glycerol. Enzyme activity is maintained for years in this storage condition.

Testing Enzyme Activity

A quick agarose gel assay can be used to determine if the extraction procedure was successful. PCR product is incubated with different amounts of CJE. The DNA duplex will degrade with increasing concentrations of enzyme, indicating that activity has been obtained. This assay can also be used as a lower cost and lower throughput alternative to fluorescent detection for enzymatic mismatch cleavage (Figure 8.2). Accuracy of this assay for mutation detection, however, may be an issue, especially for detecting SNPs [15].

1. Prepare ~500 ng of gene-specific PCR product with an amplicon length of 1–1.5 kb. It is preferable to amplify two to four samples for replication of the activity test. The PCR parameters described in "PCR Amplification" (below) can be used for this step. (*Note*: If you are attempting to use this assay for mutation detection, PCR product must be denatured and annealed to create heteroduplexed molecules before enzymatic treatment.)
2. Prepare 50 ml stock of 10 × CJE cleavage buffer. Combine 37.5 ml water, 5 ml 1 M HEPES, pH 7.5, 5 ml 1 M $MgSO_4$, 2.5 ml 2 M KCl, 100 µl 10% Triton X-100, 5 µl 20 mg/ml bovine serum albumin. Store in aliquots at $-20\,°C$.
3. Make a 1 : 10 dilution of CJE in 1 × cleavage buffer (diluted in water from the 10 × buffer stock).
4. Prepare the following enzyme master mixtures on ice. The amount of enzyme suggested is based on multiple enzyme extractions yielding similar unit activities when following the CJE extraction protocol described above.

	A	B	C	D
Water (µl)	85	84.5	84.5	80
10 × Buffer (µl)	15	15	15	15
Enzyme (µl)	0	0.5 (1:10 diluted)	0.5	5

5. Combine 20 µl of enzyme master mix with 10 µl of PCR product (~100 ng). Incubate at 45 °C for 30 min.
6. Stop the reaction by adding 5 µl of 0.225 M EDTA.
7. Electrophorese samples for ~90 min at 100 V through a 1.5% agarose gel. Stain with ethidium bromide (Figure 8.2). The degradation of full-length PCR product upon incubation with increasing amounts of enzyme is an indication that mismatch cleavage activity has been obtained. If no degradation is observed, repeat the experiment with increasing amounts of CJE.

Enzymatic Mismatch Cleavage and Fluorescent Detection

The protocol described below is for a 96-well assay. The volumes used here are suitable for 384-well assays, allowing for increased throughput of liquid handling. Volumes are also suitable for automated and semiautomated 96- and 384-channel pipettors. Throughput can be further increased by pooling genomic DNA samples up to 8-fold prior to PCR. When pooling samples together, it is important that all samples are of the same DNA concentration. Perform test reactions with different levels of sample pooling to ensure mutations can be discovered in a pool of the desired size. Both one- and two-dimensional sample pooling strategies have been described [5, 30]. Pooling samples is efficient for the discovery of rare nucleotide changes, but may not be a suitable approach when the goal is to catalog common nucleotide variation in populations, as assignments of genotypes to a specific individual will become impossible [22].

Primer Design

1. Design gene specific primers that are between T_m 67 and 73 °C, preferably with a GC content of ~50%, and producing amplicons between 750 and 1600 bp. Primers can be designed with the aid of a program such as Primer3 [38]. For TILLING applications, the CODDLe suite of programs is used to provide gene model assembly, protein homology assembly, and target amplicon choice based on the density of potentially deleterious mutations. CODDLe links to Primer3 for primer design (http://www.proweb.org/input/) [39].
2. Order one set of unlabeled primers and one set of IRD fluorescently labeled primers (e.g., MWG Biotech). The forward labeled primer should contain the IRD700 attached to the 5′ end, and the reverse should have the IRD800. (*Note*: Fluorescently labeled primers are quite expensive and a cost-saving alternative is to use a universal priming strategy [22, 27, 40]. Universal priming may take some optimization and there is some risk that amplicon contamination will destroy the ability to detect mutations. It is therefore advised to start with fluorescently end-labeled gene-specific primers and test the universal strategy once the protocol is optimized.)

PCR Amplification

1. Suspend primers in TE buffer pH 7.4 to a final primer concentration of 100 µM. To maintain fluorescence, avoid extended exposure of labeled primers to fluorescent light and aliquot samples to reduce the number of freeze–thaw cycles.
2. Prepare a mixture of fluorescently labeled and unlabeled primers by combining 2 µl unlabeled forward primer, 3 µl IRD700-labeled forward primer, 1 µl unlabeled reverse primer, and 4 µl IRD800-labeled reverse primer. (*Note*: Using a mixture of labeled and unlabeled primers is aimed at improving amplification yields that can sometimes be reduced when using only fluorescently labeled primers.)
3. Prepare a PCR master mix for a 96-well assay. Combine the following on ice:

Water	360 μl
10 × Ex Taq™ buffer (TaKaRa)	57 μl
25 mM MgCl$_2$	68 μl
2.5 mM dNTP mixture (TaKaRa)	92 μl
Primer mixture	4 μl
Hot start Ex Taq (TaKaRa)	6 μl

4. Combine 5 μl of PCR master mix with 5 μl genomic DNA per sample well. Centrifuge for 2 min at 1000 × g. The amount of genomic DNA for PCR should be optimized in advance. Perform titration experiments with varying concentrations of genomic DNA to determine the optimal concentration. A concentration yielding ∼7–10 ng/μl of PCR product is sufficient for downstream mutation discovery. Using too little genomic DNA can result in increased false-negative and false-positive errors. Using too much genomic DNA can result in increased background and a failure to detect mutations. For rice, 0.3 ng of genomic DNA is used per PCR amplification [30]. A good starting point is to scale the amount of genomic DNA according to genome size in order to maintain the molar balance between DNA and primer. For example, when screening a genome size twice the size of rice, start with 0.6 ng of genomic DNA, and test amounts lower and higher than this value. Choose the amount that produces the strongest signal with the least background.

5. Incubate samples using the following parameters: (i) 95 °C for 2 min; (ii) 94 °C for 20 s; (iii) 73 °C for 30 s, −1 °C/cycle; (iv) ramp to Step (v) at 0.5 °C/s; (v) 72 °C for 1 min; (vi) go to Step (ii), repeat 7 additional times; (vii) 94 °C for 20 s; (viii) 65 °C for 30 s; (ix) ramp to Step (x) at 0.5 °C/s; (x) 72 °C for 1 min; (xi) go to Step (vii), repeat 44 additional times; (xii) 72 °C for 5 min; (xiii) 99 °C for 10 min; (xiv) 70 °C for 20 s, −0.3 °C/cycle; (xv) go to Step (xiv), repeat 69 additional times; (xvi) hold at 8 °C. Amplification with fluorescently end-labeled primers can in some instances lead to increased mispriming and reduced yield when compared to unlabeled primers. The incubation conditions are aimed at increasing the specificity of amplification by including a touch-down (Steps ii–vi) and increasing the yield of product by increasing the number of incubation cycles. These parameters were determined empirically. Step (xiii) denatures the PCR product and Step (xiv) slowly anneals DNA to generate heteroduplexed molecules that are the substrate of the CJE cleavage assay.

Nuclease Cleavage of Mismatches and Sample Purification

1. Perform in advance. Determine the optimal amount of CJE for the assay by incubating fluorescently labeled PCR product with varying amounts of CJE following the protocol described here. The optimal amount of CJE is defined as the amount that produces the highest intensity of signal and the lowest intensity of noise (Figure 8.3).

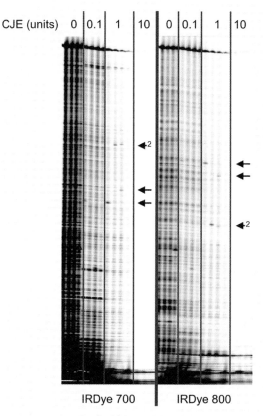

Figure 8.3 Enzyme concentration optimization for the detection of SNPs. LI-COR IRD700 (left) and IRD800 (right) images shown for an approximately 1-kb *Arabidopsis* gene target. Four samples were incubated with different amounts of CJE (listed above gel images). Previously identified single nucleotide mutations are marked by arrows [6]. The amount of enzyme that produces the best signal-to-noise ratio is defined as 1 U of activity. One unit of enzyme was also used to detect polymorphisms shown in Figure 8.2. *Arabidopsis* samples were provided by the Seattle TILLING Project (tilling.fhcrc.org) [24].

2. Prepare a CJE reaction master mix for a 96-well assay by combining 420 μl 10 × reaction buffer, the optimized amount of CJE, and water to a final volume of 2.8 ml.
3. Add 20 μl reaction mix to each sample. Centrifuge for 2 min at 2000 × g.
4. Incubate at 45 °C for 15 min.
5. Stop the reaction by adding 5 μl of 0.225 M EDTA. Process immediately or store at −20 °C for up to 2 weeks.
6. Prepare a 96-column Sephadex G50 purification plate by adding Sephadex G50 medium (Amersham Pharmacia) into a Millipore multiscreen 96-well separation plate (Fisher) using a 45-μl well Millipore multiscreen column loader (Fisher). Add 300 μl of water to each well of the plate. Allow the plate to

hydrate for a minimum of 1 h. Plates can be made in advance and stored in a sealed bag at 4 °C for up to 1 week.
7. Done in advance. Prepare loading buffer stock by combining 19.2 ml deionized formamide (Sigma), 770 μl 0.225 M EDTA, pH 8, and 1 mg bromophenol blue. Store in small aliquots at −20 °C.
8. Done in advance. Prepare a 200-bp lane marker. Design target-specific primers that amplify a 200-bp product. Amplify with IRD-labeled primers as described in "PCR Amplification". Purify samples using Sephadex G50 as described in that section, omitting the addition of formamide load buffer in Step 11. Combine all reaction wells together and load different amounts of marker on a LI-COR gel to determine optimal volume for gel loading. If signal intensity is high, the marker can be diluted in TE buffer.
9. Assemble hydrated purification plate from Step 6 with an empty 96-well assay plate using a Millipore centrifuge alignment frame (Fisher).
10. Centrifuge plate at $440 \times g$ for 2 min.
11. Prepare 96-well sample catch plate. Add 1.5 μl formamide load buffer to each well of a new 96-well plate. To aid in lane identification, add a 200-bp lane marker to the samples falling in every eighth sample gel lane starting with lane 4, then 12, 20, and so on. This lane marker distribution matches the lane markings used in the GelBuddy gel analysis program described below.
12. Assemble centrifuged Sephadex plate from Step 10 with sample catch plate and alignment frame. Load entire volume of the nuclease cleavage reaction onto the Sephadex plate, taking care to load samples over the center of the column.
13. Centrifuge at $440 \times g$ for 2 min.
14. Remove sample catch plate. Reduce sample volume to ∼1.5 μl by incubating at 90 °C for ∼40 min. Store sample at −20 °C for up to 1 week.

Millipore multiscreen plates can be reused many times to reduce assay costs. After the final spin, allow Sephadex in the plate to dehydrate completely (around 3 days). Remove dehydrated Sephadex and thoroughly rinse plate with deionized water. Allow plate to dry completely (around 3 days) before reuse. A less-expensive but more time-consuming alternative to Sephadex-based purification of samples is alcohol precipitation [33].

Preparation and Loading LI-COR Gels

1. Assemble 25-cm glass plates and 25-mm spacers according to manufacturer's instructions (LI-COR).
2. Prepare acrylamide gel mixture by combining 20 ml of 6.5% acrylamide gel matrix (LI-COR) with 15 μl TEMED and 150 μl 10% APS (ammonium persulfate).
3. Pour acrylamide into gel assembly using a disposable plastic syringe. Insert casting comb and pressure plate according to manufacturer's instruction (LI-COR), and allow 90 min for polymerization before use. After

polymerization, gels can be stored at 4 °C for up to 1 week by covering the ends of the assembly with damp paper towels and wrapping the entire assembly in plastic wrap.

4. After polymerization, clean glass plates, remove comb, and insert into LI-COR DNA analyzer. Fill buffer tanks with $0.8 \times$ TBE buffer and thoroughly clean out gel well to remove all loose acrylamide.
5. Heat the 96-well plate of samples for 5 min at 90 °C and cool to 4 °C for 3 min.
6. Load 0.25–0.5 µl of sample onto a 100-tooth membrane comb (Gel Company). Loading can be performed using a comb-loading robot (Aviso) or done manually. Manual loading is aided with the use of a multichannel pipettor and a comb loading tray (Gel Company).
7. Apply 0.25 µl of IRD size standards to the membrane comb teeth flanking the 96 sample teeth. To avoid loading mistakes that can confuse sample lane 1 with 96, apply IRD700 size standard to comb tooth 1 and IRD800 size standard to comb tooth 100. IRD size standards can be purchased commercially (LI-COR). Alternatively, size standards can be prepared by creating IRD-labeled PCR products of the appropriate size following the method described for preparing the 200-bp lane marker.
8. Prerun LI-COR gel using the following parameters: 20 min, 1500 V, 40 mA, 40 W, 50 °C, scan speed 2, and image width 1028.
9. Upon completion of prerun, thoroughly rinse the gel well. Remove buffer from upper tank and insert filter paper into gel well to remove excess buffer. Fill the gel well with 1–2 ml 1% Ficoll (prepared in water) so that a bead forms between the gel well and the top of the front glass plate.
10. Insert membrane comb into well at a 45° angle. Practice first with a blank comb if this procedure is new. After comb insertion, gently fill reservoir with $0.8\times$ TBE running buffer.
11. Run gel using the following parameters: 4 h 15 min, 1500 V, 40 mA, 40 W, 50 °C, scan speed 2, and image width 1028.
12. After 10 min, pause the run and remove the membrane comb. Gently rinse well with a 1000-µl pipettor and disposable plastic tip to remove Ficoll from well. Restart the run.

To reduce assay cost and time, the gel can be run a second time with different samples within 1 week of the first run. Membrane combs can be reused many times. Soak the membrane comb in deionized water for at least 30 min and air dry (1–3 days) before reuse.

Data Analysis

Two .tiff images are produced from a single LI-COR run – one containing data from the IRD700 channel and one from IRD800. True polymorphisms will produce a cleaved fragment in both image channels (Figure 8.1B). The sum of fragment molecular weights equals the molecular weight of the full-length PCR product. Data analysis can be performed completely manually or with the aid of data analysis software. The freely available GelBuddy program (www.gelbuddy.org) combines automated lane identification compatible with the 200-bp lane markers used in this protocol. The program provides molecular weight determination, automated band calling, haplotype identification, data reports that can be directly exported to databases, and other features for the analysis of rare induced mutations and common natural nucleotide polymorphisms [21, 22]. Alternative gel analysis software is available from Softgenetics (http://www.softgenetics.com/products.html).

For mutation discovery applications such as TILLING where knowledge of the exact nucleotide change is desired, individual samples can be sequenced. Sequence analysis is aided because the molecular weight of the cleaved products provides the location of the nucleotide change within approximately 10 nucleotides [7]. For genotyping applications where knowledge of the presence or absence of a polymorphism or haplotype is desired, sequencing need not be performed, further reducing the cost of the assay.

8.3
Applications

The most widely used application of enzymatic mismatch cleavage in plants has been for the discovery of induced point mutations in the reverse genetics strategy known as TILLING (Chapter 11). The TILLING strategy has been used to probe gene function and for developing crops with novel traits (e.g., [32, 41, 42]). In addition to providing a broad range of allele types, the strategy is appealing for crop improvement because it is a nontransgenic approach. Successes with crops such as barley, rice, wheat, and soybean have allowed the consideration of reverse genetic platforms for mutation breeding in understudied crops and understudied genotypes important to developing nations. The rising human population, changing climate, and competition from nonfood uses such as biofuels are putting an unprecedented pressure on global food security, highlighting the need for the development of hardier crops and those with diversified end uses [43, 44]. Using induced mutations to develop crops with novel traits is one approach to strengthen food security. Reverse genetic strategies promise to enhance the efficiency of the process. The Plant Breeding Unit of the Joint Food and Agriculture Organization/International Atomic Energy Agency Program on Nuclear Techniques in Food and Agriculture has developed a TILLING and EcoTILLING platform for mutation discovery in crops important to developing countries. Pilot-scale studies with three important food security crops – banana, cassava, and rice – suggest that the laboratory can serve as a mutation screening facility for understudied crops from throughout the world. The efficacy of this approach has been tested by screening DNA samples provided by laboratories in South America, Africa and Indonesia (B.J. Till and C. Mba, unpublished data). Other labs are also working to establish enzymatic mismatch cleavage for mutation discovery in understudied crops. For example, a TILLING project at the University of Bern in Switzerland is focused on developing a dwarf variety of the African crop Tef that can resist lodging (Z. Tadele, personal communication). Developing such a variety using a forward genetic strategy is impeded because the plant is tetraploid. Knowledge of the genes involved in plant dwarfing makes the task of developing a dwarf tetraploid Tef via a reverse genetic approach potentially straightforward.

8.4
Perspectives

Enzyme activity and readout platform are the main areas of focus for improving enzymatic mismatch cleavage. Alternative enzymes such as the MutS, MutY, and endonuclease V repair enzymes, and resolvases T4 endonuclease VII, and T7 endonuclease I have been adopted for enzymatic mismatch cleavage, but have not been as widely used as CEL1 or CJE [11, 45]. This may be due in part to high background over signal and the inability of some enzymes to detect all types of mismatches [11]. These and other enzymes could potentially be optimized or modified for improved mutation detection. A crude extract from *Brassica* petiols has shown enzymatic mismatch cleavage activity similar to other single-strand specific nucleases [14]. A CEL1 homolog from *Arabidopsis thaliana* termed ENDO1 has recently been reported to provide improved efficiency in enzymatic mismatch cleavage [46]. The extent to which improved or altered enzymatic activity can increase the throughput of mismatch cleavage for the discovery of unknown mutations, however, is difficult to estimate. It may be as important to focus on stabilizing DNA duplex ends from breathing in order to reduce end degradation. Alternatively, internal fluorescent labeling of the DNA may boost signal and reduce noise as has been proposed for the EMAIL capillary system [18]. The ability of alternative readout platforms to improve throughput and reduce run costs is also largely untested as side-by-side comparisons have not been performed. Some systems offer the possibility of multiplexing prior to enzymatic cleavage and others provide nonfluorescence detection promising lower consumable costs [16]. Equipment, servicing, run costs, reliability, throughput, and sensitivity should all be carefully evaluated before choosing a readout platform.

Improvements to the protocol provided in this chapter are therefore likely to be incremental and not provide the orders of magnitude increases in throughput or reduction in costs as has been lauded for advanced technologies such as massively parallel sequencing. With large governmental and private industry support, rapid advances in what is known as next-generation sequencing might turn the "high"-throughput method described in this chapter into a medium-throughput option for those not interested in the massive throughput and higher start-up costs of new technologies. Interestingly, one method for increasing throughput and reducing the cost of massively parallel sequencing has been through target enrichment [47, 48]. It is therefore intriguing to consider applications utilizing enzymatic mismatch cleavage for the selective enrichment of polymorphic sequences prior to high-throughput sequencing. For example, the free 3′-OH created upon enzymatic cleavage of mismatches can be modified to enrich cleaved products [49]. It is an exciting time to be involved in mutation discovery research, and future advances will no doubt have major impacts in the fields of functional genomics and plant breeding.

Acknowledgments

I thank Owen Huynh of the Plant Breeding Unit for providing gel images in Figures. Martin Fregene (International Center for Tropical Agriculture, Colombia), Melaku Gedil (International Institute of Tropical Agriculture, Nigeria), and Enny Sudarmonawati (Indonesian Institute of Sciences, Indonesia) provided genomic DNA samples used to test the efficacy of a global mutation screening facility at the Plant Breeding Unit. Owen Huynh, Chikelu Mba, and Rachel Howard-Till provided helpful comments on the manuscript. The protocol was developed at the Seattle TILLING Project with support from National Science Foundation grant 0 077 737 awarded to Steven Henikoff and Luca Comai. The agarose gel enzyme activity protocol and gel images were produced with support from the Food and Agriculture Organization of the United Nations and the International Atomic Energy Agency through their Joint Program of Nuclear Techniques in Food and Agriculture.

References

1 Shenk, T.E., Rhodes, C., Rigby, P.W., and Berg, P. (1975) Biochemical method for mapping mutational alterations in DNA with S1 nuclease: the location of deletions and temperature-sensitive mutations in simian virus 40. *Proc. Natl. Acad. Sci. USA*, **72**, 989–993.

2 Kroeker, W.D. and Kowalski, D. (1978) Gene-sized pieces produced by digestion of linear duplex DNA with mung bean nuclease. *Biochemistry*, **17**, 3236–3243.

3 Desai, N.A. and Shankar, V. (2003) Single-strand-specific nucleases. *FEMS Microbiol. Rev.*, **26**, 457–491.

4 Oleykowski, C.A., Bronson Mullins, C.R., Godwin, A.K., and Yeung, A.T. (1998) Mutation detection using a novel plant endonuclease. *Nucleic Acids Res.*, **26**, 4597–4602.

5 Colbert, T., Till, B.J., Tompa, R., Reynolds, S., Steine, M.N., Yeung, A.T., McCallum, C.M., Comai, L., and Henikoff, S. (2001) High-throughput screening for induced point mutations. *Plant Physiol.*, **126**, 480–484.

6 Till, B.J., Burtner, C., Comai, L., and Henikoff, S. (2004) Mismatch cleavage by single-strand specific nucleases. *Nucleic Acids Res.*, **32**, 2632–2641.

7 Greene, E.A., Codomo, C.A., Taylor, N.E., Henikoff, J.G., Till, B.J., Reynolds, S.H., Enns, L.C., Burtner, C., Johnson, J.E., Odden, A.R., Comai, L., and Henikoff, S. (2003) Spectrum of chemically induced mutations from a large-scale reverse-genetic screen in *Arabidopsis*. *Genetics*, **164**, 731–740.

8 Chaudhry, M.A. and Weinfeld, M. (1995) Induction of double-strand breaks by S1 nuclease, mung bean nuclease and nuclease P1 in DNA containing abasic sites and nicks. *Nucleic Acids Res.*, **23**, 3805–3809.

9 Kroeker, W.D., Kowalski, D., and Laskowski, M. Sr. (1976) Mung bean nuclease I. Terminally directed hydrolysis of native DNA. *Biochemistry*, **15**, 4463–4467.

10 Silber, J.R. and Loeb, L.A. (1981) S1 nuclease does not cleave DNA at

single-base mis-matches. *Biochim. Biophys. Acta*, **656**, 256–264.

11 Taylor, G.R. and Deeble, J. (1999) Enzymatic methods for mutation scanning. *Genet. Anal.*, **14**, 181–186.

12 Sokurenko, E.V., Tchesnokova, V., Yeung, A.T., Oleykowski, C.A., Trintchina, E., Hughes, K.T., Rashid, R.A., Brint, J.M., Moseley, S.L., and Lory, S. (2001) Detection of simple mutations and polymorphisms in large genomic regions. *Nucleic Acids Res.*, **29**, E111.

13 Qiu, P., Shandilya, H., D'Alessio, J.M., O'Connor, K., Durocher, J., and Gerard, G.F. (2004) Mutation detection using Surveyor nuclease. *Biotechniques*, **36**, 702–707.

14 Sato, Y., Shirasawa, K., Takahashi, Y., Nishimura, M., and Nishio, T. (2006) Mutant selection from progeny of gamma-ray-irradiated rice by DNA heteroduplex cleavage using *Brassica* petiole extract. *Breed. Sci.*, **56**, 179–183.

15 Garvin, M.R. and Gharrett, A.J. (2007) DEco-TILLING: an inexpensive method for single nucleotide polymorphism discovery that reduces ascertainment bias. *Mol. Ecol. Notes*, **7**, 735–746.

16 Suzuki, T., Eiguchi, M., Kumamaru, T., Satoh, H., Matsusaka, H., Moriguchi, K., Nagato, Y., and Kurata, N. (2008) MNU-induced mutant pools and high performance TILLING enable finding of any gene mutation in rice. *Mol. Genet. Genomics*, **279**, 213–223.

17 Caldwell, D.G., McCallum, N., Shaw, P., Muehlbauer, G.J., Marshall, D.F., and Waugh, R. (2004) A structured mutant population for forward and reverse genetics in barley (*Hordeum vulgare* L.). *Plant J.*, **40**, 143–150.

18 Cross, M.J., Waters, D.L., Lee, L.S., and Henry, R.J. (2008) Endonucleolytic mutation analysis by internal labeling (EMAIL). *Electrophoresis*, **29**, 1291–1301.

19 Perry, J.A., Wang, T.L., Welham, T.J., Gardner, S., Pike, J.M., Yoshida, S., and Parniske, M. (2003) A TILLING reverse genetics tool and a web-accessible collection of mutants of the legume *Lotus japonicus*. *Plant Physiol.*, **131**, 866–871.

20 Middendorf, L.R., Bruce, J.C., Bruce, R.C., Eckles, R.D., Grone, D.L., Roemer, S.C., Sloniker, G.D., Steffens, D.L., Sutter, S.L., Brumbaugh, J.A. *et al.* (1992) Continuous, on-line DNA sequencing using a versatile infrared laser scanner/electrophoresis apparatus. *Electrophoresis*, **13**, 487–494.

21 Zerr, T. and Henikoff, S. (2005) Automated band mapping in electrophoretic gel images using background information. *Nucleic Acids Res.*, **33**, 2806–2812.

22 Till, B.J., Zerr, T., Bowers, E., Greene, E.A., Comai, L., and Henikoff, S. (2006) High-throughput discovery of rare human nucleotide polymorphisms by Ecotilling. *Nucleic Acids Res.*, **34**, e99.

23 McCallum, C.M., Comai, L., Greene, E.A., and Henikoff, S. (2000) Targeted screening for induced mutations. *Nat. Biotechnol.*, **18**, 455–457.

24 Till, B.J., Reynolds, S.H., Greene, E.A., Codomo, C.A., Enns, L.C., Johnson, J.E., Burtner, C., Odden, A.R., Young, K., Taylor, N.E., Henikoff, J.G., Comai, L., and Henikoff, S. (2003) Large-scale discovery of induced point mutations with high-throughput TILLING. *Genome Res.*, **13**, 524–530.

25 Talame, V., Bovina, R., Sanguineti, M.C., Tuberosa, R., Lundqvist, U., and Salvi, S. (2008) TILLMore, a resource for the discovery of chemically induced mutants in barley. *Plant Biotechnol. J.*, **6**, 477–485.

26 Gilchrist, E.J., O'Neil, N.J., Rose, A.M., Zetka, M.C., and Haughn, G.W. (2006) TILLING is an effective reverse genetics technique for *Caenorhabditis elegans*. *BMC Genomics*, **7**, 262.

27 Winkler, S., Schwabedissen, A., Backasch, D., Bokel, C., Seidel, C., Bonisch, S., Furthauer, M., Kuhrs, A., Cobreros, L., Brand, M., and Gonzalez-Gaitan, M. (2005) Target-selected mutant screen by TILLING in *Drosophila*. *Genome Res.*, **15**, 718–723.

28 Cooper, J.L., Greene, E.A., Till, B.J., Codomo, C.A., Wakimoto, B.T., and Henikoff, S. (2008) Retention of induced mutations in a *Drosophila* reverse-genetic resource. *Genetics*, **180**, 661–667.

29 Cooper, J.L., Till, B.J., Laport, R.G., Darlow, M.C., Kleffner, J.M., Jamai, A., El-Mellouki, T., Liu, S., Ritchie, R., Nielsen, N., Bilyeu, K.D., Meksem, K., Comai, L., and Henikoff, S. (2008) TILLING to detect

induced mutations in soybean. *BMC Plant Biol.*, **8**, 9.

30 Till, B.J., Cooper, J., Tai, T.H., Colowit, P., Greene, E.A., Henikoff, S., and Comai, L. (2007) Discovery of chemically induced mutations in rice by TILLING. *BMC Plant Biol.*, **7**, 19.

31 Till, B.J., Reynolds, S.H., Weil, C., Springer, N., Burtner, C., Young, K., Bowers, E., Codomo, C.A., Enns, L.C., Odden, A.R., Greene, E.A., Comai, L., and Henikoff, S. (2004) Discovery of induced point mutations in maize genes by TILLING. *BMC Plant Biol.*, **4**, 12.

32 Slade, A.J., Fuerstenberg, S.I., Loeffler, D., Steine, M.N., and Facciotti, D. (2005) A reverse genetic, nontransgenic approach to wheat crop improvement by TILLING. *Nat. Biotechnol.*, **23**, 75–81.

33 Draper, B.W., McCallum, C.M., Stout, J.L., Slade, A.J., and Moens, C.B. (2004) A high-throughput method for identifying N-ethyl-N-nitrosourea (ENU)-induced point mutations in zebrafish. *Methods Cell Biol.*, **77**, 91–112.

34 Comai, L., Young, K., Till, B.J., Reynolds, S.H., Greene, E.A., Codomo, C.A., Enns, L.C., Johnson, J.E., Burtner, C., Odden, A.R., and Henikoff, S. (2004) Efficient discovery of DNA polymorphisms in natural populations by Ecotilling. *Plant J.*, **37**, 778–786.

35 Nieto, C., Piron, F., Dalmais, M., Marco, C.F., Moriones, E., Gomez-Guillamon, M.L., Truniger, V., Gomez, P., Garcia-Mas, J., Aranda, M.A., and Bendahmane, A. (2007) EcoTILLING for the identification of allelic variants of melon eIF4E, a factor that controls virus susceptibility. *BMC Plant Biol.*, **7**, 34.

36 Gilchrist, E.J., Haughn, G.W., Ying, C.C., Otto, S.P., Zhuang, J., Cheung, D., Hamberger, B., Aboutorabi, F., Kalynyak, T., Johnson, L., Bohlmann, J., Ellis, B.E., Douglas, C.J., and Cronk, Q.C. (2006) Use of Ecotilling as an efficient SNP discovery tool to survey genetic variation in wild populations of *Populus trichocarpa*. *Mol. Ecol.*, **15**, 1367–1378.

37 Wang, G.-X., Tab, M.-K., Rakshi, S., Saitoh, H., Terauchi, R., Imaizumi, T., Ohsako, T., and Tominaga, T. (2007) Discovery of single-nucleotide mutations in acetolactate synthase genes by Ecotilling. *Pestic. Biochem. Physiol.*, **88**, 143–148.

38 Rozen, S. and Skaletsky, H. (2000) Primer3 on the WWW for general users and for biologist programmers. *Methods Mol. Biol.*, **132**, 365–386.

39 McCallum, C.M., Comai, L., Greene, E.A., and Henikoff, S. (2000) Targeting induced local lesions in genomes (TILLING) for plant functional genomics. *Plant Physiol.*, **123**, 439–442.

40 Wienholds, E., van Eeden, F., Kosters, M., Mudde, J., Plasterk, R.H., and Cuppen, E. (2003) Efficient target-selected mutagenesis in zebrafish. *Genome Res.*, **13**, 2700–2707.

41 Enns, L.C., Kanaoka, M.M., Torii, K.U., Comai, L., Okada, K., and Cleland, R.E. (2005) Two callose synthases, GSL1 and GSL5, play an essential and redundant role in plant and pollen development and in fertility. *Plant Mol. Biol.*, **58**, 333–349.

42 Mizoi, J., Nakamura, M., and Nishida, I. (2006) Defects in CTP: PHOSPHORYLETHANOLAMINE CYTIDYLYLTRANSFERASE affect embryonic and postembryonic development in *Arabidopsis*. *Plant Cell*, **18**, 3370–3385.

43 United Nations (2007) *World Population Prospects, The 2006 Revision*, United Nations, New York.

44 von Braun, J. and Pachauri, R.K. (2006) *The Promises and Challenges of Biofuels for the Poor in Developing Countries*, International Food Policy Research Institute, Washington, DC.

45 Yao, M. and Kow, Y.W. (1997) Further characterization of *Escherichia coli* endonuclease V. Mechanism of recognition for deoxyinosine, deoxyuridine, and base mismatches in DNA. *J. Biol. Chem.*, **272**, 30774–30779.

46 Triques, K., Sturbois, B., Gallais, S., Dalmais, M., Chauvin, S., Clepet, C., Aubourg, S., Rameau, C., Caboche, M., and Bendahmane, A. (2007) Characterization of *Arabidopsis thaliana* mismatch specific endonucleases: application to mutation discovery by TILLING in pea. *Plant J.*, **51**, 1116–1125.

47 Albert, T.J., Molla, M.N., Muzny, D.M., Nazareth, L., Wheeler, D., Song, X., Richmond, T.A., Middle, C.M., Rodesch, M.J., Packard, C.J., Weinstock, G.M., and Gibbs, R.A. (2007) Direct selection of human genomic loci by microarray hybridization. *Nat. Methods*, **4**, 903–905.

48 Okou, D.T., Steinberg, K.M., Middle, C., Cutler, D.J., Albert, T.J., and Zwick, M.E. (2007) Microarray-based genomic selection for high-throughput resequencing. *Nat. Methods*, **4**, 907–909.

49 Li, J., Berbeco, R., Distel, R.J., Janne, P.A., Wang, L., and Makrigiorgos, G.M. (2007) s-RT-MELT for rapid mutation scanning using enzymatic selection and real time DNA-melting: new potential for multiplex genetic analysis. *Nucleic Acids Res.*, **35**, e84.

9
Mutation Scanning and Genotyping in Plants by High-Resolution DNA Melting

Jason T. McKinney, Lyle M. Nay, David De Koeyer, Gudrun H. Reed, Mikeal Wall, Robert A. Palais, Robert L. Jarret, and Carl T. Wittwer

Abstract

High-resolution melting analysis after polymerase chain reaction (PCR) allows closed-tube mutation scanning and genotyping without processing, labeled probes, real-time monitoring, or allele-specific amplification. PCR is performed in the presence of the saturating dye, LCGreen® Plus, with subsequent high-resolution melting analysis on a high-throughput 96- or 384-well LightScanner®. Heterozygotes are easily distinguished from homozygotes by the shape of the normalized melting curves. On the LightScanner, the sensitivity and specificity of detecting single nucleotide polymorphism (SNP) heterozygotes is 100% for PCR product lengths from 50–400 bp, and 96.7% and 98.4%, respectively, for 500–800 bps. In contrast to other scanning techniques, no separations are required. Scanning of highly variable segments of diploid organisms differentiates many genotypes, including SNP heterozygotes, double SNP heterozygotes, and homozygotes. In addition to scanning for unknown variants, genotyping of known variants is easily performed with unlabeled probes using the same system. In tetraploid organisms like the potato, unlabeled probes are a nice solution for genotyping with accurate allele dose determination. High-resolution DNA melting can be applied to diploid or polyploid organisms and requires only 5–10 min after PCR. Variant detection sensitivity is high and unlabeled probes allow quantitative polyploidy genotyping with a minimum of cost and effort.

9.1
Introduction

Melting analysis was introduced in 1997 as closed-tube method to characterized polymerase chain reaction (PCR) products [1]. Instead of measuring size on agarose or polyacrylamide gels, different PCR products were distinguished by their thermal

The Handbook of Plant Mutation Screening. Edited by Günter Kahl and Khalid Meksem
Copyright © 2010 WILEY-VCH Verlag GmbH & Co. KGaA, Weinheim
ISBN: 978-3-527-32604-4

denaturation profiles. The melting temperature (T_m) of a PCR product depends on the GC content, length, and sequence of the duplex. Somewhat subtler changes in the shape of the melting curve indicate heterogeneity in the sequence of the target. Similar to size discrimination, the power of melting analysis depends on the resolution of the measurement [2–4]. High-resolution scanning and genotyping are enabled by a new class of saturation dyes that overcomes previous limitations of SYBR Green I [5] and by high-throughput instrumentation (LightScanner®; Idaho Technology) for 96- or 384-well analysis. The method provides closed-tube variant scanning [6–9] and genotyping with unlabeled probes [10–12] or by direct amplicon analysis [13–16]. Owing to its simplicity, speed (less than 10 min after PCR), and low cost, high-resolution melting is becoming a favorite method for high-throughput scanning and genotyping. High-resolution melting as applied to human genetics and oncology has been recently reviewed [17, 18].

Applications of high-resolution melting in plants genetics have begun to appear. It has been used to detect and quantify RNA editing sites in chloroplast transcripts of *Arabidopsis thaliana* [19]. Different grape and olive varieties have been identified by high-resolution melting of dinucleotide microsatellite repeats [20]. In barley, high-resolution melting has been used for single nucleotide polymorphism (SNP) detection and scoring of cleaved amplified polymorphic sequence markers [21]. Starting from expressed sequence tags in apple [22], almond [23], and white lupin [24], different groups have used high-resolution melting for the discovery, genotyping, and fine genetic mapping of SNPs, insertion/deletions (Indels), microsatellites, and restriction fragment length polymorphism markers. The method is particularly useful in species with limited existing genetic information.

High-resolution melting originally focused on monoploid (bacteria) or diploid (human) species. The polyploidy nature of many plants results in unique needs for variant detection and genotyping. Studies on human tumor tissue have found that high-resolution melting can detect minor alleles down to 0.1–10%, depending on the variant [25]. This level of sensitivity suggests that even one variant allele in an octaploid genome (1 : 8) should be readily detected by scanning. In polyploid genotyping, the need is to distinguish the dosage of different alleles. For biallelic SNPs, the number of possible genotypes scales linearly with ploidy (Table 9.1) and all genotypes should be distinguishable, even in octaploid cells. However, the total number of possible SNP genotypes increases much more rapidly, posing challenges for any technique.

9.2
Methods and Protocols

High-resolution melting compares profiles from multiple PCRs. Therefore, minimizing reaction-to-reaction variability is pivotal. Different samples analyzed by high-resolution melting should have their DNA extracted by the same method to limit any ionic strength differences introduced by the final elution buffers. Oligonucleotides (primers and unlabeled probes) can be synthesized by standard methods and

Table 9.1 The affect of ploidy on the number of possible SNP genotypes.

Ploidy	Examples	Number of SNP genotypes[a]	
		Biallelic	Total
$1n$	bacteria	2	4
$2n$	human, rye, apple	3	10
$3n$	banana, seedless fruit	4	20
$4n$	peanut, potato, tobacco, cotton	5	35
$6n$	wheat, oats, sweet potato	7	84
$8n$	strawberry, sugar cane	9	165

a) The general formula for the number of possible genotypes, given ploidy (n) and the number of alleles (a) is $(n + a - 1)!/(n!(a - 1)!)$, so for $a = 2$ (biallelic), the number of SNP genotypes is $n + 1$, and for $a = 4$, the total number of SNP genotypes is $(n + 3)!/(n!3!)$.

desalted before use. Unlabeled probes must be blocked at their 3′ end. Although phosphate is often used, C3 termination is more stable [26]. Sample DNA and oligonucleotides are best quantified by absorbance at 260 nm (A_{260}).

9.2.1
LightScanner Instrument

The LightScanner® resembles the LightTyper® [27] in physical form (96- or 384-well sample trays), with extensive software and hardware changes that enable high-resolution melting. The melting/cooling cycle requires only 10 min so that many thermal cyclers can feed one LightScanner. Heating is achieved by 16-bit pulse width control of a resistive heater board attached to a metal block that receives the microtiter tray. Block temperature is monitored with an imbedded resistive thermal device and 16-bit digital conversion. Mounted above the plate are two banks of 61 superbright 470-nm LEDs that illuminate the tray through a bandpass filter. A CCD camera (1392 × 1040 pixels, 12-bit depth per pixel) monitors sample fluorescence through a long-pass filter. During melting at 0.1 °C/s, between five and 10 consecutive camera images are continuously collected and averaged, storing about 12 data points every 1 °C.

9.2.2
LightScanner for Variant Scanning

Scanning Primer Design

Use LightScanner Primer Design Software (Idaho Technology) to simultaneously design primers for multiple exons:

Primer T_m range	62–66 °C
Primer size range	20–40 bases

For very GC-rich regions, you may need to accept primers below 20 bp and/or an upper T_m limit of 70 °C.

PCR Optimization

Multiple PCRs are best optimized with an annealing temperature gradient on a gradient cycler. Use a black-shell, white-well plate (Bio-Rad HSP-9665 for most thermal cyclers, Eppendorf for the ABI 9700).

1. Aliquot reagents into 96-well plates. For 10-µl reactions, dispense the following for each target.
   - Gradients across rows (12 wells):
   - 13 µl of 2.5 µM first primer
   - 13 µl of 2.5 µM second primer
   - 13 µl of control DNA ($A_{260} = 0.2$ or 10 ng)
   - 39 µl of H$_2$O
   - 52 µl of LightScanner Master Mix (Idaho Technology HRLS-ASY-0003)
   - Mix and aliquot across a row.

   Gradient across columns (eight wells):
   - 9 µl of 2.5 µM first primer
   - 9 µl of 2.5 µM second primer
   - 9 µl of control DNA ($A_{260} = 0.2$ or 10 ng)
   - 27 µl of H$_2$O
   - 36 µl of LightScanner Master Mix
   - Mix and aliquot down a column.

   The LightScanner Master Mix includes all reagents for PCR (*Taq* polymerase, hot start antibody, dNTPs, Mg^{2+}, and LCGreen Plus dye) at optimal concentrations for high-resolution scanning. Only primers and DNA need to be added.

2. Overlay with 15 µl of mineral oil (Sigma M5904). This is required to prevent evaporation during PCR and high-resolution melting. Seal the plate with adhesive film (Nunc 232 702).
3. Spin the plate at 1600 × g for 30 s (Eppendorf 5430).
4. Amplify using the following PCR protocol:
   - 95 °C – 2 min
   - 45 cycles:
     - 95 °C – 30 s
     - 60–72 °C gradient – 30 s
   - 95 °C – 30 s
   - 25 °C – 30 s.
5. Spin the plate at 1600 × g for 2 min (Eppendorf 5430).
6. Analyze the PCR products by high-resolution melting on the LightScanner:

Start temperature	70 °C
End temperature	96 °C
Hold temperature	60 °C
Exposure	auto

Using LightScanner software, analyze each target (row or column) individually. Most PCR products <200 bp will melt completely in one transition between 76 and 96 °C. Determine the annealing temperature range over which the PCR product is pure. Some products, especially those >300 bp, melt in more than one transition (multiple domains). If the product T_m is >92 °C, additives like 5–10% dimethylsulfoxide (DMSO) and/or 1–2 M betaine can be added to improve melting analysis at high temperatures.

7. Predict the melting curves and domains for each target:
   - Access: http://www.biophys.uni-duesseldorf.de/local/POLAND/poland.html
   - Enter your amplicon sequence
   - Use 75 mM NaCl Blake and Delcourt parameters
   - Enter temperature limits of 70 and 100 °C with a step size of 0.5 °C.

8. Observed T_ms are ∼5 °C greater than these predicted T_ms, but the shape of the melting curves and the presence of domains are usually accurate. Compare the melting curves obtained in Step 6 with the melting curves predicted in Step 7.

9. Gel electrophoresis. If the experimental melting curves do not fit the predicted curves, retrieve the PCR products melted in Step 6 and perform gel electrophoresis. Standard 1.5% agarose slab gels in 0.5 × TBE are fine. Alternatively, an automated system such as the Agilent 2100 Bioanalyzer can be used. Assess the amount and purity of the PCR products. Determine the annealing temperature range over which the PCR product is pure. If undesired products are present, increase the specificity by decreasing the concentration of primers to 0.10–0.20 µM, or add 5–10% DMSO and/or 1 M betaine to lower primer T_ms. Repeat the optimization. If the yield is low or no PCR products are obtained, decrease the specificity by increasing the Mg^{2+} concentration. Repeat the optimization. It may be easier to choose alternative primers than to perform extensive optimization with difficult primer sets.

PCR

1. Plate design. Your plate layout depends entirely on how many DNA samples you need to analyze and how many PCR products you need to scan. Since scanning compares melting curves from different samples, it is always better to have more samples to compare. Common variants can be identified by analyzing 96 individual DNA samples from a normal population before you begin experimental runs. In this case, all wells need the same primers and the DNA is typically dispensed into the plate first. These "DNA plates" can be

9 Mutation Scanning and Genotyping in Plants by High-Resolution DNA Melting

Table 9.2 Format options for a 96-well plate.

DNA samples	Primer pairs	Primer priority	DNA priority
>48	1	entire plate	—
33–48	2	4 rows or 6 columns	—
25–32	3	4 columns	—
17–24	4	2 rows or 3 columns	—
13–16	5–6	2 columns	—
9–12	7–8	1 row	1 column
7–8	9–12	1 column	1 row
5–6	13–16	—	2 columns
4	17–24	—	2 rows or 3 columns
3	25–32	—	4 columns
2	33–48	—	4 rows or 6 columns

used immediately or dried and stored for later use. The PCR setup is completed by adding a single solution that includes the master mix and primers. As long as the number of DNA samples is greater than the number of primer pairs, pipette the complex pattern of the DNA first, followed by the row or column format of each primer pair (Table 9.2). When the number of primer pairs exceeds the number of DNA samples, pipette the complex pattern of the primer pairs first, followed by the row or column format of each DNA sample. Complex primer pair patterns can be dispensed, dried, and stored for later use as "primer plates". For example, this service is provided by Idaho Technology, among other vendors.

2. Aliquot reagents into the plate. For example, if you have a 96-well plate, 12 DNA samples and eight primer pairs, manually or robotically aliquot each primer pair into one of the eight rows. Then, each DNA sample should be combined with the PCR master mix and aliquoted into one of the 12 columns. The manual procedure is detailed below.

Each of eight primer pairs into one row:
- 13 µl of the first primer (2.5 µM or otherwise optimized)
- 13 µl of the second primer (2.5 µM or otherwise optimized)
- Mix and aliquot 2 µl into each well across a row.

Each of 12 DNA samples into one column:
- 27 µl of H$_2$O (with supplemental Mg^{2+}, DMSO, or betaine if optimized)
- 9 µl of control DNA (A_{260} = 0.2 or 10 ng)
- 36 µl of LightScanner Master Mix
- Mix and aliquot 8 µl into each well down a column.

If dried DNA or primer plates are used, rehydrate with the complementary nucleic acid and Master Mix, allowing for the extra volume that evaporated from drying. Such plates should be lightly covered with adhesive film (Nunc 232 702 or equivalent) and shaken on a rotary mixer (Eppendorf MixMate 5353) at 1600 rpm for 30 s. Remove the film after mixing to overlay with oil.

3. Immediately overlay with 15 μl of mineral oil (Sigma M5904) to prevent evaporation. Seal the plate with adhesive film (Nunc 232 702 or equivalent).
4. Spin the plate at 1600 × g for 30 s (Eppendorf 5430).
5. Temperature cycling. If you only have one primer pair on a plate, choose the optimal annealing temperature determined by the gradient run for that primer pair. If you have multiple primer pairs on the plate, compare the acceptable annealing temperature ranges for all primer pairs and choose an annealing temperature common to all ranges:
   - 95 °C – 2 min
   - 45 cycles:
     - 95 °C – 30 s
     - X °C – 30 s (X is determined from 60–72 °C gradients)
   - 95 °C – 30 s
   - 25 °C – 30 s.

   As the number of primer pairs on the plate increases, it becomes more difficult to find a common temperature for annealing. Although this can usually be achieved, an alternative is the following universal touchdown PCR protocol:
   - 95 °C – 2 min
   - 8 step-down cycles:
     - 95 °C – 30 s
     - 72, 71, 70, 69, 68, 67, 66, 65 °C – 30 s
   - 35 cycles:
     - 95 °C – 30 s
     - 64 °C – 30 s
   - 95 °C – 30 s
   - 25 °C – 30 s.

Melting Analysis

1. Spin the plate at 1600 × g for 2 min (Eppendorf 5430) to remove bubbles.
2. Analyze the PCR products by high-resolution melting on the LightScanner. Use the data from the optimization runs to narrow the analysis range as follows:

Start temperature	5 °C below the first melting inflection
End temperature	2 °C above complete denaturation (stable baseline)
Hold temperature	15 °C below the first melting inflection
Exposure	auto

Most product melting protocols can be narrowed to a window of 12–16 °C, decreasing the melting time to about 7 min (including instrument cool-down time) and allowing up to eight 96- or 384-well plates to be processed per hour. This allows up to eight thermal block cyclers to feed a single LightScanner instrument. Using LightScanner software, analyze each PCR product across all DNA samples.

9.2.3
LightScanner for Lunaprobe™ (Unlabeled Probe) Genotyping

Primer and LunaProbe Design

Use LightScanner Primer Design Software (Idaho Technology) to design primers and unlabeled probes. LunaProbe genotyping requires asymmetric PCR (i.e., one primer is present in excess so that single strands are produced for probe binding). The software will automatically indicate which primer should be run in excess and which primer will be limiting.

Primer T_m range	62–66 °C
Primer size range	20–40 bases
Amplicon size range	80–200 bp (favor the smaller amplicons)
LunaProbe T_m range	$\leq$ primer T_m
LunaProbe size range	18–30 bases
Variant location within LunaProbe	$\geq$3 bases from either end of probe

For very GC-rich regions, you may need to accept primers below 20 bp and/or an upper T_m limit of 70 °C.

PCR Optimization

LunaProbe genotyping can be optimized with an annealing temperature gradient on a gradient cycler. Use a black-shell, white-well plate (Bio-Rad HSP-9665 for most thermal cyclers, Eppendorf for the ABI 9700).

1. Aliquot reagents into 96-well plates. For 10 μl reactions, dispense the following for each target.
   - Gradients across rows (12 wells):
   - 13 μl of 2.5 μM excess primer
   - 13 μl of 0.5 μM limiting primer
   - 13 μl of 2.0 μM LunaProbe
   - 13 μl of control DNA ($A_{260} = 0.2$ or 10 ng)
   - 26 μl of H$_2$O
   - 52 μl of LightScanner Master Mix
   - Mix and aliquot across a row.

   Gradient across columns (eight wells):
   - 9 μl of 2.5 μM excess primer
   - 9 μl of 0.5 μM limiting primer
   - 9 μl of 2.0 μM LunaProbe
   - 9 μl of control DNA ($A_{260} = 0.2$ or 10 ng)
   - 18 μl of H$_2$O

- 36 μl of LightScanner Master Mix
- Mix and aliquot down a column.

The LightScanner Master Mix includes all reagents for PCR (*Taq* polymerase, hot start antibody, dNTPs, Mg^{2+}, and LCGreen Plus dye) at optimal concentrations for LunaProbe genotyping. Only primers, DNA, and the unlabeled probe need to be added.

2. Overlay with 15 μl of mineral oil (Sigma M5904). This is required to prevent evaporation during PCR and high-resolution melting. Seal the plate with adhesive film (Nunc 232 702).
3. Spin the plate at $1600 \times g$ for 30 s (Eppendorf 5430).
4. Amplify using the following PCR protocol:
   - 95 °C – 2 min
   - 55 cycles of three-step PCR
     - 95 °C – 30 s
     - 60–72 °C gradient – 30 s
     - 74 °C – 30 s
   - 95 °C – 30 s
   - 25 °C – 30 s.
5. Spin the plate at $1600 \times g$ for 2 min (Eppendorf 5430).
6. Analyze the PCR products by high-resolution melting on the LightScanner:

Start temperature	45 °C
End temperature	96 °C
Hold temperature	40 °C
Exposure	auto

Using the unlabeled probe analysis module in the LightScanner software, analyze each target (row or column) individually. Most LunaProbes will melt between 50 and 70 °C, while the PCR amplicon will melt at temperatures above 80 °C. Determine the annealing temperature range over which the PCR product is pure and the LunaProbe melting signal is strongest. If undesired products are present, you may increase the specificity by lowering the primer concentration or adding DMSO or betaine as mentioned above for scanning. Specificity can also be increased by rapid cycle PCR if appropriate instrumentation is available (LS-32; Idaho Technology). If the probe signal is strong, you can decrease the number of cycles. If the probe signal is weak, consider increasing the primer asymmetry. If the yield is low or no PCR products are obtained, decrease the specificity by increasing the Mg^{2+} concentration. It may be easier to choose alternative primers than to perform extensive optimization with difficult primer sets.

PCR

1. Aliquot reagents into the plate. For example, if you are using a 96-well plate to analyze 96 DNA samples with one LunaProbe, aliquot 1 µl of each DNA ($A_{260} = 0.2$ or 10 ng) into the wells first. The PCR setup is completed by adding a single solution that includes the master mix, primers, and the LunaProbe:
   - 100 µl of the excess primer (2.5 µM)
   - 100 µl of the second primer (0.5 µM)
   - 100 µl of LunaProbe (2.0 µM)
   - 200 µl of H_2O (with supplemental Mg^{2+}, DMSO, or betaine if optimized)
   - 400 µl of LightScanner Master Mix
   - Mix and aliquot 9 µl into each well.

   If dried DNA or primer plates are used, rehydrate with the complementary nucleic acid and master mix, allowing for the extra volume that evaporated from drying. Such plates should be lightly covered with adhesive film (Nunc 232 702 or equivalent) and shaken on a rotary mixer (Eppendorf MixMate 5353) at 1600 rpm for 30 s. Remove the film after mixing to overlay with oil.
2. Immediately overlay with 15 µl of mineral oil (Sigma M5904) to prevent evaporation. Seal the plate with adhesive film (Nunc 232 702 or equivalent).
3. Spin the plate at $1600 \times g$ for 30 s (Eppendorf 5430).
4. Temperature cycling. If you only have one LunaProbe set on a plate, choose the optimal annealing temperature determined by the gradient run. If you are analyzing LunaProbe sets on the same plate, compare the acceptable annealing temperature ranges for all primer pairs and choose an annealing temperature common to all ranges:
   - 95 °C – 2 min
   - 55 cycles of three-step PCR:
     - 95 °C – 30 s
     - X °C gradient – 30 s (X is determined from 60–72 °C gradients)
     - 74 °C – 30 s
   - 95 °C – 30 s
   - 25 °C – 30 s.

Melting Analysis

1. Spin the plate at $1600 \times g$ for 2 min (Eppendorf 5430) to remove bubbles.
2. Analyze the LunaProbes by high-resolution melting on the LightScanner. Use the data from the optimization runs to narrow the analysis range as follows:

Start temperature	10 °C below the T_m of the lowest LunaProbe allele
End temperature	2 °C above complete denaturation (stable baseline)
Hold temperature	5 °C below the start temperature
Exposure	auto

Using LightScanner software, analyze each LunaProbe set across all DNA samples. Although you can shorten the melting window by only melting the LunaProbe and not the PCR product, we recommend including both. Often the LunaProbe and PCR product data are complementary.

9.3 Applications

9.3.1 Sensitivity and Specificity for SNP Heterozygote Detection

The sensitivity and specificity of diploid SNP scanning was determined on the LightScanner using mixtures of engineered plasmids [28] kindly provided by Lonza. All SNP variants in three template backgrounds were tested. The same protocol previously used for the HR-1 instrument [9] was followed, studying amplicons varying in size from about 50 to 800 bp. Samples were amplified with a PTC-200 thermal cycler (BioRad) in a 384-well block and then melted in a 384-well LightScanner.

The sensitivity and specificity of SNP mutation scanning on the LightScanner are shown in Figure 9.1 and related to amplicon size. No false-positive or false-negative results were obtained in amplicons 400 bp or less ($n=720$). Even at 800 bp,

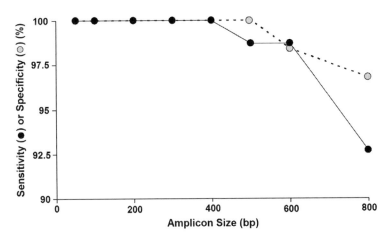

Figure 9.1 Sensitivity (filled circles) and specificity (open circles) of SNP scanning on the LightScanner instrument. Fifty, 100, 200, 300, 400, 500, 600, and 800 bp amplicons of three different plasmids were studied with the SNP near the middle (prior data suggests that the position of the SNP within the amplicon does not affect scanning accuracy [9]). Approximately half of the 1152 decisions were wild-type and half heterozygous. Melting curves were processed by fluorescence normalization and temperature overlay as previously described [34] and wild-type or heterozygous calls were made by visual inspection of blinded samples.

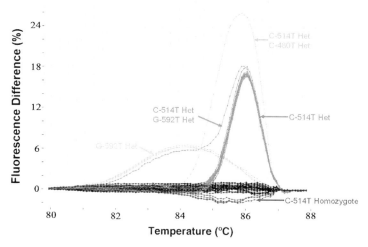

Figure 9.2 Exemplary scanning data obtained on the LightScanner instrument. A polymorphic 208-bp fragment of the 5′-UTR of hepatic lipase was amplified from 58 human genomic DNA samples. The normalized, temperature-overlaid curves were subtracted from the average wild-type curve to give difference plots. Thirty-one of the samples clustered as wild-type (horizontal lines, unlabeled), with a common variant (C-514T) resulting in 18 heterozygotes and four homozygotes. Another less-common variant (G-592T) was present in three samples, along with single C-514T/G-592T and C-514T/C-480T double heterozygotes. The C-514T/G-592T curve resembles the sum of the C-514T and G-592T curves. Note that the *difference* plots used here for mutation scanning are not the same as *derivative* plots that are commonly used in genotyping.

sensitivity was greater than 90% and specificity greater than 95%. These results are similar to a previous study reported on the HR-1 instrument [9]. Additional studies with human genomic DNA support the high sensitivity and specificity of the method [6–8, 29–33].

9.3.2
Variant Scanning by High-Resolution Melting

A 208-bp fragment of the 5′-untranslated region (UTR) of the hepatic lipase gene (GenBank accession number L77 731) was amplified from 0.3 µM of the primers CCTCTACACAGCTGGAACATTA and CCCCCAGAGGGTCCAAAT and human genomic DNA. Melting curves were displayed as difference plots after fluorescence normalization, temperature overlay, and subtraction from wild-type curves [34]. Typical data from the LightScanner is shown in Figure 9.2. In contrast to the more common derivative plots, scanning data is best displayed in difference plots that magnify curve shape differences. Deviation from wild-type is greatest for double heterozygotes, less extreme for single heterozygotes, and least for homozygous changes.

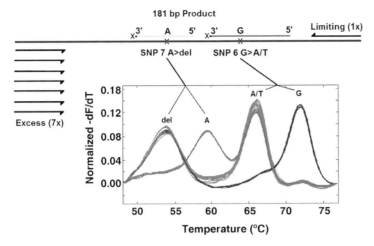

Figure 9.3 LunaProbe chlorophyll genotyping of diploid bell peppers. Two loci are interrogated by two unlabeled probes present within one 181-bp PCR. One probe (SNP7) covers an A deletion variant. The other probe (SNP8) distinguishes the G allele from A or T variants. The probes are designed so that all relevant alleles are separated by temperature. The LunaProbes are blocked at their 3' end to prevent extension. LCGreen Plus dye provides the fluorescence for genotyping – the LunaProbes not covalently labeled.

9.3.3
Bell Pepper Multiplex Genotyping with Two Unlabeled Probes

Bell pepper diploid DNA was isolated and amplified with primers TTTCTTGTCA-CACTCAGGCAGC (0.35 µM) and AGAGAACATCCTCTCCTGCAGAATAG (0.05 µM) in the presence of two unlabeled probes, SNP6 (CAAATTCA-GAACTCT**C**GGTGCCTCTAG-C3 block) and SNP 7 (AAGATGA**T**TTACCTA-CAAGG-C3 block), both at 0.15 µM. The variant sites within the probes are indicated in bold. The probes differ in stability enough so that all alleles (an A deletion under the SNP7 probe and either an A or T substitution under the SNP6 probe) are separated by T_m (Figure 9.3).

9.3.4
Potato Tetraploid Genotyping including Allele Dosage using an Unlabeled Probe

Potato DNA was isolated and a fragment of the dihydroflavonol 4-reductase gene (*dfr*) amplified by asymmetric PCR in the presence of an unlabeled probe. Figure 9.4 clearly shows variation in the dosage of the two alleles reflecting the five possible genotypes of tetraploid DNA. Hexaploid species show a maximum of seven curve clusters and octaploid species, nine clusters, corresponding to the possible biallelic genotypes.

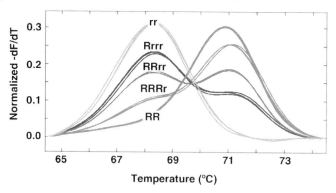

Figure 9.4 A LunaProbe assay for tetraploid bialleleic genotyping in potatoes. Variants of the *dfr* gene in the anthocyanin pathway control red pigmentation in potato tuber skin [40]. Shown are normalized melting peaks of five different *dfr* genotypes including the clones 07 506-01 (*rr*), Chieftain (*Rrrr*), Redsen (*RRrr*), Superior (*RRRr*), and W5281.2 (*RR*) in triplicate. The five different genotypes cluster into five groups.

9.4
Perspectives

High-resolution melting analysis with saturation dyes provides some very simple solutions for variant scanning and genotyping of SNPs and small Indels. Elimination of labeled probes and all processing after PCR are strong advantages of high-resolution melting over other scanning and genotyping techniques. Although most studies to date focus on human diploid DNA, the method is easily adapted to polyploidy genomes. In this chapter, we have detailed variant scanning and unlabeled probe genotyping as separate techniques. However, scanning and genotyping can be performed simultaneously by analyzing both unlabeled probe and PCR product transitions in the same melting curve [35]. Such combined analysis can be used to scan and genotype common variants in one step, to verify genotype by two independent assessments, and to identify variants within the amplicon, but not under the probe.

Although not detailed here, most SNPs can be genotyped without probes by direct PCR product analysis. All class 1 and class 2 SNPs (A–T, G–C exchanges – 84% of human SNPs) can be typed directly by PCR product melting with an average T_m difference of 1.0 °C between homozygotes when the PCR products are small [15]. However, 16% of human SNPs are class 3 or 4 and have smaller T_m differences between homozygotes. In one-quarter of these (4% of human SNPs), the alternative homozygotes have the same predicted T_m because of nearest-neighbor symmetry [36]. Even some of these base-pair neutral SNPs can be completely genotyped if internal temperature controls are included [37]. In the rare case where complete genotyping is not possible, PCR products can be mixed after PCR or DNA can be mixed before PCR and quantitative heteroduplex analysis performed [36].

Unlabeled probe analysis provides an additional layer of specificity over PCR product melting and greater detail over a more limited region. Recent advances in unlabeled probe genotyping include masking unimportant variants by incorporating deletions into the probes [38] and snapback primers [39]. In snapback primers, the unlabeled probe is attached as a 5′ primer tail and forms a hairpin after asymmetric PCR. Advantages over unlabeled probes include no need for 3′ end blocking and the ability to interrogate small sequence segments.

The power of melting analysis depends on instrument resolution, software tools, and appropriate dyes, and has come a long way since its introduction in 1997 [1]. As high-resolution melting continues to advance, we can look forward to even better performance of this simplest of all analysis tools.

Acknowledgments

This study was supported by Idaho Technology, National Institutes of Health grants GM060 063, GM072 419, and GM073 396, and the State of Utah through a Center of Excellence grant.

References

1 Ririe, K.M., Rasmussen, R.P., and Wittwer, C.T. (1997) Product differentiation by analysis of DNA melting curves during the polymerase chain reaction. *Anal. Biochem.*, **245**, 154–160.

2 Herrmann, M.G., Durtschi, J.D., Bromley, L.K., Wittwer, C.T., and Voelkerding, K.V. (2006) Amplicon DNA melting analysis for mutation scanning and genotyping: cross-platform comparison of instruments and dyes. *Clin. Chem.*, **52**, 494–503.

3 Herrmann, M.G., Durtschi, J.D., Bromley, L.K., Wittwer, C.T., and Voelkerding, K.V. (2007) Instrument comparison for heterozygote scanning of single and double heterozygotes: a correction and extension of Herrmann *et al.*, Clin Chem 2006;52:494–503. *Clin. Chem.*, **53**, 150–152.

4 Herrmann, M.G., Durtschi, J.D., Wittwer, C.T., and Voelkerding, K.V. (2007) Expanded instrument comparison of amplicon DNA melting analysis for mutation scanning and genotyping. *Clin. Chem.*, **53**, 1544–1548.

5 Dujols, V.E., Kusukawa, N., McKinney, J.T., Dobrowolski, S.F., and Wittwer, C.T. (2006) High-resolution melting analysis for scanning and genotyping, in *Real-Time PCR* (ed. M.T. Dorak), Garland, New York.

6 Dobrowolski, S.F., McKinney, J.T., Amat di San Filippo, C., Giak Sim, K., Wilcken, B., and Longo, N. (2005) Validation of dye-binding/high-resolution thermal denaturation for the identification of mutations in the SLC22A5 gene. *Hum. Mutat.*, **25**, 306–313.

7 McKinney, J.T., Longo, N., Hahn, S.H., Matern, D., Rinaldo, P., Strauss, A.W., and Dobrowolski, S.F. (2004) Rapid, comprehensive screening of the human medium chain acyl-CoA dehydrogenase gene. *Mol. Genet. Metab.*, **82**, 112–120.

8 Montgomery, J., Wittwer, C.T., Kent, J.O., and Zhou, L. (2007) Scanning the cystic fibrosis transmembrane conductance regulator gene using high-resolution DNA melting analysis. *Clin. Chem.*, **53**, 1891–1898.

9 Reed, G.H. and Wittwer, C.T. (2004) Sensitivity and specificity of single-nucleotide polymorphism scanning by high-resolution melting analysis. *Clin. Chem.*, **50**, 1748–1754.

10 Dames, S., Pattison, D.C., Bromley, L.K., Wittwer, C.T., and Voelkerding, K.V. (2007)

Unlabeled probes for the detection and typing of herpes simplex virus. *Clin. Chem.*, **53**, 1847–1854.

11 Erali, M., Palais, R., and Wittwer, C.T. (2008) SNP genotyping by unlabeled probe melting analysis, in *Molecular Beacons: Signalling Nucleic Acid Probes, Methods and Protocols* (eds O., Seitz and A., Marx), Humana Press, Totowa, NJ.

12 Zhou, L., Myers, A.N., Vandersteen, J.G., Wang, L., and Wittwer, C.T. (2004) Closed-tube genotyping with unlabeled oligonucleotide probes and a saturating DNA dye. *Clin. Chem.*, **50**, 1328–1335.

13 Graham, R., Liew, M., Meadows, C., Lyon, E., and Wittwer, C.T. (2005) Distinguishing different DNA heterozygotes by high-resolution melting. *Clin. Chem.*, **51**, 1295–1298.

14 Liew, M., Nelson, L., Margraf, R., Mitchell, S., Erali, M., Mao, R., Lyon, E., and Wittwer, C. (2006) Genotyping of human platelet antigens 1 to 6 and 15 by high-resolution amplicon melting and conventional hybridization probes. *J. Mol. Diagn.*, **8**, 97–104.

15 Liew, M., Pryor, R., Palais, R., Meadows, C., Erali, M., Lyon, E., and Wittwer, C. (2004) Genotyping of single-nucleotide polymorphisms by high-resolution melting of small amplicons. *Clin. Chem.*, **50**, 1156–1164.

16 Wittwer, C.T., Reed, G.H., Gundry, C.N., Vandersteen, J.G., and Pryor, R.J. (2003) High-resolution genotyping by amplicon melting analysis using LCGreen. *Clin. Chem.*, **49**, 853–860.

17 Erali, M., Voelkerding, K.V., and Wittwer, C.T. (2008) High resolution melting applications for clinical laboratory medicine. *Exp. Mol. Pathol.*, **85**, 50–58.

18 Reed, G.H., Kent, J.O., and Wittwer, C.T. (2007) High-resolution DNA melting analysis for simple and efficient molecular diagnostics. *Pharmacogenomics*, **8**, 597–608.

19 Chateigner-Boutin, A.L. and Small, I. (2007) A rapid high-throughput method for the detection and quantification of RNA editing based on high-resolution melting of amplicons. *Nucleic Acids Res.*, **35**, e114.

20 Mackay, J.F., Wright, C.D., and Bonfiglioli, R.G. (2008) A new approach to varietal identification in plants by microsatellite high resolution melting analysis: application to the verification of grapevine and olive cultivars. *Plant Methods*, **4**, 8.

21 Lehmensiek, A., Sutherland, M.W., and McNamara, R.B. (2008) The use of high resolution melting (HRM) to map single nucleotide polymorphism markers linked to a covered smut resistance gene in barley. *Theor. Appl. Genet.*, **117**, 721–728.

22 Chagne, D., Gasic, K., Crowhurst, R.N., Han, Y., Bassett, H.C., Bowatte, D.R., Lawrence, T.J., Rikkerink, E.H., Gardiner, S.E., and Korban, S.S. (2008) Development of a set of SNP markers present in expressed genes of the apple. *Genomics*, **92**, 353–358.

23 Wu, S.B., Wirthensohn, M.G., Hunt, P., Gibson, J.P., and Sedgley, M. (2008) High resolution melting analysis of almond SNPs derived from ESTs. *Theor. Appl. Genet.*, **118**, 1–14.

24 Croxford, A.E., Rogers, T., Caligari, P.D., and Wilkinson, M.J. (2008) High-resolution melt analysis to identify and map sequence-tagged site anchor points onto linkage maps: a white lupin (*Lupinus albus*) map as an exemplar. *New Phytol.*, **180**, 594–607.

25 Nomoto, K., Tsuta, K., Takano, T., Fukui, T., Fukui, T., Yokozawa, K., Sakamoto, H., Yoshida, T., Maeshima, A.M., Shibata, T., Furuta, K., Ohe, Y., and Matsuno, Y. (2006) Detection of EGFR mutations in archived cytologic specimens of non-small cell lung cancer using high-resolution melting analysis. *Am. J. Clin. Pathol.*, **126**, 608–615.

26 Dames, S., Margraf, R.L., Pattison, D.C., Wittwer, C.T., and Voelkerding, K.V. (2007) Characterization of aberrant melting peaks in unlabeled probe assays. *J. Mol. Diagn.*, **9**, 290–296.

27 Bennett, C.D., Campbell, M.N., Cook, C.J., Eyre, D.J., Nay, L.M., Nielsen, D.R., Rasmussen, R.P., and Bernard, P.S. (2003) The LightTyper: high-throughput genotyping using fluorescent melting curve analysis. *Biotechniques*, **34**, 1288–1295.

28 Highsmith, W.E., Jr., Nataraj, A.J., Jin, Q., O'Connor, J.M., El-Nabi, S.H., Kusukawa, N., and Garner, M.M. (1999) Use of DNA toolbox for the

characterization of mutation scanning methods. II: evaluation of single-strand conformation polymorphism analysis. *Electrophoresis*, **20**, 1195–1203.

29 De Leeneer, K., Coene, I., Poppe, B., De Paepe, A., and Claes, K. (2008) Rapid and sensitive detection of BRCA1/2 mutations in a diagnostic setting: comparison of two high-resolution melting platforms. *Clin. Chem.*, **54**, 982–989.

30 Dobrowolski, S.F., Ellingson, C., Coyne, T., Martin, R., Grey, J., Naylor, E.W., Koch, R., and Levy, H. (2007) Mutations in the phenylalanine hydroxylase gene identified in 95 patients with phenylketonuria using novel systems of mutation scanning and specific genotyping based upon thermal melt profiles. *Mol. Genet. Metabol.*, **91**, 218–227.

31 Dobrowolski, S.F., Ellingson, C.E., Caldovic, L., and Tuchman, M. (2007) Streamlined assessment of gene variants by high resolution melt profiling utilizing the ornithine transcarbamylase gene as a model system. *Hum. Mutat.*, **28**, 1133–1140.

32 Lin, S.Y., Su, Y.N., Hung, C.C., Tsay, W., Chiou, S.S., Chang, C.T., Ho, H.N., and Lee, C.N. (2008) Mutation spectrum of 122 hemophilia A families from Taiwanese population by LD-PCR, DHPLC, multiplex PCR and evaluating the clinical application of HRM. *BMC Med. Genet.*, **9**, 53.

33 Vandersteen, J.G., Bayrak-Toydemir, P., Palais, R.A., and Wittwer, C.T. (2007) Identifying common genetic variants by high-resolution melting. *Clin. Chem.*, **53**, 1191–1198.

34 Gundry, C.N., Vandersteen, J.G., Reed, G.H., Pryor, R.J., Chen, J., and Wittwer, C.T. (2003) Amplicon melting analysis with labeled primers: a closed-tube method for differentiating homozygotes and heterozygotes. *Clin. Chem.*, **49**, 396–406.

35 Montgomery, J., Wittwer, C.T., Palais, R.A., and Zhou, L. (2007) Simultaneous mutation scanning and genotyping by high-resolution DNA melting analysis. *Nature Prot.*, **2**, 59–66.

36 Palais, R.A., Liew, M.A., and Wittwer, C.T. (2005) Quantitative heteroduplex analysis for single nucleotide polymorphism genotyping. *Anal. Biochem.*, **346**, 167–175.

37 Gundry, C.N., Dobrowolski, S.F., Martin, Y.R., Robbins, T.C., Nay, L.M., Boyd, N., Coyne, T., Wall, M.D., Wittwer, C.T., and Teng, D.H. (2008) Base-pair neutral homozygotes can be discriminated by calibrated high-resolution melting of small amplicons. *Nucleic Acids Res.*, **36**, 3401–3408.

38 Margraf, R.L., Mao, R., and Wittwer, C.T. (2006) Masking selected sequence variation by incorporating mismatches into melting analysis probes. *Hum. Mutat.*, **27**, 269–278.

39 Zhou, L., Errigo, R.J., Lu, H., Poritz, M.A., Seipp, M.T., and Wittwer, C.T. (2008) Snapback primer genotyping with saturating DNA dye and melting analysis. *Clin. Chem.*, **54**, 1648–1656.

40 De Jong, W.S., De Jong, D.M., and Bodis, M. (2003) A fluorogenic 5′ nuclease (TaqMan) assay to assess dosage of a marker tightly linked to red skin color in autotetraploid potato. *Theor. Appl. Genet.*, **107**, 1384–1390.

10
In Silico Methods: Mutation Detection Software for Sanger Sequencing, Genome and Fragment Analysis

Kevin LeVan, Teresa Snyder-Leiby, C.S. Jonathan Liu, and Ni Shouyong

Abstract

Mutation detection through Sanger sequencing can be intensive to review. Manually comparing electropherograms to one another is tedious when scanning peak by peak. Additionally, mutations may be missed when comparing base calls within the electropherograms. Automated mutation detection not only increases throughput, but can also drastically increase the sensitivity of detection. This is especially useful for techniques evaluating low-frequency variations, including somatic detection and polyploid alleles. The next-generation sequencing technologies have surpassed the classical Sanger sequencing method in throughput by 100- or 1000-fold – increasing the need for automated, consistent sizing and comparison software. The flexibility to set parameters for base calling increases analysis consistency. Mutation detection through fragment analysis, as in the applications of short sequence repeats and TILLING, require consistent peak calling and comparison capabilities for large data sets. This chapter uses SoftGenetics software examples to illustrate methods and applications for *in silico* mutation detection from Sanger sequencing, genome and fragment analyses.

10.1
Introduction

A multitude of benefits exist from locating natural and induced alleles in plants. Three major applications are determining gene function/mapping, diversity/phylogeny/taxonomy, and crop improvement. *In silico* mutation detection has led to innumerable discoveries in each of these fields [1–10]. Software tools for mutation discovery enable researchers to screen large data sets; generated from individual research projects or combined from online databases (major sources include GenBank (http://www.ncbi.nlm.nih.gov/Genbank/index.html, http://www.ncbi.nlm.nih.gov/projects/SNP/), Plant Genome Center SNP database (http://www.

pgcdna.co.jp/snps/), and J. Craig Venter Institute (www.tigr.org)). Software programs for mutation detection allow researchers to establish parameters for analysis. This provides a rapid, objective, consistent mutation calling procedure.

Single nucleotide polymorphisms (SNPs) and insertion/deletions (Indels) are easily detected from DNA fragment analysis, Sanger generated sequences, and sequences generated from next-generation genomic analyzers (Solexa®, SOLiD™, and 454®; see Section 10.3). Fragment-based mutation detection has lower cost and is rapid, but does not provide specific sequence information. Sequence-based methods are higher in cost and more time-consuming, but provide specific sequence information for each mutation [11]. SNPs result from substitutions of one base for another. These substitutions are the most common type of mutation and may result in the premature termination of or the incorporation of a different amino acid in a peptide. Some SNPs may cause mRNA splicing variants, changing the function of the resulting protein. SNP markers have been associated with agronomically important phenotypes [1, 12–15] and evolutionary divergence [5, 16–18]. SNP and short sequence repeats (SSRs) are also useful for constructing linkage maps and marker-assisted breeding programs [6, 8, 13, 19]. Indels often produce reading frameshifts. This shift often makes Indels even more significant than SNPs in functional genomics, since they are more likely to cause a change in the amino acid sequence of the peptide encoded by that DNA sequence. Indels may also affect splicing and regulatory sequence functions.

DNA fragment analysis methods enable mutation detection without DNA sequencing and include methods to determine loss of heterozygosity (LOH), microsatellite instability (MSI), medium throughput SNP detection (SNPWave® and SNPlex™) and targeting induced local lesions in genomes (TILLING). All of these methods analyze data resulting from polymerase chain reaction (PCR) amplification of DNA fragments. Short tandem repeats (STRs) and SSR markers can be developed from lab validation studies or using predictive software [20]. Sanger and next-generation methods compare a DNA sequence to a reference sequence, and locate SNPs and Indels in genomic, mitochondrial, or chloroplast DNA sequences. Patterns of SNPs and Indels in DNA sequence traces can be used to analyze the distribution of genetic diversity among cultivars and within genes [9]. Large resequencing projects demand fully automated calls of SNPs and homozygous and heterozygous Indels to identify genetic relationships among wild-relative and domesticated strains. Mutation Surveyor® has been used to successfully analyze sequencing data from many crops, including sugar cane, barley, and potato, and is one of the only software packages to detect heterozygous Indels by deconvoluting DNA sequence traces [21].

10.2
Mutation Detection with Sanger Sequencing using Mutation Surveyor

Sanger sequencing is essential for many applications. However, mutation detection through sequencing can be intensive to review. Manually comparing electropherograms to one another is tedious when scanning base by base. Additionally, mutations may be missed when comparing the base calls within electropherograms. Converting nucleotide sequences into amino acid sequences can be labor-intensive. Automated mutation detection not only increases the throughput, but can drastically increase the sensitivity of detection, especially useful in techniques evaluating low-frequency variations, including somatic detection and polyploid alleles. Mutation Surveyor is a software package that looks beyond the base calls within the electropherograms to the physical traces. By evaluating the peak profiles of each sample as compared to a reference, slight variations can be observed. This stringent evaluation of the electropherograms dramatically increases the accuracy of SNP detection and can reduce the error rate due to other, more subjective, analysis techniques. This allows detection of a single allele variation in multiploid species, such as hexaploid barley. The accuracy of Mutation Surveyor when the sample is sequenced in both forward and reverse direction is over 99%, with sensitivity to greater than 5% of the primary peak (with Phred 20 quality) (http://www.softgenetics.com/mutationSurveyor.html). An accuracy of 95% and higher has been demonstrated when processing single directional sequences.

Mutation Surveyor utilizes a physical trace-to-trace comparison methodology for optimized detection of variations. Mutation Surveyor recognizes .abi, .ab1, and .scf formats from instruments including those made by Applied Biosystems and GE Healthcare. The software does a trace-to-trace comparison; the sample electropherograms are physically compared to a reference electropherogram that has been added to the project or synthetically generated. By evaluating variations at the trace level, detection sensitivity is substantially increased for several reasons. The base calls of positions that contain a mixture of alleles where one allele is at a low frequency, common with somatic mutations, and organisms that are polyploid, are often shown as only one allele. Programs that detect variations based solely on base calls would miss these subtle variations. The base callers used to call the bases within the electropherograms can sometimes make incorrect base calls, which lead to a higher rate of false-positive mutation calls when evaluating base call differences. Automated evaluation of variations by comparing the peak profiles of electropherograms not only increases the throughput, but enhances the accuracy and sensitivity of the study. To ease review, Mutation Surveyor shows a mutation electropherogram for each sample/reference pair – peaks are displayed at each location of high variation, related to SNPs and Indels (Figure 10.1).

Many sample electropherograms can be analyzed simultaneously in one project. These samples can be from multiple sets of primers (or amplicons). Mutation Surveyor automatically groups samples within the project that are from different amplicons into separate contigs. One project can therefore contain samples from multiple genes as well. For added sensitivity, two-directional coverage of a sequence is often prepared. When samples are loaded into a project that have been sequenced in both forward and reverse directions, Mutation Surveyor automatically pairs them together (Figure 10.2). The software first groups the significantly homologous samples from the same amplicons into unique contigs, then inspects these contigs

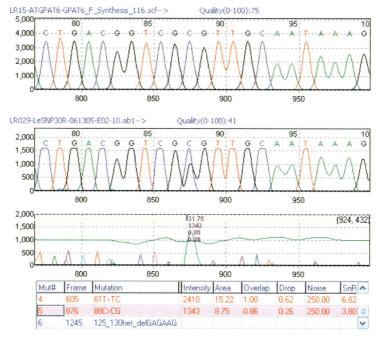

Figure 10.1 Mutation Surveyor evaluates trace differences, a SNP was automatically detected, substitutions with a minor allele frequency as low as 5% can be detected. The top electropherogram is the reference, second from top is the sample, and third from top is the mutation electropherogram showing the trace-to-trace comparison between the two previous traces. Base position 88 of the sample contains a heterozygous substitution. The table at the bottom contains a listing of all mutation calls within the sample. Position 88 is highlighted purple because this variation is reported in the reference file. Notice that the base call of both the reference and the sample is a cytosine.

for two-directional coverage by reverse complementation of the files. SNPs and Indels that are present in both forward and reverse directions are given higher weight that a true variation is present at this location. Detection sensitivity of somatic mutations is greatly improved. Electropherograms often contain extra peaks in the baseline. This noise may contribute to false-positives when attempting to increase sensitivity for analysis of somatic mutations. With two-directional coverage, the likelihood of having the complement color at the same location in the opposing direction is lower; this increases the probability that the noise at this location is in fact a mutation at low frequency.

Mutation Surveyor automates the detection of variations within sequencing samples at the trace level; sample electropherograms are physically compared to a reference electropherogram. One reference electropherogram is used for each contig and each sample electropherogram within a given contig is independently evaluated for SNPs and Indels against this same reference. An individual sample electropherogram can serve as a reference for the other samples or specific references can be selected for the trace-to-trace comparison. In addition to electropherograms serving

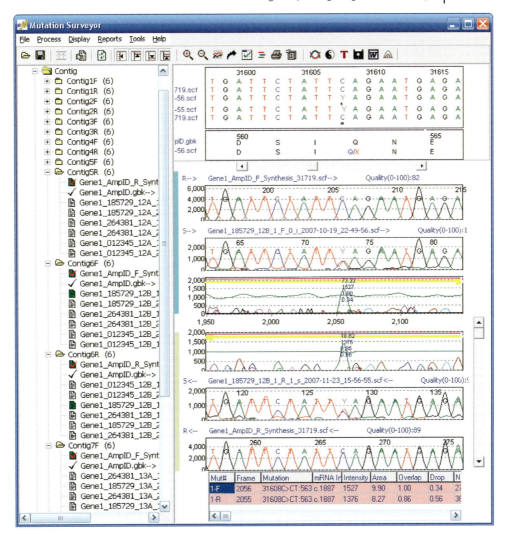

Figure 10.2 Graphic analysis display of Mutation Surveyor. On the left is a sample file tree listing samples separated into contigs (or amplicons). On the right, the active samples are shown. The top pane shows the nucleotides of each sample on the screen. The next pane displays amino acids for any nucleotide region that contains a coding sequence. The next three traces (from the top) are the synthetic reference, sample data file, and mutation electropherograms for the forward-direction sample of the pair; the bottom three traces are the mutation, sample data file, and synthetic reference electropherograms for the reverse-direction sample. Identified in the middle of the screen is a C → CT mutation that generates a stop codon.

as references, nucleotide text strings can be used as references. Several plant genomes are already sequenced and annotated, such as *Arabidopsis*, wheat, rice, and maize. By utilizing these references, many applications can be studied using Sanger sequencing, including phylogeny and crop improvement. In addition to chromosomes, reference sequences of circular genomes, such as mitochondria and chloroplasts, are available.

Mutation Surveyor conducts a physical trace-to-trace comparison to detect variations. The software synthetically creates a reference electropherogram when one is not present for the samples in the contig. For this to occur, the software utilizes the peak profiles of a high-quality sample electropherogram in the contig – without the noise – and the text string of a GenBank or other annotated sequence. Utilizing a synthetic reference can enhance the level of sensitivity for detecting low-frequency variations. Utilizing an electropherogram as the reference can remove false-positives due to repeat occurrence of some common artifacts found in electropherograms, such as dye blobs and mobility shift problems.

The general procedure for analyzing Sanger sequencing data using Mutation Surveyor is as follows:

- Load the sample electropherograms.
- Load the reference(s) – nucleotide sequences and/or electropherograms.
- Adjust the analysis settings appropriate for application.
- Run the project.
- Review the results and generate reports and outputs.

Mutation Surveyor generates a mutation electropherogram that shows the correlation between each sample/reference pair (Figure 10.2). Regions containing a high level of variance, often a result of SNPs and Indels, are represented by intense peaks. Two general formulas are graphically displayed in the mutation electropherogram – one showing peaks due to color variations between the electropherograms and another showing peaks due to spatial differences. The first identifies locations that may contain SNPs. A mutation score is assigned to each substitution mutation call, calculated using several factors that can differ between the reference and sample electropherograms; these factors are mutation peak intensity, overlap factor, and drop factor. The mutation peak intensity shows the peak height in the electropherogram, larger peaks refer to a higher color discrepancy between sample and reference. The drop factor represents the intensity decrease of the sample's normal allele with respect to the reference allele, relative to the same color peaks in the neighborhood. The overlap factor evaluates the spatial difference of a peak with respect to the reference. The mutation score indicates the reliability of the mutation calls by combining these factors: signal-to-noise ratio in the mutation electropherogram, drop factor, and overlap factor. The second formula identifies locations of homozygous Indels. In addition, Mutation Surveyor screens the samples for heterozygous Indels through a unique deconvolution of the mutation allele from the normal allele (Figure 10.3).

Sample electropherograms can be loaded into a Mutation Surveyor project without any modification. For instance, Mutation Surveyor is capable of loading the raw data contained in electropherograms, correcting some base calling errors using BasePatch, removing dye blobs, and masking (trimming) bases of defined sequence or region of the electropherogram or by base quality. The .ab1 format sequencing samples from an Applied Biosystems instrument contain both the processed,

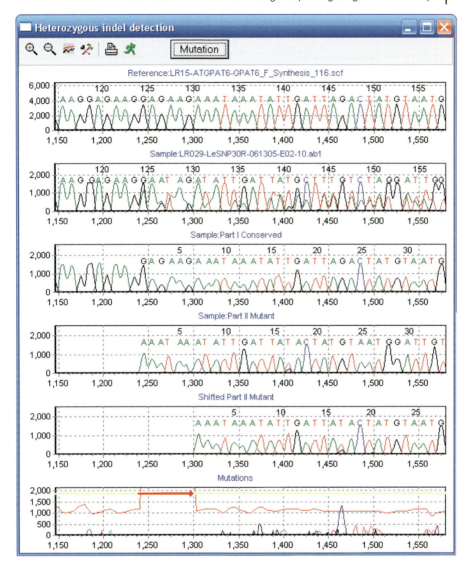

Figure 10.3 Heterozygous Indel Detection tool of Mutation Surveyor, this example from a tomato project contains a heterozygous Indel. The first trace (top) is the reference, the second trace shows a sample containing a heterozygous deletion, the third trace contains the conserved bases within the sample trace (bases that match reference sequence), and the fourth trace contains the other, mutation-containing, allele – deconvoluted at each nucleotide position. The fifth trace is the mutant allele shifted to align to the reference trace and the last trace is the mutation electropherogram of the resultant allele.

base-called results that are normally viewed, as well as the raw data generated by the sequencer. Mutation Surveyor, by default, analyzes the processed results. The software is capable of analyzing the project using the raw data contained within these files – subtracting baseline, smoothing, correcting for the different mobilities of the four dyes, and conducting its own base call.

Mutation Surveyor accepts references in a variety of types and formats. Reference files can be in the form of GenBank files that contain the nucleotide text string in addition to annotations, such as the coding sequence positions, in the form of electropherograms or both. Options are available to utilize the references in the manner appropriate for each specific application. One such application is automated methylation detection. Mutation Surveyor can determine the locations where 5-methylcytosine has been deaminated to thymine through bisulfite treatment. Mutation Surveyor generates a modified reference based on the expected methylation pattern. When annotated GenBank files are used, such as those files downloaded from National Center for Biotechnology Information (NCBI)'s Entrez Gene or RefSeq databases (Figure 10.4), Mutation Surveyor will show the amino acid

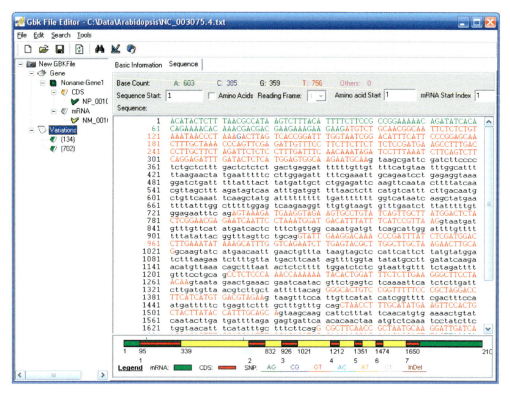

Figure 10.4 Advanced GBK file editor of Mutation Surveyor. Reference files can be in the form of annotated sequences, such as this *Arabidopsis* gene obtained from the NCBI Entrez Gene database. The noncoding mRNA is in green, coding sequence is shown in red, and reported SNPs are in blue. These GenBank files are customizable, including the ability to identify regions of interest and add custom SNPs.

No.	Sample File	Refere	Gene	Mut#	Mutation1	Mutation2	Mutation3
1	LR015-LeSNP30R-061305-G10-14	LR15-	ATGPAT6/GPAT6	1		88C>G$100	
2	LR018-LeSNP30R-061305-B01-03	LR15-	ATGPAT6/GPAT6	3	61T>C$104	88C>G$100	125_130delGAGAAG
3	LR029-LeSNP30R-061305-E02-10	LR15-	ATGPAT6/GPAT6	3	61T>TC$110	88C>CG$33	125_130het_delGAGAAG
3				2.33	66.7%	100.0%	66.7%

Figure 10.5 Many formats are available for reviewing and exporting analysis results in Mutation Surveyor. This two-direction output report shows a listing of each sample electropherogram, trace information (not all parameters are shown), and a listing of all mutation calls. These samples contain examples of both homozygous and heterozygous substitutions and deletions. Background coloring can highlight reported variations and amino acid changes. Text color represents confidence of mutation calls as well as negative SNPs. Mutation calls can be edited, added, deleted, and confirmed.

sequence for that gene and display any amino acid change that results from a SNP or Indel.

In addition to the automated detection of variations, Mutation Surveyor is capable of generating a variety of outputs, useful for simplifying annotations and importing into other applications. Mutation Surveyor has many export and viewing options available, from tables that show a listing of each variation for each sample electropherogram, to graphical reports highlighting each SNP and Indel, to files that contain the nucleotide and amino acid sequences as adjusted by the mutation calls (Figure 10.5). When annotated references are included in the project, base positions can be displayed in various formats, based on genomic numbering or relative to coding sequence and more.

Some of the other tools present within Mutation Surveyor include features to view all samples simultaneously within each contig (Figure 10.6), run multiple projects in a batch process, merge several projects together, and compare the results of multiple projects.

Mutation Surveyor is a useful tool for increasing the throughput and sensitivity of mutation detection in Sanger sequencing samples. By utilizing a physical trace-to-trace comparison, mutation detection can be automated and low-frequency variation detection is improved.

10.3
Mutation Detection with NextGENe™ and Next-Generation Sequence Technologies

The next-generation sequencing technologies have surpassed the classical Sanger sequencing method in throughput by 100- or 1000-fold. The sequence reads are often short in comparison to the Sanger sequencing method. The current sequencing systems commercially available are Genome Sequencer FLX system from 454 Life Science of Roche Applied Science, Illumina Solexa Genome Analyzer 1G system, and the SOLiD System from Applied Biosystems. These

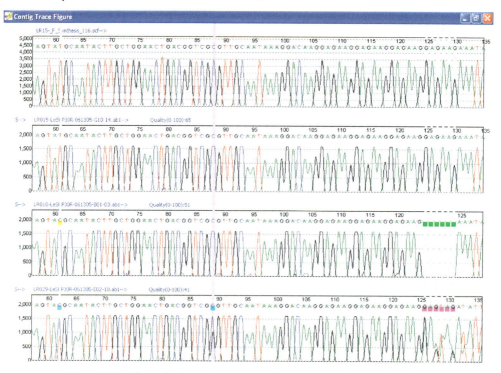

Figure 10.6 All sample electropherograms within a contig can be viewed simultaneously for easy review and comparison. Positions that contain a mutation call in at least one sample are indicated by a gray line and the individual mutation calls are highlighted.

systems have replaced the multiple steps required by the Sanger sequencing method, such as cloning of DNA libraries, PCR to amplify the DNA, and cycle sequencing, with a single instrument operable by a technician. Time and labor are reduced significantly with these next-generation genome analyzers, while the amount of information produced per run has increased dramatically. In addition, the error rates are significantly higher in relation to the reads generated by Sanger sequencing. Both of these issues result in a higher demand for software to automate the analysis.

There are a few steps in the next generation sequence technology. Genomic DNA or cDNA samples are sheared to small fragments ranging from 200 to 500 bp. These short DNA fragments are attached to a set of universal primers or adapters and amplified in a confined environment where only one allele is localized. These amplified fragments are subjected to the sequencing reaction and the resulting nucleotide sequences are exported from the instrument. These short reads are analyzed with NextGENe software to output SNP positions and copy number variations. Figure 10.7 shows a few sequence reads as they align to a reference genome.

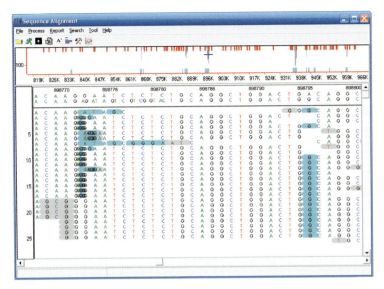

Figure 10.7 NextGENe Sequence Alignment view. In the region of aligned sequence reads mutation calls are highlighted in blue. A 7-bp insertion was identified before position 898 773 and a substitution is present at position 898 796. The Whole Genome Pane is located at the top of the display – coverage is indicated by gray lines, red tick marks indicate the breakpoints between genes, and blue tick marks identify the location of SNPs.

The 454 Genome Sequencer utilizes a pyrosequencing method to read the sequence in the flowgrams [22–24]. The luminescence intensity is proportional to the number of the consecutive duplicate nucleotides. For example, A gives an intensity and AA or TT will generate approximately twice the intensity. The temporal distributions of the intensity, allocating a portion of time to a specific type of nucleotide, will generate the sequence. The current reading length is about 250 bp. The system is able to generate 50–800 000 of such reads and 2×10^8 bases combined. The overall base-calling accuracy is about 98–99%. The accuracy for homopolymers is substantially lower in pyrosequencing.

The Illumina Solexa Genome Analyzer employs PCR on a solid surface where a template is localized. Ten million localized templates are sequenced by four-color fluorescence through sequence by synthesis. There are eight or 12 channels that may be run in parallel. The read length is often 35 bp and the newer instruments will be capable of generating reads of 70 bp. The sequencing error rate is approximately 0.5% for the first 25 bp and 3–4% for 26–35 bp.

The SOLiD System uses the technology of sequencing by ligation. DNA fragments are amplified and the resulting templates attached to beads. These bead-templates are deposited on a solid surface for sequencing. A universal primer anneals to the templates. Four fluorescently labeled dibase probes compete for each template. After successful hybridization and ligation, signal detection occurs, reading two consecutive bases as one color signal. After multiple rounds of hybridization, ligation, detection, and cleavage, a different universal primer is used for anther

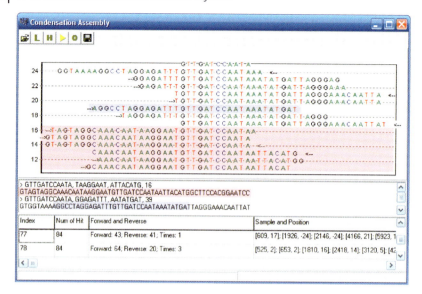

Figure 10.8 NextGENe Condensation Assembly view. Of the 84 reads containing the anchor GTTGATCCAATA (Index 77), two groups of reads were identified using 55 of these reads. The red highlighted reads are members condensed into the first group. Thirty-nine reads contained identical shoulder sequences, allowing for the blue highlighted 35-bp read to be condensed with others and generate a single read of 58 bp. The other 29 reads contain multiple sequencing errors or match more appropriately to other indexes (not shown).

cycle of sequencing by ligation. After all universal primers have been used, the template sequences are obtained. This system can generate 20 million reads with read lengths of approximately 35 bp. The data output from the instrument is in color space and the reads can be converted into base space. Sequence alignment and assembly can be done either in color space or base space.

SoftGenetics' NextGENe software has been developed to analyze data from these various platforms for a wide range of applications, including targeted sequence alignment, SNP and Indel detection, transcriptome expression analysis, and *de novo* assembly. Each instrument and application requires its own parameters and reports. NextGENe guides the users through the necessary steps to generate the anticipated results.

For the short sequencing reads, a condensation is performed to correct many of the base call errors and elongate the reads (Figure 10.8). Condensation can be repeated for several cycles. A final assembly algorithm generates large contigs for *de novo* sequencing applications. When references are loaded into the project, SNPs and expression levels can determined, reviewed, and results exported (Figures 10.9–10.11).

NextGENe is capable of aiding with the analysis of these massive amounts of data generated by the genome analyzers of today. Not only does NextGENe create reports, but it helps to improve the results by reducing the errors produced by these technologies.

10.3 Mutation Detection with NextGENe™ and Next-Generation Sequence Technologies

Mutation Output

Index	Position	Gbk letter	Coverage	A(%)	C(%)	G(%)	T(%)	(Ins%)	Del(%)
505	898972	T	4	100.00	0.00	0.00	-100.00	0.00	0.00
506	898973	T	4	0.00	0.00	100.00	-100.00	0.00	0.00
507	898974	T	4	0.00	100.00	0.00	-100.00	0.00	0.00
508	898975	G	4	0.00	100.00	-100.00	0.00	0.00	0.00
509	898977	A	4	-100.00	0.00	100.00	0.00	0.00	0.00
510	898978	A	4	-100.00	0.00	100.00	0.00	0.00	0.00
511	898980	C	4	0.00	-100.00	0.00	100.00	0.00	0.00
512	898981	C	4	0.00	-100.00	100.00	0.00	0.00	0.00
513	898982	G	4	0.00	0.00	-100.00	100.00	0.00	0.00
514	898983	C	4	0.00	-100.00	100.00	0.00	0.00	0.00
515	898984	T	4	0.00	0.00	100.00	-100.00	0.00	0.00
516	898986	T	4	100.00	0.00	0.00	-100.00	0.00	0.00
517	898987	G	4	100.00	0.00	-100.00	0.00	0.00	0.00
518	898988	G	4	0.00	0.00	-100.00	100.00	0.00	0.00
519	936949	T	4	0.00	100.00	0.00	-100.00	0.00	0.00
520	937190	C	6	0.00	-50.00	0.00	50.00	0.00	0.00
521	937193	G	8	0.00	62.50	-62.50	0.00	0.00	0.00
522	937194	A	8	-62.50	62.50	0.00	0.00	0.00	0.00
523	937202	G	20	0.00	30.00	-30.00	0.00	0.00	0.00
524	937206	C	20	0.00	-100.00	100.00	0.00	0.00	0.00
525	937236	C	22	0.00	-100.00	0.00	100.00	0.00	0.00
526	937240	T	21	0.00	95.24	0.00	-95.24	0.00	0.00
527	937243	A	20	-55.00	0.00	55.00	0.00	0.00	0.00
528	937246	G	18	0.00	0.00	-38.89	38.89	0.00	0.00
529	937251	A	14	-28.57	28.57	0.00	0.00	0.00	0.00
530	937331	C	29	24.14	-24.14	0.00	0.00	0.00	0.00
531	937378	T	5	0.00	80.00	0.00	-100.00	0.00	20.00
532	937381	G	5	0.00	0.00	-80.00	80.00	0.00	0.00
533	937403	A	20	-20.00	20.00	0.00	0.00	0.00	0.00

Figure 10.9 Mutation Output of NextGENe displaying a table of all mutation calls. On the left is a graphical representation of the selected and adjacent positions. The top chart shows the reference nucleotide and expected percentage, the middle chart shows the percentage of coverage for all nucleotides at each position, and the bottom chart shows the gain/loss of each allele.

Expression Report

Segment Index	Description	Max Count	Average Count	Reads Count
42	>gi|141802052	10	0	14
43	>gi|141802814	0	0	0
44	>gi|133892965	28	1	52
45	>gi|141803325	0	0	0
46	>gi|142345498	73	2	73
47	>gi|142349734	74	9	278
48	>gi|142351393	0	0	0
49	>gi|142362126	31	0	32
50	>gi|141802871	0	0	0

Figure 10.10 Reporting options for NextGENe – several reports can be created, including this expression report. The reference for this example included multiple genes. Information including depth of coverage and total number of reads aligned to each gene is tabulated.

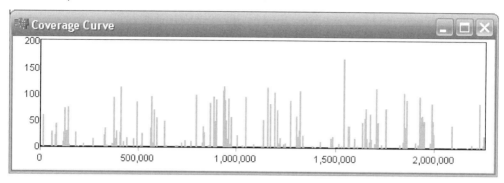

Figure 10.11 One of the charts produced by NextGENe is the Coverage Curve, showing the depth of coverage at each position of the genome.

10.4
Mutation Detection with DNA Fragments Using GeneMarker®

Allele calling is necessary prior to detecting mutations for DNA fragment analysis. Amplified fragment length polymorphism (AFLP®) [25] is the most common method for amplifying fragments. It is a PCR-based genetic fingerprinting technique developed in the early 1990s by KeyGene [39]. AFLP® uses restriction enzymes to cut genomic DNA, followed by ligation of complementary double-stranded adaptors to the ends of the restriction fragments [39]. A subset of these restriction fragments is amplified using two primers, containing amplification-selective nucleotides at their 3′ ends, complementary to the adaptor and restriction fragments [39]. These fragments are separated using denaturing polyacrylamide gels or capillary electrophoresis and visualized using either autoradiographic or fluorescence methodologies [25, 39]. Allele calls are dependent on the size of the fragment and other quality parameters, such as peak intensity and shape. The analysis is done in combination with known size standards and panels. These panels contain known alleles for each marker. SoftGenetics' GeneMarker software, with robust fragment sizing and pattern recognition technology, automatically removes chemistry and separation artifacts, such as saturated peaks, noisy data, wavelength bleed-through, instrument spikes, and stutter peaks, providing greater than 99% accuracy in allele calls (http://www.softgenetics.com/GeneMarker.html).

LOH occurs when a somatic cell contains only one copy of an allele due to nondisjunction during mitosis, segregation during recombination, or deletion of a chromosome segment. LOH becomes critical when the remaining allele contains a point mutation that renders the gene inactive. Wang *et al.* attribute accelerated hybrid fixation to LOH in rice hybrids [26]. Locating LOH after allele calling can be accomplished in one step with GeneMarker software. Samples with loss of heterozygosity are immediately apparent in two displays – an electropherogram with the lost allele marked with a light red trace and on a ratio plot [27].

MSI is a condition where repeat units are gained or lost within a locus resulting in length polymorphism. Certain repeat regions are known to be highly polymorphic and hereditable. Microsatellite markers are used extensively in many species for construction of linkage maps [28, 29]. Conversely, microsatellite instability within and around certain genes can have devastating effects due to the possibility of frameshift mutations. Test samples are compared to reference samples based on peak-to-peak comparison. Differences between the two traces are displayed with a trace comparison histogram below each electropherogram. Within the electropherogram, the test sample trace is overlain on a light red reference trace [30].

GeneMarker software contains applications for low-throughput (SNaPshot™ and single base extension procedures) and mid-throughput (SNPlex and SNPWave) SNP detection methodologies.

The techniques of TILLING [32] and EcoTILLING have been widely used since 2000 to detect SNPs [31–37]. Test samples used for TILLING may be experimentally mutagenized (ethylmethane sulfonate, radiation, etc.) or from natural populations or derived from diseased tissues. Briefly, the genes of interest are identified with gene-specific primers and PCR-amplified. The amplicon's primers are labeled with

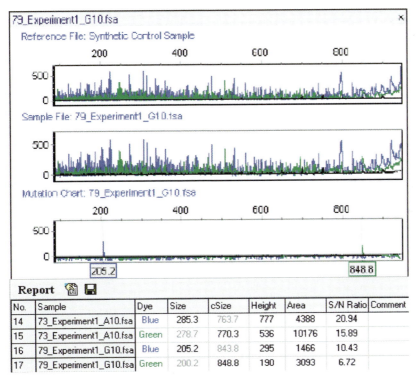

Figure 10.12 TILLING analysis using GeneMarker. The top panel shows the Synthetic Control Sample obtained from the median intensity after peak alignment. The middle panel displays the active sample. The bottom panel shows the Mutation Chart, generated by subtracting the reference from the sample, identifying individual variations. A blue peak at 205.2 bp and a green peak at 848.8 bp have been automatically identified. The original amplicon size is 1049 bp.

two fluorescent dyes. The samples are mixed, denatured, and allowed to reanneal so heteroduplexes can be formed. The hybridized fragments are cleaved at the heteroduplex site, generating multiple pairs of fragments of complementary length and dye color. The denatured fragments can be separated by gel electrophoresis or mixed with an internal size standard and separated by capillary electrophoresis. SNPs will yield two peaks of different color and the sum of the sizes will equal the amplicon length. TILLING analysis using GeneMarker software enables location of mutations within seconds from hundreds of samples. The peaks are smoothed, baseline is subtracted, and lane intensities are normalized. Low quality data is automatically rejected. A synthetic reference trace (Synthetic Control Sample) is constructed using median peak intensities from all of the high quality traces. This reference is subtracted from each sample trace generating a mutation chart that automatically identifies the sample's variations; the two fragments will equal the length of the amplicon (Figure 10.12) [38].

GeneMarker software is a useful tool that increases the speed, accuracy and sensitivity of many fragment analysis techniques. JelMarker™ software can convert many gel images into traces, similar to those generated by capillary instruments, which can be imported into GeneMarker – a versatile software package for fragment analysis.

10.5
Perspectives

Fragment analysis and Sanger sequencing are both mature techniques for detecting mutations. The next-generation sequencers offer many benefits of both previous techniques, yielding higher throughput and more detail at the same time. While all of SoftGenetics' products are continually being improved as new applications are developed and current applications are modified, team effort is devoted to the analysis of data generated by the genome analyzers. One such addition to NextGENe will be features to analyze the tSMS™ datasets produced by the Helicos™ Genetic Analysis System (Chapter 17). SoftGenetics strives to meet the immediate and evolving demands of applied and basic researchers.

References

1 Tucker, D.M., Griffey, C.A., Liu, S., Brown-Guedira, G., Marshall, D.S., and Saghai Maroof, M.A. (2007) Confirmation of three quantitative trait loci conferring adult plant resistance to powdery mildew in two winter wheat populations. *Euphytica*, **155**, 1–13.

2 Chao, S., Anderson, J., Glover, K., and Smith, K. (2006) Use of high throughput marker technologies for marker-assisted breeding in wheat and barley [abstract]. Plant and Animal Genome XIV Conference, San Diego, CA.

3 Zalapa, J.E., Brunet, J., and Guries, R.P. (2007) Isolation and characterization of microsatellite markers for red elm (*Ulmus rubra* Muhl.) and cross-species amplification with Siberian elm (*Ulmus pumila* L.). *Mol. Ecol. Notes*, **8**, 109–112.

4 Fehlberg, S.D., Ford, K.A., Ungerer, M.C., and Ferguson, C.J. (2007) Development, characterization and transferability of microsatellite markers for the plant genus *Phlox* (Polemoniaceae). *Mol. Ecol. Notes*, **8**, 116–118.

5 Wills, D. and Burke, J. (2007) QTL analysis of the early domestication of sunflower. *Genetics*, **176**, 2589–2599.

6 Riju, A., Chandrasekar, A., and Arunachalam, V. (2007) Mining for single nucleotide polymorphisms and insertions/deletions in expressed sequence tag libraries of oil palm. *Bioinformation*, **2**, 128–131.

7 Ward, J., Peakall, R., Gilmore, S.R., and Robertson, J. (2005) A molecular identification system for grasses: a novel technology for forensic botany. *Forensic Sci. Int.*, **152**, 121–131.

8 Yi, G., Lee, J.M., Lee, S., Choi, D., and Kim, B.-D. (2006) Exploitation of pepper EST–SSRs and an SSR-based linkage map Gibum. *Theor. Appl. Genet.*, **114**, 113–130.

9 Baldo, A.M., Wan, Y., Lamboy, W.F., Simon, C.J., ALabate, J., and Sheffer, S.M. (2007) NP validation and genetic diversity in cultivated tomatoes and grapes. Plant and Animal Genomes XV Conference, San Diego, CA.

10 Flanagan, N.S., Peakall, R., Clements, M.A., and Otero, J.T. (2007) Identification of the endangered Australian orchid *Microtis anqusii* using an allele-specific PCR assay. *Conserv. Genet.*, **8**, 721–725.

11 Horning, M.E. and Cronn, R.C. (2006) Length polymorphism scanning is an efficient approach for revealing chloroplast DNA variation. *Genome*, **49**, 134–142.

12 Feltus, F.A., Wan, J., Schulze, S.R., Estill, J.C., Jiang, N., and Paterson, A.H. (2004) An SNP resource for rice genetics and breeding based on subspecies *Indica* and *Japonica* genome alignments. *Genome Res.*, **14**, 1812–1819.

13 Lu, C.M., Yanga, W.Y., Zhanga, W.J., and Lu, B.-R. (2005) Identification of SNPs and development of allelic specific PCR markers for high molecular weight glutenin subunit Dtx1.5 from *Aegilops tauschii* through sequence characterization. *J. Cereal. Sci.*, **41**, 13–18.

14 Li, X., Wang, R.-C., Larson, S.R., and Chatterton, N.J. (2001) Development of a STS marker assay for detecting loss of heterozygosity in rice hybrids. *Genome*, **44**, 23–26.

15 Monteiro, M., Santos, C., Mann, R.M., Soares, A.M.V.M., and Lopes, T. (2007) Evaluation of cadmium genotoxicity in *Lactuca sativa* L. using nuclear microsatellites. *Environ. Exp. Bot.*, **60**, 421–427.

16 Kovalchuk, O., Kovalchuk, I., Arkhipov, A., Hohn, B., and Dubrova, Y. (2003) Extremely complex pattern of microsatellite mutation in the germline of wheat exposed to the post-Chernobyl radioactive contamination. *Mutat. Res.*, **525**, 93–101.

17 Le Roux, J.J. and Wieczorek, A.M. (2006) Isolation and characterization of polymorphic microsatellite markers from fireweed, *Senecio madagascariensis* Poir. (Asteraceae). *Mol. Ecol. Notes*, **7**, 327–329.

18 Hulce, D., Shouyong, N., and Liu, J.S.-C. (2007) Genetic diversity resequencing analysis with Mutation Surveyor® [Application note]. SoftGenetics Application, State College, PA.

19 Soleimani, V.D., Baum, B.R., and Johnson, D.A. (2003) Efficient validation of single nucleotide polymorphisms in plants by allele-specific PCR, with an example from barley. *Plant Mol. Biol. Rep.*, **21**, 281–288.

20 Legrendre, M., Pochet, N., Pak, T., and Verstrepen, K.J. (2007) Sequence based estimation of minisatellite and microsatellite repeat variability. *Genome Res.*, **17**, 1787–1796.

21 Hulce, D., Shouyong, N., and Liu, J.C.-S. (2007) Genetic diversity resequencing analysis with Mutation Surveyor® [Application note]. SoftGenetics Application, State College, PA.

22 Ronaghi, M., Uhlén, M., and Nyrén, P. (1998) DNA sequencing: a sequencing method based on real-time pyrophosphate. *Science*, **281**, 363–365.

23 Margulies, M., Egholm, M., Altman, W.E., Attiya, S., Bader, J.S., Bemben, L.A., Berka, J., Braverman, M.S., Chen, Y.J., and Chen, Z. *et al.* (2005) Genome sequencing in microfabricated high-density picolitre reactors. *Nature*, **437**, 376–380.

24 Huse, S.M., Huber, J.A., Morrison, H.G., Sogin, M.L., and Welch, D.M. (2007) Accuracy and quality of massively parallel DNA pyrosequencing. *System Biol.*, **8**, R143.

25 Vos, P., Hogers, R., Bleeker, M., Reijans, M., van de Lee, T., Hornes, M., Frijters, A., Pot, J., Peleman, J., and Kuiper, M. (1995) AFLP: a new technique for DNA fingerprinting. *Nucleic Acids Res.*, **23**, 4407–4414.

26 Wang, R.-C., Li, X., and Chatterton, N.J. (2006) Cytological evidence for assortment mitosis leading to loss of heterozygosity in rice. *Genome*, **49**, 556–557.

27 Serensits, T., He, H., Ning, W., and Liu, J. (2007) Loss of heterozygosity detection with GeneMarker® [Application note]. SoftGenetics Application, State College, PA.

28 Depeiges, A., Farget, S., Degroote, F., and Picard, G. (2005) A new transgene assay to study microsatellite instability in wild-type and mismatch-repair defective plant progenies. *Plant Sci.*, **168**, 939–947.

29 Burg, K., Helmersson, A., Bozhkov, P., and von Arnold, S. (2007) Developmental and genetic variation in nuclear microsatellite stability during somatic embryogenesis in pine. *J. Exp. Bot.*, **58**, 687–698.

30 Serensits, T., He, H., Ning, W., and Liu, J. (2007) Microsatellite instability analysis with GeneMarker® [Application note]. SoftGenetics Application, State College, PA.

31 Wang, D.K., Sun, Zong.-Xiu., and Tao, Yue.-Zhi. (2006) Application of TILLING in plant improvement. *Acta Genet. Sin.*, **33**, 957–964.

32 Gilchrist, E.J. and Haughn, G.W. (2005) TILLING without a plough: a new method with applications for reverse genetics. *Curr. Opin. Plant Biol.*, **8**, 211–215.

33 Gilchrist, E.J., Haugn, G.W., Ying, C.C., Otto, S.P., Zhuang, J., Cheung, D., Hamberger, G., Aboutorabi, F., Kalynyak, T., Johnson, L., Bohlmann, J., Ellis, B.E., Douglas, C.J., and Cronk, Q.C.B. (2006) Use of EcoTILLING as an efficient SNP discovery tool to survey genetic variation in wild populations of *Populus trichocarpa*. *Mol. Ecol.*, **15**, 1367–1378.

34 Yutaka, S., Shirasawa, K., Takahashi, Y., Nishimura, M., and Nishio, T. (2006) Mutant selection from progeny of gamma-ray-irradiated rice by DNA heteroduplex cleavage using *Brassica* petiole extract. *Breeding Sci.*, **56**, 179–183.

35 Nieto, C., Piron, F., Dalmais, M., Marco, C.F., Moriones, E., Gómez-Guillamón, M.L., Truniger, V., Gómez, P., Garcia-Mas, J., Aranda, M.A., and Bendahmane, A. (2007) EcoTILLING for the identification of allelic variants of melon *eIF4E*, a factor that controls virus susceptibility. *BMC Plant Biol.*, **7**, 34–42.

36 Dillon, S.L., Shapter, F.M., Henry, R.J., Cordeiro, G., Izquierdo, L., and Lee, L.S. (2007) Domestication to crop improvement: genetic resources for sorghum and saccharum (Andropogoneae). *Ann. Bot. (Lond.)*, **100**, 975–989.

37 Wang, G.-X., Tan, M.-K., Rakshit, S., Saitoh, H., Terauchi, R., Imaizumi, T., Ohsako, T., and Tominaga, T. (2007) Discovery of single-nucleotide mutations in acetolactate synthase genes by EcoTILLING. *Pestic Biochem. Physiol.*, **88**, 143–148.

38 LeVan, K., Hulce, D., Shouyong, N., Ning, W., and Liu, C.-S.J. (2008) Automated TILLING® analysis of fluorescent electrophoresis data with GeneMarker® [Application note]. SoftGenetics Application, State College, PA.

39 Riley, M. and Liu, J.C.-S. (2007) Software for amplified fragment length polymorphism (AFLP®) [Application note]. SoftGenetics Application, State College, PA.

Part III
High-Throughput Screening Methods

11
Use of TILLING for Reverse and Forward Genetics of Rice

Sujay Rakshit, Hiroyuki Kanzaki, Hideo Matsumura, Arunita Rakshit, Takahiro Fujibe, Yudai Okuyama, Kentaro Yoshida, Muluneh Oli, Matt Shenton, Hiroe Utsushi, Chikako Mitsuoka, Akira Abe, Yutaka Kiuchi, and Ryohei Terauchi

Abstract

Targeting induced local lesions in genomes (TILLING) was first demonstrated in *Arabidopsis* for the detection of allelic series and functional analysis of genes. Subsequently the technique was extended to allow discovery of polymorphism in natural populations, termed EcoTILLING. In TILLING, DNA from reference and subject genomes are mixed and polymerase chain reaction-amplified using infrared dye (IRD)-labeled primers. Mutations in the amplified region cause mismatches in the resulting heteroduplex. Such mismatches are cleaved by CEL1 endonuclease, generating IRD-labeled fragments that can be separated and detected using a LI-COR gel scanning system. In this chapter, we present TILLING and EcoTILLING protocols adapted for use in forward and reverse genetics in rice – the most important staple food crop supporting billions of people worldwide.

11.1
Introduction

In the present postgenome era, sequences are available for large numbers of genes, but a remaining task for biologists is to connect these sequences to meaningful biological functions. Reverse genetics can play an important role in the process. In the recent past, targeting induced local lesions in genomes (TILLING) has been well demonstrated in a number of species, including *Arabidopsis* [1–5], maize [6], wheat [7], rice [8], zebrafish [9], *Caenorhabditis elegans*, *Drosophila* [10], soybean [11], pea [12], and sorghum [13], and has been tried in other plant and animal species [10]. In TILLING, DNA from mutated genomes is mixed and amplified with infrared dye (IRD)-labeled PCR primers. Mismatches in the resulting heteroduplex caused by point mutations or small insertion/deletions (Indels) are cleaved by CEL1 endonuclease. The resulting IRD-labeled fragments are detected in a LI-COR gel analyzer

---

The Handbook of Plant Mutation Screening. Edited by Günter Kahl and Khalid Meksem
Copyright © 2010 WILEY-VCH Verlag GmbH & Co. KGaA, Weinheim
ISBN: 978-3-527-32604-4

system [2, 10]. This process enables rapid detection of an allelic series for a particular gene and can contribute to functional analysis of the examined sequence. TILLING is therefore one of the most potent reverse genetics tools available and can overcome some of the documented limitations of other reverse genetics tools such as T-DNA insertional mutagenesis, transposable elements-induced mutagenesis, RNA interference-based gene silencing, or use of morphono oligonucleotides [7, 10].

The TILLING technique was further developed in order to detect polymorphisms in natural populations, in a technique termed EcoTILLING [2]. Using EcoTILLING, a large germplasm may be grouped into a small number of haplotypes depending on the distribution of naturally occurring mutations in a particular locus. Thus, only representative genotypes from each of the haplotypes need to be sequenced in order to survey the mutations within the germplasm. EcoTILLING may therefore be used in order to greatly improve the speed of single nucleotide polymorphism (SNP) detection [14].

In this chapter, we present protocols for the application of TILLING and EcoTILLING to rice genetics as follows:

- Mutation detection in rice using the TILLING platform.
- Ethylmethane sulfonate (EMS) mutagenesis of rice.
- Use of EcoTILLING for rapid forward genetics in rice.

The protocols presented here may be easily applied to other crop species with minor modifications.

11.2
Methods and Protocols

Mutation Detection in Rice using the TILLING Platform

For the general protocols and methodology of TILLING, see Chapter 8 by Brad Till and the references therein. Here, we present modifications of the original protocol adapted for rice mutation detection.

In order to reduce the cost of synthesizing IRD-labeled primers, we employ a two-step polymerase chain reaction (PCR) process as shown in Figure 11.1. The first step amplifies the target locus using specific primers tailed with known noncomplimentary 20-bp sequences (UniU and UniL), but unlabeled with IRD dyes (Universal-tailed amplification primers, UniU-Tailed Upper Primer and UniL-Tailed Lower Primer, Figure 11.1a–c). The resulting PCR product is purified and diluted, then reamplified using IRD700 UniU and IRD800 UniL primers labeled with IRD700 and IRD800, respectively (Figure 11.1d and e). The target locus is now labeled at both ends with the dyes (Figure 11.1f). Subsequent steps are the same as the regular TILLING protocol. By synthesizing only two dye-labeled oligonucleotides, many different target loci could be examined, substantially reducing the cost of the protocol.

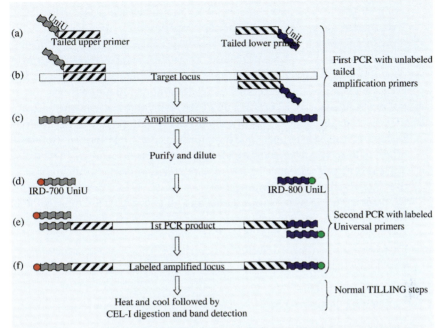

Figure 11.1 Procedure of two-step PCR to economically label DNA fragments for TILLING in rice.

DNA Extraction, Mixing, Primer Sequence, and PCR

1. DNA is routinely extracted using the Qiagen DNeasy® Plant Mini Kit. DNA quality is checked and quantified on 1.5% agarose gel by comparing with uncut λ-DNA. Subsequently, DNA is diluted to 3 ng/μl concentration. For the standard protocol, equal amounts of DNA from reference and subject individuals are mixed. The final concentration of each mix is brought to 1.5 ng/μl.
2. For the example *LM1* locus (Rakshit *et al.*, in preparation), the following primers were used: upper, *GCTACGGACTGACCTCGGAC***ACCATCATCACTGACATAATAACCA**; lower, *CTGACGTGATGCTCCTGACG***TCCTC-CTCAGATGACACTATTAGAT**. The sequences in bold type are designed to specifically amplify the target genomic region of 1.5 kb. The italic sequence 5′-GCTACGGACTGACCTCGGAC incorporated at the 5′ end of the upper primer is referred to as UniU and the italic sequence of 5′-CTGACGTGATGCTCCTGACG attached to the 5′ end of the lower primer is referred to as UniL. UniU and UniL primers were labeled with IRD700 and IRD800, respectively.
3. First PCR was carried out in a reaction volume of 20 μl containing 2.25 ng genomic DNA, 4 mM each dNTPs, 0.4 U TaKaRa Ex Taq™ polymerase and 6 μM each of tailed amplification primers (UniU-Tailed Upper primer and UniL-Tailed Lower Primer). The following thermal cycling conditions were used: 95 °C for 2 min/35 cycles of 95 °C for 1 min, 55 °C for 1 min, 72 °C for

1 min 30 s/72 °C for 7 min; 5 μl of the reaction mixture was analyzed in a 1% TAG agarose gel to check for successful amplification.

4. Remaining reaction mix was diluted with 90 μl of sterile water and purified using MultiScreen™ plates (Millipore) as per the manufacturer's instructions. Purified amplified products were eluted by dissolving in 30 μl sterile water and further diluted 20 times (final dilution ~1/40 times).

5. Second PCR was carried out in a 10-μl reaction volume consisting of 2 μl diluted amplified product, 2 mM each dNTPs, 0.2 U TaKaRa Ex Taq polymerase, 0.2 μM each of upper and lower primer mix. Upper primer mix was prepared by mixing labeled and unlabeled UniU primers in a ratio of 3 : 7, and lower primer mix was made by mixing labeled and unlabeled UniL in a ratio of 2 : 3. The following cycling program was used: 95 °C for 2 min/35 cycles of 95 °C for 1 min, 55 °C for 1 min, 72 °C for 1 min 30 s/72 °C for 7 min/99 °C for 10 min/70 °C for 20 s with touch down of 0.3 °C/cycle for 70 cycles.

6. CEL1 digestion, purification, and other steps were followed as described elsewhere [3] with the modification that Surveyor™ nuclease (Transgenomic) was used at 0.05 μl/reaction.

Bulking of DNA in Rice TILLING

The standard TILLING protocol mixes reference and subject DNA in a 1 : 1 ratio. Screening large numbers of DNAs for the purpose of mutation detection in this

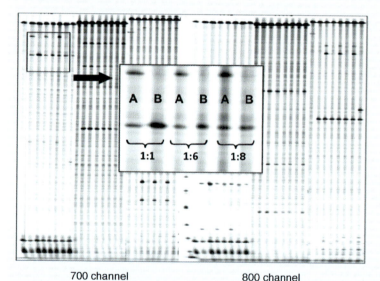

Figure 11.2 Bulking of sample DNA allows high-throughput screening of mutations. Two different subject DNAs (samples A and B) and reference DNA were mixed in the ratios 1 : 1, 1 : 6, and 1 : 8, respectively, and applied for TILLING. All the mutations detected in the 1 : 1 ratio mixture can be detected in the 1 : 8 ratio mixture.

way would be rather time-consuming. We compared the information generated by TILLING whereby subject and reference genomic DNA were mixed in 1 : 1, 1 : 6, or 1 : 8 ratios (Figure 11.2). It was clear that a mutation that could be detected in a 1 : 1 mix could also be clearly detected from a 1 : 8 mix. Therefore, in order to screen large numbers of sample we could combine eight samples into a bulk without any loss of sensitivity of the detection system. Using the LI-COR electrophoresis system, 94 samples could be analyzed in a single run, meaning that $94 \times 8 = 752$ samples could be screened in a single experiment.

EMS Mutagenesis of Rice

Both for reverse and forward genetics, a population of mutants is required. Here, we provide a protocol for rice mutagenesis with EMS. EMS is an alkylating agent that produces primarily GC → AT transitions. EMS treatment can be applied to seeds or to fertilized egg cells in immature flowers. We prefer the latter in order to reduce the chance of formation of genetic chimerism in the M1 plants.

1. To generate 10 000 M1 seeds, 100–200 rice plants are individually grown in plastic pots or, alternatively, transferred to plastic pots from the paddy field. Heading of plants is observed and the best days are selected (up to 1 week) during which maximal anthesis occurs. This depends on the cultivars and weather conditions (day length, light intensity, and temperature), so the timing should be empirically determined. After heading of panicles, only the flowers opened in the same day are kept and other flowers (old and premature ones) are removed by scissors. Since rice flowers in the morning, unnecessary flowers are removed in the afternoon. Usually ~100–300 flowers per plant are kept and used for EMS mutagenesis. Between 20 and 30 plants are treated per day (Figure 11.3).
2. In a draft chamber 0.175% EMS solution is prepared and kept in a tightly sealed plastic container. About 200 ml of diluted EMS is needed for treatment of each plant. (*Note*: EMS is highly carcinogenic, so that utmost care should be taken not to touch or inhale it. The following steps should be carried out in an open space with good ventilation. Plastic gloves and a gas mask should be worn when handling EMS solution.)
3. All the panicles of a plant are covered with a plastic bag and each plant is laid down so that EMS solution can be poured into the plastic bags.
4. In the evening after sunset, 0.175% EMS solution is poured into the plastic bag containing the panicles. The bag is tightly closed and tied with string, making sure that all the flowers are soaked in the solution. The flowers are kept in EMS solution overnight (14–16 h). EMS is unstable in ultraviolet light, so the EMS treatment should be done in the dark.
5. Next morning, the panicles are removed from the bag. Panicles are rinsed with water and plants are allowed to grow so that grains mature. EMS waste and water used for rinsing are collected in a plastic container, and detoxified

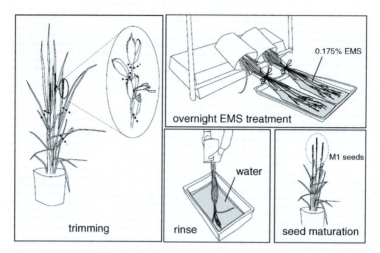

Figure 11.3 EMS mutagenesis of immature embryos of rice. Only the flowers showing anthesis are kept and other flowers are trimmed with scissors. Flowers are soaked in EMS solution in a plastic bag overnight. EMS-treated flowers are rinsed with water and plants are further grown to obtain the M1 seeds.

by adding NaOH to a final concentration of 1 N. All the plasticware used for handling EMS should also be treated with 1 N NaOH and kept under sunlight for 2–3 days.

The concentration of EMS solution in the above procedure has been optimized for our rice materials (ssp. Japonica rice cultivars like "Sasanishiki"). A higher concentration of EMS may result in higher frequency of mutations. However, viability and fertility of EMS-treated plants drop remarkably in response to a slight increase in EMS concentration, resulting in the failure of recovering seeds. Therefore, it is recommended to apply a series of EMS concentration centering at 0.175% to obtain optimum results both in mutation frequency and seed setting.

Evaluation of Mutation Frequency by TILLING

EMS-treated seeds (M1 generation) are grown and are self-pollinated to obtain the M2 seeds. In the majority of the cases, mutated loci are in a heterozygous state in the M1 and phenotypically segregate to 3 : 1 (wild : mutant) ratio in the M2 generation, so that the mutants showing phenotypes of interest should be screened after growing M2 progeny (around 10 individuals). Frequency of induced mutations can be evaluated either in the M1 or M2 generation by TILLING. For evaluation in the M1 generation, a single leaf per M1 plant is used for mutation detection; for evaluation in the M2 generation, leaves of several individuals of M2 progeny derived from a single M1 plant should be pooled. Although testing in the M1 generation is more convenient, there is an increased risk of chimerism compared to the M2. For evaluating the frequency of induced

Table 11.1 EMS-induced mutations detected in exon 5 and exon 10 of *LM1* locus in M1 rice (cv. Hitomebore) plants.

	Exon 5	Exon 10
No. of bands detected by TILLING	10	7
M1 individuals screened	1632	1414
No. mutations/1000 individuals	6.1	5.0

mutations in those EMS-treated lines using TILLING, leaf tissue from eight M1 plants are pooled as described. Table 11.1 gives an example of mutations detected in two regions in exon 5 and exon 10 of the *LM1* locus, each 1.5 kb in size, as detected by TILLING. The total number of detected "positive" bands (bands detected in both IRD700 and IRD800 channels at corresponding sizes) was 10 in exon 5 and seven in exon 10 in approximately 1500 individuals. Based on these, the mutation frequency in exon 5 and exon 10 regions of *LM1* was estimated to be 6.1 and 5.0 mutations per 1000 lines, respectively. This corresponds to 3000–2500 mutations per line assuming the rice genome is around 500 Mb.

Use of EcoTILLING for Rapid Forward Genetics in Rice

Notwithstanding the remarkable technical development in reverse genetics approaches, use of the forward genetics approach – moving from phenotype to gene and gene sequences by means of genetic linkage and association analyses – is still a major tool to connect phenotypes and genes. DNA polymorphisms detected by EcoTILLING can be useful for facilitating forward genetics. We developed an EcoTILLING-based strategy for the forward genetics study, which is efficient in detecting causal mutations in the genome.

Currently rice forward genetics is primarily carried out by:

1. Identification of mutant individuals by appropriate screens.
2. Crossing the mutant with another cultivar, which is distantly related to the mutant line, to obtain F1.
3. Self-pollinating F1 to obtain F2 progeny.
4. Checking segregation of phenotypes in F2 whether it conforms to the 3 : 1 wild-type to mutant type ratio.
5. Mapping the causal mutation by using the DNAs of F2 progeny exhibiting mutant phenotype.
6. Identification of the causal mutation by sequence comparison between the wild and mutant DNAs.

Use of EcoTILLING can substantially improve the efficiency of the Steps (5) and (6).

Use of EcoTILLING Polymorphisms as DNA Markers

In rice, simple sequence repeat (SSR) markers (alternatively called microsatellite markers) are most frequently used for mapping genes [15]. SSR markers are simple and convenient to use, and abundant in eukaryotic genomes. Since the rice genome sequence is available, the chromosomal location of each SSR marker is unequivocally determined. Using SSR markers with defined genomic positions we can carry out SSR-based mapping to narrow down the location of the mutation. In some locations, however, no SSR markers are available to further narrow down the region by linkage analysis. In such cases, use of EcoTILLING to detect SNP or Indels for use as DNA markers may help. Below is the protocol we are employing in rice mutant mapping.

1. If no SSR markers are available between the two flanking markers, 10–20 genomic locations evenly distributed between the delimited distance are selected. PCR primers are designed to amplify 1.5-kb regions for analysis by EcoTILLING. These regions are tested for polymorphism between the two parents by EcoTILLING. Detected polymorphisms are used as markers to narrow down the target region as illustrated in Figure 11.4.

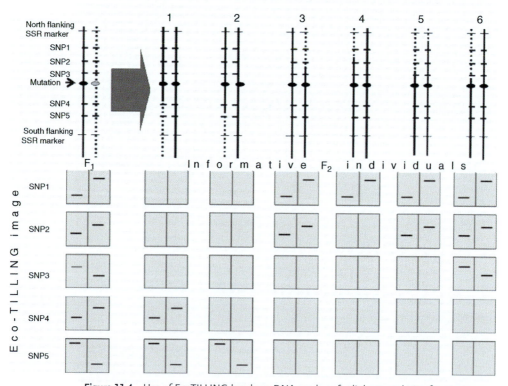

Figure 11.4 Use of EcoTILLING bands as DNA markers for linkage analysis of mutants.

2. Figure 11.4 depicts the EcoTILLING-based fine mapping strategy. The F2 mutant segregants showing heterozygosity for the flanking markers, depicted as "North" and "South" flanking SSR markers, will be informative, while the remaining will not be informative for further fine mapping purpose. This is because only the informative segregants will retain heterozygosity for the region close to the causal mutation (indicated as "Mutation" in Figure 11.4), while remaining F2 segregants have already attained homozygosity for the region in close vicinity of the causal mutation. In Figure 11.4 the first two F2 individuals are informative as they have still retained heterozygosity between them from the South side. On the other hand, F2 individuals 3–6 are informative from the North side. Now say SNPs 1–5 are SNPs between the two parents at regular intervals. From the North side the four F2 individuals (nos 3–6) are heterozygous for SNP1 and if we conduct EcoTILLING among themselves, the SNP will be revealed on the gel image in the form of band in all the four individuals. Now for SNP2 the F2 individual no. 4 has attained homozygosity by recombination so that it will not give any band in the EcoTILLING gel, thus for this locus F2 individuals nos 3, 5 and 6 will remain informative. Similarly, for SNP3 only individual no. 6 remains informative as others attained homozygosity for the region. Thus, from the North side SNP3 will remain as a flanking SNP. Similarly, SNP5 has remained informative (i.e., heterozygote) in both the F2 individuals nos. 1 and 2, and thus the same is depicted as occurrence of bands in the EcoTILLING gel, while SNP4 F2 individual no. 2 has attained homozygosity and individual no. 1 has remained informative (heterozygote). Thus, by following this strategy the region flanked by SSR markers (North and South flanking marker) may be further narrowed down to the region flanked by SNP3 (North marker) and SNP4 (South marker) by using EcoTILLING.

Use of EcoTILLING to Identify the Causal Mutation

Once a mutation has been located to within a narrow region by mapping, DNA sequencing of the region is traditionally performed. However, this process is laborious and error-prone. Therefore, we are utilizing EcoTILLING in place of sequencing to rapidly identify the causal mutation (Figure 11.5).

1. PCR primers are designed so that the entire DNA region is covered by tiles of PCR products. Each PCR product should be <1.5 kb. Two neighboring PCR products should have 0.3- to 0.5-kb overlaps to each other. Thus, in order to scan a 100-kb region, we need ~150 PCR primer sets.
2. DNAs of mutant individuals and wild-types are mixed in 1 : 1 ratio. EcoTILLING is performed using the three DNAs: wild-type DNA, mutant DNA, and mixed DNA. The causal mutation should be detected as a polymorphism that appears as a band only in the mixed DNA (3) (Figure 11.5). The mutation detected by EcoTILLING should be verified by DNA sequencing.

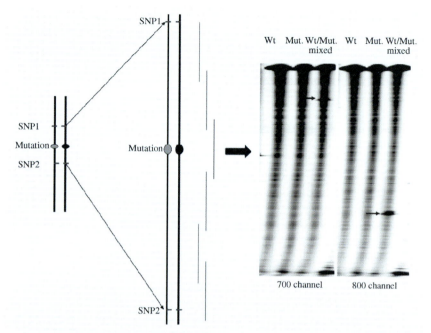

Figure 11.5 Detection of causal mutation by EcoTILLING. DNAs of mutant (Mut.) and wild-type (Wt) are mixed in a 1 : 1 ratio, and screened for mutation between the genomic region flanked by DNA markers (SNP1 and SNP2).

11.3
Perspectives

Organismal evolution including crop improvement is driven by three major forces – mutation, selection, and recombination. Therefore, efficient crop improvement can be achieved by increasing the mutation rate and efficacy of selection and recombination. A higher mutation rate can be attained by mutagenesis. Genome sequence information and knowledge on gene functions, as revealed by forward genetics studies, tell us what kind of allelic variants of which locus should be the target of artificial selection. TILLING is a suitable technology to perform efficient selection of individuals possessing such desirable alleles of interest. Once selected, these individuals can be used as parents for further breeding. We propose that the TILLING platform in conjunction with genome information will offer a tremendous opportunity to enhance breeding of crops including rice.

Acknowledgments

This work was carried out in part by support from "Program for Promotion of Basic Research Activities for Innovative Biosciences" (Japan), "Iwate University 21st

Century COE Program: Establishment of Thermo-Biosystem Research Program," Ministry of Agriculture, Forestry and Fisheries of Japan (Genomics for Agricultural Innovation PMI-0010) to R.T., and Japan Society for the Promotion of Science grants 18310136 and 18688001 to R.T. and H.M., respectively. R.T. thanks Brad Till and Steven Henikoff for kind training in TILLING.

References

1. McCallum, C.M., Comai, L., green, E.A., and Henikoff, S. (2000) Targeting induced local lesions in genomes (TILLING) for plant functional genomics. *Plant Physiol.*, **123**, 439–442.
2. Comai, L., Young, K., Till, B.J., Reynolds, S.H., Greene, E.A., Codomo, C.A., Enns, L.C., Johnson, J.E., Burtner, C., Odden, A.R., and Henikoff, S. (2004) Efficient discovery of DNA polymorphisms in natural populations by EcoTILLING. *Plant J.*, **37**, 778–786.
3. Till, B.J., Stevene, H.R., Greene, E.A., Comodo, C.A., Enns, L.C., Johnson, J.E., Burtner, C., Odden, A.R., Young, K., Taylor, N.E., Henikoff, J.G., Comai, L., and Henikoff, S. (2003) Large-scale discovery of induced point mutations with high-throughput TILLING. *Genome Res.*, **13**, 524–530.
4. Colbert, T., Till, B.J., Tompa, R., Rynold, S., Steine, M.N., Yeung, A.T., McCallum, C.M., Comai, L., and Henikoff, S. (2001) High-throughput screening for induced point mutations. *Plant Physiol.*, **126**, 480–484.
5. Green, E.A., Codomo, C.A., Taylor, N.E., Henikoff, J.G., Till, B.J., Reynolds, S.H., Enns, L.C., Burtner, C., Johnson, J.E., Odden, A.R., Comai, L., and Henikoff, S. (2003) Spectrum of chemically induced mutations from a large-scale reverse genetic screen in *Arabidopsis. Genetics*, **164**, 731–740.
6. Till, B.J., Stevene, H.R., Weil, C., Springer, N., Burtner, C., Young, K., Bowers, E., Comodo, C.A., Enns, L.C., Odden, A.R., Greene, E.A., Comai, L., and Henikoff, S. (2004) Discovery of induced point mutations in maize by TILLING. *BMC Plant Biol.*, **4**, 12.
7. Slade, A.J., Fuerstenberg, S.I., Loeffler, D., Steine, M.N., and Facciotti, D. (2005) A reverse genetic, nontransgenic approach to wheat crop improvement by TILLING. *Nat. Biotechnol.*, **23**, 75–81.
8. Till, B.J., Cooper, J., Tai, T.H., Colowit, P., Greene, E.A., Hanikoff, S., and Comai, L. (2007) Discovery of chemically induced mutations in rice by TILLING. *BMC Plant Biol.*, **7**, 19.
9. Wienholds, E., van Eeden, F., Kosters, M., Mudde, J., Plasterk, R.H.A., and Cuppen, E. (2003) Efficient target-selected mutagenesis in zebrafish. *Genome Res.*, **13**, 2700–2707.
10. Henikoff, S., Till, B.J., and Comai, L. (2004) TILLING. Traditional mutagenesis meets functional genomics. *Plant Physiol.*, **135**, 630–636.
11. Cooper, J.L., Till, B.J., Laport, R.G., Darlow, M.C., Kleffner, J.M., Jamai, A., Mellouki, T., Liu, S.M., Ritchie, R., Nielsen, N., Bilyeu, K.D., Meksem, K., Comai, L., and Hanikoff, S. (2008) TILLING to detect induced mutations in soybean. *BMC Plant Biol.*, **8**, 9.
12. Triques, K., Sturbois, B., Gallais, S., Dalmais, M., Chauvin, S., Clepet, C., Aubourg, S., Rameau, C., Caboche, M., and Bendahmane, A. (2007) Characterization of *Arabidopsis thaliana* mismatch specific endonucleases: application to mutation discovery by TILLING in pea. *Plant J.*, **51**, 1116–1125.
13. Xin, Z., Wang, M.L., Barkley, N.A., Burow, G., Franks, C., Pederson, G., and Burke, J. (2008) Applying genotyping (TILLING) and phenotyping analyses to elucidate gene function in a chemically induced sorghum mutant population. *BMC Plant Biol.*, **8**, 103.
14. Comai, L. and Hanikoff, S. (2006) TILLING: practical single nucleotide mutation discovery. *Plant J.*, **45**, 684–694.
15. Temnykh, S., William, D.P., Ayres, N., Carinhour, S., Hauck, N., Lipovich, L., Cho, Y.G., Ishii, T., and McCouch, S.R. (2000) Mapping and genome organization of microsatellite sequences in rice (*Oryza sativa* L.). *Theor. Appl. Genet.*, **100**, 697–712.

12
Sequencing-Based Screening of Mutations and Natural Variation using the KeyPoint™ Technology
Diana Rigola and Michiel J.T. van Eijk

Abstract

Reverse genetics approaches rely on the detection of sequence alterations in target genes to identify allelic variants among mutant or natural populations. Current methods such as TILLING and EcoTILLING are based on the detection of single-base mismatches in heteroduplexes using endonucleases such as CEL1. However, there are drawbacks in the use of endonucleases due to their relatively poor cleavage efficiency and exonuclease activity. Moreover, these methods do not reveal information about the nature of sequence changes and their possible impact on gene function. In this chapter, we describe KeyPoint technology – a high-throughput mutation/polymorphism discovery technique based on massive parallel sequencing of target genes amplified from mutant or natural populations. KeyPoint combines multidimensional pooling of large numbers of individual DNA samples and the use of sample identification tags with next-generation sequencing technology. We show the power of KeyPoint by identifying two mutants in the tomato eukaryotic translation initiation factor 4E (*eIF4E*) gene based on screening more than 3000 M2 families and discovery of six haplotypes of the tomato *eIF4E* gene by resequencing three amplicons in a subset of 92 tomato lines from the Eu-Sol core collection. We propose KeyPoint technology as a broadly applicable amplicon sequencing approach to screen mutant populations or germplasm collections for identification of (novel) allelic variation in a high-throughput fashion.

12.1
Introduction

Rapid, high-throughput mutation and single nucleotide polymorphism (SNP) discovery technologies are fundamental to identify allelic variants in large

The Handbook of Plant Mutation Screening. Edited by Günter Kahl and Khalid Meksem
Copyright © 2010 WILEY-VCH Verlag GmbH & Co. KGaA, Weinheim
ISBN: 978-3-527-32604-4

populations. Such analyses are very useful for functional genetics, clinical diagnostics, forensic medicine, population genetics, molecular epidemiology, and plant and animal breeding. Mutations are the basis of genetic variation and mutant populations are indispensable genetic resources in all organisms. This variation can be either naturally occurring or, in plants, animals, and lower organisms, induced by chemical or physical treatments. Mutation induction, for example, has played an important role in the genetic improvement of crop species that are of economic importance [1].

Although sequencing is considered the "gold standard" for DNA-based mutation detection because it reveals the exact location and the type of mutations, direct gene sequencing methods have rarely been used to screen large (mutant) populations because of cost limitations. Instead, in the past a number of prescreening methods have been developed and applied in order to scan amplicons in large (mutant) populations for the presence of sequence polymorphisms, prior to confirmation by Sanger sequencing. For example, the characteristic DNA properties of melting temperature and single-strand conformation have been used in techniques such as denaturing high-performance liquid chromatography (DHPLC), denaturing gradient gel electrophoresis, temperature gradient gel electrophoresis, single-strand conformational polymorphism analysis [2], and high-resolution DNA melting [3]. In addition, other procedures have been developed including the protein truncation test [4], enzymatic or chemical cleavage methods [5], the restriction site mutation method [6] and s-RT-MELT (Surveyor-mediated real-time melting) technology [7].

The targeting induced local lesions in genomes (TILLING) technique makes use of chemical mutagenesis to induce mutations throughout an entire genome and applies an enzyme-mediated detection method (e.g., based on CEL1 nuclease) combined with DHPLC or gel electrophoresis detection [8–10]. A variation of the TILLING method, known as EcoTILLING [11], aims at the detection of natural (allelic) variation in the germplasm. TILLING has been successfully applied to organisms such as zebrafish [12], *Caenorhabditis elegans* [13], and plants including *Arabidopsis* [14], rice [15], soybean [16] wheat [17], and maize [18]. However, limitations of an endonuclease such as CEL1 are its relatively poor cleavage efficiency and $5' \rightarrow 3'$ exonuclease activity. This diminishes signal/noise levels and prevents performing pooled sample analysis with more than eight samples per pool [19]. Moreover, TILLING, like other prescreening methods, does not reveal information about the nature of sequence changes and their possible impact on gene function. Consequently, there is a need for robust and low-cost amplicon sequencing methods applicable to large populations.

The recent introduction of instruments capable of producing millions of DNA sequence reads in a single experiment opened the possibility of developing a new high-throughput mutation discovery technology based on massive parallel sequencing. We describe the KeyPoint technology – a novel mutation/polymorphism screening technique using the Genome Sequencer (GS) FLX platform (454™/Roche Applied Science) – that allows massive parallel picoliter-scale amplification and pyrosequencing of individual DNA molecules [20]. Using KeyPoint, genes of interest are directly amplified by polymerase chain reaction (PCR) and sequenced. To significantly reduce sample preparation costs, KeyPoint applies a multidimensional pooling strategy of amplification templates (DNA samples) belonging to mutant or natural populations. Gene-specific PCR primers carry sample identification tags specific for each multidimensional pool in order to assign sequence reads to individual mutant plants or to assign sequence haplotypes to pooled or individual samples of a germplasm collection (Figure 12.1). Using custom developed bioinformatics tools the sequence reads are clustered, aligned, and mined for mutations or SNPs). Statistical probability calculation methods are used to distinguish true mutations and polymorphisms from amplification or sequencing errors. With the KeyPoint technology we identified two ethylmethane sulfonate (EMS)-induced mutations in exon 1 of the tomato (*Solanum lycopersicum*) eukaryotic translation initiation factor 4E (*SleIF4E*) gene by screening 3008 M2 families in a single GS FLX run. In addition, power calculations were performed to define the throughput of the KeyPoint technology. Finally, KeyPoint revealed at least six naturally occurring haplotypes defined by 15 SNPs observed in three amplicons of the *SleIF4E* gene in a subset of 92 lines of the Eu-Sol (www.eu-sol.net) tomato germplasm core collection in just one-quarter GS FLX run.

We propose KeyPoint as a generic approach to screen for induced and naturally occurring sequence variation in selected target genes of mutant and/or germplasm populations.

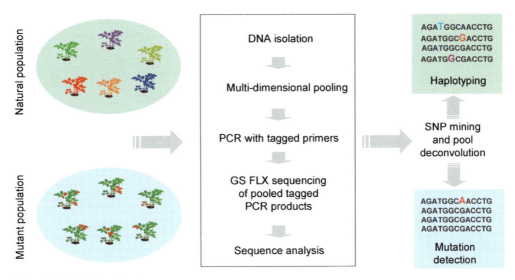

Figure 12.1 KeyPoint flowchart.

12.2
Methods and Protocols

Biological Samples

- The mutant library consisted of an isogenic library of inbred tomato cultivar M82 derived from EMS mutagenesis treatment [21]. Leaf material was harvested from five individual greenhouse-grown plants of each of 3008 M2 tomato mutant families to reduce the likelihood of not sampling a mutation as consequence of segregation to 0.1% ($0.5^{10} = 0.001$).
- The tomato natural population samples used consisted of 92 tomato lines belonging to the Eu-Sol core collection [22].

DNA Extraction, Normalization, and Pooling

1. Genomic DNA was isolated from pooled leaves of five segregating plants of each M2 family and from leaves of a single plant of each tomato line of the Eu-Sol core collection, using a modified hexadecyltrimethylammonium bromide procedure [23].
2. DNA samples were quantified using Quant-iT PicoGreen® double-stranded DNA reagent (Invitrogen) on the FLUOstar Omega (BMG Labtech) using a standard procedure. DNA samples were diluted to a concentration of 20 ng/nl and subsequently pooled.
3. The 3008 M2 DNAs were first subjected to a 4-fold pooling, resulting in 752 pooled samples contained in eight 96-well microtiter plates. These 752 4-fold pooled M2 DNA samples were then subjected to a three-dimensional (3-D) pooling strategy, such that each sample was represented once in an X-, Y-, and Z-coordinate pool. X pools were assembled by pooling all 4-fold pooled M2 DNA samples per column of eight wells (e.g., A1–H1) of the eight 96-well plates and Y pools were assembled by pooling all M2 DNA samples per row of 12 wells (e.g., A1–A12) of the eight 96-well plates. This resulted in 12 X and 8 Y pools, which each represented a maximum of 256 ($8 \times 8 \times 4$) and 384 ($12 \times 8 \times 4$) M2 families, respectively. Z pools were assembled by pooling all 4-fold pooled DNA samples of an entire 96-well pooled plate, resulting in eight Z pools each representing 384 (4×96) M2 families.
4. The 92 DNA samples of the tomato core collection lines were subjected to a two-dimensional (2-D) pooling strategy, obtaining 12 X and 8 Y pools, each representing a maximum of eight or 12 tomato lines, respectively.

KeyPoint Template Preparation

The *SleIF4E* gene was chosen as a target sequence for KeyPoint mutation and natural polymorphism detection. Several recessive mutations in this gene are associated with broad resistance to potyviruses in some plant species [24]. Specific primers were designed for PCR amplification of *SleIF4E* exon 1 (amplicon 1; 287 bp), exon 2, intron 2 and exon 3 (amplicon 2; 402 bp), and exon 4, intron 4, and exon 5 (amplicon 3; 200 bp) (Figure 12.2). *SleIF4E* amplicon 1 was chosen as a target for mutation screening (MS) and *SleIF4E* amplicons 1, 2, and 3 for natural variation screening (NVS).

1. These primers contained the following target-specific sequences (5′–3′):

ATGGCAGCAGCTGAAATGG	amplicon 1 forward primer
CCCCAAAAATTTTCAACAGTG	amplicon 1 reverse primer
TGCTTACAATAATATCCATCAC	amplicon 2 forward primer
CCTGAGCTGTTTCATTTGC	amplicon 2 reverse primer
TTAGCATTGGTAAGCAATGG	amplicon 3 forward primer
CTATACGGTGTAACGATTC	amplicon 3 reverse primer

2. Six nucleotide nontarget sequences were added at the 5′ ends of these primers. The sample identification tags all differed by at least two nucleotides to exclude the possibility that a single nucleotide substitution error could cause incorrect assignment of the sequence reads to a sample pool [22, 25].
3. The 50-μl PCR reactions were performed containing 80 ng DNA for each of the 28 3-D tomato M2 family pools and each of the 20 2-D tomato line pools, 50 ng forward tagged primer, 50 ng reverse tagged primer, 0.2 mM dNTP, 1 U Herculase® II Fusion DNA polymerase (Stratagene), and 1 × Herculase II reaction buffer. PCRs were performed with the following cycle profile: 2 min at 95 °C, followed by 35 cycles of 30 s 95 °C, 30 s 56 °C, and 30 s 72 °C, followed by cooling down to 4 °C. Equal amounts of PCR products of the 3-D or 2-D pooled samples were combined and further treated as one GS FLX fragment library sample.

GS FLX Library Preparation and Titration

1. Samples of 5 and 3.2 μg of the combined PCR fragments (i.e., pooled and tagged PCR products), obtained from mutant and natural population pools, respectively, were used as input for GS FLX library construction. The use of tagged and pooled PCR products, however, necessitated some adaptations in the published GS library construction protocol [20]. First, no shearing was carried out and, second, T4 DNA polymerase treatment, generally used to generate fragment termini suitable for blunt-end ligation, was omitted

```
  1 atggcagcagctgaaatggagagaacgatgtcgtttgatgcagct      Exon 1
    M  A  A  A  E  M  E  R  T  M  S  F  D  A  A
 46 gagaagttgaaggccgccgatggaggaggaggagaggtagacgat
    E  K  L  K  A  A  D  G  G  G  E  V  D  D
 91 gaacttgaagaaggtgaaattgttgaagaatcaaatgatacggca
    E  L  E  E  G  E  I  V  E  E  S  N  D  T  A
                        160-1        170 172
136 tcgtatttagggaaagaaatcacagtgaagcatccattggagcat
                        aa           t  g
    S  Y  L  G  K  E  I  T  V  K  H  ?  L  E  H
                           K        L  V
                            210           221
181 tcatggacttttggtttgataaccctaccactaaatctcgacaa
                                t              a
    S  W  T  F  W  F  D  N  P  T  T  K  S  R  Q
                               T              Q
226 actgcttgggaagctcacttcgaaatgtctacactttctccact
    T  A  W  G  S  S  L  R  N  V  Y  T  F  S  T
271 gttgaaaattttgggg t
    V  E  N  F  W  G
```

```
                    gcttacaataatatccatcacccaagc       Exon 2
                    A  Y  N  N  I  H  H  P  S
                       47
316 aagttaattatgggagcagactttcattgttttaagcacaaaatt
                     a
    K  L  I  M  G  A  D  F  H  C  F  K  H  K  I
                       N
361 gagccaaagtgggaagatcctgtatgtgccaatggagggacgtgg
    E  P  K  W  E  D  P  V  C  A  N  G  G  T  W
406 aaaatgagtttttcgaagggtaaatctgataccagctggctgtat
    K  M  S  F  S  K  G  K  S  D  T  S  W  L  Y
           171            193           203
451 acggtattccgaagttatttccatccagcccttaatgataggtca    Intron 2
            g                  t            c
    T
    209                                  245
    ttctagtaatgttattttcccctttgatataatttccactcttgt
    c                                     a
                266 269
    tttcttatatggaattattgtagctgctggcaatgattggacat    Exon 3
                    g  a
                        L  L  A  M  I  G  H
    caattcgatcatggagatgaaatttgtggagcagttgttagtgtc
    Q  F  D  H  G  D  E  I  C  G  A  V  V  S  V
496 cgggctaagggagaaaaaatagctttgtggaccaagaatgctgca
    R  A  K  G  E  K  I  A  L  W  T  K  N  A  A
541 aatgaaacagctcag
    N  E  T  A  Q
```

```
                                       gttagc       Exon 4
                                       V  S
                        41
586 attggtaagcaatggaagcagtttctagattacagtgattcggtt
                                       c
    I  G  K  Q  W  K  Q  F  L  D  Y  S  D  S  V
                                    S
                                84 85
631 ggcttcatatttcacgtatgaaatcttggttatcatacgccttta   Intron 4
                                         tg
    G  F  I  F  H
    attcagtttctcttcaattagcaagactcataaagaatcatcttc
    ttttgcaggacgatgcaaagaggctcgacagaaatgccaagaat   Exon 5
              D  D  A  K  R  L  D  R  N  A  K  N
676 cgttacaccgtatag 696
    R  Y  T  V  *
```

Figure 12.2 *SleIF4E* sequence, including position of discovered mutant and naturally occurring SNPs.

because PCR fragments were produced using a proofreading *Taq* polymerase yielding blunt ends.
2. After library construction, emulsion titration and bead enrichment were carried out according to the standard GS FLX protocol (Roche Applied Science).

GS FLX Sequencing

One picotiterplate (PTP) (70 × 75 mm) with two regions was used for sequencing the mutant population (EMS) library and one-quarter PTP was used for sequencing of the natural population (Eu-Sol) library. Sequencing was performed according to the manufacturer's instructions (Roche Applied Science).

KeyPoint Bio-Informatics Analysis

The mutation/SNP screening process consisted of four parts: GS FLX data processing, KeyPoint preprocessing, polymorphism detection, and SNP mining and analysis

1. GS FLX data processing was performed using the Roche GS FLX software (release 1.1.03.24). Base-called reads were trimmed and filtered for quality and converted into FASTA format.
2. The origin of the reads was identified based on the target-specific primer sequences and the six base sample identification tags. This step was implemented in a custom Perl script that used the Semi Global Smith Waterman algorithm within a TimeLogic DeCypher system (Active Motif). Furthermore, sample tag and primer sequences were trimmed, and the preprocessed reads of each multidimensional pool were saved separately to the database.
3. Each pool dataset was mapped against the reference sequence using SSAHASNP (www.sanger.ac.uk) with the *-454* and the *-NSQ* parameters set to true. Raw SSAHASNP output was parsed using a custom Perl script. SNPs derived from the SSAHASNP output were divided in "forward" and "reverse" set depending on the orientation of the read compared to the reference. Both sets were saved in a comma-separated format.
4. For the MS sequence data, the combined set of polymorphisms observed for "forward" and "reverse" orientation reads was filtered on G → A and C → T substitutions only, as these are the changes expected from EMS-induced treatment. On the contrary, for the NVS data analysis all polymorphisms were considered. Using Microsoft Excel, a matrix was built with polymorphism position on the reference sequence as rows and pools as columns. Per position, the average polymorphism error rate was calculated by dividing the sum of all polymorphisms over all pools by the total number of reads.

Next, the probability of finding the observed number of polymorphisms (k) was calculated given an H_0 of no underlying mutation for each pool at each position. This was done by taking the Poisson distribution with $\lambda = np$, where p is the estimated error rate and n is the number of all reads detected per pool. P-values below a significance value of 0.01 indicate pools with true mutants. A combination of two significant 2-D or three significant 3-D coordinate pools, one in each different coordinate, were considered as pointing at (pooled) samples harboring true SNPs in the natural population and true mutations in the mutant population, respectively.

12.3
Applications

12.3.1
EMS Mutation Screening and Validation

A total of 3008 M2 families were screened for EMS-induced mutations in exon 1 of the *SleIF4E* gene based on amplification from 28 3-D pools (12*X*, 8*Y*, and 8*Z*). Successfully trimmed and tagged reads (Table 12.1) were taken into the mutation/polymorphism mining step starting with mapping them onto the reference sequence and followed by generating pairwise sequence alignments. Next, the numbers of C → T and G → A changes from the wild-type sequence were counted for each

Table 12.1 Overview of results of GS FLX KeyPoint runs in tomato.

KeyPoint MS	no. amplicons	1
	total no. of reads after filtering (GS FLX raw sequencing reads)	667864
	reads with sample identification tag assigned	580471 (87%)
	faulty reads	87393 (13%)
	no. identified mutants	2
KeyPoint NVS	no. amplicons	3
	total no. of reads after filtering (GS FLX raw sequencing reads)	95222
	reads with sample identification tag assigned	80634 (85%)
	faulty reads	14588 (15%)
	reads with sample identification tag assigned for amplicon 1	31724 (33.3%)
	no. haplotypes	4
	reads with sample identification tag assigned for amplicon 2	11495 (12.2%)
	no. haplotypes	5
	reads with sample identification tag assigned for amplicon 3	37415 (39.3%)
	no. haplotypes	3

Counts																														
Pools																														
Pos	X1	X2	X3	X4	X5	X6	X7	X8	X9	X10	X11	X12	Y1	Y2	Y3	Y4	Y5	Y6	Y7	Y8	Z1	Z2	Z3	Z4	Z5	Z6	Z7	Z8	Total	Error rate
166	1	3			2	3	3	1					2					3	4	2	2			1		1	2		2	36	0.00006
169	1		1	2		5	3	1	2	5	1		6	3		4		4	1	2	1				4	4		2	52	0.00009
170	5	4	3		4	1	4			9		81	10	4	5	7	3	2	40	77			4	1	5	101	1	5	376	0.00065
175		2	1			3		1					4	2	1	1			3	2	12	4			4	1		3	44	0.00008
177	3	2		1	4	1	3	1	5	4	1	1	7	3	2	12	5	2	1	2	1	5		1	10	14	4	4	99	0.00017
212	3							5			2	3	1	2	2	8		3	1	2		2	2	1	2	3		5	47	0.00008
220	3	2	2		3	3	5	1	2				2	2	1	2	4	1					2		2	4	1	6	53	0.00009
221	11	4	9	5	2	22	6	1	3	1	2	63	4	2	59	8	7	6	1	7			8	4	1	13	131	3	11	394	0.00068
223	4	3	1			2	1				1	2	1	2	8	1	7	1	1	11			2	1	4	11			65	0.00011
227	3	3		1		8	6	1	2				6	2	1		2		1	1	1			1		3	3	5	50	0.00009

P-values																														# Pools	
Pools																															
Pos	X1	X2	X3	X4	X5	X6	X7	X8	X9	X10	X11	X12	Y1	Y2	Y3	Y4	Y5	Y6	Y7	Y8	Z1	Z2	Z3	Z4	Z5	Z6	Z7	Z8	P<0.001	P<0.01	P<0.05
166	0.83	0.11	1.00	0.15	0.11	0.23	0.89	1.00	1.00	0.17	1.00	1.00	0.90	1.00	1.00	1.00	0.18	0.04	0.21	0.43	0.17	0.76	0.42	1.00	0.90	0.86	1.00	0.75	0	0	1
169	0.92	1.00	0.48	0.25	1.00	0.10	0.61	0.56	0.35	0.00	0.56	1.00	0.12	0.28	1.00	0.14	1.00	0.10	0.71	0.62	0.65	1.00	1.00	1.00	0.43	0.74	1.00	0.90	1	1	1
170	1.00	1.00	0.84	1.00	1.00	1.00	1.00	1.00	1.00	0.35	1.00	0.00	1.00	1.00	1.00	0.99	1.00	1.00	0.00	0.00	1.00	1.00	1.00	0.72	0.00	1.00	1.00	1.00	4	4	4
175	1.00	0.41	0.42	1.00	1.00	0.33	1.00	0.50	1.00	1.00	1.00	1.00	0.31	0.44	0.78	0.81	1.00	1.00	0.09	0.53	0.00	0.10	1.00	1.00	0.32	0.99	1.00	0.64	1	1	1
177	0.86	0.82	1.00	0.84	0.38	0.99	0.94	0.79	0.09	0.14	0.79	0.95	0.44	0.65	0.85	0.00	0.38	0.85	0.91	0.91	0.87	0.36	1.00	0.79	0.11	0.10	0.01	0.94	1	1	2
212	0.41	1.00	1.00	0.00	1.00	1.00	0.77	0.04	1.00	1.00	1.00	1.00	0.95	0.48	0.48	0.00	1.00	0.22	0.67	0.57	1.00	0.56	0.16	0.53	0.80	0.83	1.00	0.28	2	2	3
220	0.48	0.51	0.14	1.00	0.24	0.44	0.22	0.57	0.36	1.00	1.00	1.00	0.85	0.54	0.84	0.59	0.17	0.84	1.00	0.17	1.00	0.62	1.00	0.21	0.44	0.99	0.36	0.21	0	0	0
221	0.98	1.00	0.06	0.86	1.00	0.22	1.00	1.00	1.00	0.98	0.00	1.00	1.00	0.00	0.98	1.00	0.99	1.00	1.00	0.99	0.84	1.00	1.00	0.00	0.65	1.00	3	3	3
223	0.40	0.35	0.55	1.00	1.00	0.80	0.98	1.00	1.00	0.64	0.59	0.98	0.65	0.00	0.91	0.02	0.89	0.79	0.00	0.73	1.00	0.25	0.65	0.60	0.05	1.00	1.00	1	2	3
227	0.45	0.22	1.00	0.61	1.00	0.00	0.09	0.55	0.34	1.00	1.00	1.00	0.10	0.51	0.82	1.00	0.60	1.00	0.70	0.87	0.64	1.00	1.00	0.55	1.00	0.86	0.01	0.32	0	2	2

Figure 12.3 Results of KeyPoint analysis of the mutant population. The top panel shows numbers of G → A (position 221) and C → T (position 170) sequence deviations compared to the wild-type sequence observed in each of the 3-D pools in a subset of nucleotide positions of the SleIF4E amplicon 1. The total number of observed sequence deviations and calculated average error rates are shown at the right hand side. The bottom panel shows corresponding P-values of false positives for each X, Y, and Z pool. Total numbers of pools surpassing significance thresholds $P<0.001$, $P<0.01$, and $P<0.05$ are shown at the bottom right.

position per pool (Figure 12.3) and the probabilities that they represent true EMS mutations were calculated taking into account their distribution across the 3-D sample pools. At significance threshold $P<0.01$, two mutations were identified: a C → T mutation at position 170 and a G → A mutation at position 221, which encode a proline to leucine (both hydrophobic amino acids) and arginine (positively charged and hydrophilic) to glutamine (hydrophilic) amino acids changes, respectively (Figure 12.2). The impact of the amino acid changes on gene function is unknown.

These mutations were based on significantly elevated numbers of non-wild-type nucleotides at positions 170 and 221 in four (X12, Y7, Y8, and Z5) and three pools (X12, Y3, and Z6), respectively (Figure 12.3). A complete overview of the statistical analysis is provided elsewhere [22]. Sanger sequencing confirmed the C170T mutation in one of the four M2 families located at the plate position specified by the 3-D pool coordinates X12, Y7, and Z5, and the G221A mutation in one of the four M2 families at the plate position defined by the X12, Y3, and Z6 coordinates [22].

During the development of the KeyPoint technology, we noticed that the number of sequence changes compared to the wild-type sequence, which reflects the combined total of PCR and sequence errors and genuine EMS mutations, was highly variable across all positions of the amplicon sequence. Moreover, these numbers were also rather variable between "forward" and "reverse" orientation GS FLX reads of the amplicon (data not shown). Based on this observation, which has been made previously by others [26–28], we concluded that identification of EMS mutations or SNPs could not be performed reliably based on the total number of sequence deviations from the wild-type sequence alone, without considering their distribution across the individual multidimensional pools. Consequently, a probability calculation method based on Poisson statistics was established which takes this distribution into account.

Unexpectedly for a 3-D pooling scheme, at the position C170T EMS mutation in *SleIF4E* exon 1, four positive pools were observed (X12, Y7, Y8 and Z5). This includes two Y-dimension coordinates and specifies two unique plate addresses: X12, Y7, and Z5, and X12, Y8, and Z5. In fact the mutation was confirmed in one of four M2 families of the first address, whereas the second one pointed to an empty (adjacent) well, despite the fact that 77 C → T changes were observed in Y8 compared to 40 in Y7 (Figure 12.3). We can only explain this observation by an experimental error made either during (manual) assembly of the 3-D pools or setting up the PCRs. Despite this flaw, the C170T mutation was identified and assigned to the correct position.

Based on analysis of progressively smaller subsets of the EMS data [22] we estimated that KeyPoint technology enables screening of four amplicons in 3000 M2 families per GS FLX run of approximately 500 000 sequence reads. The maximum read-length of these amplicons is defined by the specifications of GS FLX platform (approximately 240 bases). Hence, a total of $4 \times 3000 \times 240$ bases equaling 2.88 Mb of amplicon sequence can be screened in a single GS FLX run. Other combinations of sample and amplicon numbers within these boundaries can be considered as well, with appropriate adaptation of the pooling strategy.

Based on these results, we conclude that KeyPoint screening offers a sequence-based alternative to TILLING (and EcoTILLING) screening which comes with certain advantages:

- The sequence context of mutations is determined, which is important when screening for knockout mutations, especially given the fact that only a minor fraction of EMS mutations are expected to confer a stop codon or splice site errors [8].
- KeyPoint is robust and does not rely on endonucleases with known robustness issues.
- KeyPoint does not rely on visual inspection or interpretation of slab-gel data aided by image analysis software, but is based on an objective statistical analysis method which is easy to perform.
- The method is flexible with respect to changing numbers of samples and amplicons.
- Compared to unidirectional Sanger sequencing, KeyPoint is based on highly redundant sequencing of amplicons, which improves the accuracy when appropriate analysis methods are applied; this in turn reduces the likelihood of identifying false positives due to imperfections of Sanger sequencing.
- Costs are scalable as it is not mandatory required to perform an entire GS FLX (*Titanium*) run, but PTPs containing multiple compartments are available for use as well.

12.3.2
Natural Polymorphism Screening and Validation

A total of 92 tomato lines of the Eu-Sol core collection were screened for natural variation in exons 1–5 and introns 2 and 4, based on PCR amplification of three

Counts

Pos	WT	SNP	X1	X2	X3	X4	X5	X6	X7	X8	X9	X10	X11	X12	Y1	Y2	Y3	Y4	Y5	Y6	Y7	Y8	Total	Error rate
44	G	A												1	1					1			3	0.00026
47	G	A		1					43		26		67		35			10		41			223	0.01940
171	T	G			15											25							40	0.00348
193	C	T			15	1										25							41	0.00357
203	A	C	1							75										69			145	0.01261
206	T	C													1			1		1			3	0.00026
209	T	C			15											25							40	0.00348
216	A	T						1					1								1		3	0.00026
223	T	C				1				3				4	1			1		2			12	0.00104
226	T	C	1	1					3	2	1	1	1	2	3	1				1			17	0.00148
233	T	C				1		1	1	2					1				1	1			7	0.00061
245	C	A			15											25	1						41	0.00357
256	T	C						1		1						1							3	0.00026
266	A	G			7									1		12							20	0.00174
269	T	A	37		17			63										15	44	29		11	216	0.01879
346	G	A												1						2			3	0.00026

P-values

Pos	WT	SNP	X1	X2	X3	X4	X5	X6	X7	X8	X9	X10	X11	X12	Y1	Y2	Y3	Y4	Y5	Y6	Y7	Y8	P<0.001	P<0.01	P<0.05
44	G	A	1.0	1.0	1.0	1.0	1.0	1.0	1.0	1.0	1.0	1.0	1.0	0.2	0.1	1.0	1.0	1.0	1.0	0.2	1.0	1.0	0	0	0
47	G	A	1.0	1.0	1.0	1.0	1.0	1.0	0.0	1.0	0.0	1.0	0.0	1.0	0.0	1.0	1.0	0.3	1.0	0.0	1.0	1.0	4	4	5
171	T	G	1.0	1.0	1.0	0.0	1.0	1.0	1.0	1.0	1.0	1.0	1.0	1.0	1.0	0.0	1.0	1.0	1.0	1.0	1.0	1.0	2	2	2
193	C	T	1.0	1.0	1.0	0.0	1.0	1.0	1.0	1.0	1.0	1.0	1.0	1.0	1.0	0.0	1.0	1.0	1.0	1.0	1.0	1.0	2	2	2
203	A	C	1.0	1.0	1.0	1.0	1.0	1.0	1.0	1.0	0.0	1.0	1.0	1.0	1.0	1.0	1.0	1.0	1.0	0.0	1.0	1.0	2	2	2
206	T	C	1.0	1.0	1.0	1.0	1.0	1.0	1.0	1.0	1.0	1.0	1.0	1.0	0.1	1.0	1.0	0.1	1.0	0.2	1.0	1.0	0	0	0
209	T	C	1.0	1.0	1.0	0.0	1.0	1.0	1.0	1.0	1.0	1.0	1.0	1.0	1.0	0.0	1.0	1.0	1.0	1.0	1.0	1.0	2	2	2
216	A	T	1.0	1.0	1.0	1.0	1.0	0.2	1.0	1.0	1.0	1.0	1.0	1.0	1.0	1.0	1.0	1.0	1.0	1.0	0.3	1.0	0	0	0
223	T	C	1.0	1.0	1.0	0.2	1.0	1.0	1.0	0.1	1.0	1.0	1.0	0.0	0.3	1.0	1.0	1.0	1.0	0.6	1.0	0.4	0	1	1
226	T	C	0.4	0.3	1.0	1.0	1.0	1.0	0.0	0.5	0.7	0.5	0.6	0.3	0.0	1.0	0.6	1.0	1.0	1.0	0.8	1.0	0	0	2
233	T	C	1.0	1.0	1.0	1.0	1.0	0.2	0.2	0.2	1.0	1.0	1.0	1.0	0.2	1.0	1.0	1.0	1.0	0.4	0.5	1.0	0	0	0
245	C	A	1.0	1.0	1.0	0.0	1.0	1.0	1.0	1.0	1.0	1.0	1.0	1.0	1.0	0.0	0.9	1.0	1.0	1.0	1.0	1.0	2	2	2
256	T	C	1.0	1.0	1.0	1.0	1.0	1.0	1.0	0.2	1.0	0.2	1.0	1.0	1.0	0.2	1.0	1.0	1.0	1.0	1.0	1.0	0	0	0
266	A	G	1.0	1.0	1.0	1.0	1.0	1.0	1.0	1.0	1.0	1.0	1.0	0.7	1.0	0.0	1.0	1.0	1.0	1.0	1.0	1.0	2	2	2
269	T	A	1.0	0.0	1.0	0.0	1.0	1.0	0.0	1.0	1.0	1.0	1.0	1.0	1.0	1.0	1.0	0.0	0.0	0.0	1.0	0.0	5	6	7
346	G	A	1.0	1.0	1.0	1.0	1.0	1.0	1.0	1.0	1.0	1.0	0.2	1.0	1.0	1.0	1.0	1.0	1.0	0.0	1.0	1.0	0	0	1

Figure 12.4 Results of KeyPoint analysis of the tomato natural population Eu-Sol Core Collection. The top panel shows numbers of all sequence deviations compared to the wild-type sequence observed in each of the 2-D pools at a selected subset of nucleotide positions of the SleIF4E amplicon 2. The total number of observed sequence deviations and calculated average error rates are shown at the right-hand side. The bottom panel shows the corresponding P-values for each of the X and Y pools. Total numbers of pools surpassing significance thresholds $P<0.001$, $P<0.01$, and $P<0.05$ are shown at the bottom right.

amplicons of the SleIF4E gene (Figure 12.2) from 20 (12X and 8Y) 2-D pools using tagged PCR primers. Amplicon sizes were as expected 287, 402, and 200 bp, respectively, excluding nontemplate tags. Summary statistics of the GS FLX run are presented in Table 12.1. All nucleotide changes from the SleIF4E wild-type sequence were counted for each of 20 2-D pools for all positions in all three amplicons [22]. A subset of the data obtained from amplicon 2 is shown in Figure 12.4, which includes all eight positions (47, 171, 193, 203, 209, 245, 266, and 269) where SNPs were found and a number of nonpolymorphic positions for comparison purposes. As expected for a 2-D pool design, statistically significant probabilities for harboring true mutations ($P<0.01$) were observed in at least two pools (one X and one Y) for each of these eight SNPs (Figure 12.5a). Similar analyses performed for amplicons 1 and 3 revealed four and three SNPs, respectively [22]. All 15 SNPs observed in amplicons 1–3 together were confirmed by Sanger sequencing of selected PCR products of individual samples (data not shown).

The combinations of SNPs in amplicons 1, 2, and 3 defined four, five, and three sequence haplotypes, including the wild-type haplotype (Figure 12.5). Excluding the

Amplicon 1

Haplotypes	Origin	160	161	172	210
1	WT	G	T	T	C
2	F10	A	A	T	C
3	D,E,F,H/2,4,8	G	T	G	C
4	B4	G	T	T	T

Amplicon 2

Haplotypes	Origin	47	171	193	203	209	245	266	269
1	WT	G	T	C	A	T	C	A	T
2	A,F/7,11	A	T	C	A	T	C	A	T
3	B4	G	G	T	A	C	A	G	T
4	F10	G	T	C	C	T	C	A	T
5	E,F,H/2,4,8	G	T	C	A	T	C	A	A

Amplicon 3

Haplotypes	Origin	41	84	85
1	WT	T	C	A
2	B4	C	T	G
3	A,B,D,E,F,H/2,4,5,7,8,9,10,11	T	C	G

Figure 12.5 SNPs and haplotypes of the *SleIF4E* gene observed in 92 lines of the Eu-Sol tomato core collection. Haplotype 1 is wild-type (WT) sequence. Alleles different from wild-type are shown in grey. Nucleotide positions in amplicons 1, 2, and 3 are shown at the top. The 96-well plate row (A–H) and column (1–12) positions containing samples carrying haplotypes are shown in the "Origin" column.

wild-type haplotype, two amplicon 1 haplotypes, two amplicon 2 haplotypes, and one amplicon 3 haplotype could be assigned to individual 2-D pooled samples (i.e., tomato lines) in the 96-well base plate, at positions B4 and F10 for amplicons 1 and 2, and position B4 for amplicon 3, respectively (Figure 12.5). Taken together, the three amplicons therefore defined at least six *SleIF4E* haplotypes in the collection of 92 tomato lines: the wild-type haplotype (= amplicon 1 haplotype 1 – amplicon 2 haplotype 1 – amplicon 3 haplotype 1; Figure 12.5b), and similarly for amplicons 1, 2, and 3 the haplotype number combinations, 1-2-(1 or 3), 2-4-(1 or 3) (sample F10), 3-1-(1 or 3), 3-5-(1 or 3), and 4-3-2 (sample B4) as shown in Figure 12.5.

The impact on these haplotypes on gene function is unknown. As for the EMS screening, this approach can be scaled up for analysis of larger numbers of samples, with limited additional efforts and costs for amplicon preparation. Although germplasm diversity can also be revealed based on sequencing of a single pooled sample, an advantage of using a 2-D (or higher-order) pooling scheme is that a subset of identified SNPs and haplotypes can be attributed to a specific sample, or at least a subset of rows and columns/pool coordinates, while sample preparation costs are only marginally higher. This is not the case when all samples are pooled together. In fact, for rare SNPs and haplotypes in the germplasm, which are often the most interesting to discover, there is a higher probability to identify the associated sample than for more common polymorphisms, as shown by our results. A second advantage of using a multidimensional pooling strategy is that it provides a built-in quality control capable of separating genuine polymorphisms from experimental (PCR and sequencing) errors, since true mutations must be observed in at least two dimensions in case of a 2-D design. Also, this is not the case when all samples are pooled together. These features of KeyPoint technology for screening natural variation compare favorably to a number of alternative (prescreening) technologies, which may lack the power to detect low-frequency polymorphisms [6, 7].

12.4
Perspectives

The KeyPoint technology is a flexible, high-throughput sequence-based polymorphism screening technology, applicable for detection of artificially induced and natural polymorphisms in a wide variety of species. Currently, a total of $4 \times 3000 \times 240$ bases (2.88 Mb of amplicon sequences) can be screened in a single GS FLX run [22]. The KeyPoint throughput will increase with further output improvements of the GS FLX platform. For example, the GS FLX *Titanium* platform yields around 1 million sequence reads of more than 400 bases, which increases the throughput of KeyPoint screening to (the equivalent of) 4×6000 M2 plants $\times 400$ bases (9.6 Mb) of screened amplicon sequences per GS FLX *Titanium* run.

Acknowledgments

We thank our colleagues at the bioinformatics and research units at Keygene for excellent technical contributions. We are grateful to Professor Dani Zamir for construction of the EMS population and Eu-Sol for access to the Core Collection lines used in this study. Parts of the text and figures were taken from Rigola *et al.* [22]. KeyPoint technology is covered by patent applications owned by Keygene NV. Application for trademark registration for KeyPoint has been filed by Keygene NV. Other (registered) trademarks are the property of the respective owners.

References

1 Ahloowalia, B.S., Maluszynsky, M., and Nichterlein, K. (2004) Global impact of mutation-derived varieties. *Euphytica*, **135**, 187–204.

2 Hestekin, C.N. and Barron, A.E. (2006) The potential of electrophoretic mobility shift assays for clinical mutation detection. *Electrophoresis*, **27**, 3805–3815.

3 Montgomery, J., Wittwer, C.T., Palais, R., and Zhou, L. (2007) Simultaneous mutation scanning and genotyping by high-resolution DNA melting analysis. *Nat. Protocols*, **2**, 59–66.

4 Den Dunnen, J.T. and van Ommen, G.J.B. (1999) The protein truncation test: a review. *Hum. Mutation*, **14**, 95–102.

5 Taylor, G.R. (1999) Enzymatic and chemical cleavage methods. *Electrophoresis*, **20**, 1125–1130.

6 Jenkins, G.J.S. (2004) The restriction site mutation (RSM) method: clinical applications. *Mutagenesis*, **19**, 3–11.

7 Li, J., Berbeco, R., Distel, R.J., Jänne, P.A., Wang, L., and Makrigiorgos, G.M. (2007) s-RT-MELT for rapid mutation scanning using enzymatic selection and real time DNA-melting: new potential for multiplex genetic analysis. *Nucleic Acids Res.*, **35**, e84.

8 McCallum, C.M., Comai, L., Greene, E.A., and Henikoff, S. (2000) Targeted screening for induced mutations. *Nat. Biotechnol.*, **18**, 455–457.

9 Colbert, T., Till, B.J., Tompa, R., Reynolds, S., Steine, M.N., Yeung, A.T., McCallum, C.M., Comai, L., and Henikoff, S. (2001) High-throughput screening for induced point mutations. *Plant Physiol.*, **126**, 480–484.

10 Gilchrist, E.J. and Haughn, G.W. (2005) Tilling without a plough; a new method with application for reverse genetics. *Curr. Opinion Plant Biol.*, **8**, 211–215.

11 Comai, L., Young, K., Till, B.J., Reynolds, S., Greene, E.A., Codomo, C.A., Enns, L.C.,

Johnson, J.E., Burtner, C., Odden, A.R., and Henikoff, S. (2004) Efficient discovery of DNA polymorphism in natural populations by Ecotilling. *Plant J.*, **37**, 778–786.

12 Sood, R., English, M.A., Jones, M., Mullikin, J., Wang, D.M., Anderson, M., Wu, D., Chandrasekharappa, S.C., Yu, J., Zhang, J., and Liu, P.P. (2006) Methods for reverse genetic screening in zebrafish by resequencing and TILLING. *Methods*, **39**, 220–227.

13 Gilchrist, E.J., O'Neil, N.J., Rose, A.M., Zetka, M.C., and Haughn, G.W. (2006) TILLING is an effective reverse genetics technique for *Caenorhabditis elegans*. *BMC Genomics*, **7**, 262.

14 Greene, E.A., Codomo, C.A., Taylor, N.E., Henikoff, J.G., Till, B.J., Reynolds, S.H., Enns, L.C., Burtner, C., Johnson, J.E., Odden, A.R., Comai, L., and Henikoff, S. (2003) Spectrum of chemically induced mutations from a large-scale reverse-genetic screen in *Arabidopsis*. *Genetics*, **164**, 731–740.

15 Suzuki, T., Eiguchi, M., Kumamaru, T., Satoh, H., Matsusaka, H., Moriguchi, K., Nagato, Y., and Kurata, N. (2008) MNU-induced mutant pools and high performance TILLING enable finding of any gene mutation in rice. *Mol. Genet. Genomics*, **27**, 213–223.

16 Cooper, J.L., Till, B.J., Laport, R.G., Darlow, M.C., Kleffner, J.M., Jamai, A., El-Mellouki, T., Liu, S., Ritchie, R., Nielsen, N., Bilyeu, K.D., Meksem, K., Comai, L., and Henikoff, S. (2008) TILLING to detect induced mutations in soybean. *BMC Plant Biol.*, **8**, 9.

17 Slade, A.J., Fuerstenberg, S.I., Loeffler, D., Steine, M.N., and Facciotti, D. (2005) A reverse genetic, non-transgenic approach to wheat crop improvement by TILLING. *Nat. Biotechnol.*, **23**, 75–81.

18 Till, B.J., Reynolds, S.H., Weil, C., Springer, N., Burtner, C., Young, K., Bowers, E., Codomo, C.A., Enns, L.C., Odden, A.R., Greene, E.A., Comai, L., and Henikoff, S. (2004) Discovery of induced point mutations in maize genes by TILLING. *BMC Plant Biol.*, **4**, 12.

19 Till, B.J., Zerr, T., Comai, L., and Henikoff, S. (2006) A protocol for TILLING and Ecotilling in plants and animals. *Nat. Protocols*, **1**, 2465–2477.

20 Margulies, M., Egholm, M., Altman, W.E., Attiya, S., Bader, J.S., Bemben, L.A., Berka, J., Braverman, M.S., Chen, Y.J., Chen, Z., Dewell, S.B., Du, L., Fierro, J.M., Gomes, X.V., Godwin, B.C., He, W., Helgesen, S., He Ho, C., Irzyk, G.P., Jando, S.C., Alenquer, M.L.I., Jarvie, T.P., Jirage, K.B., Kim, J.B., Knight, J.R., Lanza, J.R., Leamon, J.H., Lefkowitz, S.M., Lei, M., Li, J., Lohman, K.L., Lu, H., Makhijani, V.B., McDade, K.E., McKenna, M.P., Myers, E.W., Nickerson, E., Nobile, J.R., Plant, R., Puc, B.P., Ronan, M.T., Roth, G.T., Sarkis, G.J., Simons, J.F., Simpson, J.W., Srinivasan, M., Tartaro, K.R., Tomasz, A., Vogt, K.A., Volkmer, G.A., Wang, S.H., Wang, Y., Weiner, M.P., Yu, P., Begley, R.F., and Rothberg, J.M. (2005) Genome sequencing in microfabricated high-density picolitre reactors. *Nature*, **437**, 376–380.

21 Menda, N., Semel, Y., Peled, D., Eshed, Y., and Zamir, D. (2004) In silico screening of a saturated mutation library of tomato. *Plant J.*, **38**, 861–872.

22 Rigola, D., van Oeveren, J., Janssen, A., Bonn,é, A., Schneiders, H., van der Poel, H.J.A., van Orsouw, N.J., Hogers, R.C.J., de Both, M.T.J., and van Eijk, M.J.T. (2009) High-throughput detection of induced mutations and natural variation using KeyPoint technology. *PLoS ONE*, **4**, e4761.

23 Stuart, C.N. and Via, L.E. (1993) A rapid CTAB DNA isolation technique useful for RAPD fingerprinting and other PCR applications. *Biotechniques*, **14**, 748–750.

24 Ruffel, S., Gallois, J.L., Lesage, M.L., and Caranta, C. (2005) The recessive potyvirus resistance gene *pot-1* is the tomato orthologue of the pepper *pvr2-eIF4E* gene. *Mol. Gen. Genomics*, **274**, 346–353.

25 van Orsouw, N.J., Hogers, R.C., Janssen, A., Yalcin, F., Snoeijers, S., Verstege, E., Schneiders, H., van der Poel, H., van Oeveren, J., Verstegen, H., and van Eijk, M.J.T. (2007) Complexity reduction of polymorphic sequences (CRoPS): a novel approach for large-scale polymorphism discovery in complex genomes. *PLoS One*, **2**, e1172.

26 Huse, S.M., Huber, J.A., Morrison, H.G., Sogin, M.L., and Welch, D.M. (2007) Accuracy and quality of massively parallel DNA pyrosequencing. *Genome Biol.*, **8**, R143.

27 Wang, C., Mitsuya, Y., Gharizadeh, B., Ronaghi, M., and Shafer, R.W. (2007) Characterization of mutation spectra with ultra-deep pyrosequencing: application to HIV-1 drug resistance. *Genome Res.*, **17**, 1195–1201.

28 Hoffmann, C., Minkah, N., Leipzig, J., Wang, G., Arens, M.Q., Teblas, P., and Bushman, F.D. (2007) DNA bar coding and pyrosequencing to identify rare HIV drug resistance mutations. *Nucleic Acids Res.*, **35**, e91.

Part IV
Applications in Plant Breeding

13
Natural and Induced Mutants of Barley: Single Nucleotide Polymorphisms in Genes Important for Breeding

William T.B. Thomas, Brian P. Forster, and Robbie Waugh

Abstract

Much mutation research has been conducted on barley and many different morphological mutant classes have been identified, often with multiple loci and alleles at each locus. Barley mutants have influenced not only genetic research, but also the development of commercially significant varieties. Several important cultivars that were the direct products of mutation programs have been released, notably Pallas and Golden Promise. Many current European spring barley cultivars are the indirect results of induced mutation programs due to the widespread deployment of the *sdw1* dwarfing gene. The development of a high-throughput system to assay single nucleotide polymorphisms (SNPs) in the mapped barley gene is resulting in the identification of candidate loci affecting a range of barley mutant characters. Where a known mutation is fairly frequent in a population, SNP genotyping can be combined with classification of the phenotype to identify a clearly defined region of the barley genome affecting the character. Bioinformatics approaches can then be utilized to identify syntenic regions of the rice genome and a list of potential candidate genes. We demonstrate how this approach can be deployed using either association genetics or graphical genotyping to localize candidate regions containing *vrs1* affecting the two/six-row phenotype or *rym4/5* affecting virus resistance, respectively

13.1
Brief Review of Barley Mutants

Barley has been the subject of extensive mutation studies and was an early model plant due to its relatively simple genetics. While the number and range of mutants facilitated the development of some of the first genetic maps in plant species, improvement of the crop was the main driver for much mutation work in the majority of the twentieth century. Stadler conducted pioneering work to establish that the mutation frequency in barley, and some other species, was associated with mutagen dose, and concluded that mutation for the improvement of crop plants was "much over-rated," and that induced mutation for breeding purposes was limited as useful mutants were rare and required the development of mass screening methods [1]. Nevertheless, some workers pursued the approach of mutation breeding, foremost among these being Nilsson-Ehle and Gustafsson in Sweden [2].

The use of both spontaneous and induced mutants (using physical mutagens) in barley genetics and their application in developing linkage maps was reviewed in a series of papers [3–5]. The application of chemical mutagens in barley genetics increased greatly during the 1950s, particularly in the former Soviet Union and much of this work was reviewed by Nilan [6]. Two groups acting largely independently in Scandinavia and North America generated a prodigious number of mutants of various classes. The most numerous fell into the general classes of internode length (both stem and rachis) mutants (e.g., brachytic, erectoides, and breviaristatum), surface wax (eceriferum) mutants affecting various permutations of leaves, stems, and ears, and color mutants (Figure 13.1). Another common class was sterility; however, as these mutants had little significance to plant breeding, they were largely ignored.

The erectoides mutants, defined by reduced rachis internode lengths, were isolated by various researchers in Scandinavia and were summarized by Persson and Hagberg [7]. Loci distributed across all seven barley chromosomes are described and most are represented by a series of alleles. One of the major benefits associated with reduced rachis internodes was a reduction in the length of stem internodes and thus in total plant height – a major driver in this particular area of research, because of its potential in increasing yield [7]. The brachytic class of mutants also resulted in

Figure 13.1 Four examples of common barley mutant classes: (a) brachytic, (b) erectoides, (c) breviaristatum, and (d) eceriferum.

shorter straw and thus attracted a great deal of interest; they are phenotypically quite distinct from the erectoides series as they are characterized by a relatively erect growth habit and are frequently coupled with a rounder (globose) grain shape. There is some evidence that this globose class of mutants is associated with a lack of response to exogenously applied gibberellic acid [8, 9], although other mutants also exhibit the same lack of response without the brachytic phenotype [9]. The final stature class that attracted considerable attention was the breviaristatum group of mutants, again largely the province of the Scandinavian group. This mutant class is characterized by having a reduced awn length, with some having awns less than 1.5 times the length of the ear, compared to wild-type which has awns considerably greater than 2 times ear length. Whilst this particular mutant class is not itself of great value, several were found to be associated with other valuable characteristics such as reduced height and strong chlorophyll pigmentation, similar to the phenotype of the mutant ancestor of cv. Gunilla [10].

The eceriferum mutants are also numerous and have been the subject of intense study by the Scandinavian mutation program. Like the erectoides class, loci affecting leaf waxes have been found to be distributed across the whole barley genome with most loci represented by multiple alleles. Each locus differs in the range of plant organs affected and the timing and severity of the alteration in the "waxless" or glossy appearance of the eceriferum mutants. Lundqvist and Lundqvist [11] reported 1580 known eceriferum mutants, produced from a range of mutagenic agents, located at 79 individual loci spread across the barley genome. This survey showed that different eceriferum loci had different mutabilities, which were independent of the mutagenic agent used. Interestingly, some loci differed in their sensitivities to the different agents.

In the 1970s and 1980s, the Carlsberg group subjected the flavanoid pathway to intense analysis to study the synthesis of anthocyanins and pro-anthocyanidins in barley. The pro-anthocyanidin pathway was of particular interest as pro-anthocyanidins were associated with the development of "chill-haze" in beer after storage. They adopted a mutational approach as the absence of anthocyanin pigmentation in plant tissues is relatively easily observed by the naked eye and so mutants can be readily isolated by mass screening of an M2 or M3 populations. Pro-anthocyanidins are, however, colorless compounds found in the testa of barley grain and require a chemical assay to test for their presence or absence. Despite this they were able to identify over 750 flavanoid mutants, the vast majority being pro-anthocyanidin free. These mutants have been assembled into 28 different complementation groups, each known as *Ant*, and some have been located on the barley genetic map. By conducting biochemical analyses to identify the exact stages where flavanoid biosynthesis was affected they were able to identify suitable targets for the development of pro-anthocyanidin-free malting barley lines [12]. More recently, mutation has been used to generate and isolate variants that do not produce lipoxygenase, so-called null-Lox mutants, also by researchers in the Carlsberg laboratories. Here, one null *Lox-1* mutant was produced by sodium azide mutagenesis of each of the cultivars Barke and Neruda. These mutants have important applications in brewing and are subject of a patent [13].

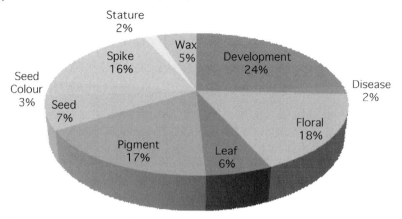

Figure 13.2 Percentages of barley mutant classes falling into 10 generalized categories.

Overall, there are over 120 different gene classes within the collection of Bowman isolines and these can be roughly grouped into 10 categories. The largest category is general plant development, followed by mutants affecting floral parts, pigment development, and spike appearance (Figure 13.2).

The targeting induced local lesions in genomes (TILLING) approach to generate a structured mutant population and then use genotypic selection to identify mutants [14] has also been applied in barley and, in fact, provides the mass screening methodology noted by Stadler [1], although one would still have to find a phenotypic association. In the United kingdom, a TILLING population comprising 23 000 M2 derived families has been developed by ethylmethane sulfonate (EMS) mutation of the popular cultivar Optic, which has been shown to carry mutations at a rate of approximately 5000 per genome [15]. Over 200 different mutant classes have been identified within the Optic TILLING population, with 18 occurring at a frequency of greater than 1%. The most frequent class was late flowering at just over 10%, followed by light green leaves and short stature at over 5%.

Subsequently, smaller TILLING populations have been developed in the North American cultivar Morex, using sodium azide [16], and the German malting cultivar Barke, using EMS [17]. One potential problem of the TILLING approach is the fact that no barley varieties can be viewed as being 100% pure. For instance, a sample from a bag of certified seed of Optic was used to construct the population described by Caldwell *et al.* [15]. Optic does not possess a resistant allele at the *Mlo* mildew resistance locus. A forward genetic screen of the mutant population, conducted to identify mildew-resistant lines, revealed that of the several mildew-resistant lines identified, most had a different haplotype to that of Optic at *Mlo* and these were therefore most likely outcrossed contaminants. When considering the numbers of seeds exposed to mutagenic treatment and the permitted levels of "off-types" in certified seed, it is not surprising that this level of contamination was detected. Clearly, to generate a TILLING population it is essential that carefully multiplied doubled haploid seed from a single plant be used as the starting material to avoid such haplotype contamination. To complement the EMS mutant population of Optic, we also developed a fast neutron population, designed to generate deletions of the order of several hundred bases. The "deletogen approach" circumvents this haplotype contamination issue as identified mutants will have a completely novel "deletion" haplotype and maximizes the chances of identifying genuine mutants.

13.2
Applications in Breeding

Barley mutants have impacted breeding programs to various degrees. Short-straw mutants have been incorporated into varieties released either directly from selections from mutant populations or by using a mutant as a crossing parent [10]. The Pallas cultivar, an erectoides mutant *ert-k^{32}* derived from cv. Bonus by the Swedish Seed Association in Svalov, was the first mutant barley released and was cultivated extensively in Scandinavia and the United Kingdom [10]. Similarly, cv. Golden Promise, a short-strawed mutant derived directly from the variety Maythorpe, dominated the United Kingdom spring barley market in the 1970s and early 1980s. The dwarfing gene in Golden Promise was mapped to barley chromosome 5H [18] and found to be an allele of the *ari-e* locus [9]. A sister mutant to Golden Promise that had the same phenotype [9] gave rise to cv. Midas was also popular in the United Kingdom in the 1970s. The dwarfing gene found in Midas was transmitted to two other notable derivatives – Goldmarker and Tyne. However, the gene is associated with reduced grain size [19] and an increase in the sieve size over which grain is traded in the United Kingdom effectively ceased development of this type, although some Tyne is still grown in some areas of the United Kingdom because of its early maturity.

Whilst the cultivars mentioned above have all been successful, mutation of the Czech cultivar Valticky to produce Diamant has arguably had the greatest impact on barley breeding in some major barley-producing areas of the world. Diamant was subsequently used in a crossing program to produce the spring barley cultivar Trumpf (reselected and released as Triumph in the United Kingdom) that combined high yield with good malting quality. Triumph features in the ancestry of at least 18 of the 20 cultivars currently on the United Kingdom recommended list (www.hgca.com) and the mutant character that has been transmitted from Diamant to all these lines is the *sdw1* dwarfing gene on barley chromosome 3H [20]. Figure 13.3 shows the percentage of United Kingdom spring barley certified seed sales that can be attributed to Triumph and its derivative *sdw1*-containing cultivars. Seed sales give an approximation of the proportion of the United Kingdom spring barley crop grown in the subsequent season that carry *sdw1*. From this it can be seen that the proportion increased from around 40% in 1983, largely due to the popularity of Triumph, to over 90% currently due to the widespread deployment of the *sdw1* dwarfing gene. This widespread deployment of *sdw1* can also be seen in other spring barley regions of North-Western Europe.

Another mutant class that has been extensively studied for commercial development is the *Mlo* mildew resistance locus. Many resistant mutants at this locus have

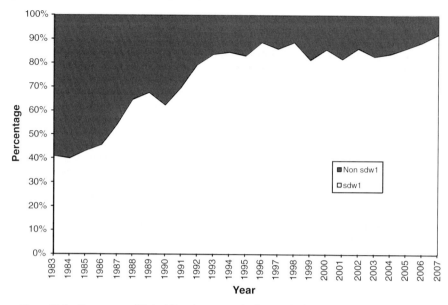

Figure 13.3 Percentage of United Kingdom spring barley certified seed production that carries the *sdw1* dwarfing gene derived from cv. Diamant.

been generated (summarized in [21]), but only the *mlo9* mutant induced in Diamant has had any great impact. It was carried by the successful cultivar Alexis, which dominated the West European malting market in the early 1990s and is also found in another successful German cultivar, Barke. Compared to the spontaneous *mlo11* mutant, found in donors such as L92, L100, and Grannenlose Zweizeilige, it has, however, had a relatively minor impact. More information on the *mlo* mutants can be found at http://www.crpmb.org/mlo.

The *Mlo* mildew resistance locus was the first disease resistance locus to be cloned in barley [22] and many of the various induced mutations producing resistance alleles have been characterized at the sequence level. Surprisingly, the coding sequence of *mlo11* does not differ from the wild-type sequence, but this naturally occurring mutant is characterized by a tandem repeat encompassing the 5'-coding and noncoding regions of the wild-type *Mlo* sequence. There are also minor differences in the sequences surrounding the *mlo11* allele in different sources, such as L92 and L100 [23]. Mutants have also played a large part in characterizing the *Mlo* mildew resistance mechanism. The *mlo5* resistance, which itself had been produced by a mutation program, had been backcrossed into the susceptible cultivar Ingrid to provide an isogenic resistant line. The isogenic Ingrid *mlo5* line was then mutated and screened to identify susceptible lines, from which two *Ror* gene loci (required for Mlo resistance) were identified [24]. This finding was a key step in characterizing the molecular mechanisms involved in host and nonhost resistance to pathogen attack [25].

13.3
Single Nucleotide Polymorphism Genotyping to Identify Candidate Genes for Mutants

13.3.1
Resources

Whilst many barley morphological mutants have been placed upon the barley genetic map, integration of these map locations with molecular marker maps is far from complete with the best summary provided by the Steptoe × Morex bin map (http://barleygenomics.wsu.edu/all-chr.pdf). The advent of a high-throughput single nucleotide polymorphism (SNP) genotyping platform using Illumina technology [26] promises not only to dramatically improve the integration of morphological mutants into molecular marker maps, but also facilitate the identification of the mutated gene loci.

Equally important is the fact that appropriate biological resources are available to progress this work and there are several key collections that are being deployed. The first is the so-called Bowman mutant collection, where Professor J. Franckowiak collected just under 1000 barley morphological mutants and began a backcrossing program to introgress them into a common recipient, the North American two-row cultivar Bowman. Whilst this was easier for some mutants than others, all range from the BC2 to BC10 generation, with the majority greater than BC4. Thus we can expect that, on average, less than 1.6% of the donor genome remains in these nearly isogenic lines. Whilst it is likely that there will be considerable "linkage drag" around the introgressed character, it does mean that these lines present a unique resource with which to localize many major mutant characteristics to specific regions of the barley genome. In a European Research Area in Plant Genomics (ERA-PG) project called "BARCODE," over 900 of the Bowman isolines have been genotyped with the first Illumina Barley Oligo Pooled Array (BOPA1, 1536 SNPs) providing a detailed genome wide survey of the amount of donor genome remaining in the Bowman isolines (A. Druka, N. Stein, and M. Morgante, personal communication).

The barley genetic stocks held by a number of the major barley gene banks around the world (e.g., the Nordic Gene Bank) contain many of the barley mutants that have been reported over the years. Many of these mutants are alleles at specific loci and thus form allelic series that can be used to detect the effects of specific sequence variants on plant phenotype. In addition, the recently established TILLING populations provide further sources of potential allelic variation at specific loci. However, most of the observed phenotypic variants have yet to undergo allelism testing. For instance, several thousand phenotypic variants are described in the DISTILLING database (http://germinate.scri.ac.uk/cgi-bin/mutantsdatabase/home.pl), but this vast number means that allelism testing of all is a considerable undertaking. Comparison of the phenotypes described in the DISTILLING database with the characteristics listed in the Barley Genetic Stocks Database (ace.untamo.net), which contains a wealth of information on most of the current known barley mutants, means that new alleles can potentially be sought in enriched subpopulations that exhibit a phenotype of interest.

13.3.2
Case Study: Two/Six-Row Locus in Barley

A mutation at the *Vrs1* locus on chromosome 2H controls the fertility of the lateral florets on the barley spikelets. Agronomically, this mutation can be commercially significant in some regions of the world as it can improve the number of grains per barley spike and thus confer a yield advantage in conditions where tillering ability is limiting (Figure 13.4). The Bowman isoline for *vrs1* is BC7 and its BOPA1 genotype showed polymorphism at six loci spanning a map distance of 5 cM on barley chromosome 2H. Further evidence for the localization of *vrs1* comes from an association genetics scan of 190 barley cultivars that represented both two- and six-row phenotypes. This germplasm set has been genotyped with 4600 SNP markers, most which have been located on a consensus linkage map. Of these, 1445 had a minor allele frequency of greater than 10% in this germplasm set. Consistent with the analysis of the Bowman isolines, association analysis located *vrs1* to a tightly defined region on barley chromosome 2H (Figure 13.5). A second significant peak was located on barley chromosome 4H, presumably reflecting the fact that six-row cultivars possess a different allele at the *int-c* locus (L. Ramsay, personal communication).

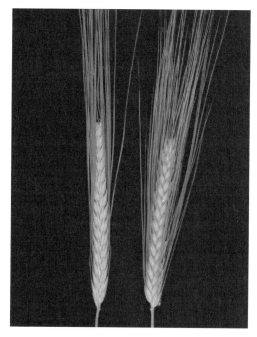

Figure 13.4 Six-row mutant of Optic (right) compared to two-row wild-type (left).

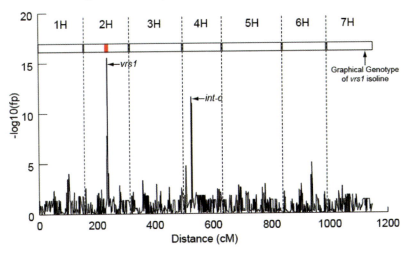

Figure 13.5 Genome scan for significant SNP marker associations with two/six-row phenotype after accounting for kinship shown with graphical genotype of Bowman isoline for *vrs1* where open and red rectangles denote absence and presence, respectively, of polymorphism with the recurrent parent.

All the SNPs on the BOPA platform were derived from publicly available expressed sequence tag (EST) sequences, which can be easily used by standard BLAST analysis to search for orthologous sequences on the publicly available rice genome sequence. The majority of the chromosome 2H markers are represented by rice chromosomes 4 and 7, with the latter inverted and inserted at the 2H "genetic centromere" (Figure 13.6). Figure 13.6 shows that relative gene order is largely conserved between these two rice chromosomes and barley chromosome 2H. However, Figure 13.6 also shows that many of the genes on rice chromosome 7 are compressed into a small region of the barley genetic map at around 60 cM. This reflects the lack of recombination in the centromeric regions of barley chromosomes. The implication of this is that many rice genes are located in regions exhibiting conserved synteny with barley centromeres. Whilst this makes strategies for gene isolation based on recombination (e.g., positional cloning) difficult in these regions, it is reassuring that the majority are located in the recombinationally active telomeric ends, facilitating candidate gene identification. Thus, we can rapidly explore synteny between the genetic location of barley gene-based SNPs and the orthologous regions in the rice genome sequence, and make use of rice gene annotations to generate a list of potential barley candidate genes.

From the association genetics study mentioned above, the SNP with the most significant association with the two/six-row locus was located at 86 cM on the barley genetic map, and this same SNP is also located in the polymorphic segment of the Bowman isoline. The *vrs1* locus has been cloned and sequenced, and found to be a homeodomain–leucine zipper I–class homeobox gene [27]. The gene containing the most significant SNP corresponds to LOC_Os04g45490 on rice chromosome 4. However, the rice homolog of the barley homeodomain–leucine zipper I–class

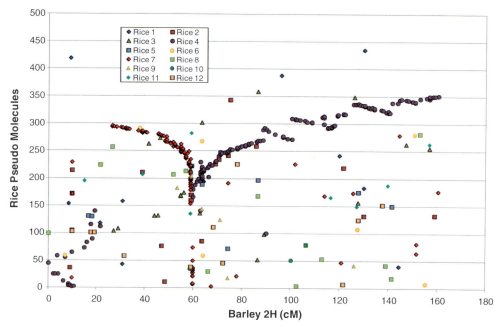

Figure 13.6 Comparison of the linear order of barley genes from chromosome 2H with the rice genome.

homeobox gene is located on rice chromosome 7. In Figure 13.6, it can be seen that barley chromosome 2H largely corresponds to rice chromosome 4 at genetic distances greater than 60 cM. A detailed survey of the genomic region around *vrs1* [28] revealed a breakdown in microsynteny in the region with an insertion of two genes from rice chromosome 7, including *vrs1*, after LOC_Os04g45600. Orthologs of genes from rice chromosome 4 that flanked this locus and that were also represented on BOPA1 exhibited conserved synteny from LOC_Os04g45820, at least 18 rice genes distal to *vrs1*. Whilst this particular example did not directly lead to the identification of the "right" gene, it demonstrates that the use of currently available biological resources, coupled with gene-based genetic mapping technologies and model genome sequences, provide a rapid means of tightly defining a genomic region likely to contain a candidate gene.

This approach will not lead directly to candidate gene identification if conserved synteny has broken down in the region of interest. In addition, as the number of genes located in the genetic centromeres of barley is high (reflected as a compression of many genes into a small genetic distance on the barley genetic map, e.g., Figure 13.6) determining gene order in these regions is difficult. This compression can be seen in all barley chromosomes and is the result of limited recombination in the centromeric regions. Consequently, this approach is less likely to be successful for genes located in barley centromeres.

Association mapping is only appropriate when the locus under consideration is relatively frequent in the gene pool under study and not associated with any population substructure issues. Association genetics approaches are normally deployed when the minor allele frequency is greater or equal to 10%. There are a number of major morphological (e.g., six/two-row case above) and developmental (e.g., photoperiod insensitivity) genes which are sufficiently frequent in germplasm collections for the association mapping approach to be deployed.

13.3.3
Case Study: Graphical Genotyping of a Disease Resistance Locus

With a suitable graphical genotyping display, it may not be necessary to conduct a full association analysis to identify markers linked to mutant characters. Resistance to the barley yellow mosaic virus (BaYMV) complex is conferred by allelic variants at a distal locus on the long arm of barley chromosome 3H. This locus has been cloned, sequenced, and identified as an eIF4E translation initiation factor [29]. By assembling a test panel of genotypes with known resistant and susceptible reactions to the complex, graphical genotyping tools (e.g., FLAPJACK (www.bioinf.scri.ac.uk)) can be deployed to sort the test panel according to the resistance locus, and then examine the surrounding graphical genotypes to confirm the association between marker and trait. Figure 13.7 illustrates two distal SNP loci on the long arm of chromosome 3H that show close association with resistance and susceptibility phenotypes. The most highly associated SNP (11_10767) wrongly identifies three lines as susceptible when they are known to be resistant. One of the unknown lines was predicted by this marker to be resistant and although its BaYMV phenotype is unknown, one of its parents carried a resistant allele. The barley EST from which SNP 11_10767 was derived has highest homology with LOC_Os01g73940, which is located distally on rice chromosome 1, which exhibits conserved synteny with barley 3H. This locus is 7 genes from LOC_Os01g73880, which encodes a translation initiation factor and shows 93.2% similarity to EMBL:AY661558, the sequence of the *Hordeum vulgare* eIF4E that corresponds to the BaYMV resistance gene on barley chromosome 3H. The other SNP that is less well associated with the resistance locus corresponds to LOC_Os01g73690, which is located proximal to LOC_Os01g73880. This association is, however, only apparent in winter barley germplasm as both alleles of the SNP are frequent in spring barley germplasm, despite the fact that none of the spring lines carry any resistance genes to the complex. Nevertheless, provided the analysis is restricted to the relevant germplasm group, the detailed genotypes revealed by marker analysis provide a rapid means of localizing genomic regions likely to contain candidate genes for major genetic loci.

13.3.4
General Protocol for using High-Throughput Genotyping to Localize Mutants

Most mutants are not, however, present in sufficient numbers in germplasm collections to permit their localization to specific genomic regions, whether by a full association study or by visual assessment of a test panel. In such cases, the use of high-throughput genotyping technologies such as that provided by BOPA1 and

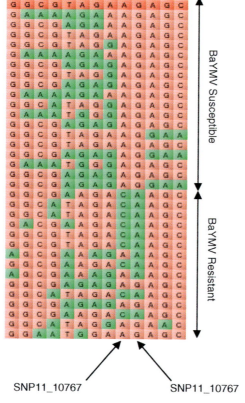

Figure 13.7 Graphical genotypes of the distal portion of barley chromosome 3HL for a range of barley genotypes with identified SNP polymorphisms showing association with a resistance locus for the BaYMV complex. Boxes in red have the same genotype as cv. Flagon and boxes in green have the alternative allele.

biparental mapping populations can quickly identify the interval in which the mutant is located. In the ERA-PG BARCODE project we have generated over 300 biparental populations segregating for mutant genes to facilitate positional isolation of the affected loci. As the developing barley physical map matures, is anchored to the genetic map, and the associated draft genome sequence emerges, the once daunting challenge of facile positional cloning in barley will become more of a reality. While it is highly unlikely that causal mutations will be represented by the SNPs on high-throughput genotyping platforms, information specifying the chromosomal position of mutant loci can be rapidly obtained by Bulk Segregant Analysis [30] prior to embarking on a standard positional cloning procedure.

In summary, high-throughput genotyping coupled with appropriate genetic resources and bioinformatics is revolutionizing the localization of potential candidate genes and can be utilized to identify mutant loci. So far, genotyping with

Illumina SNPs has refined the interval locating the *ari-e* and *sdw1* dwarfing gene, and bioinformatics has identified some likely candidates, but these important genes have yet to be cloned.

Acknowledgments

We thank Arnis Druka for producing the information on the Bowman *vrs1* isoline and the photographs in Figure 13.1, Luke Ramsay and Jordi Comadran for producing the information for the association genetics scan for two/six-row habit, and Iain Milne and David Marshall for providing the FLAPJACK graphical genotyping tool. We also thank colleagues from the Association Genetics of UK Elite Barley (AGOUEB), US Barley Coordinated Agricultural Project (CAP) and ERA-PG BARCODE projects for use of some data. The Scottish Crop Research Institute receives grant-in-aid support from the Scottish Government Rural and Environment Research and Analysis Directorate.

References

1. Stadler, L.J. (1930) Some genetic effects of X-rays in plants. *J. Hered.*, **21**, 2–19.
2. Gustafsson, A. (1986) Mutation and gene recombination principal tools in plant breeding, in *Research and Results in Plant Breeding* (ed. G. Olsson), Svalof, Stockholm.
3. Smith, L. (1951) Cytology and genetics of barley. *Bot. Rev.*, **17**, 1–51.
4. Smith, L. (1951) Genetics and cytology of barley. *Bot. Rev.*, **17**, 133–202.
5. Smith, L. (1951) Genetics and cytology of barley. *Bot. Rev.*, **17**, 285–355.
6. Nilan, R.A. (1964) *The Cytology and Genetics of Barley, 1951–1962*, Washington State University Research Studies Monographic Supplement 3, Washington State University, Washington, DC.
7. Persson, G. and Hagberg, A. (1969) Induced variation in a quantitative character in barley. Morphology and cytogenetics of erectoides mutants. *Hereditas*, **61**, 115–178.
8. Boulger, M., Sears, R., and Kronstad, W. (1981) An investigation of the association between dwarfing sources and gibberellic acid response in barley. Proceedings of the Fourth International Barley Genetics Symposium, Edinburgh.
9. Franckowiak, J. and Pecio, A. (1991) Coordinator's report: semidwarf genes: a listing of genetic stocks. *Barley Genet. Newslett.*, **21**, 116–127.
10. Gustafsson, A., Hagberg, A., Persson, G., and Wiklund, K. (1971) Induced mutations and barley improvement. *Theor. Appl. Genet.*, **41**, 239–248.
11. Lundqvist, U. and Lundqvist, A. (1988) Mutagen specificity in barley for 1580 eceriferum mutants localized to 79 loci. *Hereditas*, **108**, 1–12.
12. Jende-Strid, B. (1993) Genetic-control of flavonoid biosynthesis in barley. *Hereditas*, **119**, 187–204.
13. Breddam, K., Olsen, O., Skadhauge, B., Lok, F., Knudsen, S., and Bech Lene, M. (2009) Barley for production of flavor-stable beverage. Patent PCT/DK05/00160.
14. McCallum, C.M., Comai, L., Greene, E.A., and Henikoff, S. (2000) Targeting induced local lesions in genomes (TILLING) for plant functional genomics. *Plant Physiol.*, **123**, 439–442.
15. Caldwell, D.G., McCallum, N., Shaw, P., Muehlbauer, G.J., Marshall, D.F., and Waugh, R. (2004) A structured mutant population for forward and reverse genetics in barley (*Hordeum vulgare* L.). *Plant J.*, **40**, 143–150.
16. Talame, V., Bovina, R., Sanguineti, M.C., Tuberosa, R., Lundqvist, U., and Salvi, S. (2008) TILLMore, a resource for the

discovery of chemically induced mutants in barley. *Plant Biotechnol. J.*, **6**, 477–485.

17 Gottwald, S., Bauer, P., Altschmeid, L., and Stein, N. (2007) A TILLING population for functional genomics in barley cv. "Barke". Plant and Animal Genomes XV Conference, San Diego, CA.

18 Thomas, W.T.B., Powell, W., and Wood, W. (1984) The chromosomal location of the dwarfing gene present in the spring barley variety Golden Promise. *Heredity*, **53**, 177–183.

19 Powell, W., Thomas, W.T.B., Caligari, P.D.S., and Jinks, J.L. (1985) The effects of major genes on quantitatively varying characters in barley. 1. The *GP-ert* locus. *Heredity*, **54**, 343–348.

20 Barua, U.M., Chalmers, K.J., Thomas, W.T.B., Hackett, C.A., Lea, V., Jack, P., Forster, B.P., Waugh, R., and Powell, W. (1993) Molecular mapping of genes determining height, time to heading, and growth habit in barley (*Hordeum vulgare*). *Genome*, **36**, 1080–1087.

21 Jørgensen, J.H. (1992) Discovery, characterization and exploitation of Mlo powdery mildew resistance in barley. *Euphytica*, **63**, 141–152.

22 Buschges, R., Hollricher, K., Panstruga, R., Simons, G., Wolter, M., Frijters, A., vanDaelen, R., vanderLee, T., Diergaarde, P., Groenendijk, J., Topsch, S., Vos, P., Salamini, F., and Schulze-Lefert, P. (1997) The barley *mlo* gene: a novel control element of plant pathogen resistance. *Cell*, **88**, 695–705.

23 Piffanelli, P., Ramsay, L., Waugh, R., Benabdelmouna, A., D'Hont, A., Hollricher, K., Jorgensen, J.H., Schulze-Lefert, P., and Panstruga, R. (2004) A barley cultivation-associated polymorphism conveys resistance to powdery mildew. *Nature*, **430**, 887–891.

24 Freialdenhoven, A., Peterhansel, C., Kurth, J., Kreuzaler, F., and Schulze-Lefert, P. (1996) Identification of genes required for the function of non-race-specific Mlo resistance to powdery mildew in barley. *Plant Cell*, **8**, 5–14.

25 Collins, N.C., Thordal-Christensen, H., Lipka, V., Bau, S., Kombrink, E., Qiu, J.L., Huckelhoven, R., Stein, M., Freialdenhoven, A., Somerville, S.C., and Schulze-Lefert, P. (2003) SNARE-protein-mediated disease resistance at the plant cell wall. *Nature*, **425**, 973–977.

26 Rostoks, N., Ramsay, L., MacKenzie, K., Cardle, L., Bhat, P.R., Roose, M.L., Svensson, J.T., Stein, N., Varshney, R.K., Marshall, D.F., Grainer, A., Close, T.J., and Waugh, R. (2006) Recent history of artificial outcrossing facilitates whole-genome association mapping in elite inbred crop varieties. *Proc. Natl. Acad. Sci. USA*, **103**, 18656–18661.

27 Komatsuda, T., Pourkheirandish, M., He, C.F., Azhaguvel, P., Kanamori, H., Perovic, D., Stein, N., Graner, A., Wicker, T., Tagiri, A., Lundqvist, U., Fujimura, T., Matsuoka, M., Matsumoto, T., and Yano, M. (2007) Six-rowed barley originated from a mutation in a homeodomain–leucine zipper I–class homeobox gene. *Proc. Natl. Acad. Sci. USA*, **104**, 1424–1429.

28 Pourkheirandish, M., Wicker, T., Stein, N., Fujimura, T., and Komatsuda, T. (2007) Analysis of the barley chromosome 2 region containing the six-rowed spike gene *vrs1* reveals a breakdown of rice–barley micro collinearity by a transposition. *Theor. Appl. Genet.*, **114**, 1357–1365.

29 Stein, N., Perovic, D., Kumlehn, J., Pellio, B., Stracke, S., Streng, S., Ordon, F., and Graner, A. (2005) The eukaryotic translation initiation factor 4E confers multiallelic recessive Bymovirus resistance in *Hordeum vulgare* (L.). *Plant J.*, **42**, 912–922.

30 Michelmore, R.W., Paran, I., and Kesseli, R.V. (1991) Identification of markers linked to disease-resistance genes by bulked segregant analysis – a rapid method to detect markers in specific genomic regions by using segregating populations. *Proc. Natl. Acad. Sci. USA*, **88**, 9828–9832.

14
Association Mapping for the Exploration of Genetic Diversity and Identification of Useful Loci for Plant Breeding

André Beló and Stanley D. Luck

Abstract

Genetic diversity is essential for the continuous progress of plant breeding. The use of molecular markers and genetic mapping allowed the dissection of quantitative trait loci (QTLs) and use of new alleles in plant breeding by marker-assisted selection. Association mapping (or linkage disequilibrium mapping) is a recent method to detect associations between markers and phenotypes, capable of sampling more genetic diversity than classical QTL mapping. It provides high resolution in the identification of genes or genome regions responsible for phenotypes and can have great impact in the use of genetic diversity in plant breeding. In this chapter, we discuss the methodology of association mapping in plants, focusing mainly on genome-wide association scans and population statistical methods to identify associations. The main steps necessary to conduct association mapping, their comparison to QTL mapping in biparental populations, analysis of limitations, and perspectives provide an overview for its use in plant genetics and breeding.

14.1
Introduction

Domestication and plant breeding affect the genetic diversity of crops mostly by reducing the number of alleles in specific genomic regions [1]. While this diversity is essential for crop improvement [2], most of the germplasm diversity is not used in breeding [3]. Repositories of seeds and germplasm banks are important sources of natural occurring alleles, and the development of methods to take advantage of the newly introduced diverse alleles have to be developed [2]. Continuously developing molecular techniques allow the characterization of genetic and genomic variation that can consequently be used in plant breeding. For example, molecular markers have been successfully used as a tool to evaluate plant genetic diversity [3], map quantitative trait loci (QTLs), discover genes responsible for phenotypes [4], and track alleles in breeding programs. Genetic dissection of complex traits [5–7] and identification of genes responsible for phenotypes using different approaches [5] has become a reality. Genomic tools such as physical maps and/or whole-genome sequencing allowed positional cloning in several plant species (reviewed in [7–9]). Although positional cloning has been the most common method used to identify genes controlling QTLs in plants [7], association studies have recently become popular in human genetics [10, 11]. The ability to create experimental mapping populations probably contributed to the popularity of QTL mapping in plants. This is not possible in humans and, therefore, the development of alternative methods such as association mapping took place [10].

Association mapping is the analysis of correlation between phenotypes and DNA polymorphisms in groups of individuals. Its genetic principle is the presence of linkage disequilibrium (LD) between molecular markers and genes responsible for phenotypes. LD is the nonrandom association of alleles at different loci (reviewed in [12]). This exception of Mendel's second law of independent segregation allow us to use molecular markers to detect correlations between plant genotypes and phenotypes similarly to QTL mapping. Two advantages of association mapping are the higher mapping resolution and the ability to sample multiple alleles in the same experiment [13]. Higher resolution is obtained because LD across the genomic regions of individuals used in association mapping decays more rapidly than LD in classical biparental mapping populations, such as F2 or recombinant inbred lines (for a review of LD, saturation with molecular markers, and resolution for mapping, see [14]). This results from multiple historical recombination events that occurred in many previous generations of the individuals being currently studied. The ability to sample more alleles is a consequence of genotyping individuals of natural populations, germplasm collections, or sets of elite lines that certainly have greater numbers of alleles per locus than biparental segregant populations.

Most of the association mapping studies can be grouped in four different types [15]:

- Candidate polymorphism, which tests the association of a particular polymorphism with phenotypes.
- Candidate gene, which tests the association of several markers within a candidate gene and flanking regions with phenotypes.
- Fine mapping, which tests the association of markers spanning a region of several centiMorgans, usually identified previously by classical QTL mapping or other methods.
- Genome-wide association (GWA) mapping, which tests genotypes of markers at many locations along the entire genome for association with phenotypes.

Many human disease genes have been identified by association mapping studies, especially GWA, in recent years [10, 11]. These studies proved that GWA works and revealed important aspects related to (i) the power and resolution to detect associations, and (ii) the effect size of the identified loci [10]. Common features of these studies are the extensive use of single nucleotide polymorphism (SNP) molecular markers and large sample sizes, which are both required for adequate genome resolution and statistical power to detect associations. Based on these previous results, it is likely that this methodology will become more common, growing in both number of markers and population size [16].

In plants, the use of association mapping, in particular GWA, is a field under development. Many details of the technology have to be optimized in order to allow its use in plant genetics and breeding. In this chapter, we discuss the association mapping methodology in plants, emphasizing GWA. However, many of the concepts are also applicable to the other types of association mapping. Some limitations, comparison to classical QTL mapping, and perspectives are presented to exemplify the methods described.

14.2
Methods and Protocols

14.2.1
Population for Association Mapping

Association mapping in plants can be performed using natural populations or germplasm collections without requiring the creation of experimental mapping populations. This advantage leads to quicker identification of marker × phenotype associations compared to classical QTL mapping [13, 17]. The number of individuals, their population structure, relatedness, level of diversity, and reproductive habits affect association mapping results. Increasing the number of individuals will always provide higher power for the statistical detection of associations. Selecting individuals that maximize the genetic diversity increase the chances of having more distinct alleles per locus. A penalty of having too diverse germplasm might be the presence of many rare alleles that will result in low statistical power or require an extreme large population size to detect an association. Pre-existing characterized germplasm, phylogenetic, genotypic, or phenotypic assessments can help to define the most suitable germplasm for association mapping projects. The presence of subpopulations with different allelic frequencies or related samples can generate spurious associations (false-positives) due to nonrandom deviations in their allelic frequencies. Therefore, adequate analyses have to be performed for structured populations (see Section 14.2.4). Self-pollinating species are amenable to more straightforward statistical treatment since no heterozygous individuals are present. When outcrossing species are studied, the use of available inbred lines is an option, although proper statistical procedures can account for the allelic state comparisons in populations with heterozygous individuals.

Populations of some plant species have been created for association mapping. In *Arabidopsis*, 96 ecotypes had 876 short genomic regions sequenced [18] that were used to study LD and perform association mapping [18–20]. In maize, 302 inbred lines have been developed, phenotyped, and genotyped to correct for population structure [21]. Recently, a large collection of maize inbred lines has been developed and is available for the scientific community [22]. In sorghum, 377 inbred line accessions have been characterized for association mapping studies [23]. These panels of germplasm can save time in association studies and allow the accumulation of genetic information from independent studies carried out by the scientific community. For example, an immortalized panel of individuals, such as recombinant inbred lines or double haplotypes, genotyped in advance at high marker density can subsequently be tested for any measurable phenotype. New panels of individuals will only be needed in case the current panels do not have enough variability for the trait of interest. Even in this case it might be possible to use the available panels by adding more individuals that incorporate new alleles. Thus, association mapping germplasm collections can be extended by adding more individuals. Whether the addition of new individuals will be possible and its difficulty will vary depending on the initial choices of germplasm, molecular markers, and project goals. Extending existing biparental populations in a similar manner would usually be much more laborious.

14.2.2
Genotyping

Two important considerations for genotyping are the choice of the molecular markers to be used and the marker density needed for sufficient genomic coverage. Although several types of molecular markers can be used for association mapping, SNPs are preferred. They have codominant inheritance and are the most frequent in the genome; therefore, they provide the most detailed view of genetic variation and the highest resolution [14, 24] (for a review of available molecular markers and SNP genotyping methods, see [25] and [26], respectively). SNPs are usually biallelic markers; however, several SNPs located in the same small region (e.g., a single gene) can be used to define an allele (haplotype) [27]. In principle, n biallelic SNPs could define 2^n haplotypes. In practice, it is beneficial to analyze the data using both haplotypes of SNPs and individual SNPs.

The number of markers to be used is largely dependent on the pattern of LD [14] in the species to be analyzed [12, 13, 17, 28], the diversity of the individuals selected for the association study [29, 30], and LD of the studied loci [31, 32]. It is important to note that LD is not evenly distributed along the genome. The presence of recombination hotspots creates a haplotype block structure corresponding to regions of high LD separated by regions of low LD [12, 33]. Unfortunately, the haplotype structure is very irregular and cannot be predicted *a priori*; therefore, the number of markers necessary to provide adequate genome coverage has to be estimated based on an average value of LD. In candidate-gene association mapping projects, available simple sequence repeat markers have been commonly used to evaluate population structuring [21, 23, 34–36].

14.2.3
Phenotyping

Phenotyping is a critical part of association mapping studies. As we will see in Section 14.2.4, the statistical power to detect an association between a molecular marker and a phenotype depends mainly on the extent of the phenotypic effect and allelic frequency. Therefore, precise and accurate phenotyping will always provide better association results. If the phenotype incorporates large experimental variation, then the real associations might be missed (false-negatives). In most cases, the phenotypic variation is inherent to the trait (e.g., yield and other quantitative traits usually have larger variation and environmental interaction). For quantitative traits, proper experimental design [37] and phenotype measurement have to be used to minimize experimental error.

14.2.4
Statistical Procedures

Approaches to association mapping can be roughly divided into two types depending on the relatedness in the germplasm of interest. Family or pedigree-based approaches are commonly used for studies of human diseases where data for extended families can be obtained [38]. In plant genetics where both diverse natural populations and large managed breeding populations are available, mixed-model population-based approaches have become popular [39, 40].

The following is a brief review of the statistical basis of association tests and the complications that arise in association mapping. The simplest case is obtained when there is a biallelic marker and the genotypes are from a collection of homozygous inbred lines. The inbreds consequently form two groups. For a categorical trait the frequencies of the joint occurrences of genotype and phenotype constitute a 2×2 contingency table. A similarity in the proportions across columns in the table would indicate independence of genotype and phenotype. The statistical significance of the difference in proportions can be estimated with a Pearson's χ^2-test. For a quantitative trait the phenotype data for the two groups can be regarded as constituting two distributions. If the distributions are normal, Student's t-test can be used to determine the significance of the difference in the means [41]. There is an increase in dimensionality for multiallelic markers and heterozygous individuals because there are many more genotypes and the possibility of genetic dominance effects. In such situations the analysis of variance (ANOVA) method can be used to simultaneously estimate genotypic effects. However, the null hypothesis of the equality of means in ANOVA does not have a straightforward interpretation and should not be used for inferring the significance of effects – this is a problem for many multivariate hypotheses. Instead, it is necessary to break down the hypothesis "into smaller, more easily describable pieces" [42]. The resulting contrasts should correspond to relevant genetic effects. A genotype can be contrasted with each of the other genotypes, or various combinations of the other genotypes, which results in several comparisons for a multiallelic marker. The latter can arise in association tests for haplotypes which can be more powerful than single-marker methods in detecting variants with low frequency [43]. Thus, in association mapping, the multitude of possible genetic effects for a single marker as well as for interactions between markers will give rise to many contrasts for each marker [44].

The *t*-test and linear statistical models such as ANOVA rely on the normality of null distributions in order to estimate significance. However, significant deviation from normality is common in association mapping. Phenotype distributions are often bounded on one side with a long tail on the other. The latter can arise from either experimental or genetic effects that can be of heterogeneous origin. In this case it is necessary to apply a deconvolution and deal with the long-tailed phenomena separately. The focus in association mapping is on the central part of the distribution, and the effects for common alleles. Thus, nonparametric methods for association tests such the Kolmogorov–Smirnov test or the Mann–Whitney test might be preferred because of the robustness with respect to non-normality. These methods are able to detect shifts in the distributions as well as changes in the shape of the distribution. Consequently, while these tests are useful for detecting significant associations, it is necessary to examine the empirical distribution to confirm the nature of the effect. Outliers in the data should also be ignored. Statistical methods such as the Grubbs test can be used to identity outliers [45].

The straightforward application of association tests can result in false-positives due to the confounding effect of population stratification [46]. A simple demonstration of this effect is obtained by considering a 2×2 contingency table for a case-control study as described above. Stratification refers to condition wherein the sample is admixed, comprising individuals from different genetic backgrounds that could have arisen because of different geographic origin or heterotic groups. If allele frequencies for the marker and the frequencies of cases and controls are both different between the genetic backgrounds then the hypothesis test will be biased [46]. In a GWA scan this bias will give rise to a distribution of false associations, many lower and fewer highly significant effects. These associations will be widely distributed across the genome serving as diagnostic indicator of the presence of population stratification. In addition, the degree of genetic differentiation within a population can be estimated using Wright's F_{ST} statistic [47, 48], which can serve as another indicator of population structure.

Population-based association mapping methods are expected to have more power compared to transmission-disequilibrium tests and family-based tests [16]. This has led to the development of various methods that circumvent the confounding effects from population structure [15, 40]. The genomic control method uses a random set of markers to estimate an overall inflation factor for controlling the false-positive rate [49]. In structured populations the inflation factor would still be ineffective because the highly significant false associations would not be removed. An alternative approach is to use clustering methods to uncover hidden population structure and adapt the association tests appropriately. The approach implemented in the program STRUCTURE uses randomly chosen markers and model-based clustering to estimate admixture coefficients for membership of individuals in subpopulations; the coefficients correspond to a vector space representation of population structure [50]. The number of subpopulations is specified as an input parameter, K, ranging typically from 2 to 8 depending on the population size. Calculations are run with Markov chain Monte Carlo (MCMC) parameters of 100 000 each for burn-in and the number of iterations. The calculations for a dataset of 200 SNP markers and 1500 individuals divided in eight subpopulations take about 24 h on a Linux server with a 2.93-GHz Intel Xeon X7350 CPU. A likelihood ratio is reported at the end of each calculation. The likelihood increases monotonically with K, reflecting the corresponding increase in the degrees of freedom, and does not provide an effective way for choosing an optimal value – clustering methodology is inherently inexact [51]. Instead, biological considerations and Occam's razor should be applied. The admixture coefficients for a sequence of calculations for a range of K can be displayed using software such as DISTRUCT [52]. As K is increased, a sequential bifurcating pattern in the coefficients would provide an indication of the optimal choice of the number of populations. Fewer subpopulations are preferred because the objective is to find quantitative genetic effects that apply generally. Fine distinctions obtained at larger K might not be relevant, and reduce sample size and power unnecessarily. Additional information on the population structure can be obtained by hierarchical clustering using the marker data to produce a phylogenetic tree. The tree provides a visual display of genetic distance and branching between subpopulations. Population structure can also be analyzed using principal component analysis (PCA) without the need to identify subpopulations [20, 53]. The principal components correspond to axes of genetic variance. Computationally, PCA is much less demanding than the MCMC calculation and data for all markers can be incorporated.

In the structured association approaches to association mapping, the population structure coefficients from either STRUCTURE or PCA are used as covariates in statistical models that relate genotype to phenotype, providing the basis for estimating the significance of an association. For categorical traits, the STRAT method models the distribution of genotypes as a function of admixture, allele frequency, and phenotype, using expectation-maximization to estimate a likelihood ratio [50]. For quantitative traits the mixed-model method in TASSEL expresses phenotype as a linear function of admixture coefficients to estimate population effects [39]. Additional genetic contributions to the variance due to relatedness between individuals are included in the random effects component for kinship. Procedures are available for calculating kinship using a known pedigree or from marker data [20, 54]. For large datasets, the command-line version of TASSEL is more convenient and a significantly faster implementation of the mixed-model has been developed [54]. The principal components representation can be used in place of admixture coefficients in a mixed-model for association mapping. A comparison of the PCA and admixture approaches suggests that comparable results are obtained [20]. Finally, we note that the incorporation of epistatic effects in association mapping is a developing area [55–57].

A genome scan calculation generates a large number of statistical comparisons using a particular set of phenotypic data. In order to control the false-positive rate, a multiple testing procedure is needed to estimate an experiment wide significance level. One approach is to empirically estimate an experiment wide critical value by permutation [58]. Phenotypic data are randomly permuted and reassigned to the individuals and the genome scan association tests are recalculated. Calculations with up to 1 million permutations or more are needed to accommodate the large sample sizes and high significance levels that are often reached in genome-wide scans. For associations with significance beyond the limit of permutation the significance based on asymptotic distributions are inserted. The computational requirements could be reduced by calculating significances for all markers for each permutation of phenotypes but this could introduce correlations [58]. With existing computing capability the permutations can be performed separately for each marker in an adaptive fashion, using fewer permutations when the significance is low. The permutation procedure could also take advantage of the fact that it is only necessary to sample the genotypically distinct grouping of numbers; the order of numbers within groups is not important. For small samples it becomes possible to completely sample the permutation space and obtain an exact test. The P-values obtained for the association tests are ordered to estimate the critical value for the GWA scan. The estimation of critical values can be viewed more generally in terms of controlling the false discovery rate (FDR) [59–61]. However, naive application of these methods can give misleading results [62]. The permutation and FDR methods are both derived for a uniform, null distribution of P-values, which does not apply in association mapping because of the correlation structure introduced by LD.

14.3
Applications

Several examples of association mapping have already been reported in plants (reviewed in [13, 17, 28, 63–65]); however, only a few used GWA scans with varying number of markers and sample size. We recently reported the use of SNPs in a GWA study of oleic acid content in maize kernels [66]. A fatty acid desaturase-2 gene (*fad2*) was identified on chromosome 4 at around 1.7 kb from the most significant SNP haplotype and was validated using a classical biparental segregant population. Several features of this study illustrate how the GWA method can be applied. The inbreds chosen for this study were identified as historically important individuals or key elite inbreds in our maize breeding program. Despite the fact that the elite inbred lines were derived from a relatively small number of founders, the genetic diversity of the collection has proven to be sufficient for association mapping. The frequency of the natural occurring allele that conferred higher oleic acid content was around 0.28 among all inbred lines and around 0.82 in a main subpopulation (see Table 1 in [66]). The 8590 SNP haplotypes used incorporate almost 50 000 SNPs. The high density of markers certainly facilitated the identification of the oleic acid QTL spanning 4 cM and the *fad2* gene. The phenotypic trait studied was another important factor that contributed to the clear identification of a major QTL. Oleic acid content was precisely measured by gas chromatography in extracts of maize kernels and it is minimally influenced by environmental or experimental effects. This reinforces the importance of obtaining high-quality phenotypic data.

As described above, the germplasm was clustered in subpopulations due to the high population structure, a nonparametric test was used to avoid issues with nonnormality of the data. For every locus, each SNP haplotype was tested against a combination of all the others and permutation tests were used to estimate the P-values of the test statistics [66]. Although the significance of the major locus

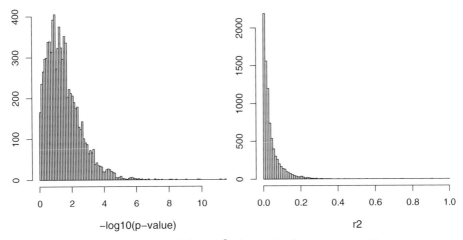

Figure 14.1 Frequencies of P-values (left) and r^2 values (right) from a genome-wide scan association mapping of kernel oleic acid content using 8590 SNP haplotypes in 553 maize inbred lines.

affecting oleic acid content reported on chromosome 4 was clearly the lowest, many other markers presented low P-values (Figure 14.1). The presence of many possible associated markers results from the high number of tests performed (see Section 14.2.4) or from the presence of LD among unlinked regions of the genome. Testing whether the distribution of P-values deviates from the expected in the null hypothesis can help in the identification of false-positives. An enrichment of low P-values is expected as these would correspond to real associations. However, this is not sufficient to indicate that all low P-values are true-positive associations. Checking the presence of LD between the associated regions and/or across the genome can also help to distinguish real associations from false-positive ones [66]. When different genomic regions are in LD and one is associated with the phenotype, then the others will display some degree of association – this LD might be a result of selection, for example. More than one statistic can be used to measure LD [67, 68]; however, when using values that range from 0 to 1 (e.g., r^2), even values as low as 0.2–0.3 can indicate the presence of LD between loci (Figure 14.1) and create false-positive associations. Observation of the same association in more than one subpopulation increases the confidence in the detected association. However, the observation of an association in only one subpopulation is not conclusive because alleles and allelic frequencies might be different in each subpopulation. A careful look at the allelic frequencies of the markers in those regions can identify whether the lack of association in some subpopulations is due to low power resulting from skewed allelic frequency distributions. Alleles with low frequency (below 5%) might have arisen recently, given little time for recombination and mutation to disrupt their haplotypes [69]. These new alleles are not informative because it is difficult to distinguish them from the haplotypes of origin using flanking molecular markers even if they have a large phenotypic effect [69].

14.3.1
QTL Mapping versus Association Mapping

Even though high sensitivity can be achieved with GWA when a phenotype, markers, and population with adequate allelic frequency are used, there is no guarantee that all QTLs will be found. An example is the positional cloning of a high-oil QTL in maize embryos using oleic acid content as a proxy for oil concentration to successfully identify an acyl-coenzyme A: diacylglycerol acyltransferase (DGAT) on chromosome 6 [70]. The authors were able to further identify the polymorphism responsible for higher oil content and concluded that this allele is ancestral, having its frequency reduced by selection. An interesting example of complementarity is the dissection of QTL that affect oleic acid content in maize kernels using different approaches (GWA mapping [66] and QTL mapping [70]). In the case of rare alleles (high-oil DGAT), QTL mapping proved to be efficient while GWA mapping was not able to identify this QTL. On the other hand, GWA mapping was able to define a QTL for oleic acid content with good resolution due to sampling a larger genetic diversity that was not identified by QTL mapping.

The combination of both approaches can provide additional and valuable information for plant genetics and breeding benefit. The F2 biparental population segregating for *fad2* and used in [66] for association validation was also segregating for the DGAT locus. ANOVA of the percentage of oleic acid among individuals of the F2 population showed no interaction between the two loci (Table 14.1), suggesting that additivity is the main component of the inheritance of these loci. Observation of the phenotypes of oleic acid content according to the genotypes from both loci showed that individuals with the DGAT and *fad2* alleles for higher oleic acid content have the highest percentage of oleic acid and those heterozygous for both alleles showed intermediate phenotypes (Figure 14.2).

Regarding costs and resources, association mapping projects on this scale [66] are more expensive than QTL mapping; however, the direct estimation of the cost per QTL or gene identified cannot be easily measured since the same population can be used repeatedly to dissect other QTLs. Another benefit of developing an association mapping population is that the individuals genotyped and markers developed (especially in case of SNPs) are available for future use. It is reasonable to think that projects of association mapping will be used for a long time and for many traits, otherwise the costs might not worth the benefits. At the same time, the costs of high-

Table 14.1 ANOVA of oleic acid content in maize kernels of a F2 population segregating for DGAT and *fad2*.

Effect on oleic acid content	Degree of freedom	Mean squared	F-test	P-value
DGAT locus	2	1456.0	61.7	$<2 \times 10^{-16}$
fad2 locus	2	1984.5	84.0	$<2 \times 10^{-16}$
DGAT × *fad2* interaction	4	37.5	1.6	0.1796
Residuals	163	23.6		

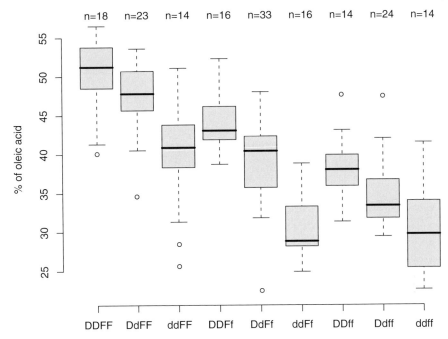

Figure 14.2 Phenotypes of oleic acid content in individuals of a F2 biparental segregant population according to their genotypes in the DGAT and *fad2* locus.

density genotyping and of marker discovery are rapidly decreasing, increasing the attractiveness of GWA. Overall, both methods seemed to be complementary (as discussed in [71]), and GWA mapping has proved to be a valuable option for QTL dissection and gene discovery in plants [19, 66].

14.3.2
Limitations

The extent of phenotypic differences and allelic frequency of the locus being tested can dramatically affect association mapping results. These factors are related to the power to detect a positive association between the markers and phenotype. If the population used for association mapping does not provide enough phenotypic variability, or if too many environmental or experimental effects are present, association mapping will probably be inefficient in detecting association. In addition, association mapping will have low power to identify alleles with low frequency even when the phenotypic effect is large. Thus, more rigorous analysis of the power of association studies is needed in different situations and experimental designs.

Experimental limitations are related to the costs of developing high-density molecular markers for whole-genome scans and statistical analysis. In humans, association mapping studies have identified association for some phenotypes and diseases thanks to the large samples and number of molecular markers available, which now exceeds 1 million SNPs. It will probably take some years until plant geneticists and breeders are able to use very-high-density genotyping for association mapping. This problem is increased by the presence of uneven LD along the genome [12, 33]; however, rapid improvement of next-generation sequencing technology continuously lowers the cost of SNP discovery. We anticipate that multiplexed genotyping assays capable of identifying multi-SNP haplotypes in every gene of common crop species, such as maize, will become available within the next few years.

The use of the candidate-gene approach for association mapping is appealing because it is cheaper than genotyping the entire genome; however, there are some significant limitations of this approach. First, significant associations can arise not only due to genetic effects, but also because of the presence of LD among unlinked regions of the genome and/or by chance (see Section 14.2.4). Conducting a genome-wide scan will allow the investigator to assess the overall distribution of P-values along the genome and have a more precise estimate of LD along the genome. Another limitation of the candidate-gene approach is that it relies on previous knowledge about the genes and their functional properties. Our current understanding of gene function is still limited, and until we have a more complete overview of gene function and pathway interactions, candidate-gene approaches will be limited to well-characterized genes and pathways. One observation from several genome-wide scans studied in humans is that very often the genes identified in association studies are of unknown function or even not annotated as genes [10] and these loci would not be properly identified by candidate-gene approaches.

14.4
Perspectives

Association mapping methodology is still under active development. Several areas needing improvement include population creation/selection, genotyping, statistical methodology, and integration with plant-breeding programs. Many breeding and research programs do not have enough resources to perform high-density association mapping, although it is likely that plant genetics will follow the same direction as human genetics in expanding the use of GWA. However, because the creation of large segregant populations is feasible in plants, QTL mapping will remain an obvious option for QTL dissection and gene discovery, and a preferred one for mapping effects associated with rare alleles. It is difficult to predict whether association mapping will become more common than QTL mapping; however, it is likely that its use will increase considerably in the next few years as the cost of genotyping decreases. Even when QTL mapping in biparental segregant population is used to first identify QTLs, association mapping can be used for fine mapping in small genome segments.

The ability to sample a larger genetic diversity in comparison to classical biparental segregant populations and the higher resolution that can be achieved will be the main reasons for the increase in popularity of association mapping. A minority of the many QTLs reported in plants were positionally cloned and molecularly characterized.

Association mapping promises to help to close this gap. With the identification of molecular markers closer to causative genes of QTLs, plant breeding can take advantage of these regions for marker-assisted breeding.

Acknowledgments

We thank John Zheng and Bo Shen for sharing unpublished data of *fad2* and DGAT phenotypes, Antoni Rafalski, Bailin Li, and Scott Tingey for frequent and challenging discussions, and Antoni Rafalski for helping to improve the manuscript.

References

1 Fu, Y. (2007) Impact of plant breeding on genetic diversity of agricultural crops: searching for molecular evidence. *Plant Genet. Resour.*, **4**, 71–78.

2 Zamir, D. (2001) Improving plant breeding with exotic genetic libraries. *Nat. Rev. Genet.*, **2**, 983–989.

3 Tanksley, S. and McCouch, S. (1997) Seed banks and molecular maps: unlocking genetic potential from the wild. *Science*, **277**, 1063–1066.

4 Martin, G.B., Brommonschenkel, S.H., Chunwongse, J., Frary, A., Ganal, M.W., Spivey, R., Wu, T., Earle, E.D., and Tanksley, S.D. (1993) Map-based cloning of a protein kinase gene conferring disease resistance in tomato. *Science*, **262**, 1432–1436.

5 Lander, E.S. and Schork, N.J. (1994) Genetic dissection of complex traits. *Science*, **265**, 2037–2048.

6 Holland, J.B. (2007) Genetic architecture of complex traits in plants. *Curr. Opin. Plant Biol.*, **10**, 156–161.

7 Salvi, S. and Tuberosa, R. (2005) To clone or not to clone plant QTLs: present and future challenges. *Trends Plant Sci.*, **10**, 297–304.

8 Morgante, M. and Salamini, F. (2003) From plant genomics to breeding practice. *Curr. Opin. Biotechnol.*, **14**, 214–219.

9 Bortiri, E., Jackson, D., and Hake, S. (2006) Advances in maize genomics: the emergence of positional cloning. *Curr. Opin. Plant Biol.*, **9**, 164–171.

10 Altshuler, D., Daly, M.J., and Lander, E.S. (2008) Genetic mapping in human disease. *Science*, **322**, 881–888.

11 McCarthy, M., Abecasis, G., Cardon, L., Goldstein, D., Little, J., Ioannidis, J., and Hirschhorn, J. (2008) Genome-wide association studies for complex traits: consensus, uncertainty and challenges. *Nat. Rev. Genet.*, **9**, 356–369.

12 Flint-Garcia, S.A., Thornsberry, J.M., and Buckler, E.S.4. (2003) Structure of linkage disequilibrium in plants. *Annu. Rev. Plant Biol.*, **54**, 357–374.

13 Yu, J. and Buckler, E.S. (2006) Genetic association mapping and genome organization of maize. *Curr. Opin. Biotechnol.*, **17**, 155–160.

14 Rafalski, A. (2002) Applications of single nucleotide polymorphisms in crop genetics. *Curr. Opin. Plant Biol.*, **5**, 94–100.

15 Balding, D.J. (2006) A tutorial on statistical methods for population association studies. *Nat. Rev. Genet.*, **7**, 781–791.

16 Kruglyak, L. (2008) The road to genome-wide association studies. *Nat. Rev. Genet.*, **9**, 314–318.

17 Abdurakhmonov, I.Y. and Abdukarimov, A. (2008) Application of association mapping to understanding the genetic diversity of plant germplasm resources. *Int. J. Plant Genomics*, **2008**, 1574927.

18 Nordborg, M., Hu, T.T., Ishino, Y., Jhaveri, J., Toomajian, C., Zheng, H., Bakker, E., Calabrese, P., Gladstone, J., Goyal, R., Jakobsson, M., Kim, S., Morozov, Y., Padhukasahasram, B., Plagnol, V., Rosenberg, N.A., Shah, C., Wall, J.D., Wang, J., Zhao, K., Kalbfleisch, T., Schulz, V., Kreitman, M., and Bergelson, J. (2005) The pattern of polymorphism in *Arabidopsis thaliana*. *PLoS Biol.*, **3**, e196.

19 Aranzana, M.J., Kim, S., Zhao, K., Bakker, E., Horton, M., Jakob, K., Lister, C., Molitor, J., Shindo, C., Tang, C., Toomajian, C., Traw, B., Zheng, H., Bergelson, J., Dean, C., Marjoram, P., and Nordborg, M. (2005) Genome-wide association mapping in *Arabidopsis* identifies previously known flowering time and pathogen resistance genes. *PLoS Genet.*, **1**, e60.

20 Zhao, K., Aranzana, M.J., Kim, S., Lister, C., Shindo, C., Tang, C., Toomajian, C., Zheng, H., Dean, C., Marjoram, P., and Nordborg, M. (2007) An *Arabidopsis* example of association mapping in structured samples. *PLoS Genet.*, **3**, e4.

21 Flint-Garcia, S.A., Thuillet, A., Yu, J., Pressoir, G., Romero, S.M., Mitchell, S.E., Doebley, J., Kresovich, S., Goodman, M.M., and Buckler, E.S. (2005) Maize association population: a high-resolution platform for quantitative trait locus dissection. *Plant J.*, **44**, 1054–1064.

22 Zhao, W., Canaran, P., Jurkuta, R., Fulton, T., Glaubitz, J., Buckler, E., Doebley, J., Gaut, B., Goodman, M., Holland, J., Kresovich, S., McMullen, M., Stein, L., and Ware, D. (2006) Panzea: a database and resource for molecular and functional diversity in the maize genome. *Nucleic Acids Res.*, **34**, 752–757.

23 Casa, A., Pressoir, G., Brown, P., Mitchell, S., Rooney, W., Tuinstra, M., Franks, C., and Kresovich, S. (2008) Community resources and strategies for association mapping in sorghum. *Crop Sci.*, **48**, 30.

24 Rafalski, J. (2002) Novel genetic mapping tools in plants: SNPs and LD-based approaches. *Plant Sci.*, **162**, 329–333.

25 Nguyen, H. and Wu, X. (2005) Molecular marker systems for genetic mapping, in *The Handbook of Plant Genome Mapping: Genetic and Physical Mapping* (eds K. Meksem and G. Kahl), Wiley-VCH, Weiheim.

26 Kahl, G., Mast, A., Tooke, N., Shen, R., and van den Boom, D. (2005) Single nucleotide polymorphisms: detection techniques and their potential for genotyping and genome mapping, in *The Handbook of Plant Genome Mapping: Genetic and Physical Mapping* (eds K. Meksem and G. Kahl), Wiley-VCH, Weiheim.

27 Abo, R., Knight, S., Wong, J., Cox, A., and Camp, N.J. (2008) hapConstructor: automatic construction and testing of haplotypes in a Monte Carlo framework. *Bioinformatics*, **24**, 2105–2107.

28 Gupta, P.K., Rustgi, S., and Kulwal, P.L. (2005) Linkage disequilibrium and association studies in higher plants: present status and future prospects. *Plant Mol. Biol.*, **57**, 461–485.

29 Ching, A., Caldwell, K.S., Jung, M., Dolan, M., Smith, O.S., Tingey, S., Morgante, M., and Rafalski, A.J. (2002) SNP frequency, haplotype structure and linkage disequilibrium in elite maize inbred lines. *BMC Genet.*, **3**, 19.

30 Jung, M., Ching, A., Bhattramakki, D., Dolan, M., Tingey, S., Morgante, M., and Rafalski, A. (2004) Linkage disequilibrium and sequence diversity in a 500-kbp region around the *adh1* locus in elite maize germplasm. *Theor. Appl. Genet.*, **109**, 681–689.

31 Palaisa, K.A., Morgante, M., Williams, M., and Rafalski, A. (2003) Contrasting effects of selection on sequence diversity and linkage disequilibrium at two phytoene synthase loci. *Plant Cell*, **15**, 1795–1806.

32 Palaisa, K., Morgante, M., Tingey, S., and Rafalski, A. (2004) Long-range patterns of diversity and linkage disequilibrium surrounding the maize Y1 gene are indicative of an asymmetric selective sweep. *Proc. Natl. Acad. Sci. USA*, **101**, 9885–9890.

33 Rafalski, A. and Morgante, M. (2004) Corn and humans: recombination and linkage disequilibrium in two genomes of similar size. *Trends Genet.*, **20**, 103–111.

34 Thornsberry, J.M., Goodman, M.M., Doebley, J., Kresovich, S., Nielsen, D., and Buckler, E.S.4. (2001) *Dwarf8* polymorphisms associate with variation in flowering time. *Nat. Genet.*, **28**, 286–289.

35 Breseghello, F. and Sorrells, M.E. (2006) Association mapping of kernel size and milling quality in wheat (*Triticum aestivum* L.) cultivars. *Genetics*, **172**, 1165–1177.

36 Matthies, I., Weise, S., and Röder, M. (2009) Association of haplotype diversity in the α-amylase gene *amy1* with malting

quality parameters in barley. *Mol. Breeding*, **23**, 139–152.

37 Piepho, H., Buchse, A., and Emrich, K. (2003) A hitchhiker's guide to mixed models for randomized experiments. *J. Agron. Crop. Sci.*, **189**, 310–322.

38 Ewens, W.J., Li, M., and Spielman, R.S. (2008) A review of family-based tests for linkage disequilibrium between a quantitative trait and a genetic marker. *PLoS Genet.*, **4**, e1000180.

39 Yu, J., Pressoir, G., Briggs, W.H., Vroh Bi, I., Yamasaki, M., Doebley, J.F., McMullen, M.D., Gaut, B.S., Nielsen, D.M., Holland, J.B., Kresovich, S., and Buckler, E.S. (2006) A unified mixed-model method for association mapping that accounts for multiple levels of relatedness. *Nat. Genet.*, **38**, 203–208.

40 Zhu, C., Gore, M., Buckler, E.S., and Yu, J. (2008) Status and prospects of association mapping in plants. *Plant Genome*, **1**, 5–20.

41 Lynch, M. and Walsh, B. (1998) *Genetics and Analysis of Quantitative Traits*, Sinauer, Sunderland, MA.

42 Casella, G. and Berger, R. (2001) *Statistical Inference*, Duxbury Press, Pacific Grove, CA.

43 Schaid, D.J. (2004) Evaluating associations of haplotypes with traits. *Genet. Epidemiol.*, **27**, 348–364.

44 Becker, T. and Knapp, M. (2004) A powerful strategy to account for multiple testing in the context of haplotype analysis. *Am. J. Hum. Genet.*, **75**, 561–570.

45 Barnett, V. and Lewis, T. (1994) *Outliers in Statistical Data*, John Wiley & Sons Ltd, Chichester.

46 Pritchard, J.K. and Rosenberg, N.A. (1999) Use of unlinked genetic markers to detect population stratification in association studies. *Am. J. Hum. Genet.*, **65**, 220–228.

47 Excofier, L. (2001) Analysis of population subdivision, in *Handbook of Statistical Genetics* (eds D. Balding, M. Bishop, and C. Cannings), John Wiley & Sons Ltd, Chichester.

48 Weir, B., (1996) *Genetic Data Analysis II*, Sinauer, Sunderland, MA.

49 Devlin, B. and Roeder, K. (1999) Genomic control for association studies. *Biometrics*, **55**, 997–1004.

50 Pritchard, J.K., Stephens, M., Rosenberg, N.A., and Donnelly, P. (2000) Association mapping in structured populations. *Am. J. Hum. Genet.*, **67**, 170–181.

51 Duda, R.O., Hart, P.E., and Stork, D.G. (2000) *Pattern Classification*, Wiley-Interscience, New York.

52 Rosenberg, N.A. (2004) DISTRUCT: a program for the graphical display of population structure. *Mol. Ecol. Notes*, **4**, 137–138.

53 Price, A.L., Patterson, N.J., Plenge, R.M., Weinblatt, M.E., Shadick, N.A., and Reich, D. (2006) Principal components analysis corrects for stratification in genome-wide association studies. *Nat. Genet.*, **38**, 904–909.

54 Kang, H.M., Zaitlen, N.A., Wade, C.M., Kirby, A., Heckerman, D., Daly, M.J., and Eskin, E. (2008) Efficient control of population structure in model organism association mapping. *Genetics*, **178**, 1709–1723.

55 Lou, X., Casella, G., Littell, R.C., Yang, M.C.K., Johnson, J.A., and Wu, R. (2003) A haplotype-based algorithm for multilocus linkage disequilibrium mapping of quantitative trait loci with epistasis. *Genetics*, **163**, 1533–1548.

56 Marchini, J., Donnelly, P., and Cardon, L.R. (2005) Genome-wide strategies for detecting multiple loci that influence complex diseases. *Nat. Genet.*, **37**, 413–417.

57 Alvarez-Castro, J.M. and Carlborg, O. (2007) A unified model for functional and statistical epistasis and its application in quantitative trait loci analysis. *Genetics*, **176**, 1151–1167.

58 Churchill, G.A. and Doerge, R.W. (1994) Empirical threshold values for quantitative trait mapping. *Genetics*, **138**, 963–971.

59 Farcomeni, A. (2008) A review of modern multiple hypothesis testing, with particular attention to the false discovery proportion. *Stat. Methods Med. Res.*, **17**, 347–388.

60 Sabatti, C., Service, S., and Freimer, N. (2003) False discovery rate in linkage and association genome screens for complex disorders. *Genetics*, **164**, 829–833.

61 Storey, J.D. and Tibshirani, R. (2003) Statistical significance for genomewide studies. *Proc. Natl. Acad. Sci. USA*, **100**, 9440–9445.

62 Churchill, G.A. and Doerge, R.W. (2008) Naive application of permutation testing leads to inflated type I error rates. *Genetics*, **178**, 609–610.

63 Mackay, I. and Powell, W. (2007) Methods for linkage disequilibrium mapping in crops. *Trends Plant Sci.*, **12**, 57–63.

64 Sorkheh, K., Malysheva, L., and Wirthensohn, M. (2008) Linkage disequilibrium, genetic association mapping and gene localization in crop plants. *Genet. Mol. Biol.*, **31**, 1415–4757.

65 Zhu, C., Gore, M., Buckler, E., and Yu, J. (2008) Status and prospects of association mapping in plants. *Plant Genome*, **1**, 5.

66 Beló, A., Zheng, P., Luck, S., Shen, B., Meyer, D.J., Li, B., Tingey, S., and Rafalski, A. (2008) Whole genome scan detects an allelic variant of *fad2* associated with increased oleic acid levels in maize. *Mol. Genet. Genomics*, **279**, 1–10.

67 Hedrick, P. (1987) Gametic disequilibrium measures: proceed with caution. *Genetics*, **117**, 331–341.

68 Lewontin, R. (1988) On measures of gametic disequilibrium. *Genetics*, **120**, 849–852.

69 Hirschhorn, J.N. and Daly, M.J. (2005) Genome-wide association studies for common diseases and complex traits. *Nat. Rev. Genet.*, **6**, 95–108.

70 Zheng, P., Allen, W.B., Roesler, K., Williams, M.E., Zhang, S., Li, J., Glassman, K., Ranch, J., Nubel, D., Solawetz, W., Bhattramakki, D., Llaca, V., Deschamps, S., Zhong, G., Tarczynski, M.C., and Shen, B. (2008) A phenylalanine in DGAT is a key determinant of oil content and composition in maize. *Nat. Genet.*, **40**, 367–372.

71 Wilson, L.M., Whitt, S.R., Ibanez, A.M., Rocheford, T.R., Goodman, M.M., and Buckler, E.S.4. (2004) Dissection of maize kernel composition and starch production by candidate gene association. *Plant Cell*, **16**, 2719–2733.

15
Using Mutations in Corn Breeding Programs
Anastasia L. Bodnar and M. Paul Scott

Abstract

It is frequently necessary to move a mutation from one genetic background to another for reasons including crop improvement and scientific studies. Two methods that can be used for this purpose are backcross breeding and forward breeding. Backcross breeding is used to move a mutation into a specific genetic background, while forward breeding is used to improve the agronomic performance of the variety carrying a mutation. These methods can be adapted and applied to a variety of situations, including both dominant and recessive mutations. They can also be adapted to take advantage of new technologies such as marker-assisted selection and doubled haploids.

15.1
Introduction

Mutations are valuable because they create the genetic variation required for crop improvement via plant breeding. Mutations are also valuable in scientific studies, especially in studies of gene function. Mutations can be naturally occurring, induced by whole-genome mutagenesis, or created with biotechnology. In this chapter, we describe breeding methods that allow a mutation to be moved into a different genetic background. The methods described here are generally applicable to all types of mutations.

It is often desirable to move a mutation into another genetic background in order to characterize the mutation. For example, whole-genome mutagenesis can result in mutations at several genetic loci in an individual and it may be necessary to isolate each mutation prior to characterization. Additionally, many mutations have different effects in different genetic backgrounds. Evaluation of the mutation in several different backgrounds can lead to a better understanding of the effect of the mutation.

The Handbook of Plant Mutation Screening. Edited by Günter Kahl and Khalid Meksem
Copyright © 2010 WILEY-VCH Verlag GmbH & Co. KGaA, Weinheim
ISBN: 978-3-527-32604-4

Finally, when comparing several mutations, the comparison should be carried out in a common genetic background.

A second reason for moving mutations into another genetic background is to improve the agronomic performance of the mutant plants. Natural mutations with potential agronomic value such as disease resistance or improved grain quality are often found in breeding stocks and these mutations must be transferred to elite germplasm prior to commercial use. Similarly, it is normally necessary to transfer transgenes into elite germplasm because only a few maize varieties, such as Hi-II [1], can be efficiently transformed and regenerated. These varieties are often agronomically inferior and highly heterozygous, which is problematic because heterozygous loci segregate in subsequent generations to give a wide range of genetic background effects.

We describe two breeding techniques that can be used to transfer a mutation from one genetic background to another. The objective of backcross breeding (Section 15.2.1) is to move the mutation into a specific genetic background, usually an inbred line. The objective of forward breeding (Section 15.2.2) is to improve the agronomic performance of the variety carrying the mutation and usually results in development of a new genotype. Plant breeders typically use a combination of these techniques, but here we describe the two methods separately for clarity. Throughout this protocol, the mutant allele will be indicated by M^* if dominant or by m^* if recessive. The wild-type allele will be indicated by M if dominant or by m if recessive.

15.1.1
Factors to Consider Before Starting a Breeding Program

The most important factor a breeder must consider when determining which type of breeding program to use is the breeding objective; however, other factors to consider include the inheritance and expression of the mutation, the population under selection, predicted response to selection, and costs and risks [2].

Before starting a breeding program, the researcher must have obtained or identified a donor parent that contains a mutation of interest. It is important to understand the inheritance of the mutation. Is it recessive or dominant? Is it transmitted equally well through either parent? Is it expressed similarly in all genetic backgrounds? The answers to these questions will determine how the breeding program is carried out. It is easiest to work with mutations that exhibit dominance without interaction and have consistent expression regardless of genetic background or environment. Recessive mutations complicate breeding programs, but can be used in either backcross or forward breeding (Section 15.2.3.2). In addition, traits that can be identified before flowering are easiest to work with. If the mutation is identifiable only after flowering, additional crosses must be to identify plants carrying the mutation. Other factors that complicate breeding and ways to adjust breeding programs accordingly are described by Fehr in *Principles of Cultivar Development* [3] and by Acquaah in *Principles of Plant Genetics and Breeding* [4].

15.1.2
Alternatives to Breeding

Backcross and forward breeding are time-consuming and expensive, and so should be avoided whenever possible. Careful planning of the mutagenesis effort can reduce or eliminate the need for backcross breeding. For example, chemical mutagenesis should be carried out in a homozygous line using a mutagen dose that minimizes the number of individuals carrying multiple mutations. Transformation of a desired inbred line may be inefficient, but because it results in the transgene being in a genetically uniform background, it might allow the effect of a mutation to be evaluated without backcrossing. The cost of transformation inefficiency should be weighed against the cost and time required for a backcrossing program.

15.2
Methods and Protocols

15.2.1
Backcross Breeding

The goal of backcross breeding is to develop varieties that contain a mutant allele and are genetically similar to existing varieties. The variety that contains the mutation at the start of the program is called the donor parent and the variety that the mutation is being transferred into, typically an inbred line, is called the recurrent parent. This is achieved using a cyclic process of crossing the donor parent to the recurrent parent, selecting for progeny containing the mutation, and crossing these progeny to the recurrent parent again to complete the cycle (Figure 15.1). A successful backcrossing program should produce lines that are near-isogenic to the recurrent parent.

The choice of recurrent parent depends on the objective of the breeding program. If the goal is to determine the effect of the mutation in a specific genetic background, then the recurrent parent should have the genetic background of interest. If the objective is simply to move the mutation into a uniform genetic background, an agronomically superior inbred line is usually chosen. Well-characterized inbred lines such as B73 are frequently used as recurrent parents because of the large amount of information available for these lines. The breeder may choose to backcross a mutation into a few different inbred lines in order to evaluate expression in different backgrounds or in preparation for further cultivar development. If the desired product is a line containing multiple mutations, the breeder may choose to backcross each mutation individually into the same recurrent parent then cross the resulting lines.

Each successive cross to the recurrent parent reduces the genetic contribution of the donor parent. Generations are typically referred to with the notation BC#, where # equals the number of crosses to the recurrent parent after the initial cross. Table 15.1 shows the average percentage of each parental genome after each backcross in the

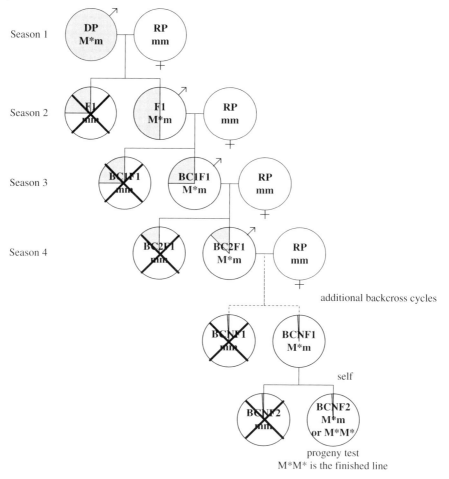

Figure 15.1 Backcross breeding with a dominant mutation. Shaded portions of circles represent the genetic contribution of the donor parent. The mutant allele is indicated by M^* and the wild-type allele is indicated by m. The donor parent is indicated by DP and the recurrent parent is indicated by RP.

absence of linkage. The number of backcrosses carried out is determined by the desired degree of similarity to the recurrent parent.

There are several drawbacks to backcross breeding. First, a large number of generations are required to recover the recurrent parent genotype to a great extent. Second, it is difficult to backcross several genetic loci simultaneously because it is necessary in each generation to select individuals with the desired mutations and the frequency of these individuals decreases exponentially with the number of independent loci being selected. Owing to linkage, it can be difficult to separate a mutation from nearby loci that may include deleterious genes. Linkage drag is a term that refers to reduction in yield caused by unwanted donor genetic material introduced by breeding. In theory, linkage drag is reduced as the proportion of the

Table 15.1 The percentage of genetic contribution of the donor and recurrent parents after each backcross, in the absence of linkage.

Generation	Donor (%)	Recurrent (%)
F1	50.00	50.00
BC1	25.00	75.00
BC2	12.50	87.50
BC3	6.25	93.75
BC4	3.13	96.88
BC5	1.56	98.44

recurrent parent genome is increased so additional cycles of backcrossing and methods that improve recovery of the recurrent parent genome, such as marker-assisted selection (Section 15.4.1.1) [5], may reduce linkage drag.

Backcross Breeding with a Dominant Mutation

1. Identify the donor and recurrent parents.
2. In Season 1, plant the recurrent and donor parents in adjacent rows. Use pollen from the donor parent (M^*m) to pollinate the recurrent parent (mm). Harvest each ear individually. The resulting seed is the F1 generation ($1/2 M^*m$ and $1/2 mm$).
3. In Season 2, plant an appropriate number of seeds (Section 15.2.3.1) from each F1 ear and an approximately equal number of recurrent parent sees in an adjacent row. Identify F1 plants carrying the mutation and use pollen from those individuals to pollinate a recurrent parent plant. Harvest each ear individually. The resulting seed is the BC1F1 generation ($1/2 M^*m$ and $1/2 mm$). This marks the end of one cycle of backcross breeding.
4. Continue with additional cycles of backcross breeding by substituting the most recent backcross generation for the F1 seed in Step 3 until satisfied with the resulting line's similarity to the recurrent parent.
5. To produce a line the breeds true for the mutation after N cycles of backcrossing, self-pollinate the BCNF1 plants to produce BCNF2 plants. Use genotyping or progeny testing (Section 15.2.3.2) to select the M^*M^* plants, which will breed true for the mutation.

Backcross with a Recessive Mutation

Backcrossing with a recessive mutation is similar to backcrossing with a dominant mutation except it requires genotyping before pollination or progeny testing after each backcross generation. If genotyping before pollination is used to identify individuals carrying the mutant allele, the protocol is the same as for backcrossing with a dominant mutation. Progeny testing can be started in the same season as a backcross by using pollen from the same donor parent to pollinate both the recurrent parent and the tester. Once the results of the progeny tests are known, the lines containing test-cross-negative parents can be removed from the backcross breeding program.

15.2.2
Forward Breeding

The goal of forward breeding is to produce an agronomically improved variety that contains a mutation. Forward breeding begins with a cross between the donor parent and an elite parent to create a breeding population. The elite parent may be an inbred line or a breeding population, such as a synthetic cultivar created by intermating several elite inbred lines. If the mutation exists in a population that is segregating for the desired traits, no cross is necessary. For simplicity, only one cross with the elite parent is used in this protocol; however, it may be desirable to make several backcrosses to the elite parent if the donor parent has very poor agronomic characteristics. After the initial cross, a typical breeding program includes repeated generations of self-pollination with selection for desired agronomic traits (Figure 15.2). One or more generations of intermating (Section 15.2.3.3) may be included in order to increase recombination between the donor and elite genomes, and to reduce the size of linkage blocks. One challenge of forward breeding is achieving the required level of genetic variability in the breeding population to allow improvement through selection.

Terminology used for each generation of forward breeding can vary depending on breeder, methods, or cultivar. In general, the progeny of the first cross between the donor and elite parents is the first filial generation (F1). Generations produced by self-pollination of plants in the breeding population are referred to as S#, where # is the number of self-pollinated generations.

Forward Breeding with a Dominant Mutation

1. Identify the donor parent and the elite parental population. Identify which traits are desired in the resulting line.
2. In Season 1, cross the donor parent (M^*m) and the elite parent (mm) to obtain F1 ears ($^1/_2$ M^*m and $^1/_2$ mm). Harvest each ear individually. The resulting seed is the F1 generation.
3. In Season 2, plant an appropriate number of seeds (Section 15.2.3.1) from each F1 ear in separate rows. Select plants that have the mutation and the desired agronomic traits. Self-pollinate each selected plant and harvest each ear individually. The resulting seed is the S1 generation.

4. In Season 3, plant an appropriate number of seeds from each S1 ear in separate rows. Select, self-pollinate, and harvest as in Season 2. The resulting seed is the S2 generation.
5. In Season 4, plant an appropriate number of seeds from each S2 ear in separate rows. The number of seeds must be sufficient to differentiate rows that are segregating for the mutation from those that are not. Observe the plants and remove entire rows that are segregating. In nonsegregating rows, select plants that have the desired agronomic traits. Self-pollinate each selected plant. Harvest each ear individually and combine equal numbers of seed from each selected ear in the row into a balanced bulk. Each bulk represents a row that contains at least one selected individual. The resulting seed is the S3 generation.
6. In Season 5, plant an appropriate number of S3 seeds from each bulk into separate rows. At this point, there should be no heterozygous plants remaining, but if rows appear to be segregating, they should be removed. Select, self-pollinate, and harvest as in Step 5. The resulting seed is the S4 generation.
7. Continue with additional cycles of selection and self-pollination by substituting the most recent self-pollinated generation for the S3 in Season 5 until satisfied with the traits in the resulting line.

Forward Breeding with a Recessive Mutation

The protocol for forward breeding with a recessive mutation is the same as for forward breeding with a dominant mutation. However, because the mutation will be expressed in only in heterozygous mutant individuals, fewer mutant plants will be available for breeding, so more seeds must be planted in each generation (Section 15.2.3.1).

15.2.3
Supplementary Protocols

There are many other factors to consider when planning a breeding program, including the number of seeds to be planted in each generation, identification of plants homozygous for a recessive mutation, and ensuring adequate recombination.

15.2.3.1 Determining How Many Seeds to Plant

The number of seeds that must be planted each season depends on several factors and was determined by Sedcole in 1977 [6]. Variables are the desired probability (p) of obtaining the desired number of mutant plants (r) and the frequency of the desired genotype (q), as shown in Table 15.2. In addition, the number of seeds needed equals the required number of plants divided by the germination rate [3]. For example, the frequency of homozygous recessive seeds resulting from a cross between two heterozygous plants is one fourth. To obtain at least two ($r = 2$) homozygous plants

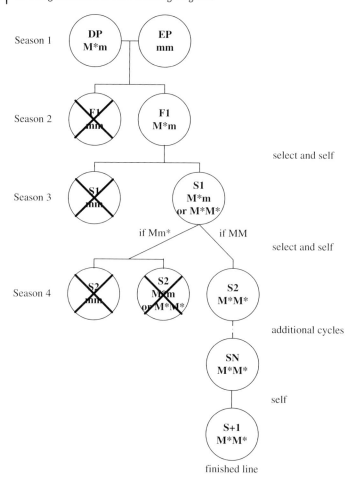

Figure 15.2 Forward breeding with a dominant mutation. The mutant allele is indicated by M^* and the wild-type allele is indicated by m. The donor parent is indicated by DP and the elite parent is indicated by EP.

Table 15.2 Number of plants required to ensure recovery of r plants with a desired genotype of frequency q and probability p [6].

p	q					r				
		1	2	3	4	5	6	8	10	15
0.95	1/2	5	8	11	13	16	18	23	28	40
0.99	1/2	7	11	14	17	19	22	27	33	45
0.95	1/4	11	18	23	29	34	40	50	60	84
0.99	1/4	17	24	31	37	43	49	60	70	96

($q = 1/4$) with a 99% probability ($p = 0.99$), 24 plants are needed. With a germination rate of 80%, at least 30 seeds must be planted per cross.

In forward breeding, a sufficient number of plants must be planted to allow for the desired selection intensity. Selection intensity (the number of individuals selected to be parents of the next generation) must be carefully controlled. If too many individuals are chosen, the population will improve slowly. If too few individuals are chosen, the genetic variability of the population may be reduced, potentially reducing future rate of gain. A selection intensity of 5–20% is typical.

15.2.3.2 Working with Recessive Mutations

If the mutation is recessive, there are two options that the breeder can use to identify plants that have the mutant allele: genotyping and progeny testing. Both of these methods allow the breeder to identify the genotype of a plant, and are suitable in different situations.

Genotyping It is often possible to use genotyping via molecular markers as an alternative to phenotyping to identify plants carrying the mutation, which can be very advantageous in certain situations. For example, a codominant molecular marker can be used to identify plants carrying a recessive mutation in the heterozygous state. If genotyping is finished before flowering, donor plants that do not contain the mutation can be removed prior to mating, reducing the number of pollinations necessary. Thus, genotyping can save time and field space, especially when used with recessive mutations or mutations that can only be identified after flowering. Genotyping requires molecular markers associated with the mutation, trained personnel, and appropriate laboratory equipment. Molecular markers may be difficult to develop for many types of induced mutations, but they can be developed relatively easily for transgenes. If a mutation will be used extensively in backcrossing programs, it may be worth characterizing the mutation molecularly for the purpose of developing markers. A less costly alternative is to map the mutation genetically relative to known molecular markers in order to identify a molecular marker that is tightly linked to the mutation. If a linked marker is used to select for the presence of the mutation, it is important to periodically verify that the mutation is still present and functional because the linkage between the marker and the mutation may be broken or expression of the mutation may be unstable.

Progeny Testing Progeny testing is used to phenotypically identify plants carrying a recessive mutation in the heterozygous state. Seeds from the plant to be tested are planted in a row and self-pollinated as in Figure 15.3 (a). The progeny are planted in the next season and scored for the mutant phenotype. The phenotypes of test-cross plants derived from heterozygous plants (Mm^*) will be one-fourth mutant and three-fourths wild-type (genotypes will be $1/4$ m^*m^*, $1/2$ Mm^*, $1/4$ MM) while the test-cross plants derived from homozygous plants (MM) will all be wild-type.

An alternative to self-pollination is to cross each plant to a "tester" that is homozygous for the mutation (m^*m^*) as in Figure 15.3(b). The phenotypes of test-cross plants derived from heterozygous plants (Mm^*) will be one-half mutant

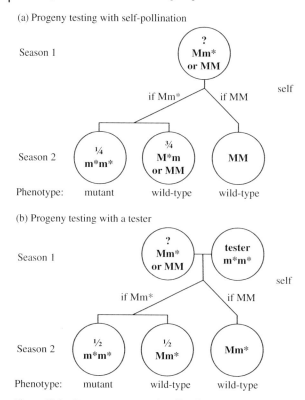

Figure 15.3 Progeny testing with self-pollination (a) or with a tester (b). The mutant allele is indicated by m^* and the wild-type allele is indicated by M.

and one-half wild-type (genotypes will be $1/2\ m^*m^*$, $1/2\ Mm^*$) while the test-cross plants derived from homozygous wild-type plants (MM) will all be heterozygous at the mutant locus, therefore phenotypically wild-type. Using a tester line may be preferred to using self-pollination because it does not require the use of an ear from the mutant. All test-cross plants can be removed after they have been identified, as they are not required for breeding

15.2.3.3 Intermating

One or more generations of intermating may be carried out to increase recombination between the elite and donor genomes. Intermating may be used to break undesired linkages with the mutation and to generate more phenotypic diversity in the breeding population. However, intermating may break desirable linkages in the elite parent. In each generation, inferior plants are removed and superior individuals carrying the mutation are intermated.

Intermating can be carried out in several different ways. The selected plants can be open pollinated if they can be isolated from other populations. However, self-pollinated seed and pollination by nearby plants result in less than optimal recombination between all plants. To avoid self-pollination, open pollination can be modified by detasseling "female" plants. Alternatively, chain sib mating can be used to optimize recombination and avoid self-pollination. In this method, seed from each parental plant is mixed in equal numbers to create a balanced bulk and then planted in rows. Inferior individuals are removed. Each plant is used to pollinate one of its neighbors, with the female plant of one cross being used as the male plant in the next cross. In this way, each plant is used as a male and as a female, and recombination is optimized because of the random placement of plants. The resulting seed from each plant are siblings of the plants before and after its female parent in the row. Selection before pollination allows the breeder to control both male and female parents of the next generation, which will result in faster line improvement than selection of female parents after pollination.

When intermating is used with a recessive mutation, genotyping or progeny testing (Section 15.2.3.2) is required. If genotyping is used with intermating, individuals in the population are tested before flowering. Progeny testing requires additional crosses that must be conducted before each generation of intermating. If greater speed in line development is desired, progeny testing can be conducted between alternating generations instead of between every generation, but this requires that more plants be used in intermating to ensure adequate frequency of the mutant allele in the population (Section 15.2.3.1).

15.2.4
Complication: Pleiotropic Effects

A single mutation can produce several phenotypes. Phenotypes that occur in addition to the phenotype of interest are termed pleiotropic effects. These effects can complicate the use of mutations for crop improvement, but can provide clues to the biological function of the mutation. An example of a mutation with pleiotropic effects is the *opaque2* (*o2*) mutation in maize – so named because mutant kernels transmit less light than wild-type kernels, giving them an opaque appearance. This mutation increases the level of the essential amino acids lysine and tryptophan in maize kernels, increasing their nutritional value [7]. Unfortunately, mutant kernels have reduced density that reduces their germination rate, and are more susceptible to attacks by insects and fungi. These pleiotropic effects limit the utility of *o2* maize. Another maize mutation with undesirable pleiotropic effects is *brown midrib 3* (*bm3*), which increases the digestibility of maize stalks, but reduces the ability of the plant to stand upright.

When breeding with mutations for crop improvement, it is frequently necessary to overcome adverse pleiotropic effects as well as to maintain the mutation. This can be done through the use of "modifier genes" that can attenuate the different phenotypes of a mutation to different degrees [8]. For example, some modifier genes play an important role in determining the severity of a number of genetic diseases in different individuals [9]. Modifier genes can decrease the opacity of *o2* maize kernels [10], including some that exhibit quantitative inheritance [11]. Thus, pleiotropic effects can offset one of the main advantages of mutation breeding, the ability to make a desired change with a simply inherited genetic locus, by requiring the manipulation of quantitatively inherited phenotypes associated with pleiotropic effects. This has been done successfully with the *o2* mutation in the 30-year long development of Quality Protein Maize (QPM) [12, 13]. From the standpoint of using mutations to understand gene function, modifier genes that alter the main or pleiotropic effects of a mutation may function in the same biochemical or genetic network as the mutant gene. Efforts to identify these modifier genes are underway [14–16].

15.3
Applications

Backcross and forward breeding have both played an important role in crop improvement and will continue to do so in the future. QPM is an example of the success of forward breeding for the development of new cultivars with a mutation and other improved traits. Transgenic plants containing Green Fluorescent Protein (GFP) are an example of a trait in a backcross breeding program moving one mutation into multiple recurrent parents.

15.3.1
Breeding with a Natural Mutation: QPM

The development of QPM is an example of a forward breeding approach in a mutation breeding program [12, 13]. The recessive *o2* mutation confers improved nutritional quality on grain by increasing the levels of nutritionally limiting amino acids [7], but also confers inferior agronomic properties. Negative pleiotropic effects are most severe in kernels with a strong opaque phenotype and are controlled in part by modifier genes that exhibit quantitative inheritance [11]. Thus, a forward breeding program is appropriate for development of new varieties that contain the *o2* mutation and have superior agronomic performance. This approach has been carried out successfully in several breeding programs, most notably at the International Maize and Wheat Improvement Center (www.cimmyt.org) [12, 13].

QPM breeding programs typically take the following approach. An *o2* donor parent is crossed to elite varieties. Kernels containing the mutation are selected and the opacity of these kernels is evaluated using a light box to identify kernels that are relatively translucent. Successive cycles are carried out with selection for agronomic traits including kernel translucence during inbreeding, with introgression of new genetic material as needed to obtain kernels with suitable agronomic characteristics. Using this approach, it is possible to produce varieties containing the *o2* mutation that have superior nutritional properties and agronomic performance. One challenge of this approach is that it becomes more difficult to visually identify the *o2* kernels as the kernels become more translucent in successive generations. For this reason, the development of molecular markers for the *o2* allele [17] has been a great benefit to QPM breeding programs.

15.3.2
Breeding with a Transgene: GFP

Modified versions of the jellyfish protein GFP have great utility as a reporter in biological experiments [18]. Transgenic maize that produces this protein in kernels has been produced [19]. As the genotype that was transformed was heterozygous, it was necessary to move these transgenes into uniform genetic backgrounds prior to use in biological studies. The backcross procedure with a dominant mutation described in Section 15.2.1 was used for this purpose. The transgenic plants that were regenerated from tissue culture were the recurrent parent. These plants were crossed to the inbred line B73 – one of the desired recurrent parents. As the GFP transgene is dominant, we were able to visually select kernels containing a functional copy of the transgene in the F1 ears produced in this cross. By visually selecting the transgenic kernels for planting and crossing these plants to the recurrent parent, we were able to complete a cycle of backcrossing in one generation. After 2–3 cycles of backcrossing, the resulting plants exhibited many phenotypic characteristics of the recurrent parent, although they exhibited more variability than the inbred recurrent parent.

15.4
Perspectives

While plant breeding has been practiced for thousands of years, the methods can be improved dramatically with new technologies. Two such technologies are marker-assisted selection and the creation of double haploids.

15.4.1
Marker-Assisted Selection

Molecular markers have the potential to save a great deal of time and money in backcross and forward breeding programs. They can be used to accelerate progress in several ways. The most obvious use of molecular markers is to track the presence of the mutation (Section 15.3.4.2), but molecular markers confer the ability to monitor the whole genome as well. This ability is useful for both backcross and forward breeding approaches.

15.4.1.1 Marker-Assisted Selection in Backcross Breeding
In backcross breeding, molecular markers can be used to identify individuals that carry an unusually large amount of the recurrent parent genome. On average, each cross to the recurrent parent reduces the amount of donor parent by 50%. However, in a large population derived from a backcross, individuals with substantially more of the recurrent parent can be identified using molecular markers that are distributed throughout the genome. This allows near-complete recovery of the recurrent parent genome in fewer generations. As the program progresses, more genetic loci become

fixed for the recurrent parent allele and the number of markers can be reduced because it is not necessary to genotype loci that are known to contain the recurrent parent allele [20].

Using molecular markers to increase the rate of recovery of the recurrent parent adds cost and requires laboratory equipment and skills not available to all breeders. An alternative method that gives a similar result is to use phenotypic selection for recurrent parent phenotypes in the course of backcrossing. This approach is based on the assumption that the plants most similar to the recurrent parent are likely to have the highest percentage of the recurrent parent genome. A convenient way to make crosses is to plant the recurrent parent in a row adjacent to the BC plants. When making crosses, simply examine the BC plants and select plants that are most phenotypically similar to the recurrent parent. Phenotypes that are easy to use are flowering date, plant height and tassel branch number. As much effort can be put into this process as desired. For example, thousands of plants could be evaluated for many different phenotypes at different developmental stages to identify the best 0.1% of the plants. Field space can be reduced by selecting seedlings germinated in a growth chamber and transplanting the selected plants to the field.

15.4.1.2 Marker-Assisted Selection in Forward Breeding

In forward breeding, the desired agronomic traits are often conferred by quantitative trait loci (quantitative trait locusQTLs). Molecular markers can be used in connection with QTL analysis to assemble desired loci from either parent in the product variety. QTLs that are known prior to the study can be selected for, but it is important to consider that the effect of a QTL is dependent on the genetic background, so that a known QTL in the elite parent may not be as favorable as the locus present in the donor parent. QTLs specific to the forward breeding population can be identified by genotyping and phenotyping the population in the course of the breeding program. It is important to consider that QTLs are often influenced by the environment, so it may be wise to carry out several generations of forward breeding with QTL identification in each generation prior to using molecular markers for selecting desired loci. As with backcross breeding, molecular markers can be used to increase the rate of inbreeding by identification of individuals with an unusually high proportion of homozygous loci.

15.4.2
Doubled Haploids

Doubled haploid technology is used extensively in commercial breeding programs to reduce the number of generations needed to produce a homozygous plant. The most commonly used process involves making a cross to "haploid inducer line" developed for the purpose of producing and identifying haploid plants. Highly efficient proprietary lines have been developed for this purpose, although public lines with lower efficiency of haploid production are available as well. The chromosome complement of the haploid plants resulting from this cross can be doubled by chemical treatment or other means, resulting in homozygous plants. While this approach would not allow recovery of a specific genotype (the goal of a backcross breeding program), it would allow production of a large number of inbred lines that could be evaluated for their agronomic performance, so could be useful in a forward breeding program [21].

References

1 Armstrong, C.L., Green, C.E., and Phillips, R.L. (1991) Development and availability of germplasm with high Type II culture formation response. *Maize. Genet. Coop. Newslett.*, **65**, 92–93.

2 Mumm, R.H. (2007) Backcross versus forward breeding in the development of transgenic maize hybrids: theory and practice. *Crop. Sci.*, **47** (Suppl. 3), S164–S171.

3 Fehr, W.R. (1991) *Principles of Cultivar Development. Volume 1: Theory and Technique*, Macmillan, London.

4 Acquaah, G. (2007) *Principles of Plant Genetics and Breeding*, Blackwell, Oxford.

5 Frisch, M. and Melchinger, A.E. (2001) Marker-assisted backcrossing for introgression of a recessive gene. *Crop. Sci.*, **41**, 1485–1494.

6 Sedcole, J.R. (1977) Number of plants necessary to recover a trait. *Crop. Sci.*, **17**, 667–668.

7 Mertz, E., Bates, L., and Nelson, O.E. Jr. (1964) Mutant gene that changes protein composition and increases lysine content of maize endosperm. *Science*, **16**, 279–280.

8 Nadeau, J.H. (2001) Modifier genes in mice and humans. *Nat. Rev. Genet.*, **2**, 165–174.

9 Dipple, K.M. and McCabe, E.R.B. (2000) Modifier genes convert "simple" Mendelian disorders to complex traits. *Mol. Gen. Metab.*, **71**, 43–50.

10 Paez, A.V., Helm, J.L., and Zuber, M.S. (1969) Lysine content of *opaque-2* maize kernels having different phenotypes. *Crop. Sci.*, **9**, 251–252.

11 Wessel-Beaver, L. and Lambert, R.J. (1982) Genetic control of modified endosperm texture in Opaque-2 maize. *Crop. Sci.*, **22**, 1095–1098.

12 Prasanna, B.M., Vasal, S.K., Kassahun, B., and Singh, N.N. (2001) Quality protein maize. *Current Sci.*, **81**, 1308–1319.

13 Krivanek, A.F., De Groote, H., Gunaratna, N.S., Diallo, A.O., and Friesen, D. (2007) Breeding and disseminating quality protein maize (QPM) for Africa. *African J. Biotech.*, **6**, 312–324.

14 Lopes, M.A., Takasaki, K., Bostwick, D.E., Helentjaris, T., and Larkins, B.A. (1995) Identification of two *opaque2* modifier loci in Quality Protein Maize. *Mol. Gen. Genet.*, **247**, 603–613.

15 Holding, D.R., Hunter, B.G., Chung, T., Gibbon, B.C., Ford, C.F., Bharti, A.K., Messing, J., Hamaker, B.R., and Larkins, B.A. (2008) Genetic analysis of *opaque2* modifier loci in quality protein maize. *Theor. Appl. Genet.*, **117**, 157–170.

16 Johal, G.S., Balint-Kurti, P., and Weil, C.F. (2008) Mining and harnessing natural variation: a little MAGIC. *Crop. Sci.*, **48**, 2066–2073.

17 Babu, R., Nair, S.K., Kumar, A., Venkatesh, S., Sekhar, J.C., Singh, N.N., Srinivasan, G., and Gupta, H.S. (2005) Two-generation marker-aided backcrossing for rapid conversion of normal maize lines to quality protein maize (QPM). *Theor. Appl. Genet.*, **111**, 888–897.

18 Chalfie, M., Tu, Y., Euskirchen, G., William, W.W., and Douglas, C.P. (1994) Green fluorescent protein as a marker for gene expression. *Science*, **263**, 802–805.

19 Shepherd, C.T., Vignaux, N., Peterson, J.M., Johnson, L.A., and Scott, M.P. (2008) Green fluorescent protein as a tissue marker in transgenic maize seed. *Cereal Chem.*, **85**, 188–195.

20 Frisch, M., Bohn, M., and Melchinger, A.E. (1999) Comparison of selection strategies for marker-assisted backcrossing of a gene. *Crop. Sci.*, **39**, 1295–1301.

21 Gordillo, G.A. and Geiger, H.H. (2008) Alternative recurrent selection strategies using doubled haploid lines in hybrid maize breeding. *Crop. Sci.*, **48**, 911–922.

16
Gene Targeting as a Precise Tool for Plant Mutagenesis
Oliver Zobell and Bernd Reiss

Abstract

Gene targeting allows precision engineering of genes, and is the method of choice for stable and heritable mutagenesis of genes. The technology has developed into an indispensible tool in reverse genetics and also forms a solid basis for the analysis of gene function with modern approaches. Unfortunately, gene targeting efficiencies are still disappointingly low in flowering plants and prevent the routine application of this tool. The situation is entirely different in the moss *Physcomitrella patens*. After the discovery that gene targeting is highly efficient in this organism, *P. patens* has developed into an important model plant within a few years. In addition, the technology has also been used to analyze an appreciable number of genes and biological processes. In this chapter, we describe the basics of gene targeting technology in *P. patens*, but also mention the problems that may be encountered in using the system.

16.1
Introduction

Gene targeting is the tool of choice for gene function analysis in reverse genetics approaches. Gene targeting, also known as targeted gene replacement, gene replacement, targeted homologous recombination, or "the production of gene knockouts," refers to a technology that allows the precise modification of any gene in the genome. In this process, an *in vitro* modified copy is introduced into the genome by transformation to replace the endogenous gene by homologous recombination. The main application of gene targeting is the analysis of gene function. The advantage of gene targeting over other methods is the production of precise, stable, and heritable mutations in any given gene of a genome. The value of this technology is illustrated best in systems with a history of application. Gene targeting is a routine technology in budding yeast (*Saccharomyces cerevisiae*), mouse [1], and chicken DT40 cells [2]. In all three systems gene targeting was either instrumental in the establishment of the system as model (mouse and chicken DT40) or has contributed substantially to the use of reverse genetics for gene function analysis (yeast) [3].

The development of gene targeting as routine tool in plants is lagging considerably behind. Although it is already 20 years ago that it was shown to be feasible in plants [4], this technology is still far from being routine [5]. In contrast, this technology is highly efficient in the moss *Physcomitrella patens*. Since its discovery about 10 years ago [6, 7], the method has become increasingly popular and is now routine. Gene targeting also led to a revival of this traditional plant model system [8–16] and *P. patens* has developed into an important model plant [17–20]. The popularity of gene targeting in *P. patens* is expected to increase even further after major drawbacks (i.e., limited availability of tools and DNA sequence information) have been overcome [21–23].

Although routine, gene targeting in *P. patens* is not without problems. Transformation of *P. patens* is not as efficient as it seems at first glance. Transformation yields stable and unstable transformants [16]. Unstable transformants remain resistant only if constant selection pressure is applied and the transgenes are lost as soon as selection is discontinued. The transgenes in this type of transformants consist of extrachromosomally replicated DNA that presumably consists of concatamers of the transformed DNA [7, 24, 25]. Unstable transformants are a particular problem with circular ends-in or insertion-type vectors – the vector type originally used for gene targeting in *P. patens*. For this vector type almost all of the transformants can be unstable and the actual yield of stable transformants is rather low. This problem is far less pronounced with the use of linearized vectors as is routine for gene replacement or ends-out vectors that are now predominantly used, but the problem still exists. An additional problem is the complexity of the modified loci. In the majority of cases, a large number of copies of the targeting vector (dozens to hundreds) integrated at the target locus [7, 25]. Therefore, single-copy integrations representing the predicted gene replacement events are rare and rather difficult to find [26, 27], even if an optimal vector design is used [28]. This problem complicates more sophisticated applications of gene targeting and is labor-intensive since a large number of transformants has to be produced and analyzed. On top of these problems, simple polymerase chain reaction (PCR) strategies proved insufficient to characterize gene replacement mutants. In addition, transformants may exhibit phenotypes that are not related to the mutation induced by gene targeting, thus making the generation and analysis of several independent transformants necessary to obtain conclusive data [26]. Another principal problem is gene redundancy. Although *P. patens* is haploid in the vegetative phase, a substantial fraction of the genome is duplicated [23, 29] and therefore gene redundancy has to be considered in any gene targeting strategy with the aim to study gene function [30].

Gene targeting is based on the transformation of vector DNA into cells. These vectors are usually referred to as targeting vectors or repair constructs, whereas the genes in the genome that are to be mutated are generally referred to as target genes. For gene targeting, two basically different strategies and, consequently, types of vectors are used – ends-in (insertional) and ends-out (replacement) vectors (Figure 16.1). Historically, the ends-in vector was the first one that was used.

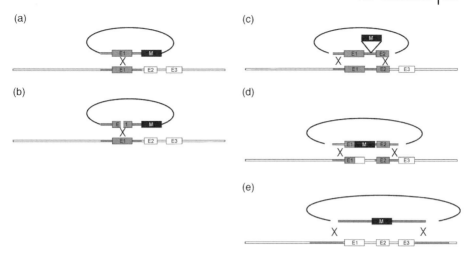

Figure 16.1 Gene targeting vector designs. Ends-in (insertion-type) vectors used in their circular form (a) or linearized before transformation within the region of homology (b). Ends-out (replacement) vectors for which the targeting fragments are released by endonuclease restriction at both ends. The targeting fragment recombines at both ends with the target gene and replaces this by the modified copy. Designs are shown in which replacement results in an insertion (c), a partial (d), and a complete (e) deletion. The crossover events theoretically mediating integration are indicated by "X".

This vector contains the portion of DNA with homology to the target gene next to a selectable marker gene cloned in a standard cloning vector (Figure 16.1a and b). Assuming that targeted gene integration follows the standard mechanisms of homologous recombination, this vector inserts into the genome via a single crossover in the region of homology, resulting in a partial duplication of the target gene sequences and at the same time a disruption of the target gene. This strategy has initially been used in *P. patens* [7, 31, 32]. The efficiency of gene targeting greatly increases by cutting the vector in the region of homology in yeast and mammalian cells [33], but in *P. patens* this strategy was not pursued since ends-in vectors were superseded very soon by ends-out vectors. In this vector type, the selectable marker gene is flanked by the regions of homology to the target gene (Figure 16.1c and d). This insert is released with restriction enzymes to generate two free ends before transformation. The DNA fragment is colinear with the target gene such that the regions of homology face outwards from the selectable marker gene towards the ends of the recombining fragment. On paper, both ends of the fragment recombine with the target gene in independent events and by two independent crossovers. Ends-out vectors are the preferred type for gene targeting since they are designed to replace the genomic region by the modified copy in a single step. However, as mentioned, this outcome is rare in *P. patens*.

There is considerable variation in the design of ends-out vectors. Although they are generally considered as replacement vectors, a vector type that has been frequently used in *P. patens* actually results in an insertion [27, 28, 34–47]. In this case, the selectable marker gene is inserted at a single position (e.g., a restriction site) into the cloned target gene fragment (Figure 16.1c). In this way gene replacement results in the insertion of the selectable marker gene at the predetermined position. Using this strategy, all genomic sequences are retained and there is a risk that the target gene remains active, at least partially. This potential problem is avoided in strategies that delete the entire target gene (Figure 16.1e). However, because large deletions are sometimes difficult to achieve, essential portions of a gene (e.g., entire exons) are often deleted instead (Figure 16.1d) [26, 31, 36, 40, 48–63]. A potential problem with the deletion strategy is that deletions could negatively impact the efficiency and precision of gene targeting, as in mammalian cells [64]. This problem could also exist in *P. patens* (B. Chrost, O. Zobell, E. Wendeler, and B. Reiss, unpublished observations).

16.2
Methods and Protocols

Basic *P. patens* Tissue Culture Methods

The *P. patens* (Hedwig) laboratory strain commonly used is Gransden Wood [65]. The strain used for the whole-genome sequence [23], "Gransden2004," was isolated from a single spore of the original isolate and can be obtained from members of the sequencing consortium (http://genome.jgi-psf.org/Phypa1_1/Phypa1_1.home.html).

P. patens can easily be cultivated in sterile Petri dish cultures in a plant growth cabinet. Continuous light conditions and temperatures of 24–26 °C will result in the strongest growth. Information on the lifecycle and also on the basic principles of tissue culture of *P. patens* can be found in several reviews [12–14, 16, 18, 19, 66–68] and on web sites (http://www.plant-biotech.net/, http://moss.nibb.ac.jp/, http://biology4.wustl.edu/moss/methods.html).

Basic Media and Solutions

- 1000 × KH_2PO_4/KOH buffer: dissolve 250 g KH_2PO_4 in H_2O. Titrate to pH 6.5 with 4 M KOH. Adjust volume to 1 l, autoclave (20 min, 120 °C), and store at 4 °C.
- 1000 × AltTES: dissolve in H_2O: 55 mg $CuSO_4 \cdot 5H_2O$, 614 mg H_3BO_3, 55 mg $CoCl_2 \cdot 6H_2O$, 25 mg $Na_2MoO_4 \cdot 2H_2O$, 55 mg $ZnSO_4 \cdot 7H_2O$, 389 mg $MnCl_2 \cdot 4H_2O$, and 28 mg KI. Adjust to 1 l, autoclave (20 min, 120 °C), and store at 4 °C.
- Minimal medium [26]: dissolve 0.8 g $Ca[NO3]_2 \cdot 4H_2O$, 0.25 g $MgSO_4 \cdot 7H_2O$ and 0.0125 g $FeSO_4 \cdot 7H_2O$ in H_2O, add dropwise while stirring 1 ml 1000 × KH_2PO_4/KOH buffer and 1 ml 1000 × AltTES and adjust volume to 1 l. Stir thoroughly until the solution is homogenous, but a slight precipitate still

remains (∼15 min is generally sufficient). Add 7 g/l Agar-agar (Merck) and autoclave (20 min, 120 °C). Dispense medium in standard 9-cm Petri dishes without vents and let solidify with the lid on (e.g., overnight in a switched-off sterile hood). Switch the sterile hood on again the next day and overlay the medium in each plate with one sterile cellophane disk (see Materials) using flame-sterilized forceps. Store the plates in plastic bags at room temperature (or at 4 °C when containing antibiotics) with the medium facing down and the lid facing up. In this way, liquid is prevented from dropping and thereby sealing off the lid, which could later impede aeration and proper growth of *P. patens*.
- Standard medium: minimal medium supplemented with 500 mg/l ammonium tartrate.
- Standard/gluc medium: standard medium supplemented with 5 g/l glucose. (*Note*: Standard/gluc medium is used to grow plant material for protoplast isolation, because it supports the most vigorous growth. It is not advised to use it for routine cultivation since the cultures rapidly age.)
- Distilled water, autoclaved (20 min, 120 °C).

Routine Propagation

The following protocol describes the weekly routine of blending, regeneration, and cultivation of protonemal tissue on standard medium that is essential to maintain this organism in tissue culture.

Materials
- Plant growth cabinet (26 °C, continuous light, intensity ∼3000 Lux).
- Sterile hood.
- Sterile 9-cm Petri dishes without vents.
- Water-permeable cellophane disks. Soak the water-permeable cellophane disks in a closable preserving jar filled with distilled water (Figure 16.2a). Rinse at least 3 times with distilled water in order to remove impurities. Wrap the jar, containing cellophane disks soaked in distilled water, in aluminum foil and autoclave (20 min, 120 °C). Refresh distilled water and autoclave again as before. The disks are now ready to use.
- Forceps with curved tip.
- Sterile 10-ml plastic pipettes.
- Sterile 50-ml plastic tube with screw cap.
- Tissue homogenizer, installed under the sterile hood. We have used a Miccra homogenizer D8 equipped with a P8 homogenizer tool (Figure 16.2c and e) and a Polytron PT 2100 homogenizer (Figure 16.2d and f).

Procedures
- For routine clonal propagation, we generally use material from four ∼1-week-old cultures grown in continuous light at 26 °C on cellophane-overlaid standard medium (Figure 16.2b). Under the sterile hood, plant material from

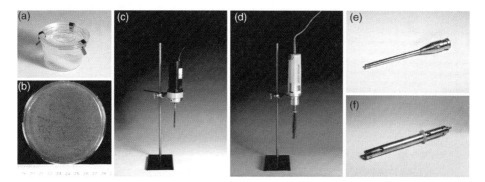

Figure 16.2 Materials used for *P. patens* tissue culture: (a) water-permeable cellophane disks soaked in distilled water in a closable preserving jar; (b) 7-day-old *P. patens* culture grown in continuous light conditions at 26 °C on cellophane-overlaid standard medium; (c) Miccra D8 homogenizer; (d) Polytron PT 2100 homogenizer; (e) Miccra P8 homogenizer tool; (f) Polytron homogenizer tool.

four plates is pushed together with flame-sterilized forceps with a curved tip and transferred to a 50-ml plastic tube filled with 10 ml sterile distilled water.
- Plant material is blended for ∼90 s at speed E with a Miccra homogenizer or at speed 11 with a Polytron PT 2100 homogenizer.
- Add 10 ml sterile distilled water to the blended plant material, to a final volume of 20 ml. Mix by pipetting. Plate a 1-ml aliquot each on one Petri dish containing cellophane-overlaid standard medium. Tilt and rotate the Petri dish to distribute the material.
- If more than one line is to be propagated, the homogenizer tip can be rinsed in between as follows. Immerse the tip in a 500-ml bottle filled with 100% ethanol and operate the homogenizer at the set speed for 5–10 s. Repeat this with 70% ethanol and, finally, with sterile distilled water. The tip is now ready for propagating the next line.
- For routine propagation, do not wrap the plates or wrap with a tape that is permeable to air (e.g., Leukosilk®). Although not absolutely necessary, the latter minimizes the risk of contamination during handling and transport of the plates.
- Cultivate plants in a growth cabinet (26 °C, continuous light, intensity ∼3000 Lux).
- After at least 1 week, cultures will be ready for a new propagation. Alternatively, plates can be wrapped with Parafilm® and stored at 4 °C.

Note: *P. patens* cultures are rather susceptible to fungal or bacterial contamination. Care has to be taken to always handle the material under the sterile hood. When using sugar-free medium such as minimal or standard media, contamination typically remains unnoticed until several rounds of propagation have passed. To monitor contamination more closely, one can routinely dispense and cultivate 1 ml of blended plant material on a plate with standard/gluc medium. When performing repeated rounds of propagation, it is further advised to regularly keep a culture on the side for storage, in order to have an uncontaminated backup. If such an uncontaminated backup is not available, a culture can always be brought back to sterile conditions by sterilization of spores.

Sexual Propagation

To induce the formation of spores, dilute cultures are first grown under optimal conditions (26 °C, continuous light, intensity around 3000 Lux) until gametophores are well established and then transferred to low-temperature (15 °C) [69], short-day (8-h photoperiod), and low-light conditions as described [70]. The culture is on minimal media with cellophane overlay since ammonium tartrate present in standard media inhibits the formation of sexual organs and hence the completion of the sexual lifecycle. This was the most efficient method in our hands and spores can be obtained in less than 4 months.

Solutions and Media
- Petri dishes (without vents) with minimal medium and cellophane overlay.
- Distilled water, autoclaved (20 min, 120 °C).

Materials
- Sterile hood.
- Sterile 50-ml plastic tube with screw cap.
- Fine forceps and scalpel.
- Plant growth cabinet (26 °C, continuous light, intensity ~3000 Lux).
- Plant growth cabinet (15 °C, 8 h photoperiod, light intensity ~1000 Lux).

Procedures
1. Take an aliquot of blended plant material as for routine clonal propagation and dilute it 5 times (somewhat depending on the density of the culture) with sterile distilled water in a sterile 50-ml plastic tube with screw cap. Mix.
2. Plate 1-ml aliquots on a Petri dish with minimal medium and cellophane overlay. Do not wrap with Parafilm.
3. Cultivate plants in a growth chamber (26 °C, continuous light, intensity ~3000 Lux) for 4–6 weeks, until a lawn of well-established gametophores has formed.
4. Water the Petri dishes with sterile distilled water. Moisten the culture thoroughly to facilitate the movement of spermatozoa, but do not drown it. Wrap with Parafilm® to prevent them from drying out and transfer the cultures to a growth cabinet (light intensity ~1000 Lux) set at low-temperature (15 °C) and short-day conditions (8-h photoperiod).
5. After at least 4 weeks, water the cultures again as before.
6. A few weeks later, the first spore capsules will start to appear. This can occasionally take longer.
7. Harvest brown spore capsules with fine, sterile forceps and, if necessary, a scalpel. Collect one or more capsules in a 1.5-ml tube containing 200 µl sterile distilled water. Alternatively, the spores can be collected in a 1.5-ml tube and air-dried under the sterile hood.

8. Store spore capsules at 4 °C in darkness.
9. For germination of spores, add sterile distilled water to a final volume of 1 ml.
10. Release the spores from the spore capsule by squeezing with the pipette tip and by pipetting. One spore capsule contains ∼4000 haploid spores [67].
11. Dispense 20 µl of spore solution onto a Petri dish containing cellophane-overlaid standard medium. Wrap with Parafilm®. Remove Parafilm® after spores have germinated and small protonemata are established.
12. Cultivate in a growth chamber (26 °C, continuous light, intensity ∼3000 Lux). Spore germination, which critically depends on light, will be visible within a couple of days.

Sterilization of Spores

Any procedure commonly used for the sterilization of seeds that does not use ethanol generally works well for *P. patens* spores. For example, 32% Klorix (household cleaner) with 0.8% (w/v) Sarkosyl (*N*-lauroylsarcosine sodium salt) added, or 0.5% hypochlorite solution (8.5–13.5% Cl) with 0.05% Triton X-100 added work well. After mixing the spores in a 1.5-ml tube with 1 ml of sterilization solution, incubate under regular shaking for 8 min with the Klorix-based solution or 2 min with the hypochlorite-based solution. Sediment the spores by brief centrifugation and wash 5 times with 1 ml sterile, distilled water. Finally resuspend the spores in a smaller volume of sterile, distilled water. Spores can be immediately plated to germinate or stored for later use.

Storage of Cultures

One-week-old Petri dish cultures may be wrapped with Parafilm® and stored in light at 15 °C for at least 2 months or at 4 °C for at least 1 year. Storage in darkness will negatively affect culture longevity. Restarting growth will be slow after both kinds of storage. At least one additional round of clonal propagation should be performed before the material is used for experiments.

Construction of Targeting Vectors

The availability of the complete genome sequence [23] has considerably facilitated the construction of gene targeting vectors. The complete sequence information is available in public databases like EMBL or GenBank. In addition, several genome browsers (http://genome.jgi-psf.org/Phypa1_1/Phypa1_1.home.html, www.cosmoss.org, and moss.nibb.ac.jp) provide convenient access to all *P. patens* DNA sequence information such that simple homology searches like BLAST now retrieve all sequence information that is required for the cloning of any kind of targeting vector and for almost any gene of interest.

Homology Length

Length and symmetry of the homologous regions both influence targeting efficiency in *P. patens* [28]. In a symmetric targeting construct, a minimum of 400 bp of homology on each arm is sufficient to achieve gene targeting. However, homology lengths of around 1 kb on each arm are optimal. Nonhomologous termini that are easily generated by releasing fragments that had been cloned into multiple cloning sites of standard cloning vectors may decrease targeting efficiency, if the length of homology is minimal. This problem is less significant with larger regions of homology. The problem can be circumvented using PCR to generate the targeting fragment [28]. However, a better strategy is to design the targeting construct, if possible, in such a manner that original restriction sites in the genomic target locus are used for cloning or cryptic sites are converted to real sites in the PCR when the fragments are amplified from genomic DNA. In this way homology will extend to the termini of the construct and targeting will not be affected by terminal heterology.

Selectable Marker Genes

A comprehensive compilation of selectable marker genes used successfully in *P. patens* is found in a recent review [18] and at PHYSCObase (moss.nibb.ac.jp). Selectable marker genes giving rise to G418 (geneticin) or hygromycin resistance were historically the first ones to be used in *P. patens*, and are still the most popular genes. There seems to be no need to specifically adapt selectable marker genes to *P. patens*. We have used selectable marker genes quite successfully in *P. patens* that were originally designed for use in flowering plants – a hygromycin resistance gene under the control of the nopaline synthase promoter [71] and a neomycin-phosphotransferase II (*nptII*) gene under the control of the cauliflower mosaic virus (CaMV) 35S promoter [26]. In addition, we have used a sulfadiazine resistance gene under the control of the CaMV 35S promoter [26] and others used a zeocin resistance gene under the control of the CaMV 35S promoter [58]. In our experience, G418 (used with the 35S::*nptII* selection marker) is the most potent antibiotic, followed by hygromycin and sulfadiazine.

Reporter Genes

The β-glucuronidase (*uidA*, *GUS*) gene, a reporter gene well established in flowering plants, also functions well in *P. patens* [35, 37, 61, 72]. More caution is needed with fluorescent reporters like Green Fluorescent Protein or *Discosoma* sp. Red Fluorescent Protein. The signals obtained with such reporters rarely exceed the level of background fluorescence (U. Markmann-Mulisch, B. Chrost, and B. Reiss, unpublished results), except when expression is either very strong and/or the proteins specifically accumulate in a cellular compartment, as described in some rare reports [59, 61].

Note

The gene duplications in the *P. patens* genome have been mentioned already; however, it is important to note here again that there is a considerable risk that functionally redundant homologs of the gene of interest exist in the *P. patens* genome. In a number of cases [26, 35–37, 57, 73], double knockouts had to be generated by two consecutive transformation experiments, using a different selection marker for targeting each gene. This is the fastest approach to create double knockout lines; however, crossing of single knockout lines is also possible, but slower. Alternatively, simultaneous targeting of multiple genes seems possible [74], but was extremely inefficient and labor-intensive in our hands.

Targeting Experiment

A 6-fold upscaled version of the original protocol [7] is presented here. Other protocols are available at several websites (www.plant-biotech.net, moss.nibb.ac.jp, and http://biology4.wustl.edu/moss/methods.html). The dimension of this experiment was chosen to generate a sufficient number of transformants in a single experiment under optimal conditions. In this protocol, 2.7×10^6 protoplasts are transformed by polyethylene glycol (PEG)-mediated transformation with 90 μg of the linearized targeting vector. The entire transformation experiment is completed in one and a half day. All material is sterilized and the transformation is carried out under sterile conditions.

Step 1: Protoplast Isolation (Day 1)

Solutions and Media
- 0.48 M mannitol (87.5 g/l).
- 2% Driselase solution: 2% (w/v) Driselase® in 0.48 M mannitol. Stir and dissolve Driselase® in 0.48 M mannitol. Stir gently for 30 min in total. Centrifuge for 10 min at 4 °C at $12\,000 \times g$ in a fixed angle rotor. Filter the supernatant sterile through a 0.22-μm filter. Prepare solution freshly for each experiment and keep on ice until used.
- MMM: 0.48 M mannitol (8.5%); 15 mM $MgCl_2$; 0.1% MES (2-[N-morpholino]ethanesulfonic acid); pH 5.6 with KOH.

Materials
- Sterile hood.
- At 5 days before protoplast isolation, blend cultures grown on standard medium as described for routine subculturing, and prepare 10 Petri dishes by plating the material on standard/gluc medium overlaid with cellophane disks. Do not wrap with Parafilm®. Culture under continuous light conditions at 26 °C (5-day-old cultures are optimal for protoplast isolation).

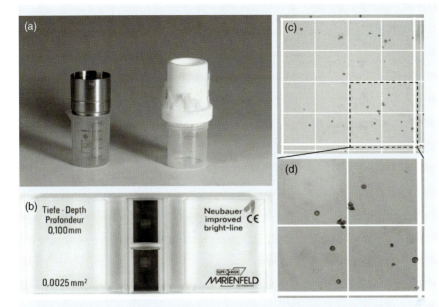

Figure 16.3 Materials used for *P. patens* protoplast isolation: (a) steel filter unit (left) and Teflon filter unit with nylon membrane (right) plus receptacle; (b) cell counting chamber; (c) Purified protoplasts in one quadrant of the cell counting chamber, observed through a binocular; (d) close-up of (c).

- Sterile 9-cm Petri dish.
- Forceps with curved tip.
- 50-μm and 100-μm filter units, such as a steel filter unit (Figure 16.3a, left) or a Teflon filter unit with nylon membrane (Figure 16.3a, right), with a receptacle like a plastic beaker (do not use glass) with a volume of at least 30 ml, completely wrapped in aluminum foil and autoclaved (20 min, 120 °C).
- Cell counting chamber (Figure 16.3b).
- Sterile 25-ml plastic pipettes.
- Sterile 50-ml plastic tube with screw cap.

Procedures
1. Fill a sterile 9-cm Petri dish under the sterile hood with 10 ml 0.48 M mannitol. Use flame-sterilized forceps to harvest plants from 8–10 Petri dishes (the exact amount depends on the quality and density of the cultures) and transfer the material to the Petri dish containing 0.48 M mannitol. Disentangle clumps with forceps.
2. Evenly dispense 15 ml 2% Driselase® solution in the Petri dish. The material is incubated at 21 °C for 30 min during which the suspension is occasionally mixed by gentle tilting and turning of the Petri dish.
3. Use a sterile 25-ml plastic pipette to transfer the material from the Petri dish onto the first, 100-μm filter unit. Pipette very gently in order not to damage the protoplasts. Wait for the entire solution to flow through. In case of a

constipated flow, slight flicking against the filter should get the flow started again.
4. Carefully pour or pipette the protoplasts from the receptacle onto the second 50-μm filter unit. Filtered protoplasts are again collected in the accompanying receptacle and then poured into a sterile 50 ml plastic tube with screw cap. Fill up gently to a volume of 50 ml with 0.48 M mannitol.
5. Protoplasts will now be washed. Centrifuge at room temperature for 5 min at $50 \times g$ or 3 min at $100 \times g$ in a swing out rotor with lowest settings of acceleration and deceleration. Discard supernatant by carefully decanting or pipetting, without disturbing the pelleted protoplasts. Gently add 0.48 M mannitol to a final volume of 50 ml, resuspend carefully protoplasts by gently tilting and turning the tube.
6. Repeat Step 5 twice, finally resuspending the protoplasts again in 50 ml 0.48 M mannitol.
7. Count protoplasts in a cell counting chamber (Figure 16.3b–d) under a microscope or binocular and calculate the protoplast concentration according to the manufacturer's instructions.
8. Centrifuge again under the same conditions as in Step 5. This time, however, resuspend the protoplasts in MMM, to a final concentration of 1.5×10^6 protoplasts/ml.

Step 2: Transformation (Day 1)

Solutions and Media
- DNA: 90 μg of sterile, linearized targeting vector DNA at 0.5 mg/ml. To sterilize DNA, precipitate with ethanol (2.5 original volume) and sodium acetate (1/10 original volume 3M solution), wash twice with ethanol (80%) and dry the pellet under a sterile atmosphere. Finally dissolve in 180 μl sterile water, to a concentration of 0.5 mg/ml.
- PEG solution: 0.38 M mannitol (69.2 g/l); 0.1 M $Ca(NO_3)_2$; 33% (w/v) PEG 4000 (SERVA); 10 mM Tris–HCl, pH 8.0. Stir and heat moderately (~50 °C) to dissolve. Filter sterile through a 0.45-μm filter. Aliquots can be stored at $-20\,°C$.
- Standard/mannitol medium: standard medium (without agar), supplemented with 66 g/l mannitol (~480 mOsmol). Autoclave (20 min, 120 °C). Store at room temperature.

Materials
- Sterile hood.
- Sterile 50-ml plastic tube with screw cap.
- Water bath (45 °C).
- Sterile plastic pipettes.

Procedures
1. Pipette 180 μl (90 μg) DNA to the bottom of a sterile 50-ml plastic tube with a screw cap.

2. Add 1.8 ml (2.7×10^6) protoplasts and mix by gentle tilting and turning.
3. Add 1.8 ml PEG solution and mix by gentle tilting and turning.
4. Heat-shock for 3 min in a 45 °C water bath.
5. Keep at room temperature for 10 min.
6. Add 4 ml standard/mannitol medium, mix by gentle tilting and turning, and let stand for 3 min. Repeat 4 times.
7. Add 8 ml standard/mannitol medium, mix by gentle tilting and turning, and let stand for 3 min. Repeat until a final volume of 50 ml is reached.
8. Store the tube overnight in darkness at room temperature (around 20 °C). Generally, storage in the lab in a dark cupboard is fine.

Step 3: Plating (Day 2)

Solutions and Media
- Standard/mannitol LM agarose: standard/mannitol medium + 1% low-melting (LM) agarose. Autoclave (20 min, 120 °C) and cool down to 42 °C in a water bath.
- Standard/mannitol agar: standard/mannitol medium with 7 g/l agar added and autoclaved (20 min, 120 °C).
  Dispense standard/mannitol agar medium in standard 9-cm Petri dishes. After solidification, overlay the medium in each plate with one sterile cellophane disk.

Materials
- Sterile hood.
- Water bath (42 °C).
- Sterile 25 ml plastic pipettes.
- Plant growth cabinet (26 °C, continuous light, intensity 3000 Lux).

Procedures
- Centrifuge for 3 min at room temperature at $50 \times g$ in a swing-out rotor with lowest settings of acceleration and deceleration.
- Remove 40 ml of the supernatant with a pipette.
- Add 10 ml standard/mannitol LM agarose. The final volume is 20 ml. Mix by gentle tilting and turning.
- Pipette the protoplast suspension in 2.5-ml aliquots on the cellophane-overlaid Petri dishes containing standard/mannitol agar medium. Act quickly to prevent agarose from gelling in the tube. Wrap with a thin layer of Parafilm®.
- Cultivate in continuous light (light intensity 3000 Lux) at 26 °C for 6 days.

Alternative Procedure
As an alternative to embedding in agarose, protoplasts can be regenerated in liquid medium as described [74].

Step 4: Selection of Transformants (~2 Months)

Two classes of antibiotic-resistant transformants are obtained in *P. patens* which cannot be distinguished directly after the transformation – stable and unstable transformants. While stable transformants are the desired ones that have the transgenic DNA inserted in the genome, extrachromosomally replicated DNA gives rise to unstable transformants which lose resistance when selection is released. To eliminate those, two rounds of selection and relaxation should be performed, after which the majority of surviving transformants are stable [75].

With the selectable marker genes described above and under our conditions, we select on 50 mg/l sulfadiazine, 15 mg/l hygromycin, and 20 mg/l G418. Selection at too high antibiotic concentrations bears the risk of selecting lines with multiple inserts, also in *P. patens* [37]. Therefore, it is advisable to determine the optimal concentration under the particular growth conditions used in the experiment. The simplest way to do this is to determine the minimal concentration of antibiotic that is lethal to untransformed cells and then use a slightly higher concentration for selection.

Solutions and Media
- Petri dishes (without vents) containing cellophane-overlaid standard medium + antibiotic.
- Petri dishes (without vents) containing cellophane-overlaid standard medium.

Materials
- Sterile hood.
- Forceps.
- Plant growth cabinet (26 °C, continuous light, intensity ∼3000 Lux).

Procedures
1. Transfer the cellophane disks together with the agarose and the embedded protoplasts to selection plates containing standard medium *with* antibiotic under sterile conditions. Wrap with a thin layer of Parafilm®.
2. Grow on selection plates in a plant growth cabinet (26 °C, continuous light, intensity ∼3000 Lux) for 3–4 weeks or until small colonies have formed, but before they grow into one another.
3. Transfer each colony with flame-sterilized forceps to cellophane-overlaid standard medium *without* antibiotic. Do not wrap with Parafilm® from here on.
4. Cultivate for ∼2 weeks in a plant growth cabinet (26 °C, continuous light, intensity ∼3000 Lux).
5. Transfer a small protonema piece of each colony with flame-sterilized forceps back to cellophane-overlaid standard medium *with* antibiotic.
6. Cultivate for ∼2 weeks in a plant growth cabinet (26 °C, continuous light, intensity ∼3000 Lux).
7. Survivors of this second selection round are scored as stable transformants.
8. Cultivate further until adequate mass has been produced, then proceed to clonal propagation by blending to generate sufficient material for isolation of nucleic acids and molecular analysis.

Analysis of Transformants

Gene targeting can produce mutants that are considerably different from the expected gene replacement and therefore a carefully designed PCR strategy and a Southern blot analysis is required to characterize the transformants. Only a combination of the data of both allows the unambiguous identification of the transformants that have the desired gene replacement. A PCR analysis commonly used exists of two PCR reactions with primer combinations designed to detect correct integration of the repair construct at the 5′ and 3′ ends. The primers for these reactions are chosen such that one primer is located in the genomic sequence outside of, but close to, the regions of homology with the repair construct (5′ and 3′). The other primers are located in the selectable marker or reporter gene in the repair construct such that an amplifiable PCR product can be generated only after correct recombination. In theory, this test is failsafe since a PCR product is detectable only after correct recombination. However, it is important to bear in mind that ectopic recombination is extremely common in plants [5] including *P. patens* (B. Chrost, O. Zobell, E. Wendeler, and B. Reiss, unpublished) and easily generates false-positives that cannot be distinguished from the expected recombination event by PCR fragment size or the DNA sequence of the product. Therefore, additional analyses are required. Since *P. patens* is haploid and thus a single copy of the target gene is present in the genome, much more conclusive analyses aim at the confirmation that this copy was altered by gene targeting. If possible, such as in targeting strategies that foresee the deletion of sequences from the genome, a PCR may be designed to confirm the absence of this portion of the gene in genomic DNA. In addition, with modern PCR technology that easily allows the amplification of long sequences from genomic DNA (e.g., Expand Long Template PCR System Kit; Roche) it is possible to analyze the modified target gene using primers located in the genomic region outside the regions of homology with the repair construct (e.g., the same ones used in the first PCR) and to amplify across the target gene. An unmodified copy of the target gene that may have persisted is easily detected and recognized by the size of the amplified fragment. However, the drawback of this approach is that a modified copy is detected only in those rare cases in which correct gene replacement has occurred. In all other cases no PCR product is detectable, either because additional copies of the repair construct have integrated into the target locus [5, 12, 26–28] or because recombination was imprecise. Therefore, and in order to exclude false-positive and false-negative PCR results, additional Southern blot data are essential. For Southern blots, genomic DNA is digested with restriction enzymes chosen such that fragments are generated that allow facile distinction of target and modified gene sequences by differences in fragment sizes. As the integration of multiple copies of the repair construct is common [5, 12, 26–28], it is advisable to chose a fragment as a probe that either detects the deleted portion [26] or that hybridizes to genomic sequences outside of the regions of homology with the repair construct ([71], and B. Chrost, O. Zobell, E. Wendeler, and B. Reiss, unpublished). In addition, probing with sequences representing the genomic regions inside the regions of homology, the marker gene, and vector sequences is necessary to determine copy numbers and additional insertions of the vector that might have occurred [26, 37]. Finally, reverse transcription-PCR can be used to confirm that the target gene is no longer transcribed.

As an example of unusual mutants that can be generated by gene targeting, we describe mutants that were generated by a larger chromosomal deletion of the target locus instead of the expected gene replacement. These mutants occurred in a significant number of experiments to target the *PpCOL2* gene [71]. The targeting vector in these experiments was designed to delete a portion of the *PpCOL2* coding region. Knockout mutants were identified using a Southern Blot analysis with a probe hybridizing to the deleted portion and a number of transformants were found in which the *PpCOL2* gene was completely deleted. However, a probe hybridizing 5′ upstream of the region of homology revealed that some of them additionally harbor a chromosomal deletion that extends beyond the target gene. Moreover, flanking chromosomal DNA is deleted in one of the transformants, but the target gene is still present. The transformants carrying the large chromosomal deletion show a significant phenotype – strongly reduced colony size with aberrant morphology (Figure 16.4a). In the absence of a careful molecular characterization, this phenotype would have been erroneously taken for the *Ppcol2* mutant phenotype. However, mutants carrying the correct deletion of the *PpCOL2* gene and transformants in which the targeting vector has integrated at random show no apparent phenotype.

Even if a careful molecular characterization had indicated a gene replacement, phenotypes exist that are not caused by the predicted loss of gene function. We have observed a number of such phenotypes that were clearly not linked to the mutated gene in our analysis of *Pprad51B* mutants [26]. One potential cause is polyploidization which is common in PEG-mediated DNA uptake in *P. patens* [26, 76] and one of our *Pprad51B* knockout mutants that exhibits a severe growth phenotype turned out later to be tetraploid (Figure 16.4b). More and unknown mechanisms causing phenotypes are likely to exist, possibly in combination with polyploidy, as demonstrated by further examples of phenotypes not linked to the *Pprad51B* mutation (Figure 16.4b). Since polyploidy is not visible in Southern blot or PCR analyses, it is advisable to perform a flow cytometric analysis [76] to exclude this as the cause of a phenotype. Moreover, it seems mandatory to create and analyze several independent knockout lines since a given phenotype seems to be reliable only if independent lines display the same.

There are as manifold methods (see references in Section 16.3) to analyze phenotypes as biological processes that are analyzed in *P. patens*. However, as a basic principle, a good way to start is the macroscopically visible colony morphology. There are two ways to inoculate a *P. patens* colony. The ideal way that yields the most standardized conditions is inoculation with a single spore. However, spore development itself might be compromised or time too limited to

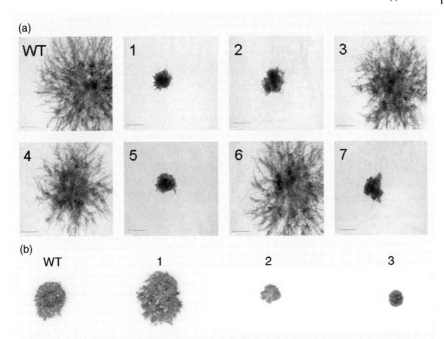

Figure 16.4 Phenotypes not linked to targeted mutations. (a) Colony morphologies of transformants obtained by transformation with a PpCOL2 targeting construct. Transformants with correct gene replacement: 4 and 6. Transformants with an additional deletion: 1, 2, and 7. Transformants with a deletion of flanking DNA but still containing the PpCOL2 gene: 5. Transformants with randomly integrated PpCOL2 targeting vector DNA: 3. Scale bars = 2 mm. (b) Polyploid P. patens lines obtained upon targeting the PpRAD51B gene [26]. The colony morphologies of wild-type (WT), a diploid line containing a mutated and wild-type PpRAD51B allele in Pprad51A background (1), a tetraploid Pprad51B knockout line (2), and a diploid line which contains a mutated and a wild-type PpRAD51B allele (3). The phenotype of all lines deviates from that of the described mutants [26]. All cultures were grown on standard medium as described [26].

wait for the completion of the lifecycle. Therefore a commonly used alternative is to start from a small piece of fresh protonemal tissue (e.g., [26, 36]; Figure 16.4). This simple analysis easily reveals defects in differentiation and development of caulonema and chloronema cells, branching of protonemal filaments, and induction and development of leafy gametophores and sporophytes, which are all reflected in colony morphology.

16.3
Applications

In the decade after the first publication of gene targeting [7], the number of genes analyzed using this methodology in P. patens is now approaching 50 and the number of publications based on gene targeting in *P. patens* is steadily increasing. Almost all biological processes important in plant sciences have been analyzed in *P. patens* by now considering that mosses are nonvascular plants and therefore it is naturally impossible to study processes that are specific for higher plants, like the development of the vascular system, the flower, or the seed. However, nearly all other facets of plant sciences are covered. These include chloroplast biology [32, 34, 37, 38, 50, 56, 77], photobiology [31, 32, 42, 52, 73, 78], plant development [35, 39, 41, 44, 46, 48, 54, 57–59, 61, 79, 80], plant hormones [49, 81], stress biology [43, 60], DNA damage repair [26, 62], sugar signaling [36, 45], water and nutrient transport [47, 53, 55], fatty acid synthesis [51, 82], and biotechnology [40]. In most of these publications fundamental, novel knowledge is reported that would not have been gained without gene targeting in *P. patens*, and this testifies to both the value of the model organism *P. patens* and the usefulness of gene targeting for systematic functional genomics studies.

16.4
Perspectives

It would be clearly desirable to have gene targeting as a routine tool for gene function analysis in flowering plants, especially since a second important application of this technology – precision engineering of transgenes – would have a multitude of applications in crops. However, this technology is still far from being routine in crops, with one important exception – rice. A highly efficient and reproducible gene targeting method is now available for this important crop plant. The rice technology is based on a highly efficient transformation system and the development of a new, highly stringent negative selection system [83]. This technology has been applied to inactivate a number of different genes [84–87], and thus has evolved into an established and reproducible methodology. Thus, rice is an alternative to *P. patens*, although gene targeting in rice is considerably more labor-intensive. Moreover, there are efforts to understand the biological mechanisms underlying the high gene-targeting efficiencies in *P. patens*. Such approaches should considerably improve the prospects to establish gene targeting as a routine tool in flowering plants in the near future.

Acknowledgments

We thank Maret Kalda for photography, and Edelgard Wendeler for providing the material used for the pictures and for carefully reading the manuscript.

References

1 Capecchi, M.R. (2005) Gene targeting in mice: functional analysis of the mammalian genome for the twenty-first century. *Nat. Rev. Genet.*, **6**, 507–512.

2 Yamazoe, M., Sonoda, E., Hochegger, H., and Takeda, S. (2004) Reverse genetic studies of the DNA damage response in the chicken B lymphocyte line DT40. *DNA Repair*, **3**, 1175–1185.

3 Gu, Z., Steinmetz, L.M., Gu, X., Scharfe, C., Davis, R.W., and Li, W.-H. (2003) Role of duplicate genes in genetic robustness against null mutations. *Nature*, **421**, 63–66.

4 Paszkowski, J., Baur, M., Bogucki, A., and Potrykus, I. (1988) Gene targeting in plants. *EMBO J.*, **7**, 4021–4026.

5 Reiss, B. (2003) Homologous recombination and gene targeting in plant cells. *Int. Rev. Cytol.*, **228**, 85–139.

6 Kammerer, W. and Cove, D.J. (1996) Genetic analysis of the effects of re-transformation of transgenic lines of the moss *Physcomitrella patens*. *Mol. Gen. Genet.*, **250**, 380–382.

7 Schaefer, D.G. and Zryd, J.P. (1997) Efficient gene targeting in the moss *Physcomitrella patens*. *Plant J.*, **11**, 1195–1206.

8 Reski, R. (1999) Molecular genetics of *Physcomitrella*. *Planta*, **208**, 301–309.

9 Reski, R., Faust, M., Wang, X.H., Wehe, M., and Abel, W.O. (1994) Genome analysis of the moss *Physcomitrella patens* (Hedw) Bsg. *Mol. Gen. Genet.*, **244**, 352–359.

10 Reski, R. (1998) *Physcomitrella* and *Arabidopsis*: the David and Goliath of reverse genetics. *Trends Plant Sci.*, **3**, 209–210.

11 Holtorf, H., Guitton, M.C., and Reski, R. (2002) Plant functional genomics. *Naturwissenschaften*, **89**, 235–249.

12 Schaefer, D.G. (2001) Gene targeting in *Physcomitrella patens*. *Curr. Opin. Plant Biol.*, **4**, 143–150.

13 Cove, D. (2000) The moss, *Physcomitrella patens*. *J. Plant Growth Regul.*, **19**, 275–283.

14 Cove, D.J., Knight, C.D., and Lamparter, T. (1997) Mosses as model systems. *Trends Plant Sci.*, **2**, 99–105.

15 Wood, A.J., Oliver, M.J., and Cove, D.J. (2000) Bryophytes as model systems. *Bryologist*, **103**, 128–133.

16 Schaefer, D.G. and Zryd, J.P. (2001) The moss *Physcomitrella patens*, now and then. *Plant Physiol.*, **127**, 1430–1438.

17 Frank, W., Decker, E.L., and Reski, R. (2005) Molecular tools to study *Physcomitrella patens*. *Plant Biol.*, **7**, 220–227.

18 Cove, D. (2005) The moss *Physcomitrella patens*. *Annu. Rev. Genet.*, **39**, 339–358.

19 Cove, D., Bezanilla, M., Harries, P., and Quatrano, R. (2006) Mosses as model systems for the study of metabolism and development. *Ann. Rev. Plant. Biol*, **57**, 497–520.

20 Quatrano, R.S., McDaniel, S.F., Khandelwal, A., Perroud, P.F., and Cove, D.J. (2007) *Physcomitrella patens*: mosses enter the genomic age. *Curr. Opin. Plant Biol.*, **10**, 182–189.

21 Richardt, S., Lang, D., Reski, R., Frank, W., and Rensing, S.A. (2007) PlanTAPDB, a phylogeny-based resource of plant transcription-associated proteins. *Plant Physiol.*, **143**, 1452–1466.

22 Rensing, S.A., Rombauts, S., Van de Peer, Y., and Reski, R. (2002) Moss transcriptome and beyond. *Trends Plant Sci.*, **7**, 535–538.

23 Rensing, S.A., Lang, D., Zimmer, A.D., Terry, A., Salamov, A., Shapiro, H., Nishiyama, T., Perroud, P.-F., Lindquist, E.A., and Kamisugi, Y. *et al.* (2008) The *Physcomitrella* genome reveals evolutionary insights into the conquest of land by plants. *Science*, **319**, 64–69.

24 Ashton, N.W., Champagne, C.E.M., Weiler, T., and Verkoczy, L.K. (2000) The bryophyte *Physcomitrella patens* replicates extrachromosomal transgenic elements. *New Phytologist*, **146**, 391–402.

25 Schaefer, D., Zryd, J.P., Knight, C.D., and Cove, D.J. (1991) Stable transformation of the moss *Physcomitrella patens*. *Mol. Gen. Genet.*, **226**, 418–424.

26 Markmann-Mulisch, U., Wendeler, E., Zobell, O., Schween, G., Steinbiss, H.H., and Reiss, B. (2007) Differential requirements for RAD51 in *Physcomitrella patens* and *Arabidopsis thaliana* development and DNA damage repair. *Plant Cell*, **19**, 3080–3089.

27 Kamisugi, Y., Schlink, K., Rensing, S.A., Schween, G., von Stackelberg, M., Cuming, A.C., Reski, R., and Cove, D.J. (2006) The mechanism of gene targeting in *Physcomitrella patens*: homologous

recombination, concatenation and multiple integration. *Nucleic Acids Res.*, **34**, 6205–6214.

28 Kamisugi, Y., Cuming, A.C., and Cove, D.J. (2005) Parameters determining the efficiency of gene targeting in the moss *Physcomitrella patens*. *Nucleic Acids Res.*, **33**, E173.

29 Rensing, S.A., Ick, J., Fawcett, J.A., Lang, D., Zimmer, A., De Peer, Y.V., and Reski, R. (2007) An ancient genome duplication contributed to the abundance of metabolic genes in the moss *Physcomitrella patens*. *BMC Evol. Biol.*, **7**, 2.

30 Markmann-Mulisch, U., Hadi, M.Z., Koepchen, K., Alonso, J.C., Russo, V.E.A., Schell, J., and Reiss, B. (2002) The organization of *Physcomitrella patens* RAD51 genes is unique among eukaryotic organisms. *Proc. Natl. Acad. Sci. USA*, **99**, 2959–2964.

31 Mittmann, F., Brucker, G., Zeidler, M., Repp, A., Abts, T., Hartmann, E., and Hughes, J. (2004) Targeted knockout in *Physcomitrella* reveals direct actions of phytochrome in the cytoplasm. *Proc. Natl. Acad. Sci. USA*, **101**, 13939–13944.

32 Hofmann, A.H., Codon, A.C., Ivascu, C., Russo, V.E.A., Knight, C., Cove, D., Schaefer, D.G., Chakhparonian, M., and Zryd, J.P. (1999) A specific member of the Cab multigene family can be efficiently targeted and disrupted in the moss *Physcomitrella patens*. *Mol. Gen. Genet.*, **261**, 92–99.

33 Langston, L.D. and Symington, L.S. (2004) Gene targeting in yeast is initiated by two independent strand invasions. *Proc. Natl. Acad. Sci. USA*, **101**, 15392–15397.

34 Strepp, R., Scholz, S., Kruse, S., Speth, V., and Reski, R. (1998) Plant nuclear gene knockout reveals a role in plastid division for the homolog of the bacterial cell division protein FtsZ, an ancestral tubulin. *Proc. Natl. Acad. Sci. USA*, **95**, 4368–4373.

35 Tanahashi, T., Sumikawa, N., Kato, M., and Hasebe, M. (2005) Diversification of gene function: homologs of the floral regulator FLO/LFY control the first zygotic cell division in the moss *Physcomitrella patens*. *Development*, **132**, 1727–1736.

36 Thelander, M., Olsson, T., and Ronne, H. (2004) Snf1-related protein kinase 1 is needed for growth in a normal day–night light cycle. *EMBO J.*, **23**, 1900–1910.

37 Yasumura, Y., Moylan, E.C., and Langdale, J.A. (2005) A conserved transcription factor mediates nuclear control of organelle biogenesis in anciently diverged land plants. *Plant Cell*, **17**, 1894–1907.

38 Hofmann, N.R. and Theg, S.M. (2005) Toc64 is not required for import of proteins into chloroplasts in the moss *Physcomitrella patens*. *Plant J.*, **43**, 675–687.

39 Hattori, M., Hasebe, M., and Sugita, M. (2004) Identification and characterization of cDNAs encoding pentatricopeptide repeat proteins in the basal land plant, the moss *Physcomitrella patens*. *Gene*, **343**, 305–311.

40 Huether, C.M., Lienhart, O., Baur, A., Stemmer, C., Gorr, G., Reski, R., and Decker, E.L. (2005) Glyco-engineering of moss lacking plant-specific sugar residues. *Plant Biol.*, **7**, 292–299.

41 Repp, A., Mikami, K., Mittmann, F., and Hartmann, E. (2004) Phosphoinositide-specific phospholipase C is involved in cytokinin and gravity responses in the moss *Physcomitrella patens*. *Plant J.*, **40**, 250–259.

42 Bierfreund, N.M., Tintelnot, S., Reski, R., and Decker, E.L. (2004) Loss of GH3 function does not affect phytochrome-mediated development in a moss, *Physcomitrella patens*. *J. Plant Physiol.*, **161**, 823–835.

43 Frank, W., Baar, K.M., Qudeimat, E., Woriedh, M., Alawady, A., Ratnadewi, D., Gremillon, L., Grimm, B., and Reski, R. (2007) A mitochondrial protein homologous to the mammalian peripheral-type benzodiazepine receptor is essential for stress adaptation in plants. *Plant J.*, **51**, 1004–1018.

44 Khandelwal, A., Chandu, D., Roe, C.M., Kopan, R., and Quatrano, R.S. (2007) Moonlighting activity of presenilin in plants is independent of gamma-secretase and evolutionarily conserved. *Proc. Natl. Acad. Sci. USA*, **104**, 13337–13342.

45 Olsson, T., Thelander, M., and Ronne, H. (2003) A novel type of chloroplast stromal hexokinase is the major glucose-phosphorylating enzyme in the moss

Physcomitrella patens. J. Biol. Chem., **278**, 44439–44447.

46 Singer, S.D. and Ashton, N.W. (2007) Revelation of ancestral roles of KNOX genes by a functional analysis of *Physcomitrella* homologues. *Plant Cell Rep.*, **26**, 2039–2054.

47 Wiedemann, G., Koprivova, A., Schneider, M., Herschbach, C., Reski, R., and Kopriva, S. (2007) The role of the novel adenosine 5′-phosphosulfate reductase in regulation of sulfate assimilation of *Physcomitrella patens. Plant Mol. Biol.*, **65**, 667–676.

48 Yasumura, Y., Crumpton-Taylor, M., Fuentes, S., and Harberd, N.P. (2007) Step-by-step acquisition of the gibberellin-DELLA growth-regulatory mechanism during land-plant evolution. *Curr. Biol.*, **17**, 1225–1230.

49 Brun, F., Schaefer, D.G., Laloue, M., and Gonneau, M. (2004) Knockout of UBP34 in *Physcomitrella patens* reveals the photoaffinity labeling of another closely related IPR protein. *Plant Sci.*, **167**, 471–479.

50 Ichikawa, K., Shimizu, A., Okada, R., Satbhai, S.B., and Aoki, S. (2008) The plastid sigma factor SIG5 is involved in the diurnal regulation of the chloroplast gene *psbD* in the moss *Physcomitrella patens*. *FEBS Let.*, **582**, 405–409.

51 Kaewsuwan, S., Cahoon, E.B., Perroud, P.F., Wiwat, C., Panvisavas, N., Quatrano, R.S., Cove, D.J., and Bunyapraphatsara, N. (2006) Identification and functional characterization of the moss *Physcomitrella patens* Δ^5-desaturase gene involved in arachidonic and eicosapentaenoic acid biosynthesis. *J. Biol. Chem.*, **281**, 21988–21997.

52 Kasahara, M., Kagawa, T., Sato, Y., Kiyosue, T., and Wada, M. (2004) Phototropins mediate blue and red light-induced chloroplast movements in *Physcomitrella patens*. *Plant Physiol.*, **135**, 1388–1397.

53 Koprivova, A., Meyer, A.J., Schween, G., Herschbach, C., Reski, R., and Kopriva, S. (2002) Functional knockout of the adenosine 5′-phosphosulfate reductase gene in *Physcomitrella patens* revives an old route of sulfate assimilation. *J. Biol. Chem.*, **277**, 32195–32201.

54 Lee, K.J.D., Sakata, Y., Mau, S.L., Pettolino, F., Bacic, A., Quatrano, R.S., Knight, C.D., and Knox, J.P. (2005) Arabinogalactan proteins are required for apical cell extension in the moss *Physcomitrella patens. Plant Cell*, **17**, 3051–3065.

55 Lienard, D., Durambur, G., Kiefer-Meyer, M.C., Nogue, F., Menu-Bouaouiche, L., Charlot, F., Gomord, V., and Lassalles, J.P. (2008) Water transport by aquaporins in the extant plant *Physcomitrella patens. Plant Physiol.*, **146**, 1207–1218.

56 Machida, M., Takechi, K., Sato, H., Chung, S.J., Kuroiwa, H., Takio, S., Seki, M., Shinozaki, K., Fujita, T., Hasebe, M. *et al.* (2006) Genes for the peptidoglycan synthesis pathway are essential for chloroplast division in moss. *Proc. Natl. Acad. Sci. USA*, **103**, 6753–6758.

57 Menand, B., Yi, K.K., Jouannic, S., Hoffmann, L., Ryan, E., Linstead, P., Schaefer, D.G., and Dolan, L. (2007) An ancient mechanism controls the development of cells with a rooting function in land plants. *Science*, **316**, 1477–1480.

58 Perroud, P.F. and Quatrano, R.S. (2008) BRICK1 is required for apical cell growth in filaments of the moss *Physcomitrella patens* but not for gametophore morphology. *Plant Cell*, **20**, 411–422.

59 Quodt, V., Faigl, W., Saedler, H., and Munster, T. (2007) The MADS-domain protein, PPM2 preferentially occurs in gametangia and sporophytes of the moss *Physcomitrella patens. Gene*, **400**, 25–34.

60 Saavedra, L., Svensson, J., Carballo, V., Izmendi, D., Welin, B., and Vidal, S. (2006) A dehydrin gene in *Physcomitrella patens* is required for salt and osmotic stress tolerance. *Plant J.*, **45**, 237–249.

61 Sakakibara, K., Nishiyama, T., Deguchi, H., and Hasebe, M. (2008) Class 1 KNOX genes are not involved in shoot development in the moss *Physcomitrella patens* but do function in sporophyte development. *Evol. Dev.*, **10**, 555–566.

62 Trouiller, B., Schaefer, D.G., Charlot, F., and Nogue, F. (2006) MSH2 is essential for the preservation of genome integrity and prevents homeologous recombination in the moss *Physcomitrella patens. Nucleic Acids Res.*, **34**, 232–242.

63 Perroud, P.F. and Quatrano, R.S. (2006) The role of ARPC4 in tip growth and alignment of the polar axis in filaments of *Physcomitrella patens*. *Cell Motil. Cytoskel.*, **63**, 162–171.

64 Russell, D.W. and Hirata, R.K. (2008) Human gene targeting favors insertions over deletions. *Hum. Gene Therapy*, **19**, 907–914.

65 Ashton, N.W. and Cove, D.J. (1977) Isolation and preliminary characterization of auxotrophic and analog resistant mutants of moss, *Physcomitrella*-patens. *Mol. Gen. Genet.*, **154**, 87–95.

66 Knight, C.D., Cove, D.J., Cuming, A.C., and Quatrano, R.S. (2002) Moss gene technology, in *Molecular Plant Biology*, vol. 2 (eds P.M. Gilmartin and C. Bowler), Oxford University Press, Oxford, pp. 285–299.

67 Cove, D.J. (1992) Regulation of development in the moss, *Physcomitrella patens*, in *Development: The Molecular Genetic Approach* (eds V.E.A. Russo, S. Brody, D. Cove, and S. Ottolenghi) Springer, Berlin, pp. 179–193.

68 Reski, R. (1998) Development, genetics and molecular biology of mosses. *Bot. Acta*, **111**, 1–15.

69 Engel, P.P. (1968) Induction of biochemical and morphological mutants in moss *Physcomitrella patens*. *Am. J. Bot.*, **55**, 438–446.

70 Hohe, A., Rensing, S.A., Mildner, M., Lang, D., and Reski, R. (2002) Day length and temperature strongly influence sexual reproduction and expression of a novel MADS-box gene in the moss *Physcomitrella patens*. *Plant Biol.*, **4**, 595–602.

71 Zobell, O. (2006) PhD Thesis, Universität zu Köln, Mathematisch-Naturwissenschaftliche Fakultät, Köln.

72 Hiwatashi, Y., Nishiyama, T., Fujita, T., and Hasebe, M. (2001) Establishment of gene-trap and enhancer-trap systems in the moss *Physcomitrella patens*. *Plant J.*, **28**, 105–116.

73 Imaizumi, T., Kadota, A., Hasebe, M., and Wada, M. (2002) Cryptochrome light signals control development to suppress auxin sensitivity in the moss *Physcomitrella patens*. *Plant Cell*, **14**, 373–386.

74 Hohe, A., Egener, T., Lucht, J.M., Holtorf, H., Reinhard, C., Schween, G., and Reski, R. (2004) An improved and highly standardised transformation procedure allows efficient production of single and multiple targeted gene-knockouts in a moss, *Physcomitrella patens*. *Curr. Genet.*, **44**, 339–347.

75 Schween, G., Fleig, S., and Reski, R. (2002) High-throughput-PCR screen of 15.000 transgenic *Physcomitrella* plants. *Plant Mol. Biol. Rep.*, **20**, 43–47.

76 Schween, G., Schulte, J., Reski, R., and Hohe, A. (2005) Effect of ploidy level on growth, differentiation, and morphology in *Physcomitrella patens*. *Bryologist*, **108**, 27–35.

77 Kabeya, Y., Hashimoto, K., and Sato, N. (2002) Identification and characterization of two phage-type RNA polymerase cDNAs in the moss *Physcomitrella patens*: Implication of recent evolution of nuclear-encoded RNA polymerase of plastids in plants. *Plant Cell Physiol.*, **43**, 245–255.

78 Uenaka, H. and Kadota, A. (2007) Functional analyses of the *Physcomitrella patens* phytochromes in regulating chloroplast avoidance movement. *Plant J.*, **51**, 1050–1061.

79 Girod, P.A., Fu, H.Y., Zryd, J.P., and Vierstra, R.D. (1999) Multiubiquitin chain binding subunit MCB1 (RPN10) of the 26S proteasome is essential for developmental progression in *Physcomitrella patens*. *Plant Cell*, **11**, 1457–1471.

80 Schipper, O., Schaefer, D., Reski, R., and Fleming, A. (2002) Expansins in the bryophyte *Physcomitrella patens*. *Plant Mol. Biol.*, **50**, 789–802.

81 Reski, R. (2006) Small molecules on the move: homeostasis, crosstalk, and molecular action of phytohormones. *Plant Biol.*, **8**, 277–280.

82 Girke, T., Schmidt, H., Zahringer, U., Reski, R., and Heinz, E. (1998) Identification of a novel delta-6-acyl-group desaturase by targeted gene disruption in *Physcomitrella patens*. *Plant J.*, **15**, 39–48.

83 Terada, R., Urawa, H., Inagaki, Y., Tsugane, K., and Iida, S. (2002) Efficient gene targeting by homologous recombination in rice. *Nat. Biotechnol.*, **20**, 1030–1034.

84 Terada, R., Johzuka-Hisatomi, Y., Saitoh, M., Asao, H., and Iida, S. (2007) Gene targeting by homologous recombination as

a biotechnological tool for rice functional genomics. *Plant Physiol.*, **144**, 846–856.
85 Iida, s., Johzuka-Hisatomi, Y., and Terada, R. (2007) Gene targeting by homologous recombination for rice functional genomics, in *Rice Functional Genomics: Challenges, Progress and Prospects* (ed. N.M. Upadhyaya), Springer, Heidelberg, pp. 273–289.
86 Iida, S. and Terada, R. (2005) Modification of endogenous natural genes by gene targeting in rice and other higher plants. *Plant Mol. Biol.*, **59**, 205–219.
87 Cotsaftis, O. and Guiderdoni, E. (2005) Enhancing gene targeting efficiency in higher plants: rice is on the move. *Transgenic Res.*, **14**, 1–14.

Part V
Emerging Technologies

17
True Single Molecule Sequencing (tSMS)™ by Synthesis
Scott Jenkins and Avak Kahvejian

Abstract

The ability to obtain DNA sequence information from single molecules has been a goal of the scientific community for decades. Helicos BioSciences has commercialized a single molecule sequencing-by-synthesis technique for detecting template-directed enzymatic incorporation of single nucleotides into billions of nascent DNA strands. Reliably detecting base incorporation events in a massively parallel manner requires a series of technological innovations in DNA chemistry, optical fluorescence, and instrument engineering to make the approach feasible. Among these is a method to anchor DNA fragments to the surface of an imaging flow cell, a cyclical chemical synthesis process to sequentially add fluorescently labeled bases to DNA templates, and the use of total internal reflection fluorescence microscopy for imaging. Single molecule sequencing now offers the potential to increase the sensitivity of mutation detection while decreasing cost and time required for sequencing. The platform housing this technology is effective for a wide range of applications, including whole-genome resequencing, digital gene expression, copy number variation, and targeted resequencing.

17.1
Introduction

The ability to obtain DNA sequence information from cells has transformed biology and medicine. Following the development of Sanger sequencing in the 1970s [1], significant effort has been devoted to increasing the speed at which DNA sequence information can be collected. In the past two decades, advances across a wide range of areas, including DNA synthesis chemistry, fluorescence microscopy, capillary electrophoresis, mass spectroscopy, instrument automation, bioinformatics, and others, have allowed researchers to generate an impressive amount of genomic sequence information. The complete genomes of humans and many other species have been sequenced, including several important biological model systems (e.g., *Arabidopsis*, fruit fly, mouse, rat, rhesus macaque) and several agriculturally relevant species (e. g., chicken, cow, honeybee, yeast) (see the National Human Genome Research Institute large-scale sequencing program web site http://www.genome.gov/10002154 and [2, 3]). In addition, millions of common human single nucleotide polymorphisms (SNPs) have been identified and a significant amount of fundamental biology uncovered.

Despite the volume of genomic sequence data generated to date, the full potential of comparative genomics and medical genomics cannot be realized until many genomes are sequenced. The cost and time required to complete full eukaryotic genomes with current technologies has been reduced dramatically since the first large-scale sequencing projects were undertaken. Costs of finished sequence for a full mammalian genome reached hundreds of millions of dollars in the early 1990s, but the cost of sequencing a full human genome had dropped to around $100 000 in 2008. Despite significant reduction in cost and time required, large-scale sequencing remains prohibitive for many types of research projects – the current sequencing expense and time horizon prevents the collection of large numbers of genomes for comparison. In the case of humans, large numbers of high-resolution sequenced genomes would enable detailed studies of genetic variability as it relates to disease, pharmaceutical efficacy, and physiology. The comparative sequencing of a large number of plant genomes, which were even more constrained by the cost and throughput barriers, would allow the elucidation of genes and variants responsible for many advantageous traits. Beyond whole-genome resequencing, rapid and low-cost sequencing can be applied to whole-transcriptome analysis of gene expression and resequencing targeted regions in all species.

Attaining the full potential of genomics for research has required new approaches to rapid, inexpensive sequencing. One approach is the direct measurement of single DNA molecules – a feat that has been a goal of the research community since at least 1989 [4]. By allowing the analysis of molecules individually, direct single molecule sequencing not only offers a high-resolution view of nucleic acid biology, but also avoids cumbersome sample preparation steps, lowers sequencing costs, and improves throughputs significantly. These advantages enable investigators to pursue new scientific enquiries not currently possible.

The potential impact of single molecule sequencing approaches will be widespread, significant and difficult to fully predict. The versatility of the method already enables a wide variety of applications including whole genome resequencing, digital gene expression (DGE) analysis, whole transcriptome resequencing (RNA-Seq), chromatin immunoprecipitation sequencing (chromatin immunoprecipitationChIP-Seq), global small RNA analysis, and copy number variation assessment. By making sequencing fast and affordable, single molecule sequencing allows individual scientists to undertake genomics research in their own laboratories, and propels multinational, multi-institutional efforts such as the 1000 Genomes Project and the Cancer Genome Atlas into a new era of cataloging the full extent of genomic variation, understanding genome function and elucidating genotype/phenotype correlations [5].

Single molecule sequencing-by-synthesis (SBS) moved beyond the theoretical in 2003, when a research team led by Stephen Quake demonstrated that sequence

Figure 17.1 The HeliScope Single Molecule Sequencer.

information could be obtained from individual DNA molecules [6]. Based on this original research, Helicos BioSciences Corporation was founded in 2003 and has since developed the first commercially available approach capable of sequencing billions of single molecules of DNA in a few days [7, 8]. Helicos® True Single Molecule Sequencing (tSMS™) is at the heart of the Helicos® Genetic Analysis System, which consists of several components – the HeliScope™ Single Molecule Sequencer (Figure 17.1), the HeliScope Sample Loader, and the HeliScope Analysis Engine.

17.2
Methods, Protocols, and Technical Principles

17.2.1
Single Molecule Sequencing Technical Challenges and Solutions

Direct measurement of individual DNA strands poses considerable technological hurdles. Helicos scientists have delivered a series of technical and manufacturing advances to overcome these challenges [7, 9–11] in developing and commercializing a single molecule SBS technology. The process begins with a unique sample preparation methodology. Genomic DNA (or DNA from any other source) is fragmented, modified with a homopolymer nucleotide tail, and hybridized on a proprietary surface within a flow cell. There, the individual DNA fragments serve as templates for SBS reactions. Helicos chemists have also developed novel and proprietary fluorescently labeled nucleotides, which are added one base type at a time, and the incorporation events are recorded with a specially designed optical microscope. The sequence of billions of individual DNA strand can be tracked through multiple cycles of labeled base addition and imaging.

The major technological challenges associated with obtaining sequence information from individual DNA molecules include the high-efficiency and high-fidelity incorporation of labeled nucleotides to a primed template using DNA polymerase, removing unincorporated nucleotides from the imaging surface, accurate detection of the incorporation events, and removal of the fluorescent label after each cycle. Additional challenges involve immobilizing individual DNA strands on a surface, rapidly acquiring thousands of images over an entire flow cell to capture billions of nucleotide incorporations, and translating terabyte-scale data output from images into DNA sequence information.

Sample Preparation Protocol

Among the major advantages of a single molecule sequencing approach is a faster and less expensive sample preparation protocol that does not require polymerase chain reaction (PCR)-based amplification of the DNA sample before sequencing. Generally, sample preparation for the Helicos single molecule sequencing approach consists of fragmenting the initial nucleic acid sample and generating a polyA oligonucleotide at the end of each fragment [7, 12]. The polyA tails on each DNA segment are designed to hybridize to surface-bound polyT oligonucleotides that will be discussed in the next section. (See Figure 17.2 for a schematic depiction of Steps 1–5.)

1. *Ultrasonic shearing of genomic DNA.* Ultrasonic shearing of the sample DNA results in its random fragmentation, permitting the sequencing of the entire length of the original sample. Typically, a few micrograms of DNA sample are sheared using a Covaris S2 instrument to generate an average fragment size of 150–200 bp.
    a. *Size selection of DNA fragments.* Although Helicos' tSMS strategy is entirely compatible with a very wide distribution of fragment sizes within a given sample, size selection is required postfragmentation to ensure uniform coverage of the original sample DNA. The sheared DNA sample is loaded onto a size-exclusion spin column, washed twice, and eluted to reduce the abundance of excessively small fragments of (<25 bp) DNA.
    b. *Calculate concentration of 3′ ends.* After fragmentation and size selection, the DNA molecules must be modified at their 3′ ends using a tailing procedure, the efficiency of which is determined by the concentration of DNA ends in the sample. This step allows the proper estimation of the concentration of 3′ ends, by taking into account both the average length of the DNA fragments, assessed by gel electrophoresis, and the total DNA concentration in the sample.

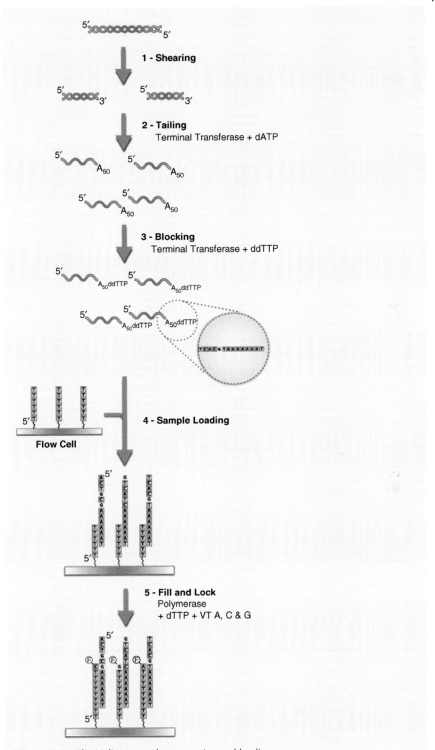

Figure 17.2 The Helicos sample preparation and loading process.

2. *PolyA tailing of the DNA.* The fragments must be modified at their 3' ends with a polyA tail to allow for efficient hybridization onto the polyT oligonucleotide-coated flow cell (discussed in Section 17.2.2). Since the length of the polyA tail is crucial for proper hybridization and subsequent sequencing, the efficiency of the tailing reaction is monitored by running a parallel tailing control reaction for every sample. dATP is added with Terminal Transferase to achieve tailing. The tailing control reactions are analyzed by gel electrophoresis to determine the success of the tailing procedure.
3. *Blocking.* During hybridization, the polyA tails on the modified templates may align imperfectly with the oligonucleotides on the Helicos Flow Cell surface. This may result in the generation of a recessed 3' end that can serve as a substrate for the SBS reaction. To prevent the incorporation of fluorescent Virtual Terminator™ nucleotides at that end of the duplexes, the 3' ends of the template molecules must be modified with a dideoxy terminator. Blocking is carried out using Terminal Transferase and a dideoxy nucleotide.
4. *Load samples onto flow cell.* The tailed DNA sample fragments are injected onto the flow cell using the HeliScope Sample Loader. The sample loader allows users to interface individually with all 25 channels of a flow cell and is equipped with a vacuum to move solutions through the channels. Temperatures are also controlled to promote proper hybridization of the tailed DNA templates to the flow cell.
5. *Fill-and-lock.* As the polyA tails on the DNA sample fragments are of variable lengths, they may align differently with the fixed-length polyT oligonucleotide capture strands on the flow cell surface. The result is an unknown and variable number of A nucleotides extending past the template/primer duplex. To ensure that the SBS reactions begin after the polyA tail, a "fill-and-lock" step adds enough free T nucleotides to extend the surface capture primer until the first non-A template base. Fluorescently labeled C, G, and A Virtual Terminator nucleotides are also added and incorporate at the first non-A template base, effectively arresting the synthesis at that position. These labeled nucleotides will be used to identify the positions of all captured DNA strands once the flow cell is on the instrument.
6. *Sequence.* The flow cell is inserted onto the HeliScope Sequencer to begin the sequencing run.

17.2.2
Flow Cell Surface Architecture

As it represents the venue for all the chemical reactions in the Helicos single molecule sequencing approach, the flow cell is critically important to generating reliable, accurate sequence data. Its design and construction help overcome many of the technical challenges to single molecule sequencing, and help enable throughput levels that cannot be achieved using amplification-based methods.

The Helicos flow cell is configured as a multichannel design consisting of an oligonucleotide primer-functionalized glass slide on which the synthesis reactions and imaging are carried out. Each flow cell contains 25 channels that can be individually loaded, with a total imageable area of around 30 mm^2.

The key characteristic of the sequencing surface are randomly deposited polyT oligonucleotides, 50 bases in length, that are covalently bound to the glass surface. The amine-functional 5' ends of the polyT oligonucleotides react with the epoxide groups on the slide's surface to establish a covalent linkage between the surface and the polyT oligos. The sequencing surface is specially designed to prevent nonspecific binding and adsorption of free nucleotides.

The surface-bound oligonucleotides serve to capture the templates DNA molecules that have been modified with polyA tails at their 3' ends. The surface-bound oligos also serve as primers for the SBS reaction. To achieve hybridization between the complementary polyT oligos on the surface on the polyA oligonucleotides on the DNA fragments, the genomic templates are incubated over the capture surface for around 60 min at a DNA template concentration of around 100 pM in buffer. Surfaces are rinsed before inserting the flow cell into the instrument.

Once the sample loading and fill-and-lock procedures (Steps 4 and 5) have been performed on the HeliScope Sample Loader, the flow cells are loaded onto the HeliScope Single Molecule Sequencer and imaged. These initial "template" images help identify locations where DNA strands are anchored to the surface as a reference against future images and also serve to estimate loading densities. Typical strand densities are greater than one strand/μm^2. In the future, techniques aimed at creating more dense ordered surfaces will allow up to a 5-fold increase in the template density on the slide, which would directly translate to a proportional increase in the system's throughput [13]. The dyes on the Virtual Terminator nucleotides are cleaved before beginning the SBS process.

17.2.3
Cyclic SBS

The availability of reference genomes enables use of sequencing technologies that employ shorter read-lengths and massive parallelism. Sequencing individual nucleic acid molecules maximizes the number of strands that can be sequenced in parallel. Helicos tSMS technology can simultaneously extract sequence information from the asynchronous growth of more than 1 billion DNA molecules.

Once the templates have been captured on the flow cell surface and imaging has determined their fixed positions, chemical synthesis of nucleotides and gathering of sequence information begins. Repeated cycles consist of adding the polymerase/labeled Virtual Terminator nucleotide mixture (containing one of the four bases), rinsing, imaging multiple positions, and cleaving the dye labels. The four bases are added sequentially in separate cycles, so only one labeled Virtual Terminator nucleotide is present in each cycle. In an example illustrating the process, Virtual Terminator "C" nucleotide analogs are added to the flow cell and the polymerase

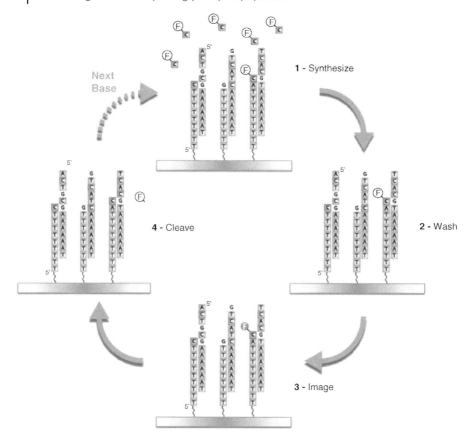

Figure 17.3 tSMS SBS. The tSMS process is a cyclical process involving multiple rounds of (1) synthesis using labeled Virtual Terminator nucleotides, (2) washing, (3) imaging, and (4) cleaving the fluorescent label until the desired read length is achieved.

catalyzes the addition of labeled Cs to those templates which have a "G" in the next available position (see Figure 17.3).

The observation of a fluorescent signal at a particular position following the addition of a base, reveals sequence information for the DNA strand at that position. For example, if a signal is observed at a given position after addition of a "C" nucleotide in the first cycle of chemical synthesis, it can be determined that the template strand has the complementary base (G) at the first position. This cycle (Synthesis → Wash → Image → Cleave) is repeated with the other three bases.

Experimental observation of the sequencing chemistry in the HeliScope Sequencer demonstrates that the rinsing between cycles is extremely efficient, the template specificity of the polymerase activity is maintained, and labels are removed effectively after each imaging cycle. Data show that label cleavage efficiency is greater than 99.5% in the Helicos system. In a typical run, 120 sequencing cycles are carried out (30 "quads," consisting of four cycles, one for each base) and read-lengths of between 25 and more than 55 bases are achieved on billions of strands.

17.2.4
Optical Imaging of Growing Strands

The optical imaging of individual nucleic acid strands is a critical component of the single molecule sequencing technology. Accurate detection of incorporated fluorescently labeled nucleotides into single strands of DNA requires two primary attributes: (i) the imaging system must have adequate resolution to discriminate among nearby strands tethered to the surface in nonordered locations, and (ii) must be sufficient signal-to-noise ratio to effectively "see" the fluorescent labels and differentiate them from the background. Reliable performance requires reduction of all nonspecific emission sources, as these nonspecific events could produce errors in the sequence. Accurately and reliably detecting fluorescence from single molecules with high signal-to-noise ratios depends on reducing the optical background interference, which can arise from Raman scattering, Rayleigh scattering, and fluorescence from impurity molecules.

The design of the HeliScope Single Molecule Sequencer addresses these requirements. Among the approaches aimed at achieving low levels of background noise and observing light emitted by single fluorophores are to avoid the use of materials that autofluoresce and to find ways to minimize the adsorption of stray fluorescent molecules onto the imaging surface. Helicos scientists have developed imaging reagents that enhance emission intensity and fluorophore detection by an order of magnitude, while reducing problems related to photobleaching and fluorophore "blinking." A specially formulated solution was developed to aid the imaging of the fluorescent nucleotide labels. The solution is an optimized mixture with oxygen-scavenging, free radical scavenger, and triplet quenching components.

To further reduce background noise, the system is designed to restrict the total volume illuminated by excitation radiation. By doing so, fluorescent emission occurs in a more localized fashion and background noise is reduced without diminishing the signal from the molecule. To achieve this with the Helicos tSMS technology, scientists at the company employed a technique known as total internal reflection fluorescence microscopy (TIRFM).

TIRFM's major advantage is to limit the fluorescence events to within a few hundred nanometers of a surface. TIRFM was developed in the 1980s by University of Michigan researcher Daniel Axelrod as a way to help resolve fluorescence of molecules interacting with cellular surfaces [14]. Total internal reflection is an optical phenomenon that occurs when light traveling through a dense material with a high refractive index encounters an interface with a less dense substance of low refractive index. The angle at which the light meets the interface determines how much light will pass through the interface (refraction) and how much will be reflected. Total internal reflection occurs when the light meets the interface at a certain critical angle, which depends on the ratio of the refractive indices of the two materials at the interface. At the critical angle, all of the light is reflected back into the denser medium.

A portion of the energy in a light beam that is totally internally reflected propagates a short distance into the material of lower refractive index, generating an exponentially decaying electromagnetic field known as an evanescent wave close to the

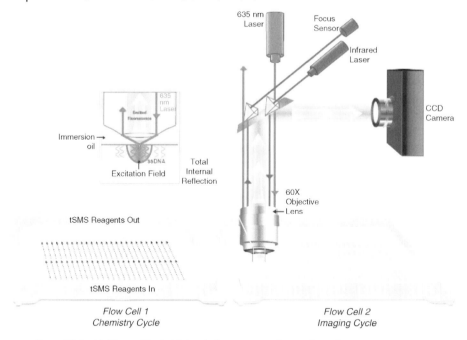

Figure 17.4 HeliScope Single Molecule Sequencer optics configuration.

interface. The energy of the evanescent wave can be exploited to excite fluorescent molecules within about 150 nm of the surface interface. As the evanescent field decays rapidly, only fluorophores in the vicinity of the surface are illuminated. This results in a dramatic reduction of noise from bulk fluids. In the context of Helicos tSMS technology, it means that only molecules bound to the surface of the flow cell will fluoresce and the signal-to-noise ratio is increased. During sequencing, rigorous washing minimizes the number of free labeled nucleotides that adhere to the surface and cause spurious detection events.

The HeliScope Sequencer utilizes an advanced, high-powered total internal reflectance microscope (see Figure 17.4). In optical microscopy, the numerical aperture (NA) is a dimensionless value representing a microscope's ability to gather light and resolve detail. Higher NA values mean a wider cone of light can be gathered by the microscope objective. For applications using TIRFM, such as single molecule sequencing, high NA values are required to detect photons emitted by a single excited fluorophore. In the single molecule technology developed by Helicos, the laser light that excites the fluorescent labels is delivered through a 60× 1.49 NA oil immersion objective microscope.

The imaging system focuses 2 W of laser power from multiple solid state red lasers at wavelength 635 nm onto an approximately 30 000 μm^2 area on the flow cell surface where the DNA templates are attached. The laser light is directed into the flow cell above the critical angle at which total internal reflection occurs. The fluorophore

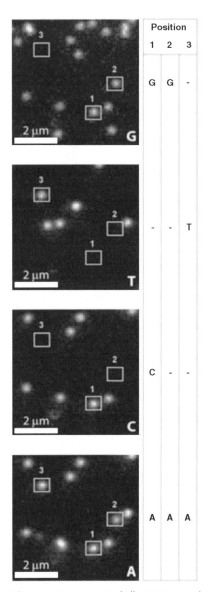

Figure 17.5 Image stack illustrating raw data. Adapted with permission from *Science*.

emits longer wavelengths of light that are imaged through the objective lens onto a four CCD camera detection system. The system can locate fluorescent objects on the flow cell surface with around 0.15-pixel precision (15 nm) and can resolve objects separated by greater than 1.2 pixels. The resulting image appears similar to the night sky, where each point of light indicates an incorporated nucleotide (see Figure 17.5).

17.2.5
Mechanical Operation

To achieve high-throughput single molecule SBS according to the Helicos method, the optical apparatus must image the entire flow cell surface for each reaction cycle. Since the area of the flow cell surface illuminated by the laser at a given time is necessarily small and the laser must remain motionless to maintain the exacting optical geometry, the flow cell must move systematically below the laser and microscope objective to allow imaging of the entire flow cell surface. To achieve this, the flow cell is mounted on a moveable stage that is capable of rapid and precise position changes.

The high-speed stage allows the HeliScope Sequencer to rapidly move the flow cell, bring the stage to rest and acquire multiple images (called "fields of view") across the flow cell channels. The moveable stage and camera setup is capable of acquiring around 10 distinct field-of-view images/s. The rate at which images can be taken and processed directly drives the number of single molecules observed, making it a major contributor to sequencing throughput as measured in megabases per hour (Mb/h). A separate infrared laser and detector are used to lock focus and maintain high-quality imaging across the flow cell.

17.2.6
System Components

The complete Helicos Genetic Analysis System includes two benchtop sample loaders (HeliScope Sample Loader), the HeliScope Single Molecule Sequencer, and the HeliScope Analysis Engine. The HeliScope Sample Loader performs all fluidic operations necessary to prepare flow cells for sequencing, including rehydration, sample loading, hybridization, and washing. The 25 independent inputs allow operators to load up to 25 unique or multiplexed samples per flow cell.

The HeliScope Single Molecule Sequencer orchestrates the tSMS SBS reactions on the flow cell surface and records the incorporation of the fluorescently labeled nucleotides into the growing DNA strands. The HeliScope Sequencer maximizes run efficiency by processing two 25-channel flow cells at once – performing strand synthesis in one flow cell while simultaneously imaging the other. Its user interface is a touch-screen monitor from which experimental runs are defined, launched and monitored. The HeliScope Sequencer can also be monitored remotely through a web interface called the HeliScope Control Center. This interface provides real-time run monitoring, data download, and system maintenance capabilities.

The HeliScope Analysis Engine is a high-performance informatics system that operates downstream of the sequencer. It works to convert the images captured by the sequencer's CCD camera into single molecule sequence data. The Analysis Engine's 48 processors and 28 TB of storage process images in real-time, and are connected to the HeliScope Sequencer via a dedicated private network.

17.2.7
Data Analysis

As images are generated by observing fluorescent emission during the addition of labeled bases, the system can begin to translate the images into sequence data. Generally, this is accomplished by spatially correlating images from subsequent chemistry cycles. The coordinates of fluorescent "spots" are determined using the intensity and distribution of the spot and subtracting background fluorescence. A correlogram is built by assembling images from subsequent chemical cycles and finding those spots that are correlated by determining whether a spot appears within a set radius from the corresponding position in a previous image. From this information, it can be determined automatically whether an incorporation event occurred at a particular position as a result of a base addition. The ultimate output is sequence data from a growing individual DNA molecule. The sequence data from individual fragments can then be aligned against a reference genome.

17.3
Applications

17.3.1
Single Molecule DGE and RNA-Seq

Gene expression studies provide key information for understanding cellular mechanisms and behavior. Large-scale sequencing of cDNA clones and differential comparisons of transcript abundance between samples have yielded extensive knowledge of genetic expression across tissue types and organisms [17].

cDNA microarrays are effective at monitoring gene expression levels of medium- to high-expressing known genes, but their ability to accurately measure low-expressing genes is poor and they are limited to previously known transcripts.

Helicos single molecule DGE offers a hypothesis-free, quantitative analysis of the entire transcriptome. DGE involves high-throughput sequencing of short cDNA fragments followed by matching reads to a reference transcriptome. Individual transcript abundances are inferred from their relative tag counts.

Researchers at Helicos BioSciences have applied single molecule sequencing to digital gene expression [15]. Helicos-driven DGE is an open-ended tool that analyzes the level of expression of virtually all genes in a sample by counting the number of individual mRNA molecules produced from each gene. While the reads are counted based on the transcriptome reference utilized, there is no requirement that genes be identified and characterized prior to conducting an experiment as novel genes may be identified by computational methods of self-clustering of reads that do not align to the chosen transcriptome reference.

Like single molecule sequencing, Helicos DGE involves the hybridization of single-stranded cDNA molecules to oligonucleotides attached to the flow cell surface. Single molecules are densely packed onto the flow cell surface, as in the sequencing application, resulting in extremely high throughput. The cDNA is sequenced by imaging the DNA polymerase addition of fluorescently labeled nucleotides incorporated one at a time into the surface oligonucleotide's growing strand.

Helicos DGE is not limited to exclusively interrogating known transcripts from the public domain. Since single molecule DGE is highly sensitive and quantitative, each signature generated in the dataset can be enumerated, which not only allows for the detection of highly expressed transcripts, but also for the detection of very rare transcripts represented by only a few molecules of mRNA per cell.

The digital nature of the DGE data output enables the comparison of expression levels of different transcripts within the same sample as well as the comparison of transcript levels between samples.

Helicos DGE generates accurate transcript counts consistent with a complete cellular dynamic range. The effectiveness of counting by single molecule DGE is driven by the fact that only a single read may be generated from each mRNA molecule, thereby maintaining a faithful one-to-one representation of transcript distribution in the data and making counting more efficient than methods based on covering the whole transcript. Single molecule DGE generates sequence reads from the 5′ ends of cDNA molecules and does not require the cDNA to be of full length. Consequently, it works equally well with short cDNAs generated as a result of mRNA degradation or incomplete reverse transcription. In fact, the variability in read start sites resulting from the presence of short cDNAs enriches the read pool for informative reads, which reduces bias and improves counting accuracy.

In contrast to serial analysis of gene expression-like approaches, DGE sample preparation does not require a PCR-based amplification step – a feature that allows faster experimental set-up and reduced cost. Freedom from biases introduced by restriction enzyme digestion and ligation steps is a further advantage of using the Helicos single molecule sequencing approach for DGE.

The effectiveness of the single molecule Helicos DGE application has been demonstrated by the complete profiling of the *Saccharomyces cerevisiae* transcriptome [16]. Helicos scientists generated millions of transcriptome-aligned reads in a single run of the sequencing platform. These were used to accurately quantify the complete range of transcripts expressed in the DBY746 strain of *S. cerevisiae*. The relative simplicity of the yeast's transcriptome allowed effective comparisons between the DGE results and classical quantification methods. Single molecule Helicos DGE demonstrated a quantitative analysis of the entire yeast transcriptome with accuracy comparable, if not more accurate, to quantitative PCR, often suggested as a gold standard for mRNA measurements.

While DGE provides the most accurate quantitative assessment of gene expression, there are times when a better assessment of RNA splicing, chimeric transcripts, or other information is desired that is best addressed via complete coverage of transcripts. Rather than initiating cDNA synthesis at the 3′ end, cDNA is generated by random priming throughout the length of the transcript. In this way, coverage of splice junctions and other useful functional regions can be identified.

17.3.1.1 DNA Sequencing Applications

The ultra-high-throughput yield of DNA sequencing provides a wealth of opportunities. In addition to resequencing of genomes, targeted regions can be sequenced either alone or in a multiplexed fashion such that many different samples can be sequenced in the same channel. Furthermore, DNA sequence information can be used to identify sites of protein binding or epigenetic modification via ChIP-based techniques. As no amplification is required, extremely quantitative mapping of the purified sequences is possible with sites of over-representation indicating areas of protein binding or modification. Similarly, structural variation in a genome can be readily ascertained by simply counting regions of coverage and examining regions of high or low coverage across the genome.

17.3.2
Single Molecule Sequencing Techniques under Development

Helicos is developing several specialized techniques to take advantage of its single molecule sequencing platform – dual-tag sequencing, and paired-read and paired-end sequencing. The dual-tag sequencing method enables the sequencing of templates with an internal priming site. Internal priming sites may be generated using a variety of ligation/circularization strategies for applications such as paired-end sequencing or high-complexity DNA barcoding. The process begins with generating a DNA template containing an internal universal priming site, and adding homopolymer "A" tails. The modified DNA fragments are hybridized to the surface and sequenced. The sequencing reaction is terminated and the universal primer, which is complementary to the internal universal priming sequence, is introduced before sequencing again.

Paired-read sequencing and paired-end sequencing enable the acquisition of sequence information from two regions or the two ends of a single molecule of DNA, respectively. These methods are tremendously beneficial for the alignment of reads to large reference sequences such as human genomes and transcriptomes. They also enable the detection of structural rearrangements, such as inversions, translocations, deletions, and amplifications in the genome [18], and alternative splice forms in the transcriptome.

In the case of paired-read sequencing, genomic DNA or cDNA fragments are tailed and hybridized to the flow cell surface and sequenced to obtain the first read. With the addition of polymerase and natural nucleotides in a controlled manner, the nascent strand is extended a given distance along the template. SBS with Virtual Terminator nucleotides is then continued to obtain the second read.

For paired-end sequencing, genomic DNA or cDNA fragments are tailed, and an adaptor containing a universal priming site is ligated to their 5' ends. The molecules are hybridized to the flow cell surface and sequenced to obtain the first read. In this case, unlike for the paired-read methodology, the complementary strand is synthesized to the end with the addition of polymerase and natural nucleotides. The original template is then melted off the surface, leaving a copy of the strand directly attached to the surface. Another SBS process is carried out using a universal primer that is complementary to the adapter sequence, to obtain a read from the 5′ end of the fragment.

17.4
Perspectives

The Helicos Genetic Analysis System has exceptional performance capabilities (see Table 17.1), enabling experiments on the scale required for understanding genome biology. The instrument is designed to accommodate future product developments, which will result in dramatic performance improvements with little or no modifications to the system hardware.

Among the several product development aims are optimizing the efficiency of the single molecule sequencing chemistry and increasing the density of captured strands on the flow cell surface. All current product development efforts are aimed at increasing instrument performance without requiring upgrades to the system. With current surface density and chemistry, the HeliScope Single Molecule Sequencer is designed to image 1.7 billion strands per run, producing a data throughput of 45 Mb/h. With the increased surface density and optimized sequencing chemistry, the instrument will be capable of producing up to 1 Gb/h.

On the data analysis side, Helicos has also set up an open-source initiative for its bioinformatics software. The company recognized early in its history that an open-source model for its informatics tools would produce the most effective software for analyzing data from the HeliScope Sequencer. The initiative benefits customers and software developers, as well as the wider scientific community by making the source code and support tools easily accessible. The community of users fostered by such an initiative will allow the widest possible access to real data sets and

Table 17.1 Routine use system performance specifications.

Strand output	12–16 million usable strands/channel[a]
Total output	420–560 Mb/channel
	21–28 Gb/run
Throughput	105–140 Mb/h
Read-length	25–55 bases in length
	30–35 average length
Accuracy	>99.995% consensus accuracy at >20 times coverage
Raw error rate	≤5% (~0.3% for substitutions)
	consistent from 20–80% GC content of target DNA independent of read-length and template size
Template size	25–2000 bases

The imaging system is designed for 1 Gb/h throughput performance. Actual sequencing performance is determined by chemistry efficiency and loading density. Routine use specifications based on genomic DNA samples loaded at recommended concentrations and 30-Quad run (8 days).
a) Usable strands are defined as those having 25 bases or greater in length and less than a defined raw error rate.

documentation from the Helicos web site. The site currently supports both "tarball" download and Subversion checkout for the source code as well as well known supporting tools, such as wiki docs and Mailman mailing lists. The initiative will also feature an open bug tracking system and will entertain patch submission from developers. To provide access to the largest possible group, the source code can be licensed through the widely used free software license GPL (general public license) for general use and through a commercial license for corporate partners.

The appetite for gene sequencing and associated gene expression studies will continue to grow, and Helicos believes its single molecule sequencing technology will enable future research. The scientific community recognizes that the completion of the first human genome reference sequence (Human Genome Project) represents only the beginning of the road to a comprehensive understanding of the human genome. The breadth of human genetic variation seems to be much wider than anticipated [19] and the implications of that realization will only begin to be understood once the sequences of a large number of genomes are known. A single molecule sequencing platform like the Helicos Genetic Analysis System will help advance the collection of that information.

One area where understanding genetic variation is crucial is in cancer treatment. The potential to personalize the treatment of cancer is directly dependent on the ability to understand genetic variation among individual patients as well as the genetic heterogeneity of tumor genomes. To fully take advantage of genetic information, researchers must combine knowledge of common variants with rare variants to reveal the true picture of biology. Examples of the importance of genetic variation in cancer include the knowledge that the effectiveness of oncology drugs varies among individuals, the ability to avoid costly and painful chemotherapy in some cases by predicting the likelihood of breast cancer recurrence, and the ability to determine medication dosage of anticoagulants based on an individual's metabolism. Connecting molecular variations with cancer phenotypes and clinical outcomes will help uncover new discoveries pertinent to disease diagnosis and treatment.

The "$1000 genome" is a theoretical benchmark for the cost of sequencing a genome that would allow comprehensive genetic information to routinely be used in preventing and treating disease. To win the race to the $1000 genome, scientists are working toward achieving a technology inflection that will allow full genome sequencing at that price level. Doing so will help realize the promise of personalized medicine – greater efficacy, safer drugs, preventive treatments, and potentially lower costs to the healthcare system.

Acknowledgments

The authors would like to acknowledge the valuable contributions of Parris Wellman, Steve Kellett, Aaron Kitzmiller, Jennifer MacArthur, and Doron Lipson. The authors would also like to thank John Thompson and Patrice Milos for critical reading of the manuscript.

References

1 Sanger, F. and Coulson, A.R. (1975) A rapid method for determining sequences in DNA by primed synthesis with DNA polymerase. *J. Mol. Biol.*, **94**, 441–448.

2 Lander, E.S. *et al.* (2001) Initial sequencing and analysis of the human genome. *Nature*, **409**, 860–921.

3 Venter, J.C. *et al.* (2001) The sequence of the human genome. *Science*, **291**, 1304–1351.

4 Jett, J.H. *et al.* (1989) High-speed DNA sequencing: an approach based on fluorescence detection of single molecules. *J. Biomol. Struct. Dyn.*, **7**, 301–309.

5 Gupta, P. (2008) Single molecule sequencing technologies for future genomics research. *Trends Biotechnol.*, **26**, 602–611.

6 Braslavsky, I. *et al.* (2003) Sequence information can be obtained from single DNA molecules. *Proc Natl Acad Sci USA*, **100**, 3960–3964.

7 Harris, T.D. *et al.* (2008) Single molecule DNA sequencing of a viral genome. *Science*, **320**, 106–109.

8 Hebert, B. and Braslavsky, I. (2007) Single-molecule fluorescence microscopy and its applications to single-molecule sequencing by cyclic synthesis, in *New High Throughput Technologies for DNA Sequencing and Genomics* (ed. K.R. Mitchelson), Perspectives in Bioanalysis Series, vol. 2, Elsevier, Amsterdam.

9 Kahvejian, A., Quackenbush, J., and Thompson, J.F. (2008) What would you do if you could sequence everything? *Nat. Biotechnol.*, **26**, 1125–1133.

10 Kahvejian, A. and Kellett, S. (2008) Making Single-Molecule Sequencing a Reality. *Am. Lab.*, **40**, 48–53.

11 Kahvejian, A. (2008) Single molecule sequencing advances: Helicos' genetic analysis system allows acquisition of billions of DNA sequences in parallel. *Genetic Eng. Biotechnol. News*, **28**, http://www.genengnews.com/articles/chtitem.aspx?tid=2360.

12 Harris, T.D. *et al.* (2008) Single molecule DNA sequencing of a viral genome. *Science*, **320**, supporting online material. http://www.sciencemag.org/cgi/data/320/5872/106/DC1/1.

13 Schwartz, J.J. and Quake, S.R. (2007) High-density single molecule surface patterning with colloidal epitaxy. *Appl. Phys. Lett.*, **91**, 083902.

14 Axelrod, D. (1984) Total internal reflection fluorescence in biological systems. *J. Lumin.*, **31–32**, 881–884.

15 Helicos BioSciences Corporation (2008) *Helicos™ Digital Gene Expression Primer*, Helicos BioSciences, Cambridge, MA.

16 Helicos BioSciences Corporation (2008) *Accurate Quantification of the Yeast Transcriptome by Single Molecule Sequencing*, Helicos BioSciences, Cambridge, MA.

17 Stranger, B.E. *et al.* (2007) Population genomics of human gene expression. *Nat. Genet.*, **39**, 1217.

18 Korbel, J.O. *et al.* (2007) Paired-end mapping reveals extensive structural variation in the human genome. *Science*, **318**, 420–426.

19 The International HapMap Consortium (2007) A second-generation human haplotype map of over 3.1 million SNPs. *Nature*, **449**, 851–861.

18
High-Throughput Sequencing by Hybridization
Sten Linnarsson

Abstract

Sequencing by hybridization was conceived as early as 1987, and was applied in its first decade to small-scale gene resequencing projects. However, more than twenty years elapsed before the method was scaled up (in our lab) to resequence whole bacterial genomes. Key to this development was the invention of a massively parallel and hybridization-friendly random array platform based on rolling-circle amplification. Surprisingly, less than two years later, a very similar technology platform is being used to resequence human genomes at unprecedented speed (R. Drmanac et al., 2009 Science is press, DOI:10.1126/science.1181498). One reason for this rapid progress is the simplicity of both the rolling-circle random array, and the hybridization chemistry. Here we present our protocol in an accessible step-by-step form.

18.1
Introduction

Recently, resequencing methods that promise drastically reduced costs and increased throughput have been developed ([1–6], reviewed in [7, 8]). In general, such methods combine a massively parallel DNA display technology (bead cloning [9], emulsion polymerase chain reaction (PCR) [10], in-gel PCR [11], solid-phase PCR [12], and single molecules [13]) with a compatible sequencing chemistry such as pyrosequencing [14, 15], sequencing by ligation [5, 9], or cyclic reversible termination [16, 17]. Conceptually distinct methods, such as nanopore sequencing, are more distant prospects [18].

Current commercially available examples include SOLiD™ (Applied Biosystems) and the Polonator (Dover Systems), which both use emulsion PCR to generate an array of beads carrying amplified single fragments, and then interrogate this array by sequential steps of ligation and cleavage of labeled nonamer probes (the Polonator lacks cleavage chemistry and therefore generates only very short reads); the Genome

The Handbook of Plant Mutation Screening. Edited by Günter Kahl and Khalid Meksem
Copyright © 2010 WILEY-VCH Verlag GmbH & Co. KGaA, Weinheim
ISBN: 978-3-527-32604-4

Analyzer II (Illumina), which uses solid-phase PCR to generate an array of *in situ* amplified fragments that are interrogated by stepwise incorporation and cleavage of fluorescently labeled nucleotides; the HeliScope™ (Helicos Biosciences), which sequences single molecules using stepwise incorporation of fluorescently labeled nucleotides; and the Genome Sequencer FLX™ (Roche/454 Life Sciences), which uses emulsion PCR and pyrosequencing.

We have developed a rapid and inexpensive DNA sequencing method termed "shotgun sequencing-by-hybridization (SBH)" [2]. The method is conceptually similar to tiling arrays [19] and to regular SBH [20–23], in that sequence is reconstructed from a complete tiling of the target sequence with short probes. However, resequencing is achieved hierarchically using a small universal set of probes compatible with any genome and proceeds in four steps: (i) *in situ* rolling-circle amplification of millions of randomly dispersed circular single-stranded DNA (ssDNA) fragments, (ii) sequential controlled hybridization of 582 pentamer probes, generating a *hybridization spectrum* for each target, (iii) alignment of hybridization spectra to the reference genome, and (iv) reconstruction of the target sequence using the combined hybridization patterns of all aligned fragments.

In this chapter, we provide a detailed, step-by-step protocol for preparing samples for shotgun SBH, followed by a discussion of potential pitfalls and future improvements.

18.2
Methods and Protocol

The general strategy is to enzymatically fragment the sample, ligate adaptors, select a narrow size range on polyacrylamide gel electrophoresis (PAGE), and then circularize using a helper oligo.

In more detail, the protocol is summarized in Figure 18.1. Genomic DNA, symbolized by the linear double helix (upper right), is fragmented by DNase I treatment. Adaptors of known sequence (blue, yellow) are then ligated and the library is selected to have insert size of 200 ± 10 bp. Strands are separated and circularized using a helper oligo, generating a library of circular, ssDNA molecules.

To make an array for sequencing, the template (library of circular ssDNA) is annealed to a capture oligo immobilized on a microscope slide. The oligo then serves as a primer for an *in situ* rolling-circle amplification, generating a covalently attached tandem-replicated product. Under suitable conditions, the products spontaneously curl up and form submicrometer fluffy structures, each containing a few thousand copies of the template molecule. There is no need for cross-linking as long as the ionic strength is kept reasonably high using, for example, NaCl or $MgCl_2$.

The array of *in situ* amplified templates is then subjected to sequential hybridizations with a set of 512 probes designed to detect every possible 5mer sequence.

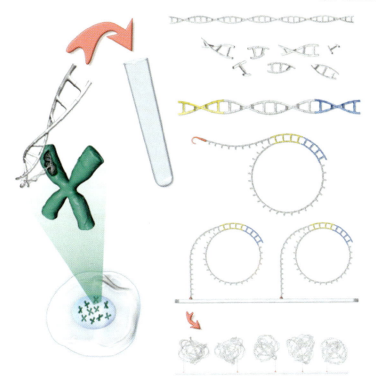

Figure 18.1 Protocol summary.

Stage 1: Sample Preparation

Input material	Genomic DNA (4 µg)
Output	Circularized ssDNA library with 200 ± 10-bp insert size and a universal adaptor sequence

Note: To decrease the loss of DNA during the procedures, due to adsorption to the centrifuge tubes, use Beckman polyallomer tubes throughout.

DNA Fractionation

- Mix the following for each reaction, *in the order indicated* (a high concentration of Mn^{2+} will precipitate proteins in the reaction, including the bovine serum albumin (BSA) in the buffer).

4 µg DNA sample	X µl
10 × DNase I buffer (= 0.5 M Tris–HCl, pH 7.5, 0.5 mg/ml BSA)	11.6 µl
Water	(to 116 µl)
10 × MnCl$_2$ (= 100 mM MnCl$_2$)	11.6 µl
Total volume	116 µl

- Just before use dilute DNase I (NEB) to 0.01 U/µl (2/400) in 1 × DNase I buffer and 1 × MnCl$_2$; 4 µl of DNase I mix is added to each reaction. Incubate at 25 °C for between 10 and 15 min. In our experience best profile for 200-bp fragments is achieved with a 12-min incubation.
- Stop the reaction by immediate purification using the QIAquick PCR Purification Kit (Qiagen). Elute in 55 µl.

  Note: We have found that enzymatic fragmentation using DNase I works reasonably well for bacterial genomes with a balanced GC content, such as *Escherichia coli*. However, more consistently reproducible results may probably be obtained using a Covaris AFA™ instrument [24].

  Samples are now ready for end repair using Klenow.

Klenow Treatment

DNA sample	55 µl
10 × NEB2 buffer	7 µl
dNTP (1 mM)	2.1 µl
Klenow (1 U/µl)	2 µl
H$_2$O	3.9 µl
Total	70 µl

- The suggested amount of Klenow is 1 U/µg of DNA. This protocol is based on 2 µg of DNA. For low-concentration/large-volume samples adjust the volumes and the enzyme amount.
- Incubate at room temperature for 10 min.

Purifying DNA from Agarose Gel

- To prevent chimeras, fragments <150 bp are removed by purification on a 2% agarose gels using QIAquick Gel Extraction Kit (Qiagen) according to the manufacturer's recommendation. Elute in with 55 µl. (*Note*: Others have noted a loss of AT-rich sequences due to the heating step in the gel extraction procedure [24]. More uniform results can probably be obtained if gel extraction is performed at room temperature.)
- Measure concentration.

Adaptor Ligation

- Prepare the following ligation mixture:

5 pmol DNA	X μl
375 pmol Adaptor (187.5 pmol of each adaptor)	15 μl
Quick ligation buffer (NEB)	50 μl
Water up to 100 μl	X μl
Quick ligase (NEB)	5 μl
Total volume	105 μl

- Samples are incubated at 25 °C for 15 min.
- Purify sample using PCR Cleanup (Qiagen) according to the manufacturer's protocol. Elute in 55 μl.
- Samples need to be treated with *Taq* polymerase (Phusion™; Finnzyme) to fill in the adaptor sequence and produce blunt-ended fragments:

DNA samples (everything)	X μl
5 × Phusion buffer HF	20 μl
dNTP (10 mM each)	2 μl
Water up to 99 μl	X μl
Total volume	99 μl

- Start PCR program 72_20SEC.cyc; when block temperature reaches 72 °C, put in the tubes for 5 min to melt apart the adaptors. Add 1 μl (2 U) Phusion polymerase and press "Start".
- Purify sample using PCR Cleanup (Qiagen) and elute samples in 30 μl elution buffer.

Purification of Ligated Template from PAGE

- Samples are purified from nondenaturing PAGE, 8%, run low voltage (250 V) overnight (∼24 h). Load four lanes per sample. Let bromphenol blue dye run out of gel; take apart the glasses and stain gel with 1 × SYBR® Gold (Invitrogen) for 15 min. Cut out 200 ± 10 bp as exactly as possible. Gel samples are collected in 50 μl of 10 mM Tris, pH 8. Leave samples for 3 h at 37 °C.

PCR on Ligated and Purified Template

- Prepare one or more of the following mixture (as standard, eight reactions are prepared for each purified fraction):

5 × Phusion buffer HF	20 μl
dNTP 10 mM	2 μl
Biotin oligo 20 μM	2 μl

5 × Phusion buffer HF	20 μl
Phospho oligo 20 μM	2 μl
Template	1 μl
Water	72 μl
Phusion *Taq*	1 μl
Total volume	100 μl

- Run 15 cycles (depending on the yield in previous steps, this may need to be increased) using the following cycle parameters: 98 °C for 10 s; 72 °C for 20 s.
- Purify the sample using PCR Cleanup (Qiagen). PCR reactions should be pooled and purified over one or two columns (depending of the number of reactions) to concentrate the sample. Just do the binding over the column several times if the initial volume is too high for one round of binding.

Purifying DNA from Agarose Gel

- To remove all contaminating products, the samples are purified from a 2% agarose gels using QIAquick Gel Extraction Kit (Qiagen) according to the manufacturer's recommendation. Elute in with 55 μl.

Purifying ssDNA

- Prepare 100 μl of Dynabeads (Dynal M280).
- Wash twice in 200 μl of B&W buffer (5 mM Tris HCl, pH 7.5, 0.5 mM EDTA, pH 8.0, 1.0 M NaCl), after final wash add 100 μl of 2 × B&W.
- Add 100 μl of purified PCR product and mix, let stand for 20 min at room temperature.
- Wash samples twice with 200 μl of B&W and twice with 10 mM Tris, pH 8, after final wash add 100 μl of 0.1 M NaOH and incubate for 3 min.
- Transfer supernatant to fresh tube and add 25 μl of 1 M Tris, pH 7.5.
- Purify sample using PCR Cleanup (Qiagen) according to manufacturer's protocol. Elute with 55 μl.
- Measure concentration.

Circularization of Template

- Annealing mix:

Single-stranded linear template	0.03–0.3 μM
5′-Biotinylated linker (Biolinker-512)	0.06–0.6 μM
in 30 μl 1 × Ligation buffer (Fermentas EL0011)	

- Ligase mix:

DNA ligase (Fermentas EL0011 5 U/μl)   0.5 μl
in 70 μl 1 × Ligation buffer (Fermentas EL0011)

- Use twice the concentration of linker compared to template.
- For annealing, the "Annealing mix" is heated at 65 °C 2 min and then cooled down to 25 °C with incubator block. The cooling takes approximately 15 min.
- Add 70 μl of ice-cold "Ligase mix" to "Annealing mix," mix, and spin.
- Incubate the reaction at 25 °C for 1 h.

Circularized Template Purification

- Prepare 25 μl of Dynabeads (Dynal M280).
- Wash twice in 100 μl of B&W buffer, after final wash add 100 μl of B&W.
- Add 100 μl of circularized template and mix, let stand for 20 min at room temperature.
- Wash beads twice in 100 μl of B&W and leave dry after final wash
- Elute circularized rolling-circle amplification templates at room temperature:
  - Add 30 μl H_2O, mix, save supernatant.
  - Add 30 μl NaOH 40 mM, mix, save supernatant.
  - Add 30 μl H_2O, mix, save supernatant.
- Pool the fractions and add 5 μl of 1 M Tris–HCl, pH 7.6 (Sigma) to stabilize the products.
- Standard dilution for array when using 0.03 μM single-stranded template is around 1/200. Prepare test arrays with 1/50, 1/100, 1/200, and 1/400 dilution.

Stage 2: *In Situ* Rolling-Circle Amplification

Input material	Circularized ssDNA with 200 ± 10-bp insert size
Output	Microscope slide with 5–10 million *in situ* amplified templates covalently bound in a 10×50-mm central region

Note: Before starting take the SAL-1 or SAL-1 Ultra slides out from the refrigerator and let them reach room temperature before taking the slide out of the box. This is to prevent water condensation on the slide surface.

Wear gloves. Prepare a chamber from two Secure-Seal™ Hybridization Chambers SA500 (Grace BioLabs). Seal the two half-chambers with a piece of tape (see Figure 18.2).

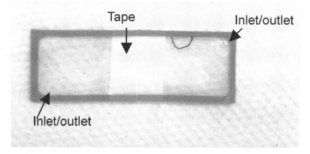

Figure 18.2 Tape used to seal the two half-chambers.

Primer Binding (for Whole Slide Volume ~1200 µl)

- Mix *in this order*:

Dimethylsulfoxide (final concentration 15%)	180 µl
RCA-512/RCA-G-Ring oligo 100 µM (final concentration 1 µM)	12 µl
400 mM Carbonate buffer, pH > 9.0 (final concentration 100 mM)	300 µl
Triton X-100 0.1% (final concentration 0.001%)	12 µl
H_2O	700 µl

- *To make carbonate buffer pH 9.00:* make 1 M Na_2CO_3 and 1 M $NaHCO_3$ solution, mix 1:1, adjust pH with concentrated glacial acetic acid and then dilute to the final concentration.
- Load the solution into the chamber and incubate for 50 min at 30 °C.
- Block unreacted groups by washing twice in 1% NH_4OH for 2 min.

Template Hybridization

- Prehybridize in 2 × sodium chloride/sodium citrate (SSC), 0.1% sodium dodecylsulfate (SDS): 65 °C for 2 min; 50 °C for 3 min; 30 °C for 5 min.
- Rinse in 2 × SSC, 0.1% Tween 20.
- Dilute circles in 2 × SSC, 0.1% SDS. Standard dilution for array when using 0.03 µM single-stranded template is around 1/200. Prepare test array with 1/50, 1/100, 1/200, and 1/400 dilutions.
- Spin the tube with template to sediment the leftover Dynabeads. Mix 100 µl of H_2O and 6 µl of template, and heat to 95 °C. Cool on ice and add 1100 µl of 2 × SSC, 0.1% SDS. Spin the tube.

Hybridization	Template at 65 °C for 2 min; 50 °C for 3 min; 30 °C for 10 min
Wash	2 × SSC, 0.1% SDS 30 °C 5 min
Rinse	Twice in 2 × SSC, 0.1% Tween 20
Rinse	1.5 mM Tris, pH 8, 10 mM $MgCl_2$

Rolling-Circle Amplification

- Mix the following for each slide (full-size chamber):

10 × phi29 Polymerase buffer	120 µl
dNTP (10 mM)	100 µl
BSA (NEB 100×)	0.9 µl
phi29 Polymerase (NEB or Fermentas)	11 µl
H_2O	890 µl

- Incubate reaction for 3 h at 30 °C

After Reaction

Rinse	1.5 mM Tris, pH 8, 10 mM $MgCl_2$
Wash	2 × SSC, 0.1% SDS 65 °C 2 min, 50 °C 3 min, 30 °C 2 min
Rinse	2 × SSC, 0.1% Tween 20
Rinse	1.5 mM Tris, pH 8, 10 mM $MgCl_2$
Rinse	1.5 mM Tris, pH 8, 10 mM $MgCl_2$
Dry	30 °C for 4 min

The slide can now be visualized by hybridization with a universal probe. A typical successful amplification is illustrated in Figure 18.3.

Figure 18.3 Typical successful amplification visualized by hybridization with a universal probe.

Stage 3: Sequencing by Hybridization

The sequencing stage is performed by an automated instrument as previously described [2]. In general, each probe is injected at elevated temperature (55 °C), the temperature is dropped to 33 °C below the T_m of the probe, the slide is washed and imaged, and the temperature is again elevated in preparation for the next probe.

As for other next-generation sequencing platforms, data is provided in both raw and analyzed forms. Raw intensities are extracted from the images for each feature and each probe. This raw data is aligned to a reference genome and a consensus sequence is called. Finally, a quality score is calculated for each position in the genome.

18.3
Discussion

The protocol presented here was used successfully to resequence viral (bacteriophage λ) and bacterial *(E. coli)* genomes. However, several challenges remain. First, we noticed a strong bias against AT-rich regions. In more general terms, there are several steps of the current protocol that may introduce bias. This could be alleviated by reducing or eliminating the PCR step, by omitting heating of gel pieces during purification (as noted above), and by using a physical rather than enzymatic fragmentation method (e.g., the Covaris AFA™ instrument). Finally, more uniform arrays may be obtained by using a larger concentration of phi29 polymerase during rolling-circle amplification (unpublished data), probably because this reduces the variation in polymerization start times. That is, at high polymerase concentration, essentially all templates begin elongation simultaneously, whereas at low concentration there may be a variable lag time before starting.

Future challenges include improvements to genome alignment algorithms, which are currently not fast enough to manage mammalian whole-genome projects, and reducing the error rate of individual probe hybridizations.

18.4
Applications

The method as currently developed shows promise in several fields. Its chief advantage is the simplicity of sample preparation, although further improvements are possible as noted above. In particular, the use of a rapid and inexpensive *in situ* amplification procedure rather than the laborious and expensive emulsion PCR is a key benefit for the user. Similarly, the simplicity of an enzyme-free sequencing chemistry leads to simplified instrument design, and relaxed demands on the optical and liquid handling systems. The method also provides long read-lengths (200 bp), although it currently suffers from a higher error rate than would ultimately be desirable.

With these characteristics in mind, one major application could be in metagenomics, particularly for deep 16S rRNA sequencing in environmental samples. Here, the long reads are necessary to cover enough variation in the gene, but some errors can be tolerated provided that enough reads can be collected to eliminate most errors during assembly. Extensions to other genes, such as the cytochrome c oxidase subunit I (COI) locus favored in genetic barcoding, would be straightforward.

Another promising application is in digital expression profiling. This is the approach where gene expression is quantified by counting the number of occurrences of mRNA molecules (or fragments of molecules) in a high-throughput shotgun sequencing experiment. The chief requirements for such a method are high throughput, low cost, and a fast genome alignment method to map reads to transcripts. With some improvements to alignment algorithms, shotgun SBH would fit this application perfectly. In particular, the simplicity of the chemistry leads to very low cost per run, which is crucial. To this end, we are currently developing a high-throughput single-cell cDNA preparation method that will allow inexpensive whole-genome expression profiling of hundreds of single cells in a single 1-week experiment.

References

1 Margulies, M. et al. (2005) Genome sequencing in microfabricated high-density picolitre reactors. *Nature*, **437**, 376–380.

2 Pihlak, A. et al. (2008) Rapid genome sequencing with short universal tiling probes. *Nat. Biotechnol.*, **26**, 676–684.

3 Bentley, D.R. et al. (2008) Accurate whole human genome sequencing using reversible terminator chemistry. *Nature*, **456**, 53–59.

4 Eid, J. et al. (2009) Real-time DNA sequencing from single polymerase molecules. *Science*, **323**, 133–138.

5 Shendure, J. et al. (2005) Accurate multiplex polony sequencing of an evolved bacterial genome. *Science*, **309**, 1728–1732.

6 Blazej, R.G., Kumaresan, P., and Mathies, R.A. (2006) Microfabricated bioprocessor for integrated nanoliter-scale Sanger DNA sequencing. *Proc. Natl. Acad. Sci. USA*, **103**, 7240–7245.

7 Shendure, J. and Ji, H. (2008) Next-generation DNA sequencing. *Nat. Biotechnol.*, **26**, 1135–1145.

8 Rothberg, J.M. and Leamon, J.H. (2008) The development and impact of 454 sequencing. *Nat. Biotechnol.*, **26**, 1117–1124.

9 Brenner, S. et al. (2000) Gene expression analysis by massively parallel signature sequencing (MPSS) on microbead arrays. *Nat. Biotechnol.*, **18**, 630–634.

10 Ghadessy, F.J., Ong, J.L., and Holliger, P. (2001) Directed evolution of polymerase function by compartmentalized self-replication. *Proc. Natl. Acad. Sci. USA*, **98**, 4552–4557.

11 Mitra, R.D. and Church, G.M. (1999) In situ localized amplification and contact replication of many individual DNA molecules. *Nucleic Acids Res.*, **27**, e34.

12 Bing, D.H. et al. (1996) Bridge amplification: a solid phase PCR system for the amplification and detection of allelic differences in single copy genes. Genetic Identity Conference: Seventh International Symposium on Human Identification, http://www.promega.com/geneticidproc/ussymp7proc/0726.html.

13 Braslavsky, I., Hebert, B., Kartalov, E., and Quake, S.R. (2003) Sequence information can be obtained from single DNA molecules. *Proc. Natl. Acad. Sci. USA*, **100**, 3960–3964.

14 Hyman, E.D. (1988) A new method of sequencing DNA. *Anal. Biochem.*, **174**, 423–436.

15 Ronaghi, M., Karamohamed, S., Pettersson, B., Uhlen, M., and Nyren, P. (1996) Real-time DNA sequencing using detection of pyrophosphate release. *Anal. Biochem.*, **242**, 84–89.

16 Metzker, M.L. *et al.* (1994) Termination of DNA synthesis by novel 3′-modified-deoxyribonucleoside 5′-triphosphates. *Nucleic Acids Res.*, **22**, 4259–4267.

17 Canard, B. and Sarfati, R.S. (1994) DNA polymerase fluorescent substrates with reversible 3′-tags. *Gene*, **148**, 1–6.

18 Branton, D. *et al.* (2008) The potential and challenges of nanopore sequencing. *Nat. Biotechnol.*, **26**, 1146–1153.

19 Hinds, D.A. *et al.* (2005) Whole-genome patterns of common DNA variation in three human populations. *Science*, **307**, 1072–1079.

20 Drmanac, R., Petrovic, N., Glisin, V., and Crkvenjakov, R. (1989) Sequencing of megabase plus DNA by hybridization: theory of the method. *Genomics*, **4**, 114–128.

21 Drmanac, S. *et al.* (1998) Accurate sequencing by hybridization for DNA diagnostics and individual genomics. *Nat. Biotechnol.*, **16**, 54–58.

22 Bains, W. and Smith, G.C. (1988) A novel method for nucleic acid sequence determination. *J. Theor. Biol.*, **135**, 303–307.

23 Lysov, Y.P., Florent'ev, V.L., Khorlin, A.A., Khrapko, K.R., and Shik, V.V. (1988) [Determination of the nucleotide sequence of DNA using hybridization with oligonucleotides. A new method]. *Dokl. Akad. Nauk SSSR*, **303**, 1508–1511.

24 Quail, M.A. *et al.* (2008) A large genome center's improvements to the Illumina sequencing system. *Nat. Methods*, **5**, 1005–1010.

19
DNA Sequencing-by-Synthesis using Novel Nucleotide Analogs

Lin Yu, Jia Guo, Ning Xu, Zengmin Li, and Jingyue Ju

Abstract

The completion of the Human Genome Project has increased the need for high-throughput DNA sequencing technologies aimed at uncovering the genomic contributions to diseases. The DNA sequencing-by-synthesis (SBS) approach has shown great promise as a new platform for deciphering the genome. Recently, much progress has been made in the fundamental sciences required to make SBS a viable sequencing technology. One of the unique features of this approach is that many of the steps required are compatible in a modular fashion, allowing for the best solution at each stage to be effectively integrated. Recent advances include the design and synthesis of novel cleavable fluorescent nucleotide reversible terminators, DNA template preparation using emulsion polymerase chain reaction and clonal clusters on immobilized single DNA molecules, and new surface attachment chemistries for DNA template immobilization. The integration of these advances will lead to the development of a high-throughput DNA sequencing system based on SBS that is able to decipher an entire human genome for $1000 in the near future.

19.1
Introduction

DNA sequencing is a fundamental tool for biological sciences. The completion of the Human Genome Project has set the stage for screening genetic mutations to identify disease genes on a genome-wide scale [1]. Recent discoveries indicate that the human genome of 3 billion base pairs is a complex interwoven network and contains very little unused sequence [2]. These new discoveries will drive the continued development of accurate, cost-effective, and high-throughput DNA sequencing technologies to decipher the functions of the complex genome for applications in clinical medicine and healthcare. Decreased cost of sequencing is critical to the comparative genomics efforts, including the ultimate goal of personalized medicine based on genetic and genomic information. Accuracy, speed, and size of the instrument are among the vital considerations for the development of new DNA analysis methods that can be implemented directly in the hospital and clinical settings, such as forensics and pathogen detection. Accuracy is essential for genetic mutation detection and haplotype analysis.

The Sanger dideoxy chain termination method [3] has been the technique of choice for large-scale DNA sequencing projects for over three decades. Widely used automated versions of this method employ either four differently end-labeled fluorescent primers or terminators to generate all the possible DNA fragments complementary to the template to be analyzed. The fragments terminating with the four different bases (A, C, G, and T) are then separated at single-base-pair resolution on a sequencing gel and identified by the four distinct fluorescent emissions [4, 5]. Application of laser-induced fluorescence for DNA sequencing is a major advancement for the automated DNA sequencing technology that makes large-scale genome sequencing initiatives possible. An "ideal" set of fluorophores for four-color DNA sequencing must consist of four different fluorophores. These fluorophores should have similar high molar absorbance at a common excitation wavelength, high fluorescence quantum yields, strong and well-separated fluorescence emissions, and the same relative mobility shift of the DNA sequencing fragments. These criteria cannot be met optimally by the spectroscopic properties of single fluorescent dye molecules and indeed are poorly satisfied by the initially used sets of fluorescent tags. Ju *et al.* overcame these obstacles imposed by the use of single dyes and developed fluorescence energy transfer dyes for DNA sequencing that fulfill the performance criteria set out above [6]. The higher sensitivity offered by these new sets of fluorescent dyes also allows the direct sequencing of large-template DNA (larger than 30 kb) with read-lengths of over 700 bases per sequencing reaction, leading to significant progress in the large-scale genome sequencing and mapping projects [7–9].

Despite Sanger sequencing's success, the electrophoresis-based sequencing technologies have some shortcomings due to the difficulty in achieving high throughput and the complexity involved in the automation. Recently, a variety of new DNA sequencing methods, including sequencing-by-hybridization [10], mass spectrometry sequencing [11–13], nanopore sequencing of single-stranded DNA [14], sequencing-by-ligation [15], and single DNA molecule sequencing [16, 17], have been investigated. In some of these reports, emulsion polymerase chain reaction (PCR), one commonly used technique for various biological assays including directed enzyme evolution [18, 19] and genotyping [20], was implemented for generating template from single DNA molecules. Margulies *et al.* used emulsion-based microreactors to amplify DNA templates in a one-tube reaction for pyrosequencing [21]. Beads containing a single amplified template were then isolated in individual wells and reagents were flowed across the well for the pyrosequencing reactions. Shendure *et al.* used emulsion PCR on 1-μm beads to prepare DNA template in the sequencing-by-ligation approach.

Among some of the novel approaches for DNA sequencing, the sequencing-by-synthesis (SBS) approach has emerged as a viable candidate for massive parallel high-throughput sequencing platforms. SBS takes advantage of the polymerase reaction – a key process for DNA replication inside cells. The basic concept of SBS is to use DNA polymerase to extend a primer that is hybridized to a template by a single nucleotide, determine its identity, and then proceed to the next nucleotide, eventually reading out the entire DNA sequence serially. In contrast to Sanger sequencing, in which fluorescently labeled DNA fragments of different sizes are all generated in a single reaction, and then separated and detected, SBS approaches have an advantage in that individual bases are detected simultaneously without the need for separations. Thus, SBS can easily scale-up over Sanger's dideoxy sequencing techniques. Currently, array scanners already exist that can easily detect over 100 000 sample spots arrayed on a glass surface [22]. Advanced array scanners enable fast screening of large areas with high resolution, allowing automated detection of hundreds of thousands and even millions of samples simultaneously.

19.2 General Methodology for DNA SBS

The concept of DNA SBS was first established in 1988 with an attempt to sequence DNA by detecting the pyrophosphate group that is released when a nucleotide is incorporated during DNA polymerase reaction [23]. Pyrosequencing, which was developed based on this concept, has been explored for DNA sequencing [24]. In this approach, each of the four dNTPs is added sequentially with a cocktail of enzymes, substrates, and the usual polymerase reaction components. If the added nucleotide is complementary with the first available base on the template, the nucleotide will be incorporated and a pyrophosphate will be released. Through an enzyme cascade, the released pyrophosphate is converted to ATP and then turned into visible light signal by firefly luciferase. On the other hand, if the added nucleotide is not incorporated, no light will be produced and the nucleotide will simply be degraded by the enzyme apyrase.

Pyrosequencing has been applied to single nucleotide polymorphism (SNP) detection and DNA sequencing [21, 25]. However, there are inherent difficulties in this method for determining the number of incorporated nucleotides in homopolymeric regions (e.g., a string of several Ts in a row) of the template. Wu *et al.* have solved this problem by using nucleotide reversible terminators (NRTs) to decipher the homopolymeric regions for pyrosequencing [26]. However, other aspects of pyrosequencing still need improvement. For example, each of the four nucleotides has to be added and detected separately, which increases the overall detection time. The accumulation of undegraded nucleotides and other components could also lower the accuracy of the method when sequencing a long DNA template. Ideally, as one examines the fundamental limitations towards miniaturization, one would prefer a simple method to directly detect a reporter group attached to the nucleotide that is incorporated into a growing DNA strand during polymerase reaction rather than relying on a complex enzymatic cascade.

Ju *et al.* have developed an integrated SBS approach for high-throughput sequencing platform as shown in Figure 19.1 [27]. This method relies on using the polymerase reaction to read out the DNA sequence through the incorporation of novel reporter nucleotides. After each nucleotide is added, the attached reporter

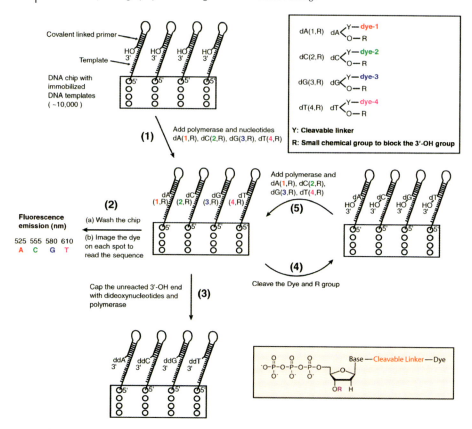

Figure 19.1 In the SBS approach, a chip is constructed with immobilized DNA templates that are able to self-prime for initiating the polymerase reaction. Four nucleotide analogs are designed such that each is labeled with a unique fluorescent dye on the specific location of the base and a small chemical group (R) to cap the 3'-OH group. Upon adding the four nucleotide analogs and DNA polymerase, only the nucleotide analog complementary to the next nucleotide on the template is incorporated by polymerase on each spot of the chip (Step 1). After removing the excess reagents and washing away any unincorporated nucleotide analogs, a four-color fluorescence imager is used to image the surface of the chip and the unique fluorescence emission from the specific dye on the nucleotide analogs on each spot of the chip will yield the identity of the nucleotide (Step 2). After imaging, the small amount of unreacted 3'-OH group on the self-primed template moiety will be capped by excess ddNTPs (ddATP, ddGTP, ddTTP, and ddCTP) and DNA polymerase to avoid interference with the next round of synthesis (Step 3). The dye moiety and the 3'-O-R protecting group will be removed chemically to generate a free 3'-OH group with high yield (Step 4). The self-primed DNA moiety on the chip at this stage is ready for the next cycle of the reaction to identify the next nucleotide sequence of the template DNA (Step 5) [27].

group is detected to determine the identity of the added nucleotide. In order to temporarily pause the sequencing reaction and to accurately sequence through homopolymeric regions, the 3'-OH group of the nucleotide must be blocked by a moiety to stop the polymerase reaction during the identification of the added nucleotide. This blocking group then needs to be easily removed to regenerate a free hydroxyl group for subsequent extension. In order to design an ideal system for SBS, new nucleotide analogs, termed cleavable fluorescent (CF)-NRTs, with the above properties must be developed. Taking this into account, the following requirements should be met to establish an entire SBS system:

- Standard cloning techniques to amplify DNA must be replaced by a high-throughput method for DNA template preparation.
- After initial amplification, DNA templates must be physically arrayed in a format that allows each template to be probed multiple times.
- Nucleotides must be reversible terminators (3'-OH is blocked) so that only a single nucleotide is added each step during SBS.
- The 3'-OH-blocking group and the fluorescent label used in SBS must be easily removed after detection for subsequent nucleotide addition.
- The entire system must allow for simple washing and reagent additions between detection cycles.

For SBS based on single fluorescent molecule detection, there is no need for the template amplification step. Emulsion PCR, which has been shown to have potential to address DNA template preparation for various sequencing platforms [15, 21], can be readily adapted to the approach shown in Figure 19.1 for SBS. A recently developed SBS system based on a similar design of the CF-NRTs has already found wide applications in genome biology [28–31]. The remainder of this chapter will be broken into several sections that describe different methodologies in the field that are elevating SBS to a viable sequencing technology.

19.3
Four-Color DNA SBS using CF-NRTs

19.3.1
Overview

In order to design the functional reporter nucleotides used in the SBS extension reaction, it is important to examine the structure of the polymerase enzyme complex with a DNA template, a primer, and an incoming nucleotide in the polymerase reaction. The three-dimensional structure of the ternary complexes of a rat DNA polymerase, a DNA template-primer, and a ddCTP is shown in Figure 19.2 [32]. It is apparent from this structure that the 5-position of the cytosine points away from the catalytic pocket of the enzyme, while the 3'-position of the ribose ring in ddCTP is in a very crowded space near the active amino acid residues of the polymerase. Any group that is attached at the 3'-position of the sugar must be small as to not interfere with polymerase reaction. Large bulky dye molecules have been attached at the 5-position of pyrmidines and the 7-position of purines and used in enzymatic incorporation reactions, especially in Sanger dideoxy sequencing [33–35]. Thus we reasoned that if a unique fluorescent dye is attached to the 5-position of the pyrimidines (T and C) and 7-position of purines (G and A) through a cleavable linker, and a small chemical moiety is used to cap the 3'-OH group, the resulting nucleotide analogs should be able

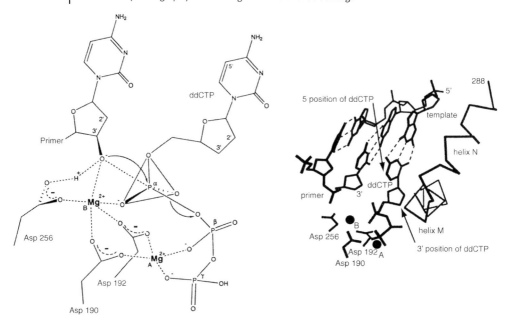

Figure 19.2 The three-dimensional structure of the ternary complexes of a rat DNA polymerase, a DNA template-primer, and a ddCTP: (a) mechanism for the addition of ddCTP and (b) active site of the polymerase in the context of the polymerase DNA complex. Note that the 3′-position of the dideoxyribose ring is very crowded, while ample space is available at the 5-position of the cytidine base.

to incorporate into the growing strand DNA. Based on this rationale, Ju et al. proposed a SBS methodology using CF nucleotide analogs as reversible terminators to sequence surface-immobilized DNA [27, 36]. In this approach, the nucleotides are modified at two specific locations so that they are still recognized by DNA polymerase as substrates: (i) a different fluorophore with a distinct fluorescent emission is linked to each of the four bases through a cleavable linker and (ii) the 3′-OH group is capped by a small chemically reversible moiety. DNA polymerase incorporates only a single nucleotide analog complementary to the base on a DNA template covalently linked to a surface. After incorporation, the unique fluorescence emission is detected to identify the incorporated nucleotide. The fluorophore is subsequently removed and the 3′-OH group is regenerated, allowing the next cycle of the polymerase reaction to proceed. As the large surface on a DNA chip can have a high density of different DNA templates spotted, each cycle can identify many bases in parallel, allowing the simultaneous sequencing of a large number of DNA molecules.

19.3.2
Design, Synthesis, and Characterization of CF-NRTs

Through previous research that established the feasibility of performing SBS on a chip using four photocleavable fluorescent nucleotide analogs [37] and the discovery of an allyl group as a cleavable linker to bridge a fluorophore to a nucleotide [38–40], Ju et al. reported the design and synthesis of nucleotide analogs containing a 3′-O-allyl group and a unique fluorophore tethered by a cleavable allyl linker for SBS [36]. The four CF-NRTs (3′-O-allyl-dCTP-allyl-Bodipy-FL-510, 3′-O-allyl-dUTP-allyl-R6G, 3′-O-allyl-dATP-allyl-ROX, and 3′-O-allyl-dGTP-allyl-Bodipy-650/Cy5) (Figure 19.3) were

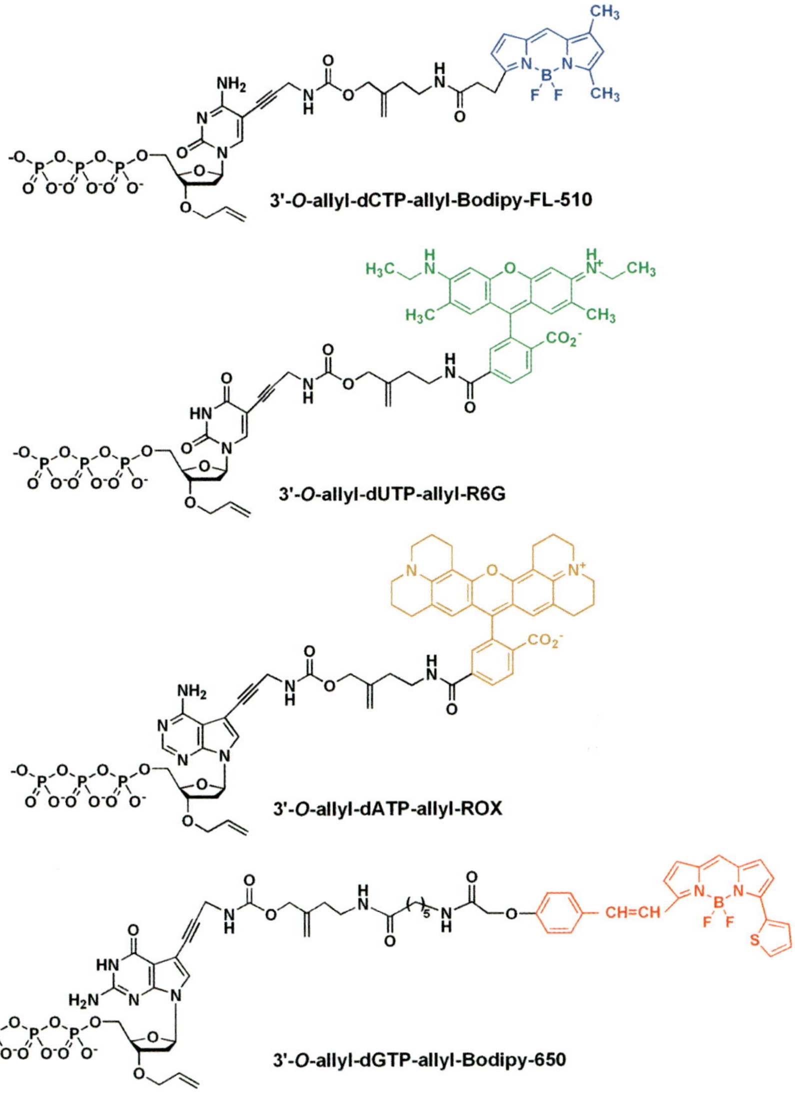

Figure 19.3 Structures of four CF-NRTs, 3′-O-ally-dNTP-allyl-fluorophores, with the four fluorophores having distinct fluorescent emissions: 3′-O-allyl-dCTP-allyl-Bodipy-FL-510 ($\lambda_{abs(max)} = 502$ nm; $\lambda_{em(max)} = 510$ nm), 3′-O-allyl-dUTP-allyl-R6G ($\lambda_{abs(max)} = 525$ nm; $\lambda_{em(max)} = 550$ nm), 3′-O-allyl-dATP-allyl-ROX ($\lambda_{abs(max)} = 585$ nm; $\lambda_{em(max)} = 602$ nm), and 3′-O-allyl-dGTP-allyl-Bodipy-650 ($\lambda_{abs(max)} = 649$ nm; $\lambda_{em(max)} = 670$ nm).

designed according to the general rationale for nucleotide modification described in the previous section. Since modified DNA polymerases have been shown to be highly tolerant to nucleotide modifications with bulky groups at the 5-position of pyrimidines (C and U) and the 7-position of purines (A and G), each unique fluorophore was attached to the 5-position of C/U and the 7-position of A/G through an allyl carbamate linker. However, due to the close proximity of the 3'-position on the sugar ring of a nucleotide to the amino acid residues of the active site of the DNA polymerase, a relatively small allyl moiety was chosen as the 3'-OH reversible capping group. After the incorporation of these nucleotide analogs and the detection of the fluorescent signal, the fluorophore and the 3'-O-allyl group on the DNA extension product are removed simultaneously in 30 s by palladium-catalyzed deallylation in aqueous solution. Such an efficient one-step dual-deallylation reaction allows the reinitiation of the polymerase reaction to incorporate the next base.

To verify that these CF-NRTs are incorporated accurately in a base-specific manner in a polymerase reaction, four continuous steps of DNA extension and deallylation were carried out in solution. This allows the isolation of the DNA product at each step for detailed molecular structure characterization by matrix-assisted laser desorption/ionization-time of flight mass spectrometry (MALDI-TOF MS) as shown in Figure 19.4. These results demonstrate that the four dual-allyl-modified CF nucleotide analogs are successfully incorporated with high fidelity into the growing DNA strand in a polymerase reaction; furthermore, both the fluorophore and the 3'-O-allyl group are efficiently removed by using a palladium-catalyzed deallylation reaction, which makes it feasible to use them for SBS on a chip.

19.3.3
DNA Chip Construction

In order to construct a DNA chip for SBS, site-specific 1,3-dipolar cycloaddition coupling chemistry was used to covalently immobilize the alkyne-labeled self-priming DNA template on the azido-functionalized surface in the presence of a Cu(I) catalyst [37]. The principal advantage offered by the use of a self-priming moiety as compared to using separate primers and templates is that the covalent linkage of the primer to the template in the self-priming moiety prevents any possible dissociation of the primer from the template during the process of SBS. To prevent nonspecific absorption of the unincorporated fluorescent nucleotides on the surface of the chip, a polyethylene glycol linker is introduced between the DNA templates and the chip surface [36]. This approach was shown to produce very low background fluorescence after cleavage to remove the fluorophore, as demonstrated by the DNA sequencing data described below.

19.3.4
Four-Color SBS using CF-NRTs

SBS on a chip-immobilized DNA template that had no homopolymer sequences was carried out using the four CF-NRTs (3'-O-allyl-dCTP-allyl-Bodipy-FL-510, 3'-O-

19.3 Four-Color DNA SBS using CF-NRTs

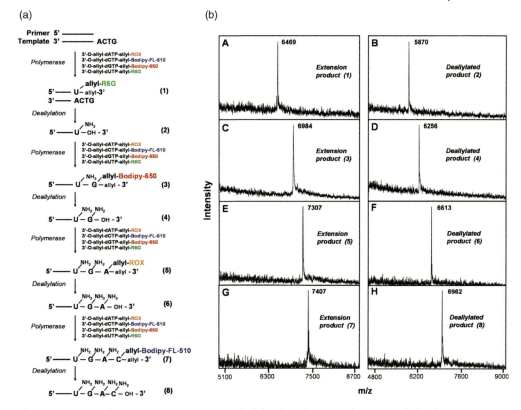

Figure 19.4 The polymerase extension scheme (a), and MALDI-TOF MS spectra of the four consecutive extension products and their deallylated products (b). Primer extended with 3′-O-allyl-dUTP-allyl-R6G (1) and its deallylated product 2; product 2 extended with 3′-O-allyl-dGTP-allyl-Bodipy-650 (3) and its deallylated product 4; product 4 extended with 3′-O-allyl-dATP-allyl-ROX (5) and its deallylated product 6; product 6 extended with 3′-O-allyl-dCTP-allyl-Bodipy-FL-510 (7) and its deallylated product 8. After 30 s of incubation with the palladium/triphenylphosphine trisulfonate mixture at 70 °C, deallylation is complete with both the fluorophores and the 3′-O-allyl groups cleaved from the extended DNA products.

allyl-dUTP-allyl-R6G, 3′-O-allyl-dATP-allyl-ROX, and 3′-O-allyl-dGTP-allyl-Cy5), and the results are shown in Figure 19.5. The *de novo* sequencing reaction on the chip was initiated by extending the self-priming DNA using a solution containing all four 3′-O-allyl-dNTP-allyl-fluorophores and a 9°N mutant DNA polymerase. To negate any lagging fluorescence signal caused by any previously unextended priming strand, a synchronization step was added to reduce the amount of unextended priming strands after the extension with the fluorescent nucleotides. A synchronization reaction mixture consisting of all four 3′-O-allyl-dNTPs (Figure 19.6), which have a higher polymerase incorporation efficiency due to the lack of a fluorophore compared with the bulkier 3′-O-allyl-dNTP-allyl-fluorophores, was used along with the 9°N mutant DNA polymerase to extend any remaining priming

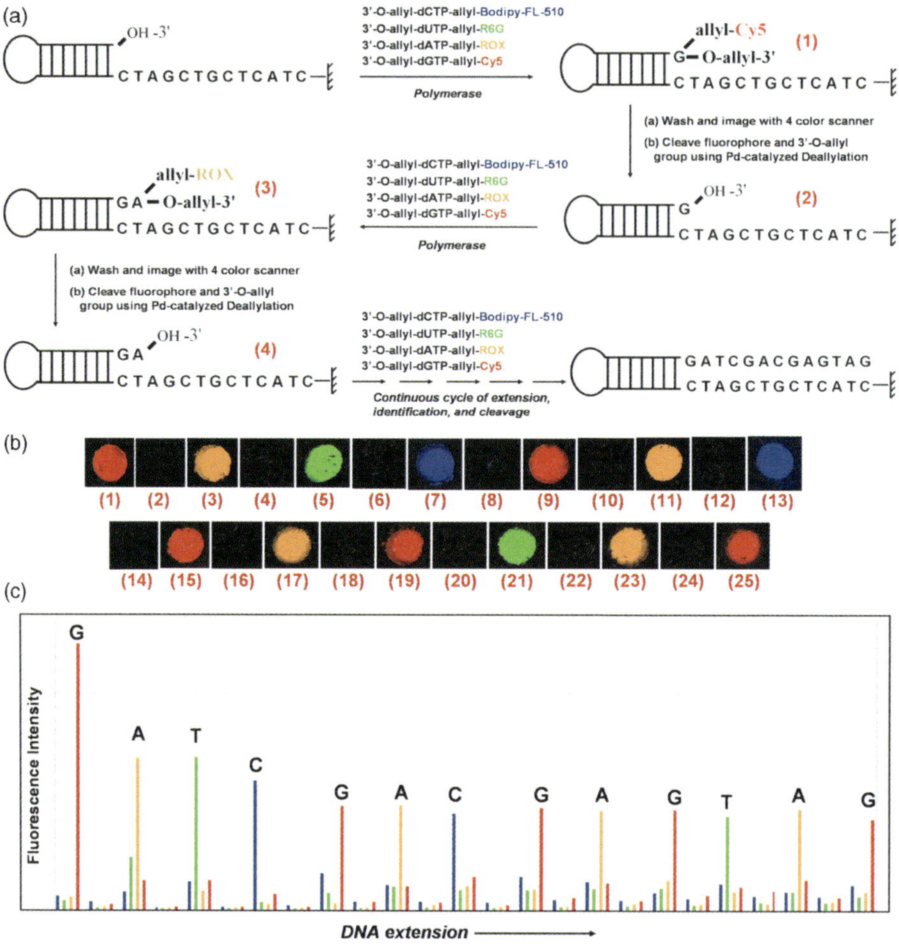

Figure 19.5 Four-color SBS data on a DNA chip. (a) Reaction scheme of SBS on a chip using four CF nucleotides. (b) The scanned four-color fluorescence images for each step of SBS on a chip: (1) incorporation of 3′-O-allyl-dGTP-allyl-Cy5; (2) cleavage of allyl-Cy5 and 3′-allyl group; (3) incorporation of 3′-O-allyl-dATP-allyl-ROX; (4) cleavage of allyl-ROX and 3′-allyl group; (5) incorporation of 3′-O-allyl-dUTP-allyl-R6G; (6) cleavage of allyl-R6G and 3′-allyl group; (7) incorporation of 3′-O-allyl-dCTP-allyl-Bodipy-FL-510; (8) cleavage of allyl-Bodipy-FL-510 and 3′-allyl group; images 9–25 are similarly produced. (c) A plot (four-color sequencing data) of raw fluorescence emission intensity at the four designated emission wavelengths of the four CF nucleotides versus the progress of sequencing extension.

strand that has a free 3′-OH group to synchronize the incorporation. The extension by 3′-O-allyl-dNTPs also enhances the enzymatic incorporation of the next nucleotide analog, since the DNA product extended by 3′-O-allyl-dNTPs carries no modification groups after the removal of the 3′-O-allyl group. The extension of the primer by only the complementary fluorescent nucleotide was confirmed by observing a red signal (the emission from Cy5) in a four-color fluorescent scanner

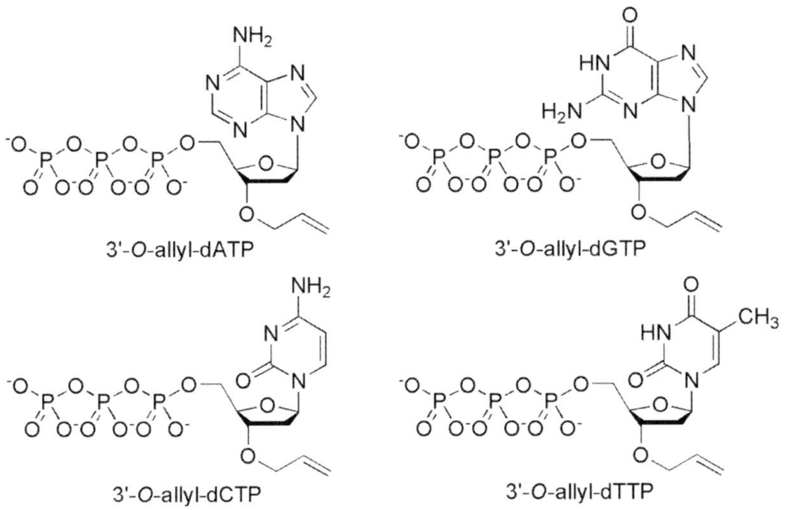

Figure 19.6 Structures of 3′-O-allyl-dATP, 3′-O-allyl-dCTP, 3′-O-allyl-dGTP, and 3′-O-allyl-dTTP.

(Figure 19.5b (1)). After detection of the fluorescent signal, the chip surface was immersed in a deallylation mixture (1 × Thermolpol I reaction buffer/Na$_2$PdCl$_4$/P (PhSO$_3$Na)$_3$) to cleave both the fluorophore and 3′-O-allyl group simultaneously. The chip was then immediately immersed in a 3 M Tris–HCl buffer, pH 8.5, to remove the palladium complex. A negligible residual fluorescent signal was detected to confirm cleavage of the fluorophore. The entire process of incorporation, synchronization, detection, and cleavage was performed multiple times to identify 13 successive bases in the DNA template. The same method was applied to sequence a DNA template with two separate homopolymeric regions as shown in Figure 19.7. All 20 bases, including the individual base (A, G, C, and T), the 10 repeated As, and the five repeated As were clearly identified. Contrarily, the pyrosequencing data of the same DNA template (Figure 19.7) displayed two large peaks that were very difficult to reveal the exact sequence.

19.4
Hybrid DNA SBS using NRTs and CF-ddNTPs

19.4.1
Overview

Based on the successful implementation of SBS using CF-NRTs [36], Guo et al. reported the development of an alternative DNA sequencing method that is a hybrid between the Sanger dideoxy chain terminating reaction and SBS [41]. In this approach, four nucleotides, modified as reversible terminators (NRTs) by capping the 3′-OH with a small reversible moiety so that they are still recognized by DNA

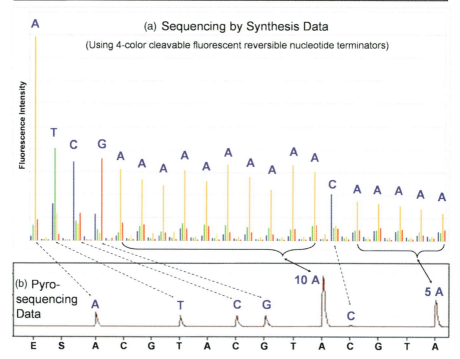

Figure 19.7 Comparison of four-color SBS and pyrosequencing data. (a) Four-color DNA sequencing raw data with our SBS chemistry using a template containing two homopolymeric regions. The individual base (A, T, C, and G), the 10 repeated As, and the five repeated As are clearly identified. The small groups of peaks between the identified bases are fluorescent background from the DNA chip, which does not build up as the cycle continues. (b) The pyrosequencing data of the same DNA template containing the homopolymeric regions (10 Ts and five Ts). The first four individual bases are clearly identified. The two homopolymeric regions (10 As) and (five As) produce two large peaks, for which it is very difficult to identify the exact sequence from the data.

polymerase as substrates to extend the DNA chain, are used in combination with a small percentage of four CF-ddNTPs to perform SBS. DNA sequences are determined by the unique fluorescence emission of each fluorophore on the DNA products terminated by ddNTPs. Upon removing the 3′-OH capping group from the DNA products generated by incorporating the 3′-O-modified dNTPs and the fluorophore from the DNA products terminated with the ddNTPs, the polymerase reaction reinitiates to continue the sequence determination. Using an azidomethyl group as a chemically reversible capping moiety in the 3′-O-modified NRTs and an azido-based cleavable linker to attach the fluorophores to the ddNTPs, Guo *et al.*

synthesized four 3′-O-N₃-dNTPs and four ddNTP-N₃-fluorophores for the hybrid SBS [41]. After fluorescence detection for sequence determination, the azidomethyl capping moiety on the 3′-OH and the fluorophore attached to the DNA extension product via the azido-based cleavable linker are efficiently removed using tris(2-carboxy-ethyl) phosphine (TCEP) in aqueous solution that is compatible with DNA. Various DNA templates, including those with homopolymer regions, were accurately sequenced with read-lengths of over 30 bases using this hybrid SBS method on a chip and a four-color fluorescent scanner.

19.4.2
Design and Synthesis of NRTs and CF-ddNTPs

Our previous research efforts have firmly established the molecular-level strategy to rationally modify the nucleotides by capping the 3′-OH with a small chemically reversible moiety for SBS [27, 36, 37, 39, 42]. Building on our successful 3′-O-modification strategy for the synthesis of NRTs, Guo et al. have explored alternative chemically reversible groups for capping the 3′-OH of the nucleotides for the preparation of the NRTs. In 1991, Zavgorodny et al. reported the capping of the 3′-OH group of the nucleoside with an azidomethyl moiety, which can be chemically cleaved under mild condition with triphenylphosphane in aqueous solutions [43]. Various 3′-O-azidomethyl nucleoside analogs following a similar procedure have been synthesized subsequently [44].

Guo et al. synthesized and evaluated four 3′-O-azidomethyl-modified NRTs (3′-O-N₃-dNTPs) (Figure 19.8) for use in the hybrid SBS approach. The 3′-O-modified NRTs containing an azidomethyl group to cap the 3′-OH on the sugar ring were synthesized based on a similar method to that reported by Zavgorodny et al. [41]. The 3′-O-azidomethyl group on the DNA extension product generated by incorporating each of the NRTs is efficiently removed by the Staudinger reaction using aqueous TCEP

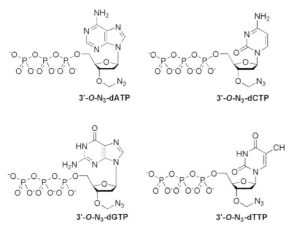

Figure 19.8 Structures of 3′-O-N₃-dATP, 3′-O-N₃-dCTP, 3′-O-N₃-dGTP, and 3′-O-N₃-dTTP.

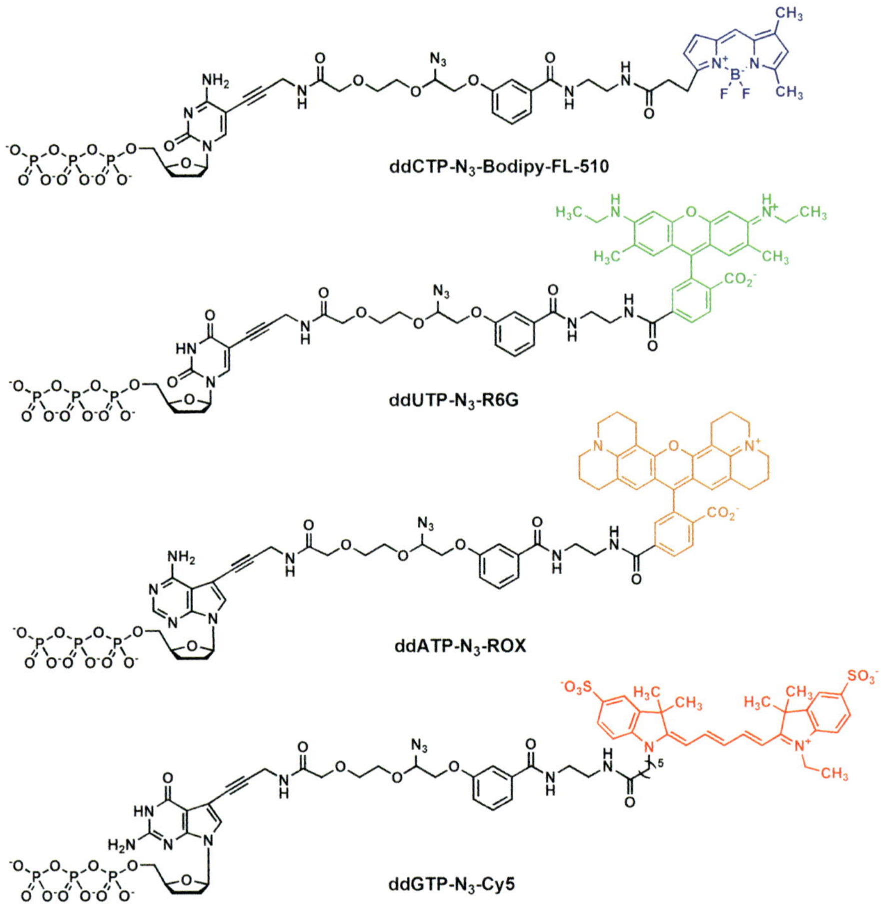

Figure 19.9 Structures of CF-ddNTP terminators ddNTP-N$_3$-fluorophores, with the four fluorophores having distinct fluorescent emissions: ddCTP-N$_3$-Bodipy-FL-510 ($\lambda_{abs\ (max)} = 502$ nm; $\lambda_{em\ (max)} = 510$ nm), ddUTP-N$_3$-R6G ($\lambda_{abs\ (max)} = 525$ nm; $\lambda_{em\ (max)} = 550$ nm), ddATP-N$_3$-ROX ($\lambda_{abs\ (max)} = 585$ nm; $\lambda_{em\ (max)} = 602$ nm), and ddGTP-N$_3$-Cy5 ($\lambda_{abs\ (max)} = 649$ nm; $\lambda_{em\ (max)} = 670$ nm).

solution [45] followed by hydrolysis to yield a free 3'-OH group for elongating the DNA chain in subsequent cycles of the hybrid SBS.

To demonstrate the feasibility of carrying out the hybrid SBS on a DNA chip, Guo et al. designed and synthesized four CF-ddNTP terminators, ddNTP-N$_3$-fluorophores (ddCTP-N$_3$-Bodipy-FL-510, ddUTP-N$_3$-R6G, ddATP-N$_3$-ROX, and ddGTP-N$_3$-Cy5) (Figure 19.9). According to a similar rationale for designing CF-NRTs [34, 36], each unique fluorophore was attached to the 5-position of C/U and the 7-position of A/G through a cleavable linker. The cleavable linker is also based on an azido-modified moiety [46] as a trigger for cleavage – a mechanism that is similar to the removal of the

3′-O-azidomethyl group. The ddNTP-N_3-fluorophores were found to efficiently incorporate into the growing DNA strand to terminate DNA synthesis for sequence determination. The fluorophore on a DNA extension product, which is generated by incorporation of the CF-ddNTP analogs, is removed rapidly and quantitatively by TCEP from the DNA extension product in aqueous solution. The ddNTP-N_3-fluorophores were used in combination with the four NRTs (Figure 19.8) to perform the hybrid SBS.

19.4.3
Four-Color Hybrid DNA SBS

In the four-color hybrid SBS approach, the identity of the incorporated nucleotide is determined by the unique fluorescent emission from the four fluorescent ddNTP terminators, while the role of the 3′-O-modified NRTs is to further extend the DNA strand to continue the determination of the DNA sequence. Therefore, the ratio between the amount of ddNTP-N_3-fluorophores and 3′-O-N_3-dNTPs during the polymerase reaction determines how much of the ddNTP-N_3-fluorophores incorporate and thus the corresponding fluorescent emission strength. With a finite amount of immobilized DNA template on a solid surface, initially the majority of the priming strands should be extended with 3′-O-N_3-dNTPs, while a relative smaller amount are to be extended with ddNTP-N_3-fluorophores to produce fluorescent signals that are above the fluorescent detection system's sensitivity threshold for sequence determination. As the sequencing cycle continues, the amounts of the ddNTP-N_3-fluorophores need to be gradually increased to maintain the fluorescence emission strength for detection. Following these guidelines, Guo *et al.* performed the hybrid SBS on a chip-immobilized DNA template using the 3′-O-N_3-dNTP/ddNTP-N_3-fluorophore combination and the results are shown in Figure 19.10. The general four-color sequencing reaction scheme on a chip is shown in Figure 19.10(a). The *de novo* sequencing reaction on the chip was initiated by extending the self-priming DNA using a solution containing the combination of the four 3′-O-N_3-dNTPs and the four ddNTP-N_3-fluorophores, and 9°N DNA polymerase. In order to negate any lagging fluorescence signal that is caused by a previously unextended priming strand, a synchronization step was added to reduce the amount of unextended priming strands after the initial extension reaction. A synchronization reaction mixture consisting of just the four 3′-O-N_3-dNTPs in relative high concentration was used along with the 9°N DNA polymerase to extend any remaining priming strands that retain a free 3′-OH group to synchronize the incorporation. The extension with the 3′-O-N_3-dNTP/ddNTP-N_3-fluorophore mixture does not have a negative impact on the enzymatic incorporation of the next nucleotide analog, because after cleavage to remove the 3′-OH capping group, the DNA products extended by 3′-O-N_3-dNTPs carry no modification groups. Previous designs of CF-NRTs left small traces of modification (propargyl amine linker) after the cleavage of the fluorophore on the base of the nucleotide [36]. Successive addition of these NRTs into a growing DNA strand during SBS leads to a newly synthesized DNA chain with, at each base site, a

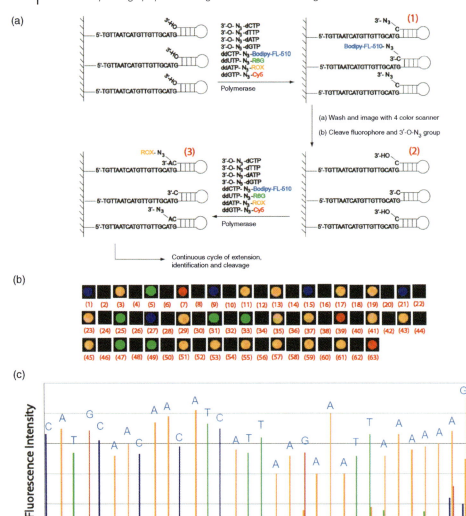

Figure 19.10 (a) Hybrid SBS scheme for four-color sequencing on a chip using the NRTs (3′-O-N$_3$-dNTPs) and CF-ddNTP terminators (ddNTP-N$_3$-fluorophores). (b) Four-color fluorescence images for each step of SBS: (1) incorporation of 3′-O-N$_3$-dCTP and ddCTP-N$_3$-Bodipy-FL-510; (2) cleavage of N$_3$-Bodipy-FL-510 and 3′-CH$_2$N$_3$ group; (3) incorporation of 3′-O-N$_3$-dATP and ddATP-N$_3$-ROX; (4) cleavage of N$_3$-ROX and 3′-CH$_2$N$_3$ group; (5) incorporation of 3′-O-N$_3$-dTTP and ddTTP-N$_3$-R6G; (6) cleavage of N$_3$-R6G and 3′-CH$_2$N$_3$ group; (7) incorporation of 3′-O-N$_3$-dGTP and ddGTP-N$_3$-Cy5; (8) cleavage of N$_3$-Cy5 and 3′-CH$_2$N$_3$ group; images 9–63 are similarly produced. (c) Plot (four-color sequencing data) of raw fluorescence emission intensity obtained using 3′-O-N$_3$-dNTPs and ddNTP-N$_3$-fluorophores at the four designated emission wavelengths of the four CF-ddNTPs.

small leftover linker. This may interfere with the ability of the enzyme to efficiently incorporate the next incoming nucleotide, which will undoubtedly lead to loss of synchrony and, furthermore, reduction in the maximal read-length. This challenge might potentially be overcome with further research efforts to re-engineer new DNA polymerases that efficiently recognize and accept the modified DNA strand. With the hybrid SBS approach, DNA products extended by ddNTP-N_3-fluorophores, after fluorescence detection for sequence determination and cleavage, are no longer involved in the subsequent polymerase reaction cycles because they are permanently terminated. Therefore, further polymerase reaction only occurs on a DNA strand that incorporates the 3'-O-N_3-dNTPs, which subsequently turn back into natural nucleotides upon cleavage of the 3'-OH capping group, and should have no deleterious effect on the polymerase binding to incorporate subsequent nucleotides for growing the DNA chains.

The four-color images from a fluorescence scanner for each step of the hybrid SBS on a chip is shown in Figure 19.10(b). The first extension of the primer by the complementary fluorescent ddNTP, ddCTP-N_3-Bodipy-FL-510 was confirmed by observing a blue signal (the emission from Bodipy-FL-510) (Figure 19.10b (1)). After detection of the fluorescent signal, the surface was immersed in a TCEP solution to cleave both the fluorophore from the DNA product extended with ddNTP-N_3-fluorophores and the 3'-O-azidomethyl group from the DNA product extended with 3'-O-N_3-dNTPs. The surface of the chip was then washed, and a negligible residual fluorescent signal was detected, confirming cleavage of the fluorophore (Figure 19.10b (2)). This was followed by another extension reaction using the 3'-O-N_3-dNTP/ddNTP-N_3-fluorophore solution to incorporate the next nucleotide complementary to the subsequent base on the template. The entire process of incorporation, synchronization, detection, and cleavage was performed multiple times to identify 32 successive bases in the DNA template. The plot of the fluorescence intensity versus the progress of sequencing extension (raw four-color sequencing data) is shown in Figure 19.10(c). The DNA sequences were unambiguously identified with no errors from the four-color raw fluorescence data without any processing. Similar four-color sequencing data were obtained for a variety of DNA templates [41].

19.5
Perspectives

A substantial number of advances have been made toward the goal of making DNA SBS a viable technology for genomic research. These include the rapid large-scale amplification of genomic libraries through emulsion PCR [15, 21] and the generation of clonal clusters from immobilized single DNA molecules [31], new developments in DNA attachment chemistries that allow increased array densities, and the invention of novel reporter nucleotides as reversible terminators for the polymerase reaction. These nucleotide analogs allow the enzymatic addition of a single nucleotide, direct detection to determine its identity, and efficient removal of the reporter fluorophore and the 3'-blocking group to allow for subsequent nucleotide additions.

The integration of these developments will make SBS a high-throughput DNA sequencing platform for the era of personalized medicine.

Acknowledgments

This work was supported by National Institutes of Health grants R01 HG003582, R01NS060762, and R01HG004774.

References

1 Collins, F.S., Green, E.D., Guttmacher, A.E., and Guyer, M.S. (2003) A vision for the future of genomics research. *Nature*, **422**, 835–847.
2 ENCODE Project Consortium (2007) Identification and analysis of functional elements in 1% of the human genome by the ENCODE pilot project. *Nature*, **14**, 799–816.
3 Sanger, F., Nicklen, S., and Coulson, A.R. (1977) DNA sequencing with chain-terminating inhibitors. *Proc. Natl. Acad. Sci. USA*, **74**, 5463–5467.
4 Smith, L.M. *et al.* (1986) Fluorescence detection in automated DNA sequence analysis. *Nature*, **321**, 674–679.
5 Prober, J.M. *et al.* (1987) A system for rapid DNA sequencing with fluorescent chain-terminating dideoxynucleotides. *Science*, **238**, 336–434.
6 Ju, J., Ruan, C., Fuller, C.W., Glazer, A.N., and Mathies, R.A. (1995) Fluorescence energy transfer dye-labeled primers for DNA sequencing and analysis. *Proc. Natl. Acad. Sci. USA*, **92**, 4347–4351.
7 Marra, M., Weinstock, L.A., and Mardis, E.R. (1996) End sequence determination from large insert clones using energy transfer fluorescent primers. *Genome Res.*, **6**, 1118–1122.
8 Lee, L.G. *et al.* (1997) New energy transfer dyes for DNA sequencing. *Nucleic Acids Res.*, **25**, 2816–2822.
9 Heiner, C.R., Hunkapiller, K.L., Chen, S., Glass, J.I., and Chen, E.Y. (1998) Sequencing multimegabase-template DNA with BigDye terminator chemistry. *Genome Res.*, **8**, 557–561.
10 Drmanac, S. *et al.* (1998) Accurate sequencing by hybridization for DNA diagnostics and individual genomics. *Nat. Biotechnol.*, **16**, 54–58.
11 Fu, D.J. *et al.* (1998) Sequencing exons 5 to 8 of the *p53* gene by MALDI-TOF mass spectrometry. *Nat. Biotechnol.*, **16**, 381–384.
12 Roskey, M.T. *et al.* (1996) DNA sequencing by delayed extraction-matrix-assisted laser desorption/ionizaton time of flight mass spectrometry. *Proc. Natl. Acad. Sci. USA*, **93**, 4724–4729.
13 Edwards, J.R., Itagaki, Y., and Ju, J. (2001) DNA sequencing using biotinylated dideoxynucleotides and mass spectrometry. *Nucleic Acids Res.*, **29**, E104.
14 Kasianowicz, J.J., Brandin, E., Branton, D., and Deamer, D.W. (1996) Characterization of individual polynucleotide molecules using a membrane channel. *Proc. Natl. Acad. Sci. USA*, **93**, 13770–13773.
15 Shendure, J. *et al.* (2005) Accurate multiplex polony sequencing of an evolved bacterial genome. *Science*, **309**, 1728–1732.
16 Harris, T.D. *et al.* (2008) Single-molecule DNA sequencing of a viral genome. *Science*, **320**, 106–109.
17 Eid, J. *et al.* (2008) Real-time DNA sequencing from single polymerase molecule. *Science*, **323**, 133–138.
18 Tawfik, D.S. and Griffiths, A.D. (1998) Man-made cell-like compartments for molecular evolution. *Nat. Biotechnol.*, **16**, 652–656.
19 Ghadessy, F.J., Ong, J.L., and Holliger, P. (2001) Directed evolution of polymerase function by compartmentalized self-replication. *Proc. Natl. Acad. Sci. USA*, **98**, 4552–4557. 4527.
20 Dressman, D., Yan, H., Traverso, G., Kinzler, K.W., and Vogelstein, B. (2003)

Transforming single DNA molecules into fluorescent magnetic particles for detection and enumeration of genetic variations. *Proc. Natl. Acad. Sci. USA*, **100**, 8817–8822.

21 Margulies, M. *et al.* (2005) Genome sequencing in microfabricated high-density picolitre reactors. *Nature*, **437**, 376–380.

22 Schena, M., Shalon, D., Davis, R.W., and Brown, P.O. (1995) Quantitative monitoring of gene expression patterns with a complementary DNA microarray. *Science*, **270**, 467–470.

23 Hyman, E.D. (1988) A new method of sequencing DNA. *Anal. Biochem.*, **174**, 423–436.

24 Ronaghi, M., Karamohamed, S., Pettersson, B., Uhlen, M., and Nyren, P. (1996) Real-time DNA sequencing using detection of pyrophosphate release. *Anal Biochem.*, **242**, 84–89.

25 Ronaghi, M., Uhlen, M., and Nyren, P. (1998) A sequencing method based on real-time pyrophosphate. *Science*, **281**, 363–365.

26 Wu, J. *et al.* (2007) 3′-O-modified nucleotides as reversible terminators for pyrosequencing. *Proc. Natl. Acad. Sci. USA*, **104**, 16462–16467.

27 Ju, J., Li, Z., Edwards, J., and Itagaki, Y. (2003) Massive parallel method for decoding DNA and RNA. US Patent 6,664,079.

28 Mikkelsen, T.S. *et al.* (2007) Genome-wide maps of chromatin state in pluripotent and lineage-committed cells. *Nature*, **448**, 553–560.

29 Johnson, D.S., Mortazavi, A., Myers, R.M., and Wold, B. (2007) Genome-wide mapping of *in vivo* protein–DNA interactions. *Science*, **316**, 1497–1502.

30 Barski, A. *et al.* (2007) High-resolution profiling of histone methylations in the human genome. *Cell*, **129**, 823–837.

31 Bentley, D.R. *et al.* (2008) Accurate whole human genome sequencing using reversible terminator chemistry. *Nature*, **456**, 53–59.

32 Pelletier, H., Sawaya, M.R., Kumar, A., Wilson, S.H., and Kraut, J. (1994) Structures of ternary complexes of rat DNA polymerase beta, a DNA template-primer, and ddCTP. *Science*, **264**, 1891–1903.

33 Zhu, Z., Chao, J., and Waggoner, A.S. (1994) Directly labeled DNA probes using fluorescent nucleotides with different length linkers. *Nucleic Acids Res.*, **22**, 3418–3422.

34 Rosenblum, B.B. *et al.* (1997) New dye-labeled terminators for improved DNA sequencing patterns. *Nucleic Acids Res.*, **25**, 4500–4504.

35 Duthie, R.S. *et al.* (2002) Novel cyanine dye-labeled dideoxynucleoside triphosphates for DNA sequencing. *Bioconjug. Chem.*, **13**, 699–706.

36 Ju, J. *et al.* (2006) Four-color DNA sequencing by synthesis using cleavable fluorescent nucleotide reversible terminators. *Proc. Natl. Acad. Sci. USA*, **103**, 19635–19640.

37 Seo, T.S. *et al.* (2005) Four-color DNA sequencing by synthesis on a chip using photocleavable fluorescent nucleotides. *Proc. Natl. Acad. Sci. USA*, **102**, 5926–5931.

38 Bi, L., Kim, D.H., and Ju, J. (2006) Design and synthesis of a chemically cleavable fluorescent nucleotide, 3′-O-allyl-dGTP-allyl-Bodipy-FL-510, as a reversible terminator for DNA sequencing by synthesis. *J. Am Chem Soc.*, **128**, 2542–2543.

39 Ruparel, H. *et al.* (2005) Design and synthesis of a 3′-O-allyl photocleavable fluorescent nucleotide as a reversible terminator for DNA sequencing by synthesis. *Proc. Natl. Acad. Sci. USA*, **102**, 5932–5937.

40 Meng, Q. *et al.* (2006) Design and synthesis of a photocleavable fluorescent nucleotide 3′-O-allyl-dGTP-PC-Bodipy-FL-510 as a reversible terminator for DNA sequencing by synthesis. *J. Org. Chem.*, **71**, 3248–3252.

41 Guo, J. *et al.* (2008) Four-color DNA sequencing with 3′-O-modified nucleotide reversible terminators and chemically cleavable fluorescent dideoxynucleotides. *Proc. Natl. Acad. Sci. USA*, **105**, 9145–9150.

42 Li, Z. *et al.* (2003) A photocleavable fluorescent nucleotide for DNA sequencing and analysis. *Proc. Natl. Acad. Sci. USA*, **100**, 414–419.

43 Zavgorodny, S. *et al.* (1991) 1-Alkylthioalkylation of nucleoside hydroxyl functions and its synthetic applications: a

new versatile method in nucleoside chemistry. *Tetrahedron Lett.*, **32**, 7593–7596.

44 Zavgorodny, S., Pechenov, A.E., Shvets, V.I., and Miroshnikov, A.I. (2000) S,X-acetals in nucleoside chemistry. III. Synthesis of 2′- and 3′-O-azidomethyl derivatives of ribonucleosides. *Nucleosides Nucleotides Nucleic Acids*, **19**, 1977–1991.

45 Saxon, E. and Bertozzi, C.R. (2000) Cell surface engineering by a modified Staudinger reaction. *Science*, **287**, 2007–2010.

46 Milton, J., Ruediger, S., and Liu, X. (2006) Nucleosides/nucleotides conjugated to labels via cleavable linkages and their use in nucleic acid sequencing. US Patent Application US20060160081A1.

20
Emerging Technologies: Nanopore Sequencing for Mutation Detection

Ryan Rollings and Jiali Li

Abstract

The emerging nanopore-based DNA sequencing methods combine high-throughput and single molecule detection without the need for tagging or optical detection. They present a fascinating route to significantly reduce the cost of sequencing human-sized genomes. In this chapter, we review the history of the development of nanopores for DNA sequencing along with the early results that show the method's promise. We also discuss strategies that move away from the tagless approach as well as the progress on developing these "third-generation" nanopore-based sequencing and mutation detection techniques.

20.1
Introduction

The technological advances that allowed the Human Genome Project to create the first working draft of the human genome in 2001 opened researcher's eyes to the benefits of high-speed sequencing technologies and the transformation they could bring to the field of genetics. This vision has been supported by the National Institute of Health's Human Genome Research Institute's funding initiatives to develop methods to sequence a human-sized genome for $100 000 in the near future with the eventual cost being less than $1000. A second generation of massively parallel sequencing platforms are becoming commercially available that are capable of producing a human-sized genome for $100 000 and are reviewed in a previous handbook in this series [1]. The eventual goal of a $1000 genome has sparked research to develop so-called "third-generation" sequencers that use massively parallel single molecule techniques to drastically reduce reagent cost and sample preparation time, and increase sequencing accuracy. A particularly promising technique is nanopore-based sequencing.

This introduction to nanopore-based sequencing methods accompanies several recent reviews of the field [2–4]. These references are indispensable to the reader interested in the historical development of nanopore technology and cover the state-of-the-art methods in great detail.

20.1.1
Nanopore Detection Principle

Nanopores, biological or synthetic, with diameters less than 10 nm and membrane thicknesses of the order of 10 nm have been used as the main sensing component for DNA sequencing. Figure 20.1(a) illustrates a typical nanopore sensing system as developed in our lab. In this system, a single pore is used as the sole electrical and fluidic connection between two electrolyte filled fluid chambers (cis and trans). When a constant DC voltage is applied across Ag/AgCl electrodes connected to each chamber, a flow of ions through the nanopore creates a stable open pore ionic current on the order of pico- to nanoamperes. For solid-state nanopores, this current has been shown to be linear and ohmic over a wide range of conditions, with an anomalous increase in conductance for very low salt concentrations [2]. The open pore current is proportional to the bulk conductivity, voltage, and area of the nanopore, yet inversely proportional to the thickness of the membrane containing the nanopore [3].

The pore probes single DNA molecules using the Coulter counter principle – also known as the resistive pulse sensing method – discovered by Walter Coulter in the 1940s (http://www.whcf.org/About/about-wallace.html). When the electric field generated by the applied voltage drives the negatively charged DNA molecule through the nanopore, the DNA molecule partially blocks the flow of the ions. Thus, the passing DNA molecule causes a transient resistance increase and resulting current decrease as shown in Figure 20.1(b). The transient current decrease contains

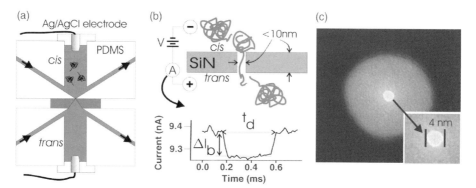

Figure 20.1 The nanopore concept. The nanopore acts as the sole connection between conducting electrolyte solution as shown in (a). An electric field is applied by Ag/AgCl electrodes driving the negatively charged DNA through the nanopore, causing a momentary decrease in ionic conduction through the nanopore (b). The TEM image shows an approximately 4-nm diameter pore fabricated in our lab by ion beam sculpting (c).

information about the physical properties of the translocating DNA molecule such as its geometric dimensions and electrical charge density. In addition to the DNA molecule itself, the physical properties of the solution such as pH, conductivity, viscosity, and the dimensions of the nanopore also contribute to the characteristics of the current blockage signal. Figure 20.1(c) shows an example of the transmission electron microscopy (TEM) image of an approximately 4-nm diameter silicon nitride pore fabricated in our lab.

20.1.2
Important Parameters and Nanopore Sensing Resolution

In solution at pH 7, the phosphate backbone of DNA is negatively charged to a charge density of 1 e/phosphate. In an electrolyte solution far from container walls the positively charged cations condense around the DNA to effectively screen the charge to around 0.33 e/phosphate over a wide range of ionic strengths [4]. The negative charge of a DNA molecule makes it possible for one to electrophoretically drive the DNA molecule by an electric field. For situations where the nanopore size and ionic strength of the solution are high enough such that the translocating molecule can translocate without significant interaction with the pore boundary, the amplitude of the current blockage or drop produced by a transiting DNA molecule increases with an increase in solution conductivity, voltage, and cross-sectional area excluded by the molecule [5]. The duration of a current blockage event increases with increasing length of a DNA molecule and solution viscosity, and decreases with an increase of charge and applied voltage [6–8].

The thickness of the membrane supporting the nanopore is a complicating factor. For a given bias voltage, the current drop amplitude increases with a decrease in nanopore thickness, potentially increasing the nanopore detection resolution. In principle, a nanopore effectively senses the average cross-section of the molecule over the length of the pore. This suggests that the thinner the pore, the higher the fidelity of the current drop to the axial cross-section variations caused by different bases. For DNA, a pore thinner than the around 3.4-Å base-pair spacing of a DNA molecule would be ideal to avoid averaging the current drop over several bases. However, for both biological and solid-state nanopores, there is an access resistance region, as shown in Figure 20.2, caused by the electric field lines extended above and below the nanopore, roughly equal to the radius of the nanopore that is also sensitive to the blocking DNA molecule [9–11]. Even for an ideal monolayer thickness nanopore, the sensing region is limited by the radius of the pore. For a pore that has a diameter of around 3 nm, we thus have a region around 3 nm long, forcing us to integrate over at least 9 bases even in this ideal situation.

20.1.3
Biological Nanopore History

The earliest published work using protein pores to probe polymers in solution was with the alemethicin ion channel in 1994 [12]. In this study, Bezrukov *et al.* detected a

20 Emerging Technologies: Nanopore Sequencing for Mutation Detection

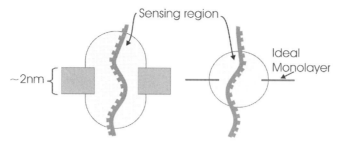

Figure 20.2 Hypothesized nanopore sensing region and its limits on nucleotide discrimination. The ionic current through the nanopore is reduced by the partial length of a DNA molecule that enters into a region that extends approximately one pore radius on either side of the nanopore. The shorter the sensing region is, the shorter the length of DNA that is being probed and the higher the sequence resolution at each moment. An ideal, single-monolayer-thick nanopore would still have a sensing region of about the diameter of the nanopore.

change in pore conductance with the addition of polyethylene glycol (PEG) molecules that were driven by diffusion through the pore. Concurrently, the Bayley lab was developing α-hemolysin ion channels as biosensors [13]. Kasianowicz *et al.* demonstrated in 1996 that single-stranded DNA (ssDNA) and RNA homopolymers translocated through the α-hemolysin channel [14]. They hypothesized that if each nucleotide along the molecule created a characteristic current blockage, the sequence could be detected as shown in Figure 20.3. Several groups began using α-hemolysin to detect the differences between different ssDNA and RNA homopolymers [15, 16]. Their research showed that α-hemolysin could readily detect the difference between polyA and polyC RNA, and even detect the difference between polyA and polyC regions on the same synthetically prepared molecules. However, the difference between polydA and polydC ssDNA was much smaller. It was concluded that the large variation in coiled secondary structure between polyA and polyC RNA, which the ssDNA homopolymers lacked, caused the large variations in current rather than the sequence itself. Their research, and the research reviewed here and else-

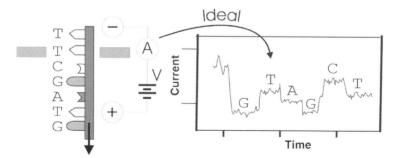

Figure 20.3 Original idealized nanopore sequencing concept. As ssDNA electrophoreses through the nanopore, differences in the nucleotides cause differences in the ion conductance through the nanopore that directly yield the sequence of the molecule.

where [17–19], has proven the ability of biological pores to detect the secondary structure of biomolecules.

20.1.4
Solid-State Nanopore History

The development of solid-state nanopores was driven by the desire to bring the promise shown by the α-hemolysin pore to a platform that could be fabricated with robust solid-state materials. The first solid-state nanopore that was able to detect single DNA molecule translocation was developed in the Harvard Nanopore Group using a unique process called ion beam sculpting [5, 20]. In this process, a free-standing silicon-rich silicon nitride or silicon dioxide membrane is perforated by a high-energy (50 keV) focused ion beam mill or by electron beam lithography creating a single approximately 100-nm hole through the entire membrane. Then, the surrounding surface is bombarded by a low-energy noble gas ion beam, causing a very thin layer of matter to flow towards the hole [21–23]. Ions also pass through the hole and are detected after exiting the pore. The rate of ions passing through the shrinking pore decreases, allowing the pore shrinking process to be monitored and the beam to be shut off when the pore reaches the desired area. By controlling the noble gas ion beam species, flux, and energy, the pore's dimensions can be sculpted to a radius of a few nanometers with a resolution of around 1 nm and a nanopore thickness of around 10 nm.

Shortly after the development of ion beam sculpted pores, high-energy electron beam sputtered silicon nitride and silicon oxide pores were developed [24]. This method has developed such that subnanometer radial resolution can be achieved [25, 26]. In this method, a commercial TEM beam is focused on a freestanding membrane and beam conditions can be set to controllably open or close a nanopore. Just as in the ion beam sculpting method, a thin, freestanding membrane is produced using standard photolithographic techniques, but is then drilled open and shaped using a combination of electron beam sputtering and heating. The TEM-based fabrication methods have been the most often used, perhaps due to the availability of TEM equipment and the resulting nanopore's high radial resolution.

20.1.5
Nanopore Promise

Perhaps the greatest attraction of nanopore-based sequencing methods is that they require minimal sample preparation. To illustrate this point, Branton et al. in their recent review of nanopore technology estimated that approximately 10^8 DNA molecules would be adequate for 6-fold sequence coverage [18, 27]. For medical analysis, these molecules could be procured from a human patient with a 20-ml blood sample that could be processed using readily available technology to yield genomic DNA. Theoretically, there is no limit on the length of a DNA to be sequenced by a nanopore, but mechanical shearing or other methods could then be used to fragment the DNA into easily translocatable fragments on the order of tens of kilobases if necessary.

A single nanopore allows high serial throughput with translocation rates on the order of 10^6 nucleotides/s depending of the biological or synthetic pore used [18]. At this rate the 10^9 bases of a mammalian genome can pass through a pore single file in less than 1 h. To overcome noise present in nanopore measurements and the bandwidth limitations of measurement equipment, nanopores can be placed in parallel on the same physical platform to increase the sensing throughput and accuracy [28]. Working in parallel, nanopores can use the modern massively parallel methods of second-generation sequencers to sequence more than one fragment at a time and reconstruct the genome computationally. The computation time would reduced compared to present methods since the read-lengths would be at least 1–2 orders of magnitude longer than at present. This is particularly appealing for solid-state nanopores since the photolithographic methods used to fabricate them are well suited for parallelization, cheap fabrication, and integration into electronic detection and data processing platforms.

20.2
Current Developments in Nanopore Sequencing

So far, only ionic current blockage sequencing has been presented. To avoid the problems inherent to this method, two routes are being pursued: (i) modification and improvement of the nanopore sensing resolution, and (ii) the tagging of the DNA to make the differences between bases larger. While the former method is more true to the original promise and eventual goal of nanopore sequencing, the latter may be able to leverage well-characterized biochemical techniques with the nanopore's abilities.

20.2.1
Improving Biological Nanopores

Several engineered variants of the α-hemolysin protein pores have been developed to increase their sensitivity. By placing a single cysteine amino acid on the opening of an α-hemolysin pore, a 5′-thiol-modified oligomer was attached by Howorka *et al.* [29]. This experimental set-up allowed his group to sequence a codon at the end of a covalently bound oligomer as it dangled into the narrowest constriction of the nanopore. By passing known sequences of DNA with varying end sequences, they found that exact matches had dwell times much longer than even single base mismatched hybrids. Their method was limited to sequencing only the codon of the covalently bound oligomer, but their attachment chemistry could allow the attachment of processing enzymes to the pore to help slow the translocation rate and modify the DNA to aid in its detection.

Astier, Wu and others of the Bayley group covalently attached an aminocyclodextrin molecule to the inner constriction of the α-hemolysin pore [30, 31]. They have shown that the current of this mutant pore is modulated to four different levels as the four different dNMPs are driven through it one by one and in preliminary tests have been able to discriminate between bases at least 93% of the time. They envision attaching an exonuclease near the entrance of the pore that would cleave the DNA into a sequence of dNMPs that would be driven electrophoretically through the pore one at a time as they are removed. Progress has gone so as for the creation of the company Oxford Nanopore Technologies Ltd (www.nanoporetech.com), which is working to commercialize this exonuclease sequencing method.

20.2.2
Improving Solid-State Nanopores

Solid-state nanopores have lagged behind biological nanopores because of the lack of precise control over pore dimensions and surface properties. However, solid-state nanopores have not only pushed closer to the dimensional control of biological pores and developed surface modification techniques, but work has been done to make them a platform for electrodes that may be capable of truly tagless and enzyme-free sequencing.

The two fabrication methods predominantly oriented towards DNA sequencing – ion beam sculpting and electron beam sputtering/annealing – have improved their control over the diameter of the nanopores. Diameter control has been achieved at the single nanometer (for ion beam sculpted pores) and subnanometer (for electron beam fabricated pores) level. The thickness of the nanopores has not been controlled to single nanometer precision. However, better characterization of the nanopore thicknesses has been reported [21, 25, 32].

Wanunu *et al.* have extended surface functionalization methods that are well-established for planar surfaces to the confined region of the nanopore by reproducibly coating the pore and surrounding area [33]. Iqbal *et al.* modified the nanopore surface chemistry to bind hairpin DNA oligomers to the pore surface and surrounding membrane [34]. They investigated the dwell times of oligomers electrophoretically driven through the nanopore with different degrees of complementarity to the bound hairpins, and found that there was a reduction in dwell time with increasing complementarity between the free and bound oligomers. Their results show the ability to detect a single mismatch between the free and bound oligomers.

Perhaps the most technically challenging goal has been to develop tunneling electrodes placed perpendicular to the axis of the pore as illustrated in Figure 20.4. These electrodes are designed to probe the electronic structure of the passing molecule, similar to scanning tunneling microscopy. A large amount of theoretical work has been done on this subject, but despite continued effort, very little of the experimental work has provided results [35–37]. Theoretical work so far has suggested larger differences between bases than ionic current blockage sequencing, with tunneling projected currents in the nanoampere range. This would reduce the need to slow the molecule and allow higher bandwidth for detection, but is limited by the requirement that each base be well oriented as it moves past the transverse electrode. Each base on a ssDNA molecule passing through the pore undergoes thermal motion that could rotate the base, which theoretically produces wildly

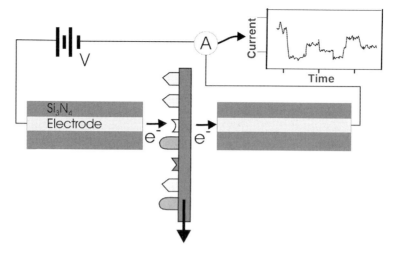

Figure 20.4 Tunneling electrodes in solid-state nanopore. The DNA is driven through the nanopore electrophoretically similar to other nanopore methods. Electrodes are imbedded in the nanopore membrane and a voltage is applied perpendicular to the DNA molecule. An electron tunneling current is modulated by each base in the translocating DNA in a manner that would be used to determine the nucleotide sequence.

different tunneling currents. This motion by even slow moving bases could make the sequence unintelligible.

An experimental approach to surmount this problem has been the attempt to use ssDNA–carbon nanotube binding to orient the nucleotides and reduce thermal fluctuations during translocation. Nanotubes and ssDNA will bind spontaneously in solution with a tendency to orient each base in a regular fashion [38]. Using the electric field and appropriate solution conditions, the ssDNA could then be pulled along a stationary nanotube, reducing the velocity of each base as well as reducing the variability in base orientation relative to the electrode. Nanopores with imbedded tunneling electrodes have been constructed, but a working device capable of detecting single molecule translocation has been elusive due to the variability in research-scale production [18]. Metal electrodes encapsulated in silicon nitride adjacent to a nanopore have been demonstrated to measure admittance and phase changes in AC measurements made by the transverse electrodes in the presence of translocating charged gold nanoparticles, but the bandwidths used were too low to detect single translocation events [39].

Another possible nucleotide orientation scheme is the functionalization of transverse electrodes that use hydrogen bonding to orient the translocating molecule. To do this each electrode would have a separate chemistry. The molecule on the end of one electrode would hydrogen bond to the nucleobase and a different molecule on the opposing electrode would hydrogen bond to the phosphate backbone, thus orienting the nucleotide for analysis. This method has a precedent in the demonstration of chemical modification as a means to orient the phosphate backbone of DNA molecules on a metal electrode for scanning tunneling microscopy (STM) analysis [40]. Similarly, functionalizing a STM tip with a thiol derivative of a single nucleotide species has been demonstrated to increase the tunneling current between the electrode and the complimentary nucleotide on a surface-bound ssDNA molecule [40, 41].

20.2.3
Slowing Translocation and Trapping

Slowing DNA translocation is a concern for all nanopore-based methods, with programmable control of translocation being the ultimate goal to allow sequencing and resequencing of the same molecule on the fly to increase resolution. A variety of techniques have been employed to slow the translocation of DNA molecules, including varying solution viscosity, temperature, and applied voltage, for a 10-fold increase in translocation time [3]. Still others have demonstrated the slowing and trapping of DNA in a nanopore by coupling DNA to a bead controlled by optical tweezers [42, 43]. In these experiments, researchers were able to repeatedly insert the tethered DNA molecule into the pore and pull it back out the same side it entered, demonstrating the possibility of resequencing a molecule with ultimate control.

Dynamic control over the DNA translocation rate has been demonstrated by varying the voltage during the translocation [44–46]. In this technique the driving voltage is reduced or shut off as a DNA hairpin is partially driven into a pore that is only large enough to allow ssDNA to translocate. The voltage is then increased at a constant rate and unzips the hairpin, pulling it through the pore. An interesting extension of this method is the ability to drive a molecule through the nanopore and, after a complete translocation, reverse the polarity of the driving field and drive the same molecule back through the pore [47]. Although this method has not yet been used to slow the molecule, it presents a method for reprobing a molecule several times to increase the signal-to-noise ratio, with the possibility of reprobing sequences within a section of DNA without it leaving the pore.

20.2.4
Modification of the DNA

In addition to developing a better detector, some researchers have begun work on amplifying the base differences between DNA molecules to make them easier to detect using biochemical labeling. The simplest method presented so far is hybridization-assisted nanopore sequencing as illustrated in Figure 20.5, patented by Ling et al. [48]. In this method, short oligos hybridize with the unknown strand of DNA and then translocate through the nanopore. Taking advantage of the solid-state nanopore's ability to detect the cross-sectional difference between a single- and double-stranded DNA molecule, the measurement should be able to determine the location of the hybridized portion of the molecule, thus revealing the sequence at that region [49]. By copying the sample DNA many times, each variation of the known oligomer can be hybridized to a copy of the sample DNA and translocated, yielding a current blockage profile that gives the sequence at a specific position along the

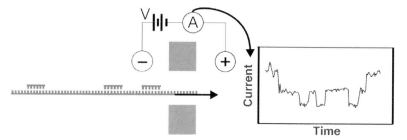

Figure 20.5 Hybridization assisted nanopore sequencing. Short oligos of known sequence are hybridized with ssDNA and drawn through the nanopore. The ionic current blockage profile caused by the single- and double-stranded sections would yield the position of the oligos, thus revealing the sequence at that position. To sequence the molecule, the interrogated DNA would be copied and annealed to many different sets of oligomers. The information from each blockage profile would then be combined by software to reconstruct the entire sequence.

molecule. These sequences for each of the known regions can then be computationally combined into the genomic sequence.

Since the hybridized region is much longer than a single nucleotide, hybridization will increase the time that a detectable portion of the molecule spends in the nanopore, letting the detection system afford a higher translocation velocity while still yielding useful information. By increasing the length of the complimentary oligomer, the hybrid region is easier to detect; however, the number of unique oligomers required to sequence the molecule increases exponentially. Another complication is that the unhybridized ssDNA regions will randomly self-hybridize and form a complicated secondary structure that could create undesirable current blockades. The same temperatures, ionic strengths, and denaturants that remove the unwanted secondary structure would also remove the short oligomers.

Another proposed method using DNA hybridization would use optically tagged oligos that would fluoresce as the hybridized DNA molecule translocates through the nanopore, as illustrated in Figure 20.6 [50]. In this method, a synthetic "designer polymer" version of the DNA molecule under study would be created, effectively replacing each nucleotide by a unique single-stranded sequence of about 24 nucleotides. The expanded DNA sequence would then be hybridized to two 12mer optically tagged oligos. Each oligo would have a fluorophore at the 5' end and a quencher at the 3' end so that as the oligos hybridize with the DNA molecule, the quencher would line up with the fluorophore, quenching its fluorescence and leaving only the final oligo visible. This hybridized construct would be driven through a nanopore too small to allow double-stranded DNA, but large enough for ssDNA, causing the final oligo to unzip from the molecule which would allow the next oligo to fluoresce. Unzipping of hairpin DNA molecules has been shown to be a slow process on the order of milliseconds and can be controlled by applied voltage, allowing enough time for optical techniques to record the transient events [51, 52]. The "designer polymer" can be designed such that the oligomer will self-hybridize, bringing its own fluorophore and quencher in close proximity, and removing its contribution to the background fluorescence.

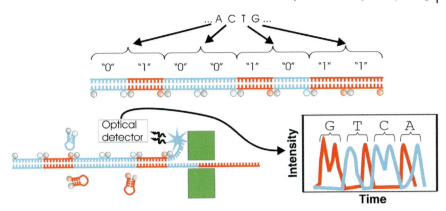

Figure 20.6 Optical detection of bound oligomers. First, a native DNA sequence is expanded into a designer DNA polymer. Each base will be expanded into two 12-base sequences that will make a single-stranded molecule 24 times the length of the original DNA. Second, each 12-base sequence will be hybridized to an oligo functionalized with one of two fluorophores (shown in blue and red) at the 5′ end and a broad-spectrum quencher at the 3′ end. As the oligos anneal to the polymer, the 5′ fluorophore and 3′ quencher line up, quenching their fluorescence, except for the first fluorophore. The fluorescence wavelength of each fluorophore can be used to encode each base into a series of two binary states. Third, the designer DNA polymer is driven through a nanopore smaller than the width of the double-stranded molecule where the nanopore sequentially unzips each hairpin, briefly unquenching each fluorophore in sequence. Every two fluorescence signals would correspond to a single base on the original molecule.

In order to avoid using four fluorophores to correspond to the four bases, a clever use of the "designer polymer" concept would allow each base to be amplified to a 24mer sequence made of two sequential 12 nucleotide sequences, each conceptually corresponding to a binary "1" or "0" as shown in Figure 20.6. In this way, each nucleotide species can be encoded as two binary digits rather than one quaternary digit, thus making each base correspond to a sequence of two flashes of light in sequence, and drastically simplifying the detection set-up and data processing. Perhaps the greatest benefit of any optical detection method is the ability to massively parallelize the detection process allowing only one CCD detector to observe a large array of nanopores. This ability alone is expected to make up for the time required for the construction of the "designer polymer" and attachment of complimentary oligomers.

20.2.5
Resequencing Applications

Aside from the *de novo* sequencing discussed so far, nanopores offer the ability to search for mutagenesis on the single molecule level, retaining the possible benefits of small sample size, high throughput, and low cost. Recent examples of polymorphism detection include solid-state nanopores that were used to detect the presence and location of two adjacent mismatched bases of oligomers hybridized to longer ssDNA molecules, while modern informatics techniques have been used to detect single base mismatches in hairpin stems threading α-hemolysin pores [53, 54]. As mentioned previously, solid-state nanopores coated in ssDNA have also shown single base discrimination in the translocation times of complimentary ssDNA [34]. These reports suggest the possibility of a nanopore for replacing the detecting mechanism used in allele-specific nucleotide-based polymorphism detection. A mutation in a DNA sequence can be detected by adding a known complimentary oligomer spanning the region of interest. As the hybrid complex passes through the nanopore, the dwell time – or some other computationally derived parameter – reflects the mismatch number. Instead of amplifying the interrogated DNA and using radioactive or optical detection methods to detect an appropriately tagged hybridized molecule, the physical binding force of the unmodified oligomer is detected one molecule at a time. Essentially any method that uses a short complimentary oligomer to detect a polymorphism can be replaced with a nanopore-based technique. The nanopore would simplify the detection of the oligomer since there is no tagging of the oligomer while retaining single molecule sensitivity.

20.3
Work Done in Our Lab

In our lab we work on both improving nanopore sensing resolution and making the DNA we wish to study easier to sequence. By studying the noise characteristics and surface properties, we are improving the current resolution of our nanopores by changing the fabrication parameters available to ion beam sculpting (i.e., varying incident ion species and energy). We have found limitations, however, in our ability to control the thickness of our using only ion beam methods. The competition between lateral mass flow to close the pore and the sputtering processes does not seem capable of creating an ideal approximately 1 nm thickness of the membrane adjacent to the pore, so we are looking at thinning the membrane supporting the nanopore by chemical wet etching of the silicon nitride. Although nanometer-resolution thinning of stoichiometric silicon nitride thin films is well established in the literature, we are investigating etch rates of the less well known surface chemistry adjacent to the opening of the pore [53]. This is more difficult than standard optical or profilometric thin film metrology techniques since the region of interest is confined to an area much smaller than an optical beam waist and too small to find with macroscopic profilometry. We are currently pursuing electron energy loss spectroscopy (EELS) thickness profiling and three-dimensional TEM tomography similar to work done previously [25].

We are simultaneously investigating detecting DNA hybridization in solid-state nanopores, and attaching biotin labels to individual bases as a method to increase the difference between any one type of nucleotide and the other three. For example, Figure 20.7 illustrates one such experiment that demonstrates the ability of our solid-state nanopore to size single- and double-stranded sections of hairpin DNA molecules. Using a pore around 4 nm in diameter, we were able to detect the presence of

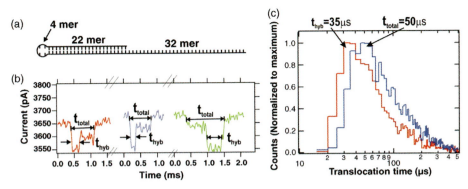

Figure 20.7 Proof-of-concept study performed to detect the current drop caused by the translocation of a 80mer hairpin structure (a) in an approximately 4 nm pore. Events show a characteristic double-level translocation profile (b) with the peak duration of the lower level and upper level being distinguishable (c).

both double- and single-stranded sections of a 80mer hairpin molecule. These results, although preliminary, have encouraged our group to further pursue hybridization and tagging DNA molecules for ionic current blockage sequencing.

20.4 Perspectives

It is difficult to predict which of the many different nanopore sequencing methods presented will prove to be the most successful. Biological nanopores are the most sensitive, have the most accurate measurements, and have the highest signal-to-noise ratio. With recent in-membrane stabilization and the use of very small lipid bilayers, biological pores using the exonuclease method may become the sensing elements of the first single molecule sequencers [54, 55]. The exonuclease method, however, still has large hurdles to overcome to ensure that the dNMPs pass in sequential order and translocate through to the trans side, rather than return to the cis side where they can later translocate to cause erroneous signals.

Biological nanopores are inherently less stable than solid-state structures and cannot be easily mass manufactured as disposable chips with the appropriate onboard electronics. Using transverse electrodes, solid-state nanopores potentially offer the only method capable of measuring the DNA sequence without any tags, but because of the difficulties in fabrication it is likely that the first solid-state nanopore to sequence DNA will rely upon a different method and require some form of tag.

In conclusion, both biological and solid-state nanopores have demonstrated promise to sequence DNA at the single molecule level with high throughput. Although the focus is typically on *de novo* sequencing, nanopores have already performed proof-of-concept experiments showing that they can be used to replace conventional detection mechanisms in mutagenesis detection. The original concept of sequence detection by ionic conductance blockage has been modified to a wide array of competing techniques, with biological and solid-state nanopores being developed along side each other. Biological nanopores still show the highest sensitivity and have recently been demonstrated with improved stability. However, they will always lack long-term stability and are difficult to integrate with detection electronics using existing fabrication techniques. No insurmountable challenges are seen for nanopore sequencing and the continued advancement of the art makes them a strong candidate for "third-generation" sequencers.

Acknowledgments

The authors would like to acknowledge Brad Ledden for helpful comments. This work is supported from grants 1R21HG003290 and 5R21HG004776 from the National Human Genome Research Institute (National Institutes of Health).

References

1 Nutter, R.C. (2008) New frontiers in plant functional genomics using next generation sequencing technologies, in *The Handbook of Plant Functional Genomics: Concepts and Protocols* (eds G. Kahl and K. Meksem), Wiley-VCH Verlag GmbH, Weinheim.

2 Smeets, R.M., Keyser, U.F., Krapf, D., Wu, M.-Y., Nynke, H.D., and Dekker, C. (2006) Salt dependence of ion transport and DNA translocation through solid-state nanopores. *Nano Lett.*, **6**, 89–95.

3 Fologea, D., Uplinger, J., Thomas, B., McNabb, D.S., and Li, J. (2005) Slowing DNA translocation in a solid state nanopore. *Nano Lett.*, **5**, 1734–1737.

4 Manning, G.S. (1978) The molecular theory of polyelectrolyte solutions with applications to the electrostatic properties of polynucleotides. *Q. Rev. Biophys. II*, **2**, 179–246.

5 Li, J., Gershow, M., Stein, D., Brandin, E., and Golovchenko, J.A. (2003) DNA molecules and configurations in a solid-state nanopore microscope. *Nat. Mater.*, **2**, 611–615.

6 Storm, A.J., Chen, J.H., Zandbergen, H.W., and Dekker, C. (2005) Translocation of double-stranded DNA through a silicon oxide nanopore. *Phys. Rev. E*, **71**, 051903.

7 Fologea, D., Brandin, E., Uplinger, J., Branton, D., and Li, J. (2007) DNA conformation and base number simultaneously determined in a nanopore. *Electrophoresis*, **28**, 3168–3192.

8 Wanunu, M., Sutin, J., McNally, B., Chow, A., and Meller, A. (2008) DNA translocation governed by interactions with solid-state nanopores. *Biophys. J.*, **95**, 4716–4725.

9 Hille, B. (2001) *Ion Channels of Excitable Membranes*, 3rd edn, Sinauer, Sunderland, MA.

10 Aksimentiev, A., Heng, J.B., Timp, G., and Schulten, K. (2004) Microscopic kinetics of DNA translocation through synthetic nanopores. *Biophys. J.*, **87**, 2086–2097.

11 King, G.M. and Golovchenko, J.A. (2005) Probing nanotube–nanopore interactions. *Phys. Rev. Lett.*, **95**, 216103.

12 Bezrukov, S.M., Vodyanoy, I., and Parsegian, V.A. (1994) Counting polymers moving through a single ion channel. *Nature*, **370**, 279–281.

13 Braha, O., Walker, B., Cheley, S., Kasianowicz, J.J., Song, L., Gouaux, J.E., and Bayley, H. (1997) Designed protein pores as components for biosensors. *Chem. Biol.*, **4**, 497–505.

14 Kasianowicz, J.J., Brandin, E., Branton, D., and Deamer, D.W. (1996) Characterization of individual polynucleotide molecules using a membrane channel. *Proc. Natl. Acad. Sci. USA*, **93**, 13770–13773.

15 Akeson, M., Branton, D., Kasianowicz, J.J., Brandin, E., and Deamer, D.W. (1999) Microsecond time-scale discrimination among polycytidylic acid, polyadenylic acid, and polyuridylic acid as homopolymers or as segments within single RNA molecules. *Biophys. J.*, **77**, 3227–3233.

16 Meller, A., Nivon, L., Brandin, E., Golovchenko, J., and Branton, D. (2000) Rapid nanopore discrimination between single polynucleotide molecules. *Proc. Natl. Acad. Sci. USA*, **97**, 1079–1084.

17 Zwolak, M. and Di Ventra, M. (2008) Colloquium: physical approaches to DNA sequencing and detection. *Rev. Mod. Phys.*, **80**, 141–165.

18 Branton, D., Deamer, D.W., Marziali, A., Bayley, H., Benner, S.A., Butler, T., Ventra, M.D., Garaj, S., Hibbs, A., Huang, X., Jovanovich, S.B., Kristic, P.S., Lindsay, S., Ling, X.S., Mastrangelo, C.H., Meller, A., Oliver, J.S., Pershin, Y.V., Ramsey, J.M., Riehn, R., Soni, G.V., Tabard-Cossa, V., Wanunu, M., Wiggin, M., and Schloss, J.A. (2008) The potential and challenges of nanopore sequencing. *Nat. Biotechnol.*, **26**, 1146–1153.

19 Healy, K. (2007) Nanopore-based single-molecule DNA analysis. *Nanomedicine*, **2**, 459–481.

20 Li, J., Stein, D., McMullan, C., Branton, D., Aziz, M.J., and Golovchenko, J.A. (2001) Ion-beam sculpting at nanometre length scales. *Nature*, **412**, 166–169.

21 Cai, Q., Ledden, B., Krueger, E., Golovchenko, J.A., and Li, J. (2006) Nanopore sculpting with noble gas ions. *J. Appl. Phys.*, **100**, 024914.1–024914.6.

22 Stein, D., Li, J., and Golovchenko, J.A. (2002) Ion-beam sculpting time scales. *Phys. Rev. Lett.*, **89**, 276106.1–276106.4.

23 Stein, D.M., McMullan, C.J., Li, Jiali, and Golovchenko, J.A. (2004) A feedback-controlled ion beam sculpting apparatus. *Rev. Sci. Instrum.*, **75**, 900–905.

24 Storm, A.J., Chen, J.H., Ling, X.S., Zandbergen, H.W., and Dekker, C. (2003) Fabrication of solid-state nanopores with single-nanometre precision. *Nature Mater.*, **2**, 537–540.

25 Kim, M.J., McNally, B., Murata, K., and Meller, A. (2007) Characteristics of solid-state nanometre pores fabricated using a transmission electron microscope. *Nanotechnology*, **18**, 205302.1–205302.5.

26 Ho, C., Qiao, R., Heng, J.B., Chatterjee, A., Timp, R.J., Aluri, N.R., and Timp, G. (2005) Electrolytic transport through a synthetic nanometer-diameter pore. *Proc. Natl. Acad. Sci. USA*, **102**, 10445–10450.

27 Meller, A. and Branton, D. (2002) Single molecule measurements of DNA transport through a nanopore. *Electrophoresis*, **23**, 2583–2591.

28 Deamer, D. and Branton, D. (2002) Characterization of nucleic acids by nanopore analysis. *Acc. Chem. Res.*, **35**, 817–825.

29 Howorka, S., Cheley, S., and Bayley, H. (2001) Sequence-specific detection of individual DNA strands using engineered nanopores. *Nat. Biotechnol.*, **19**, 636–639.

30 Astier, Y., Braha, O., and Bayley, H. (2006) Toward single molecule DNA sequencing: direct identification of ribonucleoside and deoxyribonucleoside-monophosphates by using an engineered protein nanopore equipped with a molecular adapter. *J. Am. Chem. Soc.*, **128**, 1705–1710.

31 Wu, H.-C., Astier, Y., Maglia, G., Mikhailova, E., and Bayley, H. (2007) Protein nanopores with covalently attached molecular adapters. *J. Am. Chem. Soc.*, **129**, 16142–16148.

32 Wu, M.-Y., Krapf, D., Zandbergen, M., Zandbergen, H., and Batson, P.E. (2005) Formation of nanopores in a SiN/SiO_2 membrane with an electron beam. *Appl. Phys. Lett.*, **87**, 113106–113109.

33 Wanunu, M. and Meller, A. (2007) Chemically modified solid-state nanopores. *Nano Lett.*, **7**, 1580–1585.

34 Iqbal, S.M., Akin, D., and Bashir, R. (2007) Solid-state nanopore channels with DNA selectivity. *Nat. Nanotechnol.*, **2**, 243–248.

35 Zwolak, M. and Ventra, M.D. (2005) Electronic signature of DNA nucleotides via transverse transport. *Nano Lett.*, **5**, 421–424.

36 Meunier, V. and Krstic, P.S. (2008) Enhancement of the transverse conductance in DNA nucleotides. *J. Chem. Phys.*, **128**, 041103.

37 Lagerqvist, J., Zwolak, M., and Ventra, M.D. (2006) Fast DNA sequencing via transverse electronic transport. *Nano Lett.*, **6**, 779–782.

38 Hughes, M.E., Brandin, E., and Golovchenko, J.A. (2007) Optical absorption of DNA-carbon nanotube structures. *Nano Lett.*, **7**, 1191–1194.

39 Gierhart, B.C., Howitt, D.G., Chen, S.J., Zhu, Z., Kotecki, D.E., Smith, R.L., and Collins, S.D. (2008) Nanopore with transverse nanoelectrodes for electrical characterization and sequencing of DNA. *Sensor Actuat. B Chem.*, **132**, 593–600.

40 He, J., Lin, L., Zhang, P., Spadola, Q., Xi, Z., Fu, Q., and Lindsay, S. (2008) Transverse tunneling through DNA hydrogen bonded to an electrode. *Nano Lett.*, **8**, 2530–2534.

41 He, J., Lin, L., Zhang, P., and Lindsay, S. (2007) Identification of DNA basepairing via tunnel-current decay. *Nano Lett.*, **7**, 3854–3858.

42 Trepagnier, E.H., Radenovic, A., Sivak, D., Geissler, P., and Liphardt, J. (2007) Controlling DNA capture and propagation through artificial nanopores. *Nano Lett.*, **7**, 2824–2830.

43 Keyser, U.F., Koeleman, B.N., Dorp, S.V., Krapf, D., Smeets, R.M.M., Lemay, S.G., Dekker, N.H., and Dekker, C. (2006) Direct force measurements on DNA in a solid-state nanopore. *Nat. Phys.*, **2**, 473–477.

44 Mathe, J., Visram, H., Viasnoff, V., Robin, Y., and Meller, A. (2004) Nanopore unzipping of individual DNA hairpin molecules. *Biophys. J.*, **87**, 3205–3212.

45 Bates, M., Burns, M., and Meller, A. (2003) Dynamics of DNA molecules in a membrane channel probed by active control techniques. *Biophys J.*, **84**, 2366–2372.

46 Dudko, O.K., Mathe, J., Szabo, A., Meller, A., and Hummer, G. (2007) Extracting kinetics from single-molecule force spectroscopy: nanopore unzipping of DNA hairpins. *Biophys J.*, **92**, 4188–4195.

47 Gershow, M. and Golovchenko, J.A. (2007) Recapturing and trapping single molecules with a solid-state nanopore. *Nat. Nanotechnol.*, **2**, 775–779.

48 Ling, X.S., Ready, B., and Pertsinidis, A. (2007) Hybridization assisted nanopore sequencing. US Patent Application 20070190542.

49 Fologea, D., Gershow, M., Ledden, B., McNabb, D.S., Golovchenko, J.A., and Li, J. (2005) Detecting single stranded DNA with a solid state nanopore. *Nano Lett.*, **5**, 1905–1909.

50 Soni, G.V. and Meller, A. (2007) Progress toward ultrafast DNA sequencing using solid-state nanopores. *Clin. Chem.*, **53**, 1996–2001.

51 Sauer-Budge, A.F., Nyamwanda, J.A., Lubensky, D.K., and Branton, D. (2003) Unzipping kinetics of double-stranded DNA in a nanopore. *Phys. Rev. Lett.*, **90**, 238101.1–238101.4.

52 McNally, B., Wanunu, M., and Meller, A. (2008) Electromechanical unzipping of individual DNA molecules using synthetic sub-2 nm pores. *Nano Lett.*, **8**, 3418–3422.

53 van Gelder, W. and Hauser, V.E. (1967) The etching of silicon nitride in phosphoric acid with silicon dioxide as a mask. *J. Electrochem. Soc.*, **114**, 869–872.

54 White, R.J., Ervin, E.N., Yang, T., Chen, X., Daniel, S., Cremer, P.S., and White, H.S. (2007) Single ion-channel recordings using glass nanopore membranes. *J. Am. Chem. Soc.*, **129**, 11766–11775.

55 Jeon, T.-J., Malmstadt, N., and Schmidt, J.J. (2006) Hydrogel-encapsulated lipid membranes. *J. Am. Chem. Soc.*, **128**, 42–43.

Glossary

This glossary lists a number of relevant terms for the reader who may not be so familiar with the topics of the present *Handbook of Plant Mutation Screening (Mining of Natural and Induced Alleles)*.

A

Abasic site: Any gap in a nucleic acid sequence that originates from the loss of a base.

Acentric fragment: A chromosome fragment that is the result of a chromosome breakage. Since it does not contain a centromere, it is lost during mitosis.

Acridine dye: Any one of a series of mutagenic heterocyclic compounds, including acridine and its derivatives. At low concentrations, aminoacridines (e.g., quinacrine) intercalate between the two strands of dsDNA. Higher concentrations cause the binding of acridines to the outside of dsDNA, ssDNA, and ssRNA. Acridines interfere with DNA and RNA synthesis, cause frameshift mutations, and addition or deletion of bases.

Acridine orange (3,6-bis-[dimethylamino]-acridinium chloride; euchrysine): A basic acridine dye that binds to double-stranded nucleic acids by intercalation, or to single- and double-stranded nucleic acid by electrostatic interaction with the phosphate backbone. Ultraviolet irradiation absorbed at 260 nm by a dye–dsDNA complex can be re-emitted as fluorescence at 530 nm (green) or by ssDNA or RNA at 640 nm (red). Acridine orange also functions as a mutagen. Sublethal concentrations of the dye are used for curing plasmids.

Acriflavine (euflavine; 3,6-diamino-10-methylacridinium chloride): An acridine dye producing reading frameshift mutations.

Activator–dissociator system (*Ac–Ds* system): A group of the two interacting transposable elements *Ac* and *Ds* in maize (*Zea mays*). *Ac* is a 4.6-kb autonomous element, carrying a transposase gene, whose encoded protein binds to the terminal inverted repeat ends of both the *Ac* and the *Ds* elements, catalyzing their transposition to new locations in the genome. *Ds* is most often a derivative of *Ac* that no longer produces a functional transposase and therefore is unable to transpose by itself. *Ds* is consequently nonautonomous. Upon *Ac*-mediated activation, however, *Ds* may change the expression rate of flanking genes and the timing of gene expression, and may also cause chromosome breakage. *Ac* determines the time period during morphogenesis when *Ds* acts. *Ac/Ds* loci are recognized and mapped by their action on neighboring genes.

Adaptive mutation (stress-inducible mutation; stationary-phase mutation): Any spontaneous mutation, or also genome-wide hypermutation, that occurs in bacteria (e.g., *Escherichia coli* cells) after a prolonged period of incubation (3–7 days or longer) on nonlethal selective medium, while the cells are starving and not dividing or are dividing very slowly. Under these stress conditions, the cells activate the stress protein σ^{38} that in turn activates the expression of the DNA polymerase IV gene. As a consequence, polymerase IV expression is quadrupled (from about 250 to 1000 polymerase IV copies per cell). This error-prone enzyme introduces mutations into replicating DNA. In addition, the SOS response leads to increased levels of RecA and RecF' that also are needed for adaptive mutation, which is under control of the SOS-controlled PsiB inhibitor, and the stress-response sigma factor, RpoS. Adaptive mutation then is a response to a stressful environment.

Agrobacterium chromosomal DNA (AchrDNA):

a. The DNA of *Agrobacterium tumefaciens* that is organized in chromosomes, rather than plasmids (e.g., tumor-inducing plasmid).
b. Any fragment of chromosomal DNA of *A. tumefaciens* that is transferred and integrated into a recipient (usually dicot) plant during T-DNA-induced transformation. Normally, only the T-DNA on the so-called tumor-inducing plasmid (Ti-plasmid) bracketed by the right and left border sequences, respectively, is mobilized into the plant cell after sensing of plant phenolics (usually at a wound), and subsequently integrated into the plant nuclear genome. In rare cases (about 0.4% of all infections), additionally chromosomal sequences are cotransferred and cointegrated, more frequently associated with the right border. The AchrDNA may comprise more than 18 kb and harbor several bacterial genes.

Alkylating agent: Any chemical compound that transfers alkyl groups (e.g., methyl or ethyl moieties) onto the bases in DNA.

Allele (allelomorph): One of two or more alternate forms of a given gene occupying the same locus on homologous chromosomes. An allele may differ from other alleles of the same locus at one or more mutational sites, whose number per gene ranges from 10^2 to 10^3.

Allele frequency: The number of copies of a specific allele in a given population of organisms. Erroneously called gene frequency.

Allele shift: Any change in the frequency of a specific allele in a population of organisms that is driven by selection. In extreme cases, allele shifts may cause the complete loss of an allele.

Allele-specific amplification (ASA): A technique for the detection of specific alleles, which uses high stringency in a conventional PCR to allow the annealing of only a matching (i.e., perfectly complementary) primer to the target DNA. This primer is designed to bind only to one single allele.

Allele-specific associated primer (ASAP): A synthetic oligodeoxynucleotide of some 20–30 bp in length that is complementary to the 3′ or 5′ end of a random amplified polymorphic DNA marker and used in PCR as a primer to amplify a specific allele. In short, an informative PCR product is excised from a gel, cloned, and sequenced. Then primers of about 20 bp in length are designed to the ends of the product that allow us to amplify the specific genomic region from which the original amplification product originates. If such regions from two genomes differ, the ASAP products vary in size (either by deletion or insertion events within or between the primer binding sites) and then can be used as molecular markers.

Allele-specific expression: The transcription of only one allele or the transcription of both alleles of a genetic locus to different extents. Allele-specific expression may be detected by, for example, allele-specific amplification or allele-specific PCR.

Allele-specific hybridization (ASH; allele-specific oligonucleotide hybridization): A technique for the detection of SNPs, small deletions, or insertions in a specific DNA sequence that allows us to discriminate wild-type and mutant alleles. In short, a restriction fragment length polymorphism fragment is first cloned into an appropriate vector, sequenced, and locus-specific primer oligonucleotides for conventional PCR techniques designed. These primers are then used to amplify the corresponding locus with genomic DNAs from different, closely related organisms as templates. The resulting amplicons, differing by, for example, a SNP, are then sequenced and used as allele-specific probes (labeled with a fluorochrome) to detect allelic differences by hybridization techniques at high stringency and fluorescence detection.

Allele-specific methylation: The methylation of cytosyl residues in either the paternal or maternal allele of a gene. Allele-specific methylation regulates the expression of, for example, imprinted genes in a wide range of eukaryotes.

Allele-specific oligonucleotide ligation: A technique for the detection of SNPs in genomic DNA. In short, the target DNA is first amplified by conventional PCR, then allele-specific oligonucleotides complementary to the target sequence and with the allele-specific base at either their 3′ or 5′ ends annealed to the amplicon just adjacent to the polymorphic site. Only if the oligonucleotide fully matches the target will it be ligated. If there is a mismatch, ligation is impossible.

Allele-specific oligonucleotide (ASO) probe: A synthetic, approximately 20-nucleotide oligodeoxynucleotide designed to locate single base mismatches in complex genomes and to discriminate between two alleles. Such probes are long enough to detect unique sequences in the genome, but sufficiently short to be destablilized by a single internal mismatch during their hybridization to a target sequence. The technique involves the immobilization of target DNA, hybridization with oligonucleotide probes, and finally washing under carefully controlled conditions, which allows us to discriminate sequences with a single nucleotide mismatch from their wild-type genomic counterparts on the basis of different hybridization behavior.

Allele-specific polymerase chain reaction (AS-PCR; ASP; PCR amplification of specific alleles (PASA); allele-specific amplification (ASA)): A variant of the conventional PCR that allows the amplification of specific alleles (or DNA sequence variants). For example, a single base difference, which discriminates both alleles, can be detected by using two primers for amplification – one possessing a 3′ end specific for allele A, the other one a 3′ end specific for allele A′. A third primer is designed to bind to sequences downstream of the allelic polymorphic site (that are identical for both genomes). Since *Taq* DNA polymerase does not have a 3′ exonuclease activity, it cannot use or degrade a primer with a mispaired 3′ terminus. Therefore, the allele-specific primers only amplify the allele to which they pair precisely. The amplification products can then be visualized by using primers of different lengths or carrying different fluorochromes.

Allele-specific probe: Any defined, radioactively or nonradioactively labeled DNA sequence that is complementary to a specific allele and allows its detection by for example, hybridization and autoradiography.

Allele-specific RNA interference (allele-specific RNAi): The silencing of only one allele of a specific gene by an allele-specific interference RNA. Allele-specific silencing starts with the cloning of the target gene into a plasmid vector and its mutation *in vitro* to create a variant of the gene, whose mRNA is not subject to RNA interference by the dsRNA able to knockdown the endogenous protein.

Allelic deletion: The deletion of one complete allele of a gene or an exon (or part of it) of one allele. Allelic deletions can be detected by comparative quantitative real-time PCR.

Allelic exclusion: The expression of only one allele of a parental immunoglobulin gene in immunoglobulin-producing cells (e.g., B lymphocytes). The allele on the other homologous chromosome in a diploid cell cannot rearrange.

Allelic imbalance (AI):

 a. The presence of one nonfunctional allele at a specific locus or the complete loss of an allele (loss of heterozygosity).
 b. The differential transcription of the two alleles at a particular locus such that different levels of transcripts (and finally proteins) are present in a cell. Imbalanced allelic expression is characteristic for some cancer cell lines and can be detected by a comparison of the allelic ratios measured, for example, for heterozygous SNPs both in genomic DNA and cDNA. Ratios differing from 1:

1 are indicative for allelic imbalance. In certain cancers complete loss of the expression of one allele is common.

Allelic ladder: Laboratory slang term for any display in which, for example, sequence-tagged microsatellite sites of different alleles of a population are arranged according to their molecular weight (in increasing or decreasing size order).

Allelic mining: The search for novel alleles of a gene in an organism or a population of organisms using comparative genetics. The mining process starts with a known gene sequence for organism A, the design of gene-specific primers, and the use of these primers to amplify corresponding sequences from the genome of organism B. Allelic variants are then detected by, for example, sequencing of the amplification products and discovery of sequence polymorphisms.

Allelic recombination: Any recombination event occurring between sequences located at similar or identical positions on homologous chromosomes or sister chromatids.

Alternative transposition: A variant of the transposition reaction in which two separate, rather than one single transposable, elements are involved. The first step in alternative transposition leads to the synapsis of complementary ends of separate transposable elements to form a so-called hybrid element. The subsequent steps of excision, insertion (of the hybrid element into the target site), and repair of the double-strand breaks resembles traditional transposition. Such alternative transpositions occur in bacteria (IS10/*Tn10* elements), maize, tobacco, and snapdragon (*Ac/Ds* and *Tam*3 elements, respectively), and *Drosophila* (P elements).

Amber mutant: Any mutant that synthesizes mRNAs containing the codon UAG as a consequence of a point mutation in the corresponding gene.

Amber mutation (am): A mutation generating the stop codon UAG in the coding region of a gene, thus leading to the synthesis of a truncated message and protein. Amber mutations can be suppressed by specific mutant tRNAs with a UAC anticodon that allow the incorporation of an amino acid in spite of the UAG stop codon and therefore the synthesis of a complete protein. Amber mutations are included in λ phage cloning vectors (charon phages) so that they can only be propagated in host cells that suppress the amber mutation (amber suppressor). As such suppressor cells normally do not occur outside the laboratory, the use of such amber suppressor cells is a measure of biological containment. The term "amber" is derived from a CALTECH student, named "Bernstein," which is the German word for amber. The notations ochre and opal for other stop codons were chosen arbitrarily.

Amber suppressor (Am): A mutant gene that codes for a tRNA with an anticodon recognizing the UAG stop codon. As a consequence, the growing polypeptide chain is not terminated at the stop codon, but instead elongated beyond it. Thus, the normal stop signal of the UAG (amber) codon is suppressed.

Amplification refractory mutation system (ARMS): A technique for the detection of SNPs between two (or more) genomes, based on the fact that oligodeoxynucleotides with a mismatched 3′ residue will not prime their extension in a PCR, whereas primers with a perfect match at their 3′ ends will. ARMS works with an allele-specific primer in two forms – a wild-type primer (3′ terminal nucleotide matching) and a mutant primer (3′ terminal nucleotide not matching). In separate pairs of reactions each genomic DNA is either coamplified with the wild-type or mutant primer coupled with a common primer (specific for the respective locus). The amplification products are then separated by agarose gel electrophoresis and visualized by ethidium bromide fluorescence. ARMS requires that the employed DNA polymerase does not possess any 3′ exonucleolytic proofreading activity.

Anonymous single nucleotide polymorphism (anonymous SNP): Any one of the most frequently occurring SNPs that has no known effect on the function of a gene.

Amplification fragment length polymorphism (AFLP; amplified sequence polymorphism, ASP; amplified fragment length polymorphism, AMP-FLP, PCR-RFLP): The variation in the length of DNA fragments produced by the polymerase chain reaction (PCR) using either one or several specific or arbitrary oligodeoxynucleotide primers (amplimers) and genomic DNA from two or more individuals of a species. AFLPs arise from

a. restriction site polymorphisms, where a specific restriction endonuclease recognition sequence is either present or missing at a given site (restriction fragment length polymorphism),

b. sequence length polymorphisms, where the number of tandemly arranged repetitive sequences at a given site varies (variable number of tandem repeats, simple sequence length polymorphism),

c. DNA base pair changes not associated with restriction sites.

AFLPs are used e.g. to discriminate between closely related individuals, to localize specific genes in complex genomes (linkage analysis) and to establish genome maps.

Anti-microRNA (anti-miRNA; anti-miRNA oligonucleotide AMO): Any small RNA molecule (also artificial oligonucleotide) that is complementary to a specific small interfering RNA (siRNA) or microRNA (miRNA), hybridizes to it, and thereby blocks its action and prevents the suppression of a specific mRNA such that the corresponding protein can be synthesized (i.e., no translational repression occurs). Anti-miRNAs are about 22 nucleotides long. Modifications in the ribose moiety of the oligonucleotides, especially a 2′-O-ethyl group or 2′-O-4′-C-methylene bridge, stabilize the anti-miRNAs and increase their efficacy and affinity to the target.

Antimutator DNA polymerase: Any DNA-dependent DNA polymerase that shows an unusually high degree of fidelity in proofreading. For example, a mutant T4 DNA polymerase edits all types of base substitution errors and reduces the frequency of AT → GC transitions substantially.

Antimutator gene (antimutator): Any gene that decreases the rate of spontaneous mutations of one or more other genes. Most probably, the antimutator gene product somehow increases the efficiency of the proofreading/editing function of the DNA polymerase during normal replication.

AP endonuclease (apurinic/apyrimidinic endonuclease): One of a series of generally small (25–40 kDa) monomeric endonucleases that recognizes alkylated or deaminated bases or AP sites in DNA and catalyzes the hydrolytic cleavage of the phosphodiester bond immediately 5′ to the AP site. AP endonucleases fall into two protein family classes. One family in *Escherichia coli* consists of endonuclease IV and Apn1 proteins, the other is mainly composed of exonuclease III (that exhibits 3′-phosphatase and 3′-phosphodiesterase activities in addition to its 3′ → 5′ exonuclease function). AP endonucleases and AP lyases are ubiquitous components of pro- and eukaryotic excision repair systems.

AP site (apurinic/apyrimidinic site): Any mutation in DNA caused by the spontaneous or induced hydrolysis of the relatively labile *N*-glycosyl bond, the subsequent loss of the corresponding base (abasic site), and a disturbance of the information content of the DNA. Such losses may be as high as 10 000 bases per day in humans. AP sites can result from exposure of DNA to various DNA-binding chemicals or radiation. The majority of AP sites originate from purine loss, since many DNA-affine compounds target purines and since purine loss by far exceeds pyrimidine loss. The structures of the different AP sites differ. For example, the regular AP site derives from simple *N*-glycosyl bond hydrolysis and is in equilibrium with an open-ring form. The so-called oxidized AP sites are produced by oxygen radical attack at the C-1′, C-2′, and C-4′ positions of the deoxyribose moiety of DNA. For example, the radiomimetic glycopeptide bleomycin reacts selectively with the C-4′ position and generates a 4′-oxidized AP site. Cu(II) phenanthroline abstracts protons from the C-1′ position and produces 1′-oxidized AP sites. AP sites are potential promutagenic lesions, because DNA polymerase may bypass the site during replication and insert a wrong base on the complementary strand (= base substitution mutation). AP sites also inhibit DNA polymerase transit. Such sites are corrected by AP endonucleases, or cleavage at their 3′ sides via a β-elimination reaction, catalyzed by an AP lyase, or the removal of the AP site by incising the DNA backbone on either side of the lesion, catalyzed by the multiprotein nucleotide excision repair machinery. The open-ring form of the AP site exposes an aldehyde group that can be detected with an aldehyde-reactive probe containing a biotin moiety. This biotin is then reacted with avidin or streptavidin bound to a fluorochrome.

Apurinic site: Any gap in a nucleic acid molecule created by the removal of a purine.

Apyrimidinic site: Any gap in a nucleic acid molecule created by the removal of a pyrimidine.

Array comparative genomic hybridization (aCGH; matrix comparative genomic hybridization; matrix CGH; matrix-based comparative genomic hybridization; array CGH; array-based comparative genomic hybridization; CGH chip): A variant of the conventional CGH technique that works with genomic fragments in the range from 30 to 200 kb rather than metaphase chromosome spreads. These fragments (in the form of bacterial artificial chromosome (BAC) clones, also cDNAs, or selected PCR products to reduce complexity) are spotted onto chip supports (e.g., surface-modified glass or plastics), and the resulting chips cohybridized to differentially fluorescently labeled cDNAs from two sources (e.g., normal versus tumorous tissues, cyanin 3 (Cy3)- versus cyanin 5 (Cy5)-labeled). Subsequently a laser beam generates fluorescence signals on the chip, which are detected by a CCD camera of a fluorescence scanner. The amount of emitted fluorescence from each dye is then quantified for each spot on the array, the different fluorescence ratios are determined, and the resulting data analyzed. A Cy5/Cy3 fluorescence ratio of more than 1.0 indicates a duplication or amplification in the test individual's genomic DNA spotted on the array; a ratio of less than 1.0 can be interpreted as a reduction in copy number or a deletion. Additionally, DNA from three different sources, each one labeled with a different fluorochrome (e.g., first sample: cyanin 3; second sample: cyanin 5; third sample: fluorescein) can be compared simultaneously. Matrix CGH increases the relatively poor resolution of conventional CGH (with chromosomes) by several orders of magnitude. The technique can also be combined with cDNA microarray experiments, so that genes expressed under certain conditions (e.g., in tumors) can first be selected and then spotted onto the matrix CGH array. For example, the combination of both techniques led to the molecular differentiation of two subtypes of liposarcomas (differentiated liposarcomas versus polymorphic liposarcomas). Matrix CGH arrays covering the whole human genome, consisting of about 30 000 BAC clones, are used to detect mutations in a genome, ranging from small mutated regions (e.g., SNPs or single-copy changes) over larger copy number variations to large deletions.

Arrayed primer extension (APEX): A technique for the detection of mutations (e.g., SNPs) in genomic DNA. In short, 15–25mer oligonucleotide primers complementary to multiple target sequences (e.g., the various parts of a gene) are immobilized via their 5′ termini on a solid support (e.g., a glass slide) to produce a DNA chip. Then a target DNA is amplified by conventional PCR, the amplicons restricted with appropriate restriction endonucleases, or otherwise fragmented, and annealed to the primers on the chip. Subsequently, *Taq* DNA polymerase is used to extend the 3′ ends of the immobilized primers by one single base only, employing fluorescently labeled dideoxynucleotides ("terminators"), where each base is labeled with a different fluorochrome. After extension, the target DNA together with excess ddNTPs and the polymerase are washed off at 96 °C, and the target-complementary primers marked by fluorescence detected by a CCD camera. Usually, primers are arrayed in duplicates or triplicates for better discrimination of signal and background noise. By comparison of two target genomes, SNPs can be discovered if the gene fragment of genome A can be extended and the homologous fragment of genome B cannot.

Artificial transposon (AT): Any transposon that is both synthesized and mobilized from its resident plasmid to another (acceptor) plasmid *in vitro*. For example, an artificial transposon can be constructed by inserting any foreign DNA into a multiple cloning site (MCS) of a suitable plasmid vector, that additionally contains a bacterial origin of replication and the yeast *URA3* gene. The MCS is flanked by *Ty*1 U3 cassettes, which are incorporated into *Xmn*I cleavage sites. Since two such sites exist (one on either side of the MCS), digestion with the restriction endonuclease *Xmn*I both cleaves the fragment from the plasmid vector and creates precise *Ty*1 U3 termini at the ends of the liberated fragment, which are substrates for the *Ty*1 integrase. The integrase is part of the *Ty*1 virus-like particles (VLPs), that can be isolated from yeast cell cultures. The VLPs are added and the integrase catalyzes the integration of the AT into the recipient plasmid *in vitro*. In essence, this construct acts as a transposon. ATs allows to generate recombinant molecules of any desired structure and sequence.

ATM- and Rad3-related (ATR) checkpoint pathway: A multiprotein complex of eukaryotic cells, that recognizes and signals the presence of multiple DNA damage events (e.g., double-strand breaks (DSBs), but also ssDNAs) and stalled replication forks. In short, the specific topological structure of a DNA lesion is recognized by several proteins, that bind to it and recruit additional repair proteins to the site. For example, during mismatch repair or repair of DSBs by, for example, nonhomologous end-joining, the lesion is recognized by Msh2–Msh6 (mismatches) and Ku70/80 (DSBs). Also, replication protein A (RPA, in fact, an ssDNA-binding protein complex) coats ssDNA in a complex with ATRIP (i.e., ATR-interacting protein; Mec1 and Ddc2 in yeast), that in turn recruits ATR. Once recruited, ATR efficiently phosphorylates a subunit of a second checkpoint protein-sensing complex Rad17, but only when Rad17 is itself independently recruited to RPA–ssDNA complexes. Additionally, ATR phosphorylates p53, BRCA1, and Chk1 protein kinase with the consequence of inhibition of DNA replication and mitosis, and promotion of DNA repair, recombination, or apoptosis. RPA together with ATR and ATRIP localize to nuclear foci after DNA damage. In prokaryotic cells, the ssDNA generated by damage is coated with RecA and serves as signal for the so-called SOS response.

Autonomous transposon: Any one of a class of transposons that encode a set of proteins catalyzing its transposition, or the transposition of nonautonomous elements.

B

Base excision repair (BER): A prokaryotic DNA excision repair system that is encoded by genes *ada* and *alkA*, and cuts out deaminated cytosine (= uracil) and deaminated adenine (= hypoxanthine) bases modified by, for example, alkylating mutagens via DNA glycosylases. These enzymes split the *N*-glycosyl bonds and create apurinic and apyrimidinic (AP) sites. After base removal, AP endonucleases cut out the deoxyribose phosphates, upon which DNA polymerase I fills-in and ligases complete the BER process.

Base excision sequence scanning (BESS): A technique for the screening of large genomes for mutations (e.g., deletions, transitions, transversions). In short, genomic

DNAs from two (or more) organisms are isolated and a sequence-specific, labeled primer, or two differentially labeled primers used to amplify a distinct region of the genomes (e.g., a gene in which mutation(s) are to be detected), employing conventional PCR techniques. PCR is performed in the presence of limiting amounts of dUTP, which is incorporated into the newly synthesized DNA products instead of dTTP. Then uracil DNA glycosylase is added, which cleaves the DNA at the sites of incorporated deoxyuridines, producing a set of nested DNA fragments that are subsequently separated by sequencing gel electrophoresis. The fragment patterns, similar to T lane sequencing patterns, are then detected by autoradiography. Polymorphisms between the target sequences of two genomes are caused by mutations, that can easily be classified.

Base mismatch (mismatch (MM); mispairing): The occurrence of incorrectly paired (mismatched) bases in DNA duplex molecules. Such mismatches arise by errors of the replication and/or repair (DNA repair) systems and are sources of mutations.

Base substitution: The replacement of one nucleotide by a different nucleotide in DNA or RNA.

Benign copy number variant (benign CNV): Laboratory slang term for any genomic region with a structural variation (e.g., a copy number variation) that is not linked to a disease.

Bisulfite mutagenesis: A special type of chemical mutagenesis (substitution mutagenesis) of ssDNA molecules that uses sodium bisulfite ($NaHSO_3$) for the deamination of cytosine residues to yield uracil. Subsequent synthesis of a complementary strand leads to the incorporation of an adenosine where a uracil is located on the template strand. In short, supercoiled circular double-stranded plasmid DNA is nicked at random with DNase I in the presence of ethidium bromide. The nicks are then extended with exonuclease III. Afterwards the DNA is treated with 1–3 M $NaHSO_3$ that deaminates cytosine to uracil in the previously generated gaps of the DNA double strand. The resulting mutagenized plasmid is then transferred into host bacteria where its replication leads to the filling-in of the gap by DNA polymerase and to the replacement of an original GC by TA. A method of substitution mutagenesis.

Break (double-strand break (DSB)): The disruption of a phosphodiester bond between adjacent nucleotides in both strands of a DNA duplex molecule. Such DSBs are induced by ionizing radiation and certain chemical compounds (e.g., chemotherapeutics such as bleomycin). Spontaneous DSBs can be caused by cellular metabolites such as reactive oxygen species. Also, if a replication fork encounters a single-strand break, a DSB can be produced. Moreover, torsional stress in DNA can lead to DSBs, which can also be generated during meiosis to initiate recombination between paired homologs.

Breakpoint mapping: The localization of breakpoints (i.e., sites of breaks as a prerequisite for chromosomal alterations such as deletions, inversions, or translocations) along a chromosome. For example, translocation breakpoint mapping in humans can be achieved with a combination of flow cytometry and array-based comparative genomic hybridization. The technique involves the sorting of about 150 000 metaphase chromosomes (chromosome sorting) from normal and diseased patients, the labeling of the sorted chromosomes from patients with cyanin 3 (Cy3)-dUTP and from controls with cyanin 5 (Cy5)-dUTP and their simultaneous hybridization to microarrays containing thousands of well-characterized genomic clones (e.g., bacterial artificial chromosome (BAC) clones) covering the whole genome at about 1-Mb resolution. Usually the BAC clones are spotted in triplicate. After hybridization, the array is scanned and the so-called fluorescence test over reference (T/R) ratio for each clone determined. High T/R values in the Cy3 channel indicate clones that contain breakpoint-spanning genes. As a result, a breakpoint map can be established for all human chromosomes. Since, for example, in cancerous diseases more than 100 recurring chromosomal translocations are known to trigger activation of various oncogenes, breakpoint mapping serves to localize the underlying chromosomal breakpoints.

C

Candidate single nucleotide polymorphism (candidate SNP): Any SNP in an exon of a gene that can be expected to have an impact on the function of the encoded protein.

Cassette mutagenesis: A technique for site-specific mutagenesis. In short, a specific region flanked by two restriction sites is first removed from the target molecule. Then a DNA fragment consisting of a chosen sequence (e.g., a synthetic oligonucleotide) is inserted in its place, which causes various amino acid replacements in the encoded protein.

Causative single nucleotide polymorphism (causative SNP; causal SNP; etiological SNP): Any SNP that is in linkage disequilibrium to a disease phenotype and therefore a responsible candidate for the disease.

Chemical cleavage method (CCM): A technique for the detection of mismatched or unmatched cytosine and thymidine residues in DNA:DNA, DNA:cDNA, and DNA:RNA heteroduplexes. These mismatched bases are more reactive to modifications by hydroxylamine or osmium tetroxide than fully matched bases. Any heteroduplex containing the modified residues can be cleaved with piperidine. Usually the DNA is radioactively labeled before the chemical reactions take place, so that target DNA and cleavage products can both be separated from each other by denaturing polyacrylamide gel electrophoresis and visualized by autoradiography.

Chemical mutagen: Any chemical agent, that increases the frequency of mutations in DNA above the spontaneous background level. For example, alkylating mutagens such as methyl or ethylmethane sulfonate ($CH_3SO_2OCH_2CH_3$) alkylate guanine O^6 and C^7. N-nitroso compounds, such as nitrosamines, nitrosoureas, and methyl-nitro-nitrosoguanidine ($C_2H_5N_5O_3$), and acridine dyes, such as proflavin, acriflavin, and acridine orange, as intercalating and frameshift-causing agents are also such chemical mutagens.

Chemical mutagenesis: The introduction of mutations into ssDNA or dsDNA fragments by mutagens (e.g., nitrous acid, which deaminates A, G, and C to hypoxanthine, xanthine, and uracil. This in turn causes the substitution of an A/T pair by hypoxanthine/C and of C/G by U/A (xanthine pairs with C); potassium permanganate, which substitutes N for T; hydrazine, which substitutes N for C and T), where N symbolizes any nucleotide.

Chromosomal instability (CIN; chromosome number instability): The loss or gain of whole chromosomes (or parts of them) predominantly during the early stages of tumor development in humans. For example, a tumor could lose its maternal chromosome 8, while duplicating the paternal chromosome 8, leaving the cell with a normal chromosome 8 karyotype, but an abnormal chromosome 8 allelotype. CIN may be caused by mutation(s) in genes controlling chromosome condensation, sister chromatid cohesion, kinetochore structure and function, centrosome-microtubule formation and dynamics, and the progression of the cell cycle.

Chromosomal numerical aberration (CNA): The occurrence of an abnormal number of chromosomes in a karyotype. For example, the inheritance of three (instead of two) chromosomes 13, 18, or 21, that are relatively frequent (trisomy 13 : 1 : 5000; trisomy 18 : 1 : 3000; and trisomy 21 : 1 : 650) are such numerical aberrations.

Chromosome aberration (chromosomal aberration): Any deviation from the normal number or the normal structure or composition of chromosomes of an organism. Such aberrations include, for example, deletions (interstitial or terminal deletions), duplications, insertions, inversions, and translocations. The most frequently occurring aberrations in chromosome numbers (aneuploidy) are monosomy (only a single chromosome of a chromosome pair is present) and trisomy (three instead of the normal two chromosomes are present).

Chromosome mutation:

  a. Any structural change of a chromosome that is caused by radicals (generated for instance by ionizing irradiation), chemicals (e.g., intercalating agents), or transposons and involves terminal or intercalary deletions, insertions, translocations, inversions, or chromosome breakage.
  b. A change in the normal number of chromosomes, leading to an aneuploid cell or organism. This may, for instance, be due to the nondisjunction of chromosomes during meiosis or mitosis.

Chromosome rearrangement: Any change in the normal order of genes on a chromosome caused by inversions or translocations.

Clustered lesion (locally multiply damaged site): Any ionizing radiation-induced damage of DNA in which several base lesions, abasic sites, and/or strand breaks occur within a few helical turns along the track.

Cold spot (recombinational cold spot; mutational cold spot): Any sequence within a gene or a chromosome at which mutations occur at a significantly lower frequency than usual.

Collateral mutation: Any random second-site mutation that is introduced into the target DNA during site-directed mutagenesis.

Common single nucleotide polymorphism (common SNP): Any SNP, whose minor allele occurs in more than 10% of the genomes of a population.

Compensatory mutation: Any mutation in an exon of a gene that neutralizes the effect of another mutation in the same exon (or another exon of the same gene).

Comutagenesis: The occurrence of two or more mutations at closely linked loci within a genome.

Conditional lethal mutation: A mutation that is either tolerated under permissive conditions, or leads to the death of a cell or the destruction of a virus under nonpermissive conditions.

Conditional mutation: Any mutation that is only expressed under specific conditions (e.g., high temperature).

Conserved intron-scanning polymorphism (CISP): Any SNP or small Indel polymorphism detected by PCR with conserved intron-spanning primers in two (or more) individuals of the same species or different species that allows their discrimination on a molecular level. Do not confuse with conserved intron-spanning primer (also abbreviated CISP).

Constitutive mutation: A mutation that leads to the permanent expression of a gene which is normally tightly regulated.

Constitutive mutant: Any mutant organism with a mutation in a regulatory gene that normally encodes an RNA or a protein suppressing target genes, but in the mutated state has lost suppressive capacity, so that the target genes become constitutively expressed (constitutive genes). As a consequence, the encoded RNAs or proteins accumulate in excess.

Conversion: A technique to detect refractory mutations in human genomes (e.g., for diagnostic purposes). Patient cells are first fused with a specifically designed rodent cell line to produce hybrids, which stably retain a subset of the human chromosomes. Every fourth hybrid contains only a single copy of a human chromosome, which converts the normal diploid state to a haploid state. This facilitates the detection of mutations (the normal sequence of the wild-type allele is absent) by, for example, PCR amplification.

Copy number: The number of a particular plasmid per cell or of a particular gene per genome.

Copy number change (CNC): Any increase or decrease in the number of a particular gene or its alleles, of a repeat such as a microsatellite, of a retrotransposon, or of a single nucleotide polymorphic site in a genome, caused by duplications or deletions, respectively. For example, the formation of certain cancers is frequently accompanied by a loss of heterozygosity, characterized by the continuous loss of an allele. Different CNCs in different genomes lead to the phenomenon of copy number variation.

Copy number polymorphism (CNP; gene copy number polymorphism): Any difference between two (or more) genomes that reflects the number of copies of a particular gene.

Copy number variant (CNV): Any DNA segment of 1 kb or larger that varies in copy number between individuals. About 12% of the human genome (corresponding to greater than 360 million bases) or more exists as CNVs, 100-fold more variable DNA than is accounted for by SNPs. The copy number variation is a consequence of duplications in one, but not a second individual.

Copy number variation (CNV): The presence of different numbers of specific sequences (e.g., genes, repeats such as microsatellites, retrotransposons, or SNPs) in two (or more) genomes. In diploid organisms, normally two alleles of a single locus are present. However, the number of alleles can vary as a consequence of mutations (deletion of one allele, copy number = 1) or duplications (duplication of one allele, copy number = 3 or duplication of two alleles, copy number = 4).

D

Damage avoidance mechanism (DA mechanism): A process by which the DNA replication machinery avoids copying any damaged (mutated) template. Basically two DA mechanisms are in operation – the daughter-strand gap repair involves the repair of a gap in the newly synthesized daughter strand (created by the dissociation of the replicating DNA polymerase in the vicinity of the lesion) through homologous recombination (recombinational repair) and the DNA polymerase template switching (the replicating polymerase uses the newly synthesized strand to transiently detour from the lesion, before returning to the original template strand downstream of the lesion).

Deep intronic mutation: Any mutation located within an intron, but not adjacent to an intron–exon border. Such mutations are frequently overlooked, since conventionally only intronic sequences flanking exons are targeted by primers and amplified in a PCR.

Deletion (del; d): Any loss of a nucleotide, an oligonucleotide, a segment containing one or several genes, a part of a chromosome, or a whole chromosome. Deletions may be terminal (i.e., occur at the end of a chromosome) or intercalary (i.e., occur within a chromosome). If the number of lost nucleotides is not divisible by 3, a reading frameshift mutation ensues.

Deletion hotspot: Any region of a genome in which deletions occur more frequently than in the rest of the genome.

Deletion map: A graphical description of the precise location of deletions on a linear or circular DNA molecule. A deletion map is the result of deletion mapping and serves as an important tool for functional genomics (here, the identification of the function of genes by evaluation of the resulting deletion mutant phenotype).

Deletion mapping:
   a. The localization of the positions of deletions in the DNA of an organism.
   b. The localization of a specific, yet unidentified gene on a chromosome by using overlapping deletions.

Deletion mutagenesis: The progressive unidirectional removal of sequences from the 5′ or 3′ end, or from internal regions of a target DNA. In short, one deletion mutagenesis technique uses oligonucleotide cassettes to introduce or modify specific restriction endonuclease recognition sites in a plasmid to facilitate subsequent deletion with, for example, exonuclease III and the generation of a series of deletion mutants. Exonuclease III sequentially removes nucleotides from the 3′ end of dsDNA that contains either a 5′ overhang or blunt ends, but does not attack a 3′ overhang. The use of suitable restriction endonuclease sites therefore allows unidirectional deletion of DNA. The effect(s) of the deletion(s) are then analyzed by, for example, an *in vitro* transcription system coupled to an *in vitro* translation system and an *in vitro* or *in vivo* characterization of the properties of the mutated protein (e.g., by studies involving binding of a mutated transcription factor to its cognate target sequence on DNA).

Deletion mutant: Any mutant that has arisen by the removal of one or more base pairs from its DNA.

Deletion mutation: Any mutation that is generated by the removal of one or more base pairs from a particular genome.

Deletion-TILLING (de-TILLING): A variant of the conventional TILLING technique that uses γ-rays, X-rays, and fast neutron bombardment instead of chemicals as inducing mutagens to produce a mutant population.

Deletogen: Any small chemical compound that induces deletions in a genome. For example, 1,3-butadiene diepoxide, trimethylpsoralene in combination with UV light, and 4′-aminomethyltrioxsalene with UV light are such deletogens.

Digital karyotyping (DK): A technique for the quantitative and high-resolution detection of copy number changes (amplified and deleted chromosomal regions) on a genome-wide scale that uses short sequence tags derived from specific genomic loci at approximately 4-kb intervals along the entire genome (corresponding to about 800 000 specific loci distributed throughout the human genome) by enzymatic digestion. Individual tags are linked into ditags, concatenated, cloned and sequenced. Tags are matched to reference genome sequences and digital enumeration of groups of neighboring tags provides quantitative copy number information along each chromosome. In short, genomic DNA is first isolated, sequentially digested with the so-called mapping enzyme *Sac*I, the resulting fragments ligated to biotinylated linkers, and then restricted with the so-called fragmenting enzyme *Nla*III. DNA fragments containing biotinylated linkers are isolated by capture on streptavidin-coated magnetic beads. Captured DNA fragments are ligated to linkers containing *Mme*I recognition sites, tags released with *Mme*I, and self-ligated to ditags, which are further ligated to form concatemers, which in turn are cloned into an appropriate plasmid vector. Clones are sequenced, the sequences matched to the genome, and increased or decreased genomic tag densities evaluated with specific software packages. More than six tag copies per diploid genome are interpreted to derive from an amplified chromosomal region. Generally, the analysis of sequence tag densities in sliding windows throughout each chromosome allows the identification of potential amplifications and deletions at high resolution.

Direct repair: The removal of pyrimidine dimers (cyclobutane dimers) from DNA by photolyases activated by visible light ("light repair"). Absorption of the activating light is mediated by enzyme-bound flavin adenine dinucleotide ($FADH_2$) chromophores.

Diversity array technology (DArT): A technique for the detection of mutations in genomic DNAs from two (or more) cells or organisms. In short, so-called diversity panels are first established. To that end, genomic DNA is isolated, restricted with an appropriate restriction endonuclease (e.g., *Eco*RI, *Pst*I, or *Msp*I), and adapters ligated to the ends of the resulting fragments. Then primers complementary to the adapters are used to amplify the fragments in a conventional PCR. The genome complexity is reduced about 100- to 1000-fold with primers containing one, two or three selective overhang bases (comparable to the amplified fragment length polymorphism technique). The amplified fragments from such diversity panels ("representations") are cloned into a topoisomerase I cloning vector, the cloned inserts amplified with vector-specific primers and arrayed on a solid support (e.g., glass slides coated with poly-L-lysine). This array is called a diversity array (DArT array, "diversity panel"). Then two genomic samples are converted to representations by the above technique, labeled with different fluorochromes (e.g., cyanin 3 (Cy3) and cyanin 5 (Cy5), respectively), mixed, and hybridized to the diversity array. Finally the ratio between Cy3 and Cy5 fluorescence intensity is measured for each spot. Significant differences in the signal ratios (i.e., above 2.5) identify spots (i.e., genomic fragments), which are different in the two samples. DArT is employed for the genetic fingerprinting of any organism or group of organisms (but mostly in plants, even plants with big genomes or polyploids) and detects single-base-pair changes within the restriction sites or at one of the selective bases of the PCR primer, insertions, deletions, and genomic rearrangements. The resulting dominant markers are partly transferable between species of the same genus, and only few beyond the genus.

DNA damage response gene (DDR gene): Any one of a series of nuclear genes that encode proteins with various functions in the repair of DNA damage triggered by various, mostly environmental impacts. In wild-type cells, the DNA repair and DNA damage checkpoint pathways (collectively called DDR) induce a transient cell-cycle arrest to provide the necessary time for DNA repair and a variety of DNA repair pathways correct the various types of DNA lesions (e.g., nucleotide excision repair, mismatch repair, base excision repair, nonhomologous end-joining, and homologous recombination). In metazoans, checkpoint pathways can also induce apoptosis, and thereby eliminate compromised cells. Mutations in DDR genes underlie many cancer phenotypes, as, for example, in xeroderma pigmentosum or hereditary nonpolyposis colon cancer.

DNA deletion: The removal of DNA sequences of various lengths, parts of chromosomes, or whole chromosomes from a genome in evolutionary times or during developmental programs. For example, DNA deletions normally occur in all ciliate species, including the oligohymenophorans (e.g., *Tetrahymena thermophila* and *Paramecium*) and the hypotrichs (e.g., *Oxytricha nova, O. fallax, O. trifallax, Stylonychia lemnae, S. pustulata,* and *Euplotes crassus*). In these ciliatae, so-called interstitial DNA deletions lead to loss of sequences from the genome.

DNA melting analysis (DMA): A technique for the detection of mismatched bases in DNA (SNP). In short, target DNA is first amplified with conventional PCR techniques from both wild-type and mutant. Then both samples are melted and reannealed to allow recombination of both homoduplexes (wild-type–wild-type and mutant–mutant) and a mismatched heteroduplex. The mixture is again heated. The mismatches will melt at lower temperatures than perfect matches. The rate of melting can be monitored by a fluorochrome that emits fluorescent light only when bound to dsDNA. During the melting process the fluorescence decreases and the decrease can be accurately recorded. Single-tube DMA avoids gel separation of homo- and heteroduplexes, and the labor-intensive and expensive use of columns and solvents.

DNA photolyase: An enzyme, that catalyzes the repair of so-called *cis, syn*-cyclobutane pyrimidine dimers (CPDs) generated by a photochemical [2 + 2] cycloaddition of two neighboring pyrimidine bases. The energy for the activation of the rather inert cyclobutane rings is derived from the absorption of a light quantum (λ 320–500 nm) by a deazaflavine. The energy is then transferred onto the CPD via a flavin adenine dinucleotide (FAD) that forms hydrogen bonds between the amino group of its adenine and the C4-carbonyl groups of the thymine. As a consequence of the direct transfer of an energetic electron the CPD is cleaved in a radical reaction.

DNA polymorphism: The difference in the base sequence of a distinct region between two (or more) different genomes. Such polymorphisms are generated by deletions, insertions, inversions, or generally sequence rearrangements. These mutations lead to, for example, the existence of different alleles for a specific locus. In case of repetitive DNA, variations in the number of repeats may lead to restriction fragment length polymorphisms. DNA polymorphisms may be detected by various DNA fingerprinting techniques or by DNA sequencing.

DNA radical: Any one of a series of highly reactive radicals of purines, pyrimidines, or sugars in DNA, that are generated by, for example, UV irradiation, or copper ions (e.g., $Cu(II)$-H_2O_2) as intermediates of oxidation. In some cases, the initial product of a reaction between singlet oxygen and guanine is not a free radical, but the 4,8-endoperoxide of guanine that rearranges to form 8-oxo-7,8-dihydro-2'-deoxyguanosine. DNA radicals can be detected by, for example, electron spin resonance, since they are paramagnetic, or immuno-spin trapping.

DNA rearrangement: Any structural change in a nucleotide sequence, a gene, or a chromosome.

DNA repair: The enzymatic correction of errors in the nucleotide sequence of a DNA duplex molecule. DNA repair mechanisms protect the genetic information of an organism (its genetic identity) against damage by environmental mutagens (e.g., UV light, ionizing radiation, chemicals) and replication errors. More than 20 genes in *Escherichia coli* code for repair proteins (e.g., repair nucleases) that catalyze various steps in the prominent DNA repair processes, as for example, excision repair, light repair, mismatch repair, recombination repair, short patch repair, and SOS repair.

DNA transposable element: Any one of a series of distinct DNA sequences, that move from their original site in a genome to another site by excision and reintegration. For example, activator-dissociator system elements, bacterial insertion elements, P elements and transposons are such DNA transposable sequences.

Domain: Any part or specific two- or three-dimensional structure of a macromolecule, usually a protein, that forms a structural or functional niche within the remainder of the molecule. For example, DNA-binding proteins possess specific features (DNA-binding domains, e.g., helix–turn–helix, also helix–loop–helix configurations or Zn^{2+} fingers), which enable them to recognize and bind to specific structures or sequences on their target DNA with high specificity and affinity.

Domain fusion: The combination of two (or more) naturally unrelated or synthetically produced sequences in a single DNA molecule that encode specific protein domains. The shuffling of domains generates new proteins, of which a minor fraction also has novel or improved function(s).

Domain fusion analysis: The search for and detection of functionally related proteins by analyzing the sequences for patterns of domains. For example, protein-coding genes in one organism, which are separate from each other in the genome, are often found to be fused into a single gene encoding a single polypeptide chain in another organism. The two *Escherichia coli* gyrase subunits GyrA and GyrB are present as fused homologs in topoisomerase II of *Saccharomyces cerevisiae*.

Domain mapping: The identification of specific domains of a protein that possess structural or functional features. For example, domain mapping defines regions of a protein that interact with other proteins, as, for example, the Ran-binding domain and the nuclear pore complex-binding domain of importin – a mediator of the transport of many proteins between the cytoplasm and nucleus. The Ran-GTP-binding domain maps to the 282 N-terminal amino acids, whereas the pore complex-binding domain localizes between residues 152 and 352 of the protein.

Domain number variation: The decrease or increase of the number of domains in a protein by the deletion, duplication, or insertion of domain-encoding exons in a synthetic gene or a novel gene produced by *in vitro* exon shuffling.

Double-hit single nucleotide polymorphism (double-hit SNP): Any SNP for which each allele is present in two (or more) samples from a distinct population.

Double-knockout mutation ("knock-knock mutation"): Any mutation that simultaneously occurs in two (or more) genes within the same genome and knocks out the function of both.

Double mutant: Any organism whose DNA has suffered two independent mutations.

Double-strand break-induced homologous recombination: Any repair process, that starts at a double-strand break (DSB) between two adjacent direct repeat sequences in DNA and leads to the complete restoration of the site. It involves a $5' \rightarrow 3'$ exonuclease, whose activity generates complementary overhangs on both strands at the break site that subsequently anneal to each other. Noncomplementary intervening sequences form $3'$ flaps that are removed by specific endonucleases. The gaps are filled by DNA synthesis, after which the ends are ligated. Effective DSB-induced homologous recombination requires at least 15–20 bp (optimally 70 bp) of sequence homology between the strands, and occurs in bacteria, plants, insects, and mammalian cells.

Down promoter mutation (down mutation): Any mutation in a promoter sequence that decreases the affinity of DNA-dependent RNA polymerase to the promoter and leads to less frequent transcription of the adjacent gene.

Duplicate gene: Any gene originating from a gene duplication event. Duplicate genes are characteristic for all multicellular organisms.

Duplication: The amplification of a gene, a region of the genome, or whole genomes or chromosomes such that the duplicated sequence is represented twice (or more). For example, a gene duplication results in the appearance of a paralogous gene that can be slightly modified in evolution without serious or even lethal consequences for the organism. Several types of duplications occur within chromosomes. The duplicated chromosome segments are located side by side (in tandem, tandem duplications), or the duplicated segment switches around 180°, but stays in tandem (reverse tandem duplications), or the duplicated segments are located on another chromosome (in trans, displaced duplication, "interchromosomal duplication").

E

EcoTILLING (ecotype TILLING): A variant of the original TILLING procedure that detects naturally occurring allelic variants of a given gene by amplification of the same part of a given gene of different individuals from naturally occurring plant accessions in the same PCR reaction mixture. During the final elongation step of the PCR reaction, either homo- or heteroduplexes form between similar amplification products from the different accessions, depending on whether one or the other plant contained an allelic variant of the gene region or not. As in the original TILLING protocol, the amplified gene fragments are subjected to CEL1 digestion and denaturing polyacrylamide gel electrophoresis. If heteroduplexes form during the last PCR cycle due to the presence of allelic variants of the gene in one or more individuals, one or more band(s) of differing size(s) are visible in the gel. Pools containing such allelic variants are rescreened by amplifying each individual allele of the pool together with that of a well-characterized standard in the same PCR reaction. Thus, the allelic variant of the gene is identified and sequenced. Natural allelic variations detected by this procedure can then be introduced into data bases, and oligonucleotides discriminating between the different SNPs to be detected in a large set of germplasm can be fixed on a chip and used to screen a complete population (e.g., a germplasm collection) for the presence of the particular variant. Moreover, when such variants are present in amplicons from parental lines of segregating populations, the respective genes can easily be mapped and compared to phenotypic data. EcoTILLLING is perfectly suited for high-throughput approaches, since it is 7–8 times more efficient as even the best sequencing approaches.

Ectopic recombination (nonallelic homologous recombination): Any recombination occurring between homologous sequences located at different positions in a genome. Ectopic recombination is initiated by the misalignment of nonallelic, but homologous DNA sequences of widely separated DNA elements, and frequently leads to extensive copy number variation, changes of the gene number in gene families, and segmental duplications and deletions, often with pathological consequences. For example, ectopic recombination between a pair of 24-kb repeat elements on chromosome 17 induces the duplication or deletion of the intervening 1.4-Mb DNA segment, resulting in the inherited disorders Charcot–Marie–Tooth type 1A and hereditary neuropathy with liability to pressure palsies. Ectopic recombination rates are dependent on the lengths of high sequence similarity shared between homology blocks, and certain nuclear features.

Enhancer mutation: Any mutation occurring in an enhancer sequence. Enhancer mutations (e.g., deletions, point mutations) can be detected by their effect(s) on the transcription of associated genes.

Enhancer single nucleotide polymorphism (enhancer SNP): Any polymorphism between two genomes that is based on a single nucleotide exchange, small deletion, or insertion within enhancer sequences. An enhancer SNP may be neutral, but may also effectively prevent the binding of activator protein(s). As a result, the corresponding gene is less efficiently transcribed.

Epigenetic code: The specific distribution of methylated cytosines along the DNA of a chromosome and/or the specific side-chain modifications of histones in the chromatin of this chromosome. Since both the cytosine methylation patterns as well as histone side-chain modifications (e.g., acetylation, methylation, phosphorylation) in a specific region of the genome varies with time, so does the epigenetic code.

Epigenetic context: The entirety of all epigenetic mechanisms (e.g., cytidine methylation, histone modifications, nucleosome–lexosome interchanges, chromatin packaging, and others) and pathways in a cell at a given time.

Epigenetic drug (epigenetic therapeutic): Any agent, that acts on either the epigenome or the chromosomal proteins (as e.g., histones). For example, drugs reversing the methylation of cytosyl residues in DNA ("demethylating" or "hypomethylating" agents, e.g., the azacytidine Vidaza or Dacogen for the treatment of myelodysplastic syndromes – a group of diseases affecting bone marrow and blood), or drugs influencing the histone code, that is, restoring acetylation of amino acid residues in distinct histones, or inhibiting deacetylation of histone residues such as the drug Vorinostat, or combinations of various such compounds are such epigenetic drugs.

Epigenetic haplotype: The specific arrangement of relatively few adjacent methylated cytosines in a given region of a genome, that is characteristic for this part of the genome and appears as a "block." Usually it is sufficient to estimate the extent of strategic cytosine methylations in such regions to infer the general epigenetic pattern of the surrounding flanks.

Epigenetic marker: Any cytosine residue in a specific region of a genome, that is characteristic for this region and discriminates this region from others in the same genome and from the identical region in another genome. Such cytosyl residues are also called methylation variable positions (MVPs).

Epigenetic modification: Any reversible, but inherited alteration of DNA beyond the level of the base sequences. For example, the methylation of cytosyl residues in CpG dinucleotides of the target DNA is reversible (e.g., by demethylases), but the methylation pattern is heritable (e.g., genomic imprinting).

Epigenetic reprogramming: The reversion of hypermethylation ("demethylation") of promoter sequences by administration of 5-azacytidine (5-AZA), that is incorporated into replicating DNA. The incorporated 5-AZA covalently binds ("traps") DNA methyltransferase 1 (DNMT1) that becomes depleted in, for example, myeloid leukemia cells and as a consequence leads to reduced C-methylation.

Epigenetics: A discipline of general and molecular genetics that focuses on the study of the heritable or also acquired, nonhereditable differences in gene expression patterns not caused by changes in the primary sequence of the DNA, but rather on changes in DNA methylation (cytosine methylation) and chromatin modifications (e.g., histone side-chain modifications, nucleosome patterns). Epigenetics aims at deciphering the mechanisms that generate the phenotypic complexity of an organism. Some epigenetic changes are passed from one generation to the next (transgenerational inheritance).

Epigenetic signature: The characteristic pattern of cytosine methylation in a specific region of a promoter (or a gene) at a given time. The epigenetic signatures vary with the state of a cell, and changes in response to environmental, also intrinsic factors. Methylation of strategic cytosines in promoters recruits proteins binding to the methylated sites, prevents the binding of activating transcription factors, and silences the adjacent gene.

Epigenetic switch: Any change from predominant histone methylation to cytosine methylation or vice versa. Such epigenetic switches occur frequently in, for example, prostate cancer cells.

Epigenetic therapy: The administration of drugs (e.g., 5-azacytidine) to modulate epigenetic markers (e.g., the hypermethylation of distinct promoters).

Epigenetic variation: The occurrence of differences between the genomes of two related organisms that are based on nonhereditary and also reversible modifications, as, for example, the methylation of cytidyl residues in genomic DNA.

Epigenome:

a. The complete set of genes involved in genetic imprinting.
b. The pattern of methylated cytosines and of modified histones in a genome in a cell, a tissue or an organ at a given time.

Epigenomic profiling: The genome-wide analysis of cytosine methylation, DNA replication, distribution of DNA-binding proteins, and histone modification patterns.

Epigenomics:

a. The whole repertoire of techniques that allow us to analyze epigenetic parameters, such as the methylation pattern of cytosine residues in genes or promoters during the activation or silencing of these genes, in different developmental stages, or after specific treatment of cells.
b. The whole repertoire of techniques for the identification of genes involved in genetic imprinting.

Epigenotype: The normally stable and heritable genotype (based on the four-base genetic code), upon which the so-called epigenetic code is superimposed (i.e., the specific pattern of methylated or otherwise modified bases in the genome). For example, the genotype in all cells of a multicellular organism is identical, but the different types of cells have different distribution patterns of 5-methylcytosine (i.e., have different epigenotypes).

Epigenotyping: The process of establishing an epigenotype.

Epimutation: Any alteration in either the histone code or the overall (or also specific) cytosine methylation pattern in a particular stretch of DNA. Such epimutations are blamed for the outbreak of diseases, such as diabetes, schizophrenia, or bipolar disorder, but do not involve base changes.

Error-prone polymerase chain reaction (error-prone PCR (epPCR)): A variant of the conventional PCR that is used for the introduction of random mutations into target DNA. For example, the replacement of Mg^{2+} by Mn^{2+} or an imbalanced nucleotide concentration in the PCR reaction mixture increase the intrinsic error rate of *Taq* DNA polymerase (i.e., wrong deoxynucleotides are incorporated with increased frequency during PCR).

Error rate: The number of nucleotides erroneously incorporated into a new DNA strand by DNA polymerase, leading to mutations.

Excision:

a. The enzymatic removal of a nucleotide, an oligonucleotide, or a polynucleotide fragment from a nucleic acid molecule.
b. The breakage of the peptide bond at an intein–extein junction.
c. The removal of intein(s) from a precursor protein. Intein excision is catalyzed by cleavage endoproteinases that precisely cut at the intein–extein junction.

Excision repair (cut and patch repair; dark repair): The precise enzymatic substitution of damaged or altered bases on one strand of a DNA duplex molecule. Such damage arises from thermal fluctuations (e.g., leading to depurination) or UV irradiation (e.g., formation of pyrimidine dimers). In *Escherichia coli*, the repair starts with the recognition and excision of the incorrect base together with some 10–20 nucleotides (short patch repair) or up to 1500 nucleotides (long patch repair) by an endonuclease. Subsequent insertion of the correct bases ("patch") is directed by the complementary strand and catalyzed by a DNA polymerase. Finally, the inserted bases are ligated to the 3' end of the adjacent bases by a DNA ligase.

Exon duplication: The duplication of a specific exon of a gene. In many cases such duplicated exons are retained in the pre-mRNA and even acquire new functions. Exon duplication accompanies alternative splicing frequency. For example, singletons (single genes) possess significantly more exons than paralogous members of a gene family and at the same time produce more splicing variants.

Exon shuffling: The generation of new genes through intron-mediated recombination of coding sequences (exons) that were previously specifying different proteins or different parts of one and the same protein.

Exon skipping (exonS): The elimination of one (or more) exons from a transcript during splicing such that the combination of residual exons results in a new mRNA and consequently a protein with a new arrangement of domains. For example, the deletion ("skipping") of an exon B, which was originally linking two other exons A and C, allows to recombine exons A and C, creating a new exon combination (exon shuffling). Exon skipping is a route to the generation of new genes. Out of many examples, only one is mentioned here. If exon 7 of the pre-mRNA encoding the so-called survival motor neuron (SMN) II protein is skipped ($\Delta 7$), the resulting protein is nonfunctional and cannot compensate for a defect SMNI protein (generated by a mutation in the *SMNI* gene). The defect leads to spinal muscular atrophy – an autosomal recessive disease accompanied by a degeneration of α motor neurons responsible for the innervation of musculature. As a consequence, there is paralysis of the musculature of the respiratory tract concomitantly with airway infections that frequently leads to the death of the patient. The SMNI protein is part of a protein machine ("SMN complex") that loads Sm proteins onto U small nuclear RNAs – a prerequisite for the formation of an active spliceosome.

Exonic single nucleotide polymorphism (exon SNP): Any SNP that is present in an exon of a gene. Synonymous with expressed SNP.

Exonic variation: The occurrence of sequence polymorphisms (e.g., SNPs or Indels) in homologous exons of the same gene in two (or more) different individuals of the same species.

Expressed sequence tag polymorphism (ESTP): Any difference in DNA sequence between two (or more) expressed sequence tags (ESTs) that can be detected by either restriction digestion of the ESTs or by separation of polymorphic sequences using denaturing gradient gel electrophoresis. ESTPs can be used to screen for DNA polymorphisms in populations or serve as markers in mapping and comparative mapping procedures.

Expressed sequence tag polymorphism mapping (ESTP mapping; EST locus polymorphism mapping): A technique for the conversion of expressed sequence tags (ESTs) into molecular markers that can be integrated into a genetic map, based on SNPs in coding and noncoding regions adjacent to the EST. In short, isolated genomic DNA is digested with either a 4-bp cutter (e.g., *Alu*I), 6-bp cutter (e.g., *Dra*I, *Ssp*I), or less frequent 6-bp cutters (e.g., *Eco*RV), then vectorette-like adaptors ligated onto the termini using DNA ligase, and the resulting adaptored fragments amplified in a conventional PCR using an EST-specific and an adaptor-specific primer. The EST primers are usually designed within the 5′ or 3′ untranslated regions (or near to the start or stop codon within the coding region) such that amplification occurs towards the noncoding region, either 5′ or 3′ of the EST. For fluorescent detection of the amplified fragments, the adaptor-primer can be labeled by a fluorochrome (e.g., 6-carboxyfluorescein or hexachlorofluorescein) or the fragments can be visualized simply by ethidium bromide fluorescence. A second amplification can be necessary with nested EST-specific and nested adaptor-specific primers. The resulting amplicons are separated on agarose or polyacrylamide gels. EST polymorphisms between the two parents can be exploited for the estimation of the segregation pattern in the progeny of a cross and mapped. The resulting EST map therefore is based on genic markers rather than anonymous molecular markers generated by, for example, amplified fragment length polymorphism or similar techniques. Moreover, the positions of expressed genes can be fixed on a genetic map.

Expressed sequence tag simple sequence repeat (EST-SSR): Any microsatellite repeat (simple sequence repeat) that is part of an expressed sequence tag (EST). Sequence polymorphisms in EST-SSRs allow to localize the corresponding gene on a genetic map.

Expressed single nucleotide polymorphism (eSNP): Any SNP that is present in exons (i.e., expressed sequences).

Expression variation (EV): The variation in the expression of distinct genes in the different cells, tissues, or organs of the same organism, or between the same cells, tissues, or organs of different organisms.

F

5-Bromouracil (5-BU): A mutagenic thymine analog that can be mistakenly incorporated into nascent DNA. In its prevalent keto form, 5-BU replaces thymine and consequently pairs with adenine. The bromine atom, however, causes a redistribution of electrons such that 5-BU can also adopt the (rare) ionized state in which it pairs with guanosine, mimicking the pairing of cytosine. During subsequent replication, this base exchange cause mutations in the target DNA.

5-Iodo-deoxycytidine: A halogenated derivative of deoxycytidine that is used for incorporation into an oligonucleotide, where it can be activated by light and cross-links the oligonucleotide to DNA, RNA, or protein. Halogenated nucleosides allow crystallographic studies of oligonucleotide structure.

5-Iodo-deoxyuridine: A halogenated derivative of deoxyuridine that is used for incorporation into an oligonucleotide, where it can be activated by light and cross-links the oligonucleotide to DNA, RNA, or protein. Halogenated nucleosides allow crystallographic studies of oligonucleotide structure.

Fluorouracil (FU; 5-fluorouracil): A base analog that contains fluorine at position 5. This antagonist is incorporated into mRNA instead of uracil and changes the coding properties of the messenger. This faulty mRNA translates into a faulty protein.

Forensic single nucleotide polymorphism (forensic SNP): Any SNP that unequivocally identifies an individual (e.g., a victim or his/her murderer, a victim, or one or more suspects). Forensic SNPs can be located in nuclear or mitochondrial DNA, and usually are detected by sequencing of amplicons from the various individuals and sequence alignment.

Forensically informative nucleotide sequencing (FINS): A technique for the determination of specific sequences of the DNA of a distinct animal (or plant or human individual) in specimens potentially containing DNAs from several animals (plants, humans). FINS uses evolutionary conserved primers to amplify a distinct region of the mitochondrial cytochrome *b* gene and then employ one of these primers to sequence the amplification product. The resulting nucleotide sequence is diagnostic of the species from which it originated and can be used to discriminate it from other species (e.g., in mixed meat samples).

Forward mutation: Any mutation that inactivates a gene. Such forward mutations occur at a rate of about 10^{-6} per locus per generation.

Founder mutation: Any mutation that occurs in the genome of one single individual and is subsequently transferred to many other individuals of the same species, and may ultimately be a component of the genomes of members of whole populations. Founder mutations, usually single nucleotide exchanges, are embedded in flanking DNA, which is highly conserved in the different progeny individuals, since it is transmitted to the progeny as a block. If the mutation occurred at a hotspot, the flanking DNA differs from individual to individual in the progeny. Founder mutations in strategic regions of a genome (mostly a gene) may cause diseases (e.g., sickle cell anemia).

Wild-type sequence	5′-CTACT**G**CTCGAATC**T**ATCCGTTCA**A**TCGCAT**T**-3′
Founder mutation (↑)	5′-CTACT**G**CTCGAATC**A**ATCCGTTCA**A**TCGCAT**T**-3′
Progeny of founder	5′-CTACT**G***CTCGAATC* **A** *ATCCGTTCA***A***TCGCAT***T**-3′
	5′-CTACT**G**CTCGAATC**A**ATCCGTTCA**A**TCGCAT**T**-3′
	5′-CTACT**G**CTCGAATC**A**ATCCGTTCA**A**TCGCAT**T**-3′
	5′-CTACT**G**CTCGAATC**A**ATCCGTTCA**A**TCGCAT**T**-3′
	5′-CTACT **T** *CTCGAATC* **A** *ATCCGTTCA* **C** *TCGCAT* **G** -3′
Progeny of hotspot mutation	5′-CTACT**T**CTCGAATC**A**ATCCGTTCA**C**TCGCAT**G**-3′
	5′-CTACT**A**CTCGAATC**A**ATCCGTTCA**G**TCGCAT**C**-3′
	5′-CTACT**C**CTCGAATC**A**ATCCGTTCA**T**TCGCAT**C**-3′
	5′-CTACT**A**CTCGAATC**A**ATCCGTTCA**G**TCGCAT**A**-3′

Fragile site: Any one of multiple regions within a human (or mammalian) genome, that form gaps, constrictions, and breaks of chromosomes exposed to replication stress at an increased frequency, and represent chromatin failing to compact during mitosis. Fragile sites are classified as rare or common, depending on their frequency within a population. Further subdivision is based on their specific induction chemistry (e.g., as folate-sensitive (FRA10A, FRA11B, FRA12A, FRA16A, FRAXA, FRAXE, and FRAXF) or non-folate-sensitive rare fragile sites, or as aphidicolin- (FRA2G, FRA3B, FRA4F, FRA6E, FRA6F, FRA7E, FRA7G, FRA7H, FRA7I, FRA8C, FRA9E, FRA16D, and FRAXB), bromodeoxyuridine- or 5-azacytidine-inducible common fragile sites (FRA10B and FRA16B)). Rare fragile sites are associated with expanded CGG/CCG trinucleotide repeats or AT-rich minisatellite repeats, composed of interrupted runs of AT dinucleotides, that adopt stable secondary non-B-DNA conformations (intrastrand hairpins, slipped strand DNA, or tetrahelical structures) perturbing DNA replication and interfering with higher-order chromatin folding. Such sites segregate in a Mendelian codominant fashion (in afflicted families) and occur in 1/3000 down to 1/20 individuals. The molecular basis of common fragile sites is unknown, but such sites are ubiquitous in human populations. The folate-sensitive rare fragile site is clinically most important, as it is associated with the fragile X syndrome – the most common form of familial mental retardation, affecting about 1/4000 males and 1/6000 females. FRAXA mental retardation probably results from the abolition of FMR1 gene expression due to hypermethylation of the CpG islands adjacent to the expanded methylated trinucleotide repeat. FRAXE is associated with X-linked nonspecific mental retardation and FRA11B with Jacobsen syndrome. In particular, common fragile sites are consistently involved in *in vivo* chromosomal rearrangements related to cancer.

Fragilome: The entirety of all rare and common fragile sites in a chromosome or a complete genome at which chromosomal breaks occur at higher than usual frequency. The human genome harbors about 120 such sites.

Functional polymorphism:

a. Any polymorphism in a gene or a promoter (or noncoding regulatory sequence) that changes the underlying codon and hence the amino acid composition of the encoded protein or alters the sequence of the recognition site of a transcription factor in a promoter such that its binding and the activation of the adjacent gene are prevented.
b. Any sequence polymorphism (e.g., a SNP, transition, or transversion) between two genomes (originating from two different individuals), that is linked to a particular phenotype (e.g., a disease). Linkage of SNP and phenotype is generally taken as indication for a function of the SNP (e.g., by changing the amino acid composition of a protein domain with subsequent functional consequences).

G

Gain-of-function mutation (GOF): Any mutation that converts a previously inactive (noncoding) sequence into an active (coding) sequence (e.g., a gene).

Gapped duplex mutagenesis: The introduction of mutations in a DNA molecule by using "gapped duplex" DNA. Such DNA molecules can be generated by hybridization of a single-stranded vector DNA (e.g., pBR 322) carrying an insert with a homologous single-stranded vector DNA lacking the insert. Reannealing of these two single strands yields a double-stranded molecule that is single-stranded in the insert region to be mutagenized. An appropriate oligodeoxynucleotide primer 16–18 bases in length and carrying one or more mismatched bases (acting as the mutagen) can now be hybridized to the gap region. It is then filled-in using DNA polymerase I and covalently closed by DNA ligase. After transfection into bacteria, the mutants are identified by appropriate selection and screening techniques. In another approach, base analogs are supplied during the repair synthesis (e.g., in gap misrepair mutagenesis).

Gene-based single nucleotide polymorphism (gene-based SNP): Any SNP that is located in either an exon, an intron, or a promoter of a gene.

Gene defect: The mutation-induced failure of a gene to encode a functional protein. Such gene defects underlie many human (and animal) diseases (e.g., diabetes mellitus type I and II, β-thalassemia, sickle cell anemia, familial hypercholesterolemia, cystic fibrosis, Tay–Sachs disease, α_1-antitrypsin deficiency, classical phenylketonuria, Duchenne–Griesinger muscular dystrophy, classical hemophilia, Lesch–Nyhan syndrome, metachromatic leukodystrophy, and galactosemia, to name only the most prevalent human disorders).

Gene duplication: A process by which an ancestral gene is copied ("duplicated"), so that the corresponding genome contains two identical gene sequences. One of these genes subsequently undergoes mutation(s) that may convert it to a pseudogene or its functions may be retained in spite of changed sequence composition. or the copy may be mutated such that a novel function of the encoded protein evolves. Such gene duplications are the result of unequal crossing over, reverse transcription of mRNAs, or the duplication of segments of a genome or the whole genome.

Gene excision: The removal of a gene (generally DNA sequences) from a target DNA that is achieved by site-specific recombination mediated by Cre recombinase in Cre/Lox or analogous systems, or by meganuclease-induced recombination (also double-strand break-induced homologous recombination). Gene excision shows potential for plant genetic engineering (e.g., for the removal of antibiotic resistance genes that are necessary for selection of plant transformants containing the transferred desirable gene(s), but not required for the performance of the transgenic plant in the field and all the more blamed to increase resistance in soil bacteria by horizontal gene transfer) and therapy (e.g., the excision of an integrated virus in human cells).

Gene knockdown: The reduction of a gene's activity to very low levels through various mechanisms (e.g., RNA interference), such that it can be conditionally expressed. Gene knockdown is the method of choice if gene knockout would be lethal for the carrier organism.

Gene knockdown potency: The efficiency of a single small interfering RNA (siRNA) or a set of siRNAs covering the target mRNA to knockdown the cognate gene. Gene knockdown potency is expressed as percentage of mRNA concentration left after addition of the siRNA to a target cell.

Gene knockin: Laboratory slang term for the disruption of a gene by the insertion of a sequence or mutation(s) that either activates the gene or restores its activity (if it was previously knocked out).

Gene knockout: Laboratory slang term for the disruption of a gene by the insertion of a DNA sequence or mutation(s) that abolish gene function.

Gene mutation: Any mutation occurring within the coding region of a gene (leading to the synthesis of a defective polypeptide) or the promoter (leading to an aberrant regulation of the adjacent gene).

Gene repair: The correction of mutations (preferably point mutations) in a gene within living cells, using bifunctional oligonucleotides or chimeric oligonucleotide-directed gene targeting.

Genetic load (mutational load; genetic burden): The accumulation of unfavorable or deleterious mutations in the gene pool of a specific population.

Genome mutation (genomic mutation): An incorrect term for a change in the number of a specific chromosome (leading to aneuploidy) or the whole chromosome set of an organism (leading to polyploidy).

H

Haplotype ("half-type"):

a. The linear arrangement of alleles along a region in DNA (e.g., a bacterial artificial chromosome clone, a restriction fragment, or a chromosome). In laboratory slang, a haplotype can also be an individual with a specific arrangement of alleles in a given piece of its DNA (e.g., a gene; also called "block" or "haplotype block"). Such haplotypes can, for example, be defined on the basis of specific SNPs on a chromosomal segment (in diploids, on the corresponding segments of homologous chromosomes) that requires repeated sequencing of the target region. If the target sequence from different individuals is then compared, the haplotypic organization becomes apparent. Haplotype analysis is used for the establishment of genetic risk profiles and the prediction of clinical reaction of an individual towards pharmaceutical compounds (e.g., drugs).

Haplotype patterns (haplotype block: underlined)

Individual	SNP		SNP		SNP
A	5'-ATTGAT<u>C</u>GGAT	....	<u>C</u>CATCGGA	....	CT<u>A</u>AC-3'
B	5'-ATTGAT<u>A</u>GGAT	....	<u>C</u>CAGCGGA	....	CT<u>C</u>AC-3'
C	5'-ATTGAT<u>C</u>GGAT	....	<u>C</u>CATCGGA	....	CT<u>A</u>AC-3'
D	5'-ATTGAT<u>A</u>GGAT	....	<u>C</u>CAGCGGA	....	CT<u>C</u>AC-3'
E	5'-ATTGAT<u>C</u>GGAT	....	<u>C</u>CATCGGA	....	CT<u>A</u>AC-3'
F	5'-ATTGAT<u>A</u>GGAT	....	<u>C</u>CAGCGGA	....	CT<u>C</u>AC-3'

b. The complete set of genes inherited from one parent.

Haplotype block (hapblock; linkage disequilibrium block (LD block)):

a. A specific arrangement of adjacent alleles in a given region of genomic DNA (usually in the range from 10 000 to 100 000 bp) that is inherited as a "block," probably because its recombination frequency is lower than in other parts of the genome. In practice, a haplotype block is characterized by a series of SNP in linkage disequilibrium.

b. Any one of relatively large genomic regions (from 1 to 180 kb pr more), where defined sequences (e.g., specific genes, but also SNPs) are associated.

Haplotype map: A variant of a genome map in which the haplotype blocks of the genome of an organism are depicted.

Haplotype mapping: The process of establishing a haplotype map.

Haplotype shuffling: The rearrangement of parental haplotypes in the progeny by recombination.

Haplotype signature: Any characteristic configuration of specific alleles in an organism. Haplotype signatures can be established for a collection of organisms (e.g., human patients suffering from the same disease) and a disease allele be identified.

Haplotype single nucleotide polymorphism (htSNP; haplotype tag SNP): Any SNP contained within a haplotype block. Do not confuse with high-throughput SNP.

Haplotype-tagged single nucleotide polymorphism (haplotype tagging SNP): Any SNP that is identified (or "tagged") from larger SNP databases (dbSNP), located in a specific genomic region, and used to define haplotypes and haplotype structure in that region.

Heterologous transposon: Any transposon originating from a donor organism A and transformed into an acceptor organism B previously lacking this transposon.

Hetero-mismatch: Laboratory slang term for any base mismatch, in which two different noncomplementary bases are juxtaposed to each other in DNA (e.g., A and G, A and C, C and A, C and T, G and A, G and T, T and C, and T and G, respectively).

Homeotic mutant: Any mutant (e.g., a *Drosophila* mutant) in which a normal organ is replaced by another organ ("transdetermined organ") during development. For example, a specific homeotic mutant carries a leg in place of a wing in the normal wing location.

Homeotic mutation: A mutation that leads to the exchange of one specific part of the body by another. For example, mutations in the *Drosophila* homeotic gene *Antennapedia* lead to the replacement of antennae by legs or of parts of the antennae by the corresponding parts of the legs. In *Drosophila Ultrabithorax* mutants the second thorax segment is substituted by a first one giving rise to flies with two pairs of wings. In a speculative hierarchy of developmental genes, the homeotic genes are therefore clearly distinguished from maternal effects and segmentation genes which are needed during earlier stages of ontogenesis.

Host range mutant (hr mutant):

  a. Any bacteriophage mutant that is able to infect and lyse bacterial hosts that are different from its natural host.
  b. Any mutant virus that is able to replicate in cells that are different from its natural host cells.

Host range mutation: Any mutation that changes the properties of a bacteriophage (host range mutant) so that it may infect and lyse bacteria that were previously resistant.

Hotspot (recombinational hotspot): Any sequence within a gene or a chromosome at which mutations occur at a significantly higher frequency than usual. In case of *Tn10* (*transposon* 10) mutagenesis, insertion occurs at a hotspot with a symmetrical 6-bp consensus sequence (5′-GCTNAGC-3′) where the internal 5-methyl group at the third position of the pyrimidine is necessary for strong recombination. In the human genome, hotspots of 1–2 kb, at which the recombination rate is at least 10 times higher than in the surrounding regions, occur once every 50–200 kb.

Generally, the recombination rate in humans is about 60% greater in female than in male meiosis.

Human single nucleotide polymorphism probe array (HuSNP array): A DNA chip that allows the parallel interrogation of 1500 SNPs covering all 22 autosomes and the X chromosome of humans in a single experiment.

Hypermutation: The process of dramatically increasing the mutation rate of a distinct DNA molecule, a genomic region, or a whole chromosome above background. For example, the protein PMS2 removes nucleotides from the parental rather than the newly synthesized DNA strand in immunoglobulin genes of B cells, inserts a nucleotide complementary to the mismatch, and thereby eternalizes the mutation. PMS2 recognizes the parental target strand, since it is methylated at cytosyl residues. Hypermutation creates a vast array of new antibody-encoding genes and consequently new antibodies against a multitude of foreign antigens.

Hypervariable regions (HVR): Highly polymorphic sequences scattered throughout the human genome that consist of arrays of short, usually GC-rich, tandemly repeated units to which no specific function can yet be attributed. HVRs are thought to be hotspots of recombination. Unequal exchange at meiosis or mitosis, or slippage during DNA replication (slipped strand mispairing) may result in allelic differences in the number of repeated units present at an HVR site and, consequently, in length polymorphism. HVRs may be used for the establishment of a DNA fingerprint.

I

Identity-testing single nucleotide polymorphism (identity-testing SNP): Any one of a set of SNPs that allows us to differentiate between people and is therefore used in forensic analyses. Elite identity-testing SNPs have the highest heterozygosity possible (i.e., 50% heterozygosity for a biallelic system) and a low coefficient of inbreeding (i.e., low population heterogeneity). Identity-testing SNP panels comprise from 19 to 52 SNPs.

Illegitimate recombination (nonhomologous recombination): Any recombination of two DNA molecules without any, or only very little, homology.

Indel: Abbreviation for insertion–deletion – a mutation in a target DNA caused by a combined or separate deletion or insertion event. For example, in an original sequence 5′-GATTCGTTTTACCGTTATCATCGGGTA-3′, an Indel would create the mutated sequence: (deletion) 5′-GATTCGCGTTATCATCGGGTA-3′ and (insertion) 5′-GATTCGTACGGTCTTTACCGTTATCATCGGGTA-3′.

Indel bias: The relatively higher rate of small deletions (1–400 bp) as compared to small insertions of similar size in protein-coding genes and nongenic regions. The Indel bias leads to DNA loss and reduction of genome size in evolutionary times.

Induced mutation: A mutation that is generated by a mutagen, as opposed to spontaneous mutation.

Informative single nucleotide polymorphism (informative SNP): Any SNP that is located in an exon, a promoter, an enhancer, or a silencer region of a gene, or within regulatory sequences in a genome, and therefore potentially influences the activity of the corresponding gene. The term is also used for a SNP as component of a haplotype.

Insertion:

  a. The incorporation of one or more base pairs into a DNA sequence (insertion mutation).
  b. The process of integration of foreign DNA into a cloning vector molecule.

Insertion mutagenesis: The introduction of an insertion mutation in a DNA sequence.

Insertion mutation (addition mutation): The interruption of a DNA sequence by the insertion of additional DNA. Single-base-pair insertions may be caused by certain chemicals (e.g., acridine dyes), while the integration of transposons or insertion sequences is equivalent to a longer insertion mutation. Any insertion of bases in other numbers than three or multiples thereof may result in a reading frameshift mutation. In any case, insertions may either lead to the loss of function of the original DNA (insertional inactivation) or to the restoration of a previously defect DNA (insertional activation). Insertional mutations are given three-letter designations, consisting of the designation of the mutated gene, allele numbers, and (following a double colon) the name of the inserted sequence (e.g., *his* C 527 : : Tn7).

Insertion preference: The predominant insertion of a transposon into specific regions of a genome.

Insertion sequence (IS; IS element; insertion element; simple transposon): Any member of a group of small transposons (0.7–1.5 kb in length) widely distributed throughout pro- and eukaryotic DNA that contain only a few genes encoding transposition functions and whose termini consist of inverted repeat sequences of about 30 bp. Usually these flanking DNA regions are also transposed. IS elements can insert into different regions of the chromosome or into coresident plasmids, leaving a copy of themselves at the donor site, and causing a 3- to 9-bp duplication at their integration site (direct repeat). Transposition of IS elements (denoted IS1, IS2, IS3, etc.) into genes may, and usually does, destroy the function of these genes (insertional inactivation).

Insertion sequence fingerprinting (IS fingerprinting): A technique for the detection of sequence polymorphisms between different bacteria of the same species that uses ISs as probes. Such radioactively labeled ISs are hybridized to Southern blots of restricted genomic DNA fragments generated by digestion with appropriate restriction endonucleases.

Insertion site:

a. Any unique restriction site of a cloning vector molecule (cloning site) into which foreign DNA can be inserted.
b. The integration site of transposons or insertion sequences.

Insertion-site-based polymorphism (ISBP): Any sequence polymorphism generated by the insertion of one transposon into another transposon ("nested transposon"). Such nested transposons frequently form whole sets of transposons, covering large areas of a genome (especially in plants, e.g., wheat, corn). Since the ISs vary from insertion event to insertion event, one such nested transposon island harbors many sequence polymorphisms, which can be used as genetic markers for genotyping (discovering genetic diversity) and genetic and physical mapping.

In silico **single nucleotide polymorphism (in silico SNP; isSNP):** Any SNP that is identified *in silico* by mining overlapping sequences in expressed sequence tag or genomic databases. Since isSNPs represent "virtual" polymorphisms, they have to be validated by resequencing the region in which they occur.

Intercalating agent (base intercalator): Any molecule that inserts between two complementary base pairs in a double-helical DNA or RNA molecule. Intercalation causes changes in DNA topology (e.g., unwinding), leads to mutations, and influences DNA functions (intercalated DNA cannot be transcribed or replicated). Intercalators are mutagenic and cancerogenic. Experimentally, they are used to detect DNA or RNA by staining and to separate different topological forms of DNA in density gradient centrifugation.

Intercalation: A process whereby atoms or molecules are inserted into pre-existing structures (e.g., ethidium bromide intercalates between two strands of DNA duplexes or a protein intercalates into the fluid matrix of a membrane).

Intercalator: Any usually low-molecular-weight and typically planar heterocyclic molecule of approximately the size and shape of a DNA base pair that intercalates into the DNA double helix or double-stranded parts of an RNA molecule and stabilizes the double-stranded region. Intercalation mostly follows the so-called "nearest-neighbor exclusion principle," which demands a maximum loading of one intercalator per 2 bp (i.e., alternate base pairs are not linked by the intercalator).

Interchromosomal duplication: The addition of one (or several) segment(s) from one chromosome to another chromosome (usually by faulty crossover). Such interchromosomal duplications lead to a functional imbalance of genes in the involved region and usually are the basis for genetic disorders. The term is also used for any genomic segment that is duplicated among nonhomologous chromosomes. For example, in the human genome, a 9.5-kb sequence containing the adrenoleukodystrophy locus from chromosome Xq28 has been duplicated, and now appears around pericentric regions of chromosomes 2, 10, 16, and 22.

Interposon: A recombinant DNA fragment that is used for *in vitro* insertional mutagenesis Typically, an interposon carries one (or more) antibiotic resistance gene

(s) (e.g., Sm^r/Spc^r) flanked by short inverted repeats that include transcription termination signals (e.g., from bacteriophage T4 gene 32), translational stop signals (e.g., synthetic DNA with stop codons in all three reading frames), and polylinkers. An interposon can be cloned into a linearized plasmid vector and can be easily selected on the basis of drug resistance, and its position can be precisely mapped after integration into a chromosome by the restriction sites in the flanking polylinker. The use of interposons (e.g., Ω interposon) avoids the disadvantages of transposon mutagenesis: bias for the position of integration, transcription of adjacent DNA, and DNA rearrangements (e.g., deletions, inversions) accompanying transposon integration.

Interposon mutagenesis: A method to introduce insertion mutations at specific sites of a target DNA, using interposons (e.g., Ω interposon).

Intrachromosomal duplication: Any genomic segment, that is duplicated within a particular chromosome or chromosome arm. Such duplications mediate chromosomal rearrangements that are associated with diseases. For example, recombination of such duplications on chromosome 17 give rise to contiguous gene syndromes such as Smith–Magenis syndrome or Charcot–Marie–Tooth syndrome 1A.

Intragenic single nucleotide polymorphism (intragenic SNP): Any sequence polymorphism between two (or more) genomes that is based on a single nucleotide exchange, small deletion, or insertion and occurs within a gene.

Intragenome duplication: The occurrence of identical sequences (e.g., genes, gene families) on different chromosomes of the same nucleus.

Intron intrusion: The disruption of a functional gene by the insertion of an intron.

Intronic single nucleotide polymorphism (intronic SNP; intron SNP): Any SNP that occurs in introns of eukaryotic genes. Intron SNPs are more frequent than SNPs in coding regions.

Intronic variation: The occurrence of sequence polymorphisms (e.g., SNPs or Indels) in homologous introns of the same gene in two (or more) different individuals of the same species.

Inversion: The disruption of the normal arrangement of sequences within a chromosome or chromatid by the excision of a fragment, its rotation by 180° (reversal), and its reinsertion at the excision site or another position ("shift") in the reverse orientation ("breakage-reunion"). Principally, two types of inversions occur – single inversions (where only one segment of a chromosome is inverted) and multiple or complex inversions (involving several chromosomal segments).

***In vitro* mutagenesis:** The alteration of the base sequence of DNA in the test tube (*in vitro*).

Isocoding mutation: A point mutation that changes the nucleotide sequence of a codon without changing the amino acid specified. This means both the wild-type and the mutated triplets code for the same amino acid. Due to the degenerate code, a change, for example, from GCC to GCA or GCG leaves the coding quality of the triplet unaltered (in this case all three triplets code for alanine).

K

Knockabout mutation: Laboratory slang term for any mutation in a gene that abolishes its transcription almost totally, but not completely (i.e., it leaves a residual leaky transcription).

Knockdown mutation: Any mutation that reduces the expression of a gene, but does not abolish it.

Knockin: Laboratory slang term for the insertion of a functioning gene within a mutated, and therefore inactive, copy of the same gene.

Knock-knock mutation: See Double-knockout mutation.

Knockon mutation: Any mutation in genomic DNA caused by the insertion of T-DNA that additionally carries a constitutive promoter (e.g., the cauliflower mosaic virus 35S promoter). This promoter drives ("knocks on") the expression of genes in immediate vicinity of the insertion site.

Knockout (KO; knockout mutation): Laboratory slang term for the inactivation of a gene by the insertion of a DNA sequence (by e.g., gene transfer techniques or site-specific recombination), that disrupts the coding context of the gene.

Knockout animal: Any animal in whose genome a normally active gene has been silenced ("knocked out") experimentally by either random mutation or gene targeting. If the knockout process leads to an altered phenotype, then the function of the knocked-out gene can easily be revealed by complementation (i.e., the substitution of the knocked-out gene by an intact gene).

Knockout/knockin vector: Any vector plasmid into which both the coding region of a target gene under the control of a regulated promoter and a sequence encoding a small interfering RNA (siRNA) or microRNA (miRNA) are inserted. The siRNA or miRNA sequence is designed such that it only destroys the mRNA of the endogenous gene, but not the transgene. Such a vector allows the downregulation of an endogenous gene product, while at the same time expressing, for example, a mutated replacement product.

Knockout mouse (KO mouse): A laboratory mouse in whose genome a normally active gene has been silenced ("knocked out") experimentally. In short, the generation of such knockout mice starts with the production of embryonic stem (ES) cell clones from mouse blastocysts in which the target gene is inactivated by, for example, the electroporation of a specially designed recombination vector. This vector carries isogenic DNA (originating from the same mouse strain from which the ES cells have been isolated) with two sequences homologous to the target gene, flanking a selectable marker gene (e.g., the neomycin phosphotransferase *npt* gene). The homologous sequences span 0.5–2.0 kb (at the 5′ end of the construct) and 5–8 kb at its 3′ end. Usually a herpes simplex virus thymidine kinase (HSV-*tk*) gene is fused to this construct as negative selectable marker. If the vector is not integrated into the genome of the electroporated ES cells, these will be killed in a selection medium containing geneticin (no *npt* gene mediating geneticin resistance). If the vector integrates randomly in the target genome, the transgenic ES cells will survive. However, if ganciclovir is additionally present in the selection medium, the ES cell will die (the HSV-*tk* phosphorylates the drug, which is then integrated into the newly synthesized DNA, leading to chain termination). In rare cases, the vector integrates into the target gene via homologous recombination, so that it is inactivated (integration of the npt gene into one of its exons). The *tk* gene is, however, not inserted, so that the transformants now grow on both geneticin and ganciclovir. These transgenic ES cells are now microinjected into the blastocoel of 3.5-day-old mice, and the manipulated embryos implanted into the uterus of falsely pregnant mice and left to develop there into transgenic chimeric mice. The skin color serves as a visual marker: ES cells and blastocysts originate from mice with different color, so that a rough estimate of the proportion of ES cells in the skin of the chimeric mouse can be made easily. For the production of completely transgenic animals, the chimeric mice are mated with wild-type mice and progeny selected with the skin color of the ES donor mice. Now, animals heterozygous for the transgene are selected (detected by either PCR or Southern blot hybridization) and mated among each other. About 25% of the resulting progeny is now homozygous for the transgene (i.e., both chromosomes carry the same allele; $-/-$ homozygotes). If the transgene produced a knockout mutation, then a knockout mouse has been created, which allows us to characterize the function of the knocked-out gene. An alternative technique for the generation of knockout mice is the morula aggregation method, which is based on the enzymatic removal of the zona pellucida from embryos at the morula stage (2.5 days after fertilization), their culture in paraffin oil, the addition of transgenic ES cells, and the generation of embryo–ES cell chimeras (blastocysts). These are then transplanted into falsely pregnant mice.

Knock-worst mutation: Any T-DNA insertion into a target genome that leads to chromosomal rearrangement(s).

L

Large-scale copy variation (LSC; large-scale copy number variation (LCV); large-scale copy number polymorphism (lsCNP): Any DNA polymorphism between two (or more) individuals, that comprises hundreds of thousands of base pairs (in humans greater than 100 kb). Originally, the term sequence polymorphism was reserved for smaller Indels or transition-transversion-type SNPs. LSCs, on the contrary, represent large polymorphisms, which represent genetic variations in populations, and may be diagnostic for a specific disease or sensitivity towards a drug in human beings. LSCs can be detected by, for example, representational oligonucleotide microarray analysis.

Large-scale duplication: The duplication of a whole genome, a chromosome, or a large chromosomal fragment in evolutionary times. Whole-genome duplication is a consequence of either autopolyploidy (i.e., the doubling of every set of homologous chromosomes in a genome) or allopolyploidy (the creation of a genome with doubled chromosome number through interspecific hybridization). Duplication of individual chromosomes (aneuploidy) leads to an abnormal chromosome number in a karyotype (e.g., trisomy). The duplication of a chromosomal fragment occur through DNA transposition or translocation followed by meiosis. In comparative mapping, such regional duplications manifest themselves as segments enriched for paralogous pairs in genome self-comparisons.

Leaky mutant: A mutant carrying a leaky mutation.

Leaky mutation: Any gene mutation that does not completely abolish gene function and allows the synthesis of a protein which still partly functions.

Lesion-specific DNA repair protein: Any one of many nuclear proteins that specifically recognizes a particular primary lesion in DNA, binds there, and initiates repair processes. For example, MutS proteins bind to mismatched bases, the Ku heterodimer to double-strand breaks, and the xeroderma pigmentosum group C protein (XPC) involved in nucleotide excision repair is among several proteins selectively recognizing UV-induced DNA photoproducts.

Lethal allele (lethal gene): Any usually heavily mutated gene whose expression inevitably leads to the death of the carrier organism.

Lethal mutation: Any mutation that changes a normal gene to a gene encoding a faulty protein, which does not function and leads to the death of the carrier organism.

Ligation-mediated mutation screening (ligation-dependent mutation screening): A method for the detection of mutations in DNA that employs a thermostable DNA ligase with extreme hybridization stringency and ligation specificity such that only perfectly matching probes are ligated to each other. A frequently used DNA ligase catalyzes the NAD^+-dependent ligation of adjacent 3'-OH and 5'-phosphorylated termini in duplex DNA. Since the enzyme is derived from a thermophilic bacterium, it is highly heat-stable and more active at higher temperatures than conventional DNA ligases, so that stringency of hybridization can be extraordinarily rigid and specificity of ligation be absolute.

Lineage informative SNP: Any one of a set of tightly linked SNPs mostly residing on the mitochondrial genome or the Y chromosome that function as haplotype markers for the identification of missing individuals and therefore are also informative for kinship analyses. For example, a set of 59 SNPs organized in eight different multiplex panels targets 18 specific common Caucasian HVI/HVII hypervariable region types.

Linker mutagenesis (linker scanning mutagenesis): The introduction of mutations into a DNA molecule by the insertion of linkers. First, a circular DNA molecule is treated with DNase I under conditions that allow random cutting of the duplex. Such treatment leads to the generation of a set of linear molecules with different termini. Then linkers are ligated to these ends and cut with the restriction endonuclease whose recognition site is specified by the linker, which in turn generates single-stranded overhangs that are used to recircularize the molecules. This procedure then leads to the accumulation of DNA molecules with insertion mutations at random positions that can easily be localized by restriction mapping, since the specific restriction site of the linker is known.

Loss-of-expression mutation: Any mutation in a gene that silences the gene (i.e., leads to the disappearance of its transcript). A loss-of-expression mutation represents a loss-of-function mutation.

Loss-of-function mutation (lf): Any mutation that completely abolishes the function of the encoded protein.

Loss of heterozygosity (LOH; allele imbalance): The disappearance of one of two heterozygous loci in specific cell types (e.g., tumor cells). For example, a microsatellite marker closely linked to a putative colorectal tumor suppressor gene is represented as two equivalent, heterozygous loci – a microsatellite site of shorter and one of longer size, but both at the same concentration. In contrast, in colorectal cancer cells the shorter microsatellite allele is either reduced in concentration (i.e., is under-represented) or lost, probably a consequence of mutation(s) in the microsatellite flanking regions.

Loss of imprinting (LOI): The reversal of the methylation of cytosine residues at strategic sites in a gene (i.e., in exons and also introns) or its promoter, leading to the cessation of epigenetic silencing and the activation of transcription of the gene. For example, LOI in the gene encoding the insulin-like growth factor II (*IGF2*), an important tumor growth factor, leads to the activation of the normally silenced gene. Therefore, LOI of the *IGF2* gene is associated with a family history of colorectal cancer (CRC) and a personal history of colon adenomas and CRC. LOI is inherited or acquired early in life and LOI at the *IGF2* locus serves as biomarker for a distinct risk for CRC.

"Low cop" mutation: Any chromosomal mutation that leads to a decrease in the copy number of plasmids per cell. Not desired in recombinant DNA experiments. The "low cop" mutants can be counterselected by high antibiotic concentrations. Under certain conditions, however, low copy number plasmid vectors are favored.

Low copy number plasmid (single-copy plasmid; stringent plasmid): A plasmid that is present in one or only a few copies per bacterial cell (e.g., pSC 101). Derivatives of pSC 101 carrying three antibiotic resistance markers and unique restriction sites are favored vectors for the cloning of genes that disturb the cell's normal metabolism if present in high copy number (e.g., genes encoding surface membrane proteins). Low copy number plasmids are replicated under stringent control.

M

Melting temperature (T_m; melting point): The temperature at which 50% of existing DNA duplex molecules is dissociated into single strands. For measurement of T_m, a DNA solution is heated and its absorbance at 260 nm is continuously monitored. Transition from dsDNA to ssDNA occurs over a narrow temperature range and shows a characteristic increase in absorbance at 260 nm, so that a sigmoidal (S-shaped) curve results. T_m is defined as the temperature at the midpoint of the absorbance increase (i.e., the temperature at which 50% of the molecule(s) are dissociated). T_m calculation:

  a. Simplified calculation: $T_m = [2\,°C \times (\#A + \#T)] + [4\,°C \times (\#G + \#C)]$. For example, the melting temperature of the 10mer oligonucleotide ACG TAC GTA C is: $[2\,°C \times (3 + 2)] + [4\,°C \times (2 + 3)] = 30\,°C$
  b. Alternative calculation: $T_m = 81.5\,°C - 16.6 + [41 \times (\#G + \#C)]/\text{oligonucleotide length} - (500/\text{oligo length})$. For example, the melting temperature of the 10mer ACG TAC GTA C is: $81.5\,°C - 16.6 + [41 \times (5)]/10 - (500/10) = 35.4\,°C$.

Melting temperature-shift genotyping (T_m-**shift genotyping**): A single-tube technique for the detection of SNPs in genomic DNA that is based on the discrimination of SNP alleles by the different melting temperature profiles of their amplification products. In short, genomic DNA is first amplified in a conventional PCR with two allele-specific primers, of which either only one, or both, contain a GC-rich tail at the 5′ end. If only one allele-specific primer is tailed, then the tail comprises 26 bp. In case both allele-specific primers are tailed, then the 5′ end of one primer extends by 6 bases only, that of the other primer by 14 bases:

Allele-specific primer tail 1:	5′- GCGGGC-3′
Allele-specific primer tail 2:	5′- GCGGGCAGGGCGGC-3′

This difference of only 8 bp discriminates the melting profiles between the two allelic products, but only marginally influences the priming and amplification procedures. In addition, the primers differ by the 3′-terminal base that corresponds to one of the two allelic variants. Therefore, for each SNP two 15- to 22-base forward allele-specific primers (optimized T_m: 59–62 °C) with the 3′ base of each primer matching one of the SNP allele bases, and a common 22- to 27-base reverse primer (optimal T_m: 63–70 °C) are employed in PCR. The common primer typically binds no more than 20 bp downstream of the SNP, thereby producing relatively short PCR products with a good amplification efficiency. Amplification is catalyzed by the Stoffel fragment of DNA polymerase to enhance discrimination of 3′ primer/template mismatches. Samples homozygous for allele 1 are amplified with the short GC-tailed primer (6 bases) and produce one product with lower temperature peak in the melting profile. Samples homozygous for allele 2 will be amplified with the long GC-tailed primer (14 bases) and present only one higher temperature peak. Heterozygous samples are amplified with both GC-tailed primers and correspondingly the melting curves exhibit two temperature peaks. Depending on the SNP configuration in two (or more) genotypes, either one or the other, or both allele-specific primer(s) is (are) extended. Since the allele-specific primers differ by their GC-rich tails, the corresponding PCR products also differ by their distinct T_ms, that in turn depend on which of the two primers is used for amplification. Genotypes can finally be determined by inspection of melting curves on a real-time PCR instrument.

Methyl single nucleotide polymorphism (methylSNP): Any methylation-dependent DNA sequence variation between two (or more) individual genomes, in which a specific cytosine methylation status superimposes a SNP. MethylSNPs can be converted into common SNPs of the C/T type by sodium bisulfite treatment of the DNA, which then can be subjected to conventional SNP typing.

Microhaplotype: Any haplotype, that comprises only two SNPs within 10–20 bp of a particular allele in genomic DNA. Other alleles or alleles from another individual may possess a different microhaplotype. Therefore, microhaplotyping is one approach to genotype various organisms (e.g., patients suffering from the same disease).

Micro-Indel: A more general term for any Indel that comprises only 20 bp or less.

Micro-insertion: Any insertion that comprises 20 bp or less.

Misinsertion: The incorporation of bases into a growing polynucleotide chain (DNA or RNA) that have no complementary counterparts in the template strand. Such mismatched bases are normally excised by mismatch repair systems and replaced by the matching bases.

Mismatch repair (MMR; postreplication repair): The detection and replacement of incorrectly paired (mismatched) bases or small Indel mispairs in newly synthesized DNA. For example, in *Escherichia coli* a mismatch repair system consisting of 11 proteins, encoded by the genes *mutH*, *mutL*, *mutS*, *uvrD*, and *uvrE*, screens the newly synthesized DNA strand for mismatched bases. Proteins MutS, MutL, and MutH recognize mismatches and incise the newly synthesized unmethylated DNA strand ("initiation"). The mispaired bases and a short region surrounding them are excised by one of four exonucleases (ExoI, ExoVII, ExoX, or RecJ), that catalyze 5′ or 3′ excision from the DNA strand break in concert with UvrD helicase ("excision"). Finally, the DNA polymerase III holoenzyme catalyzes the repair process, which is completed by DNA ligase. This repair mechanism acts before the newly replicated DNA is methylated. Only after its completion is the *de novo* synthesized strand modified, for example, by dam methylase according to the methylation pattern of the complementary strand (maintenance methylation). In eukaryotes, the initial recognition of mismatches is accomplished by a complex of the two proteins MSH2 and MSH6 (MutSα), or to a limited extent by MSH2–MSH3 (MutSβ), which binds to mismatched bases. A second complex, consisting of proteins MLH1, PMS2, proliferating cell nuclear antigen, replication protein A, ExoI, HMGB1, replication factor

C, and DNA polymerase δ, then joins the mismatch-MSH2/6 complex, and catalyzes excision and repair of the mismatch. The MutLα complex promotes termination of the ExoI-catalyzed excision upon mismatch removal by dissociating ExoI from the DNA. DNA ligase I finally catalyzes the ligation step. Hereditary deficiencies in the MMR system result in gene mutations and subsequent susceptibility to specific types of cancers, including hereditary colorectal cancer. Loss of the MMR leads to the so-called mutator phenotype that exhibits increased mutation rates.

Mispriming: An undesirable artifact generated by the annealing of amplimers to nontarget sequences and the extension of these amplimers by *Taq* DNA polymerase in the PCR. The generation of such artifactual products can be circumvented by the hot-start technique.

Missense mutant: A mutant carrying one or more missense mutations.

Missense mutation: Any gene mutation in which one or more codon triplets are changed so that they direct the incorporation of amino acids into the encoded protein, which differ from the wild-type (e.g., UUU, encoding phenylalanine, mutates to UGU, encoding cysteine). The replacement of a wild-type amino acid by a missense amino acid in the mutant potentially produces an unstable or inactive protein.

Missense single nucleotide polymorphism (missense SNP): Any SNP that occurs in the coding region of a gene and changes the amino acid sequence of the encoded protein. Such missense SNPs, if responsible for a functional change of, for example, a protein domain, may cause diseases.

Mu (mutator): Any one of a class of transposable elements in the maize (*Zea mays*) genome that increases the frequency of mutation of various loci by more than an order of magnitude. *Mu* elements, present in the genome in 10–100 copies, comprise maximally 2 kb and are flanked by 200-bp inverted repeats with adjacent 9-bp direct repeats. Basically two size classes prevail, of which the shorter ones are derived from longer ones by internal deletions. *Mu* elements transpose by a replicative mechanism and can also occur in circular extrachromosomal state (e.g., *Mu*1 (1.4 kb) and *Mu*1.7 (1.7 kb)). Methylation of inserted *Mu* sequences prevents their transposition and stabilizes the mutation, whereas less than complete methylation leads to transpositional activity.

Mu array: A glass chip or a nylon membrane, onto which thousands of mutator transposon flanking regions are spotted at high density and which serves to identify specific genes with mutator insertions. The various mutator flanking regions are isolated from individual *Mu* active plants with the mutator amplified fragment length polymorphism technique, so that each spot on the array represents the *Mu* flanks of an individual plant. Hybridization of these arrays with, for example, cyanin-labeled or radiolabeled gene probes (e.g., cDNAs) identifies plants with *Mu* insertions in specific genes.

Multi-exon deletion: The deletion of more than one (usually two or three) exons from a multi-exon gene. For example, exons 1–4 of the breast cancer gene 2 (*BRCA2*) are frequently deleted in patients with breast cancer.

Multiple gene disruption: The insertion of DNA sequences into two or more genes within the same genome with the result of a knockdown or complete knockout of all the genes. The function(s) of all the disrupted genes in concert is then deduced from a changed phenotype. For example, the knockout of only one strategic gene of the parasite *Plasmodium berghei* (e.g., the "upregulated in infectious sporozoites gene 3" gene *UIS3*), whose encoded protein is necessary for the establishment of the parasite in the human body, is not sufficient for a long-term and efficient protection against malaria. In fact, sporozoites lacking *UIS3* do not fully develop the liver cycle, but enduring resistance of a host is only expected from the disruption of more genes, so that the parasite cannot replace them all in a short time period.

Multiple nucleotide polymorphism (MNP): Any polymorphism between two (or more) genomes that is based on more than one SNP. For example, many human diseases are probably caused by single base exchanges at strategic sites of several genes (e.g., coding for functional domains of different proteins) that are not present in the wild-type genomes and act in concert to cause a disease. These altogether are MNPs.

Multiplex ligation-dependent probe amplification (MLPA): A technique for the detection of mutations or, more precisely, exon duplications and deletions, deletions of whole genes, SNPs, or chromosomal aberrations (e.g., in tumor cell lines or samples). In short, MLPA starts with the hybridization of target-specific probes to denatured and fragmented genomic DNA (usually 20–100 ng). Each probe consists of two oligonucleotides A and B that bind to adjacent nucleotides of the target sequence via their 50- to 70-nucleotide long DNA-binding sequence (DBS) at the 3′ end. Oligonucleotide A additionally contains a flanking universal primer-binding sequence (PBS), whereas in oligonucleotide B DBS and PBS are separated by a stuffer fragment of variable length ("variable fragment"). If both oligonucleotides hybridize to the target DNA, they can be covalently ligated by a thermostable DNA ligase (e.g., the mismatch-sensitive, NAD^+-requiring ligase-65). The resulting, usually 130- to 480-bp strand can then be amplified by conventional PCR using one fluorescently labeled and another nonlabeled primer directed to the PBSs. Since all ligated probes share identical 5′ end sequences, they can be amplified with only one single primer. The difference in length of the different probes allows their separation and quantification in high-resolution capillary gel electrophoresis (or also 6.5% polyacrylamide gel electrophoresis). In case the target sequence is deleted, the ligation is prevented and the fragment cannot be amplified by the universal primer. Should the target DNA be absent in both homologous chromosomes, the corresponding fragment cannot be detected. If the target sequence is deleted in only one of the alleles, then the peak area of the eluting fragment is reduced to about 50% of the control. Up to 40 probes with different stuffer lengths (or sequences) and targeting at 40 different genes can simultaneously be run in a single reaction.

Multiplex quencher extension (multiplex-QEXT): A single-step technique for the simultaneous real-time detection and quantification of several different SNPs that is based on the direct measurement of fluorescence changes in a closed tube. In short, the target DNA (e.g., a gene) is first amplified by specific primers in a conventional PCR, and the amplified fragment treated with shrimp alkaline phosphatase and exonuclease I to inactivate the nucleotides and to degrade residual PCR primers. Then different probes detecting different SNPs in the amplified fragment are 5'-labeled with different reporter fluorochromes (e.g., one is labeled with 6-carboxyfluorescein, the second one with tetrachlorofluorescein, the third one with hexachlorofluorescein, the fourth one with Texas Red, the fifth one with cyanin 5, and so on). These probes are subsequently extended by a single TAMRA-labeled ddCTP if the respective SNP alleles are present. TAMRA may function as fluorescence acceptor (quencher-based detection) or donor (fluorescence resonance energy transfer (FRET)-based detection), depending on the 5'-fluorescent reporter. The extension generates increased reporter fluorescence, a result of FRET, if TAMRA serves as energy donor. If TAMRA functions as energy acceptor, then the reporter fluorescence is quenched.

Multisite mutation: Any mutation that either involves alteration of two or more contiguous nucleotides, or occurs repeatedly at many loci in a given genome.

Mutagen (mutagenic agent): Any physical or chemical agent that increases the frequency of mutations above the spontaneous background level. Such mutagenic agents include ionizing irradiation, UV irradiation, alkylating compounds, and base analogs.

Mutagenesis: The induction of mutations in DNA, either in the test tube (*in vitro* mutagenesis) or *in vivo*. For example, by irradiation (irradiation mutagenesis), chemicals (chemical mutagenesis), or by the deletion, inversion or insertion of DNA sequences (insertion mutagenesis).

Mutagenically separated polymerase chain reaction (MS-PCR): A technique for the detection of point mutations in a known DNA sequence that relies on conventional PCR. It allows us to amplify normal and mutant alleles of a gene simultaneously in the same reaction, using allele-specific primers of different lengths. Additionally, the allele-specific primers differ from each other at several nucleotide positions, and therefore introduce new and discriminating mutations into the allelic PCR products (thereby reducing cross-reactions between amplification products during the PCR process). Since both products possess different lengths, MS-PCR "separates" both amplified alleles that can then be identified by agarose gel electrophoresis and ethidium bromide staining.

Mutant: An organism harboring a mutant gene whose expression changes the phenotype of the organism.

Mutant allele-specific amplification (MASA): Any one of a series of PCR-based techniques, allowing the specific amplification of an allele that has undergone a mutation (e.g., a deletion, insertion, inversion, transition, transversion). MASA techniques are presently employed in clinical screening and diagnosis.

Mutated promoter: Any promoter sequence into which a mutation(s) is (are) introduced either naturally or artificially. Such mutations may not at all affect the binding of transcription factors to their cognate sequence motifs, but can also lead to either a stronger affinity of the transcription factor to its binding sequence, or the partial or total loss of binding of the transcription factor.

Mutation: Any structural or compositional change in the DNA of an organism that is not caused by normal segregation or genetic recombination processes. Such mutations may occur spontaneously, or may be induced by mutagens such as ionizing radiation or alkylating chemicals. The change of a nucleotide base, for example, may cause the conversion of one codon into another one. It is silent if the codon change does not cause any detectable phenotypic change (e.g., if both codons stand for the same amino acid).

Mutation analysis: The detection and characterization of a mutation in DNA (e.g., deletion, insertion, inversion, mismatch mutation, point mutation, or translocation). A multitude of techniques for mutation analysis are available.

Mutation breeding: The development of plants with improved characteristics (e.g., resistance against pathogens or environmental stress, increased agricultural productivity) through physically or chemically induced mutations. Since such mutations are totally at random, no directed genetic change is possible. This method is still used, but will be replaced by directed genetic engineering in future.

Mutation cluster region (MCR): Any region of a gene or genome where various types of mutations are present at a higher frequency than in the rest of the genome. MCRs represent extended hotspots of mutations.

Mutation delay: The time lag between a mutation event and its phenotypic expression. For example, recessive mutations may only be apparent if they become homozygous.

Mutation detection electrophoresis (MDE) gel: A gel made of modified polyacrylamide with slightly hydrophobic properties that selectively alters the electrophoretic mobility of heteroduplexes such that even single mismatched bases in 1 kb of duplex DNA can be visualized by a mobility shift.

Mutation rate (μ): The number of mutations occurring per unit DNA (e.g., kilobase or gene) per unit time.

Mutator gene (mutator): Any gene (*mut* gene) that increases the rate of spontaneous mutations of one or more other genes. Such mutators may themselves originate from normal genes by mutation. If, for example, a gene is mutated whose product normally functions in DNA repair or replication, the mutant protein encoded by the mutated gene may introduce multiple errors (i.e., mutations) during these processes. High mutator gene activity probably increases evolutionary adaptation in bacteria.

Mutator phage: Any phage that is able to increase the rate of mutation in its host cell (e.g., the *Mu* phage).

Mutator polymerase: A mutated, nucleus-encoded mitochondrial genomic DNA polymerase that gives rise to the accumulation of frameshifts, point mutations, and deletions in the mitochondrial genome. One of the mutations converting the wild-type DNA polymerase to the mutator polymerase is a point mutation that changes a highly conserved tyrosine at position 955 (part of the binding pocket responsible for selection of deoxyribonucleotides against ribonucleotides) to cysteine (Y955C). This simple base exchange neither changes the catalytic rate nor the intrinsic $3' \rightarrow 5'$ exonuclease proofreading activity, but decreases the fidelity of DNA replication, which in turn leads to the mitochondrial DNA mutations. These mutations cause a series of diseases. For example, progressive external ophthalmoplegia (PEO) is the consequence of several-kilobase-long deletions primarily between short, direct repeats of 10–13 b. These deletions are associated with point mutations caused by T·dTMP mispairing, which occurs 100 times more frequent with mutator as compared to wild-type genomic polymerase. The disease appears in patients at the age of 30–40 and causes a weakness of muscles in general, the eye muscles in particular. As a consequence, the muscles moving the eye (especially the lateral rectus) deteriorate gradually, so that the patients can only follow a moving object by turning their heads. The muscle weakness is a result of impaired electron transport chain activity (depletion of ATP). Another cause of PEO is a mutant mitochondrial helicase encoded by gene *twinkle*.

MutS: Any one of a family of *Escherichia coli* methyl-directed mismatch repair enzymes that recognize and bind to mismatched bases in target DNA. MutS is part of a system for the correction of replication errors.

MutS mismatch detection: A technique for the detection of single base mismatches in a target DNA that exploits the affinity of MutS to recognize and bind mismatched bases. In short, the target DNA is first amplified using appropriate, radioactively end-labeled primers and conventional PCR techniques. The PCR products are then heat-denatured and reannealed, which results in four different DNA duplexes (in case the target DNA is heterozygous at locus A (A/a): two homoduplexes (*AA* and *aa*) and two heteroduplexes (*Aa* and *aA*)). If the heteroduplexes contain, for example, a single base-pair mismatch, the added MutS protein will bind to this mismatch and the mutant allele can be detected by mobility-shift DNA-binding assays in polyacrylamide gels with subsequent autoradiography.

N

Neutral insertion: The insertion of a nucleotide or oligonucleotide into a coding sequence of a gene without changing the function of the encoded protein.

Neutral mutation: Any mutation that has no selective advantage or disadvantage for the organism in which it occurs, for example a mutation in a cryptic gene or other noncoding DNA.

Neutral substitution: An exchange of one (or more) amino acid(s) in a protein without any change of its function.

Next-generation sequencing (NGS; next-generation sequencing technology): A generic term for novel DNA and RNA sequencing technologies with the potential to sequence a human genome for $100 000, or even only $1000, that are not based on the conventional Sanger (dideoxy) sequencing procedure. Next-generation sequencing relies on extremely high-throughput procedures, mostly based on massively parallel reactions, as, for example, in sequencing by oligonucleotide ligation and detection (SOLiD™), where each run produces at least 40 million reads, covering 1 billion bases. NGS technologies fall into two broad categories – clonal cluster sequencing and single molecule sequencing.

Nonallelic homologous recombination (NAHR): Any crossover with subsequent genetic recombination mediated by DNA base mispairing and resulting in duplication and/or deletion of DNA sequences.

Noncoding single nucleotide polymorphism (ncSNP): A misleading term for any SNP that occurs in a noncoding region of the genome (e.g., an intron). ncSNPs are the most frequent types of SNPs in eukaryotic organisms.

Nonfunctional polymorphism: Any sequence polymorphism that has no consequences for the function of a protein and is therefore selectively neutral.

Nonhomologous end-joining (NHEJ): A mechanism ("pathway") for the repair of double-strand breaks (DSBs) in DNA, that rejoins the two ends of this break. NHEJ frequently leads to error-prone repair of DSBs, because the ends are only incompletely processed. NHEJ is catalyzed by the concerted action of ligase IV, Xrcc4, Ku70 and 80, DNA-PKcs, Artemis, and Nej1/Lif2 in rodent cells.

Nonhomologous random recombination (NRR): The random recombination of DNA fragments in a length-controlled manner without the need for sequence homology. For example, NRR is used to evolve DNA aptamers that bind streptavidin. Aptamer development starts with two parental sequences of modest affinity towards streptavidin, and repeated cycles of NRR evolve aptamers with 15- to 20-fold higher affinity. Therefore, NRR enhances the effectiveness of nucleic acid evolution.

Nonhomologous recombination: See Illegitimate recombination.

Nonhomologous synapsis: The indiscriminate association of nonhomologous chromosomes during meiosis. Normally, only homologous chromosomes pair with each other, assisted by proteins of the so-called synaptonemal complex. However, in certain mutants, the homolog pairing is not functioning. For example, in maize the poor homologous synapsis (*phs*) 1 gene encodes a protein that coordinates chromosome pairing, recombination, and synapsis. A simple mutation in the gene results in the synthesis of a mutated PHS1 protein that fails to form chiasmata. Nonhomologous synapsis leads to a random segregation of chromosomes.

Nonproductive base-pairing: The imperfect pairing of bases in DNA that are not complementary to each other and therefore cannot form hydrogen bonds. For example, AA, AG, AC, GA, GG, GT, CC, CA, CT, TG, TC, and TT are such nonproductive base pairs, which, for example, destabilize hybrids and reduce the melting temperature of a hybrid through reducing the force of interaction between the two strands.

Nonsense mutation: Any mutation in a coding sequence that converts a sense codon into a nonsense codon (a stop codon) or a stop codon into a sense codon. As a consequence, the encoded protein will either be truncated (premature termination) or too long, which in turn hampers or abolishes protein function.

Nonsense suppression: A secondary mutation occurring at a chromosomal site separate from the site of a nonsense mutation and correcting the phenotype associated with the latter.

Nonsense suppressor: A tRNA that is mutated in its anticodon and recognizes a nonsense (stop) codon so that the synthesis of a specific polypeptide can be extended beyond the stop codon. As a consequence, the nonsense codon is ignored (suppressed).

Nonsynonymous sequence change: Any alteration in the nucleotide sequence of a coding region, that changes the amino acid sequence (and possibly the function) of the encoded protein.

Nonsynonymous single nucleotide polymorphism (nsSNP): Any SNP that occurs in a coding region of a eukaryotic gene and changes the encoded amino acid. nsSNPs may cause the synthesis of a nonfunctional protein, and therefore be involved in diseases.

Nonsynonymous/synonymous mutation rate ratio: The ratio of nonsynonymous versus synonymous mutations in a genome over evolutionary times, expressed as d_N/d_S (or ω). Negative selection is characterized by $d_N/d_S < 1$, no selection (H_0) by $d_N/d_S = 1$ and positive selection (H_A) by $d_N/d_S > 1$.

Nucleotide excision repair (NER): A prokaryotic DNA repair system, encoded by genes *uvr*A (encoding an ATPase subunit of endonucleases), *uvr*B and *uvr*C (encoding the endonuclease subunits of *Escherichia coli* excinuclease), and *uvr*D (coding for a helicase removing the excised stretch of DNA), that repairs from few to more than several thousands of nucleotides. It is particularly active in the removal of UV photoproducts, alkylated adducts, and oxidized DNA. First, the ABC excinuclease recognizes damaged sites ("damage recognition"), cuts at two flanking sites, and removes the intervening sequences ("dual incision excision"). Then, DNA polymerase I catalyzes repair synthesis, gaps are filled by any of the four DNA polymerases, and the ends ligated. Eukaryotic NER protein machines more or less process DNA damage sites the same way. The initial step is damage recognition by XPC and (in humans) hHR23B (homolog of *Saccharomyces cerevisiae* Rad23), that concertedly recruit other repair proteins. XPB and XPD helicases mediate strand separation at the lesion site, and XPA identifies the damaged area in an open DNA conformation. The unwound DNA is stabilized by RPA, that also positions XPG and ERCCI-XPF endonucleases. These nucleases catalyze the incision around the lesion. Once the lesion is removed, the gap is filled by replication proteins and the repair process is complete. NER systems are also active in, for example, mammalian organisms (more than 30 different proteins), and their failure causes rare autosomal recessive disorders such as xeroderma pigmentosum.

Nucleotide replacement site: Any position in a codon where a point mutation has occurred.

Nucleotide substitution: The exchange of one nucleotide in a DNA molecule for another one. Such substitutions are neutral if the genetic code is not changed, but have massive consequences if the genetic code is altered (e.g., result in the synthesis of a nonfunctional protein).

Null mutation: Any mutation that leads to a complete loss of function of the sequence in which it occurs.

O

Oligo-mismatch mutagenesis (oligonucleotide-primed mutagenesis; oligonucleotide-directed mutagenesis; oligonucleotide-directed double-strand break repair): The introduction of site-specific mutations into a target DNA molecule by annealing a specifically designed synthetic oligodeoxynucleotide (7–20 nucleotides long). The oligo is complementary to the region to be mutated except for one or more "wrong" bases that lead to specific mismatches. After hybridization of the oligonucleotide to the denatured target DNA (usually inserted in a cloning vector), the Klenow fragment of DNA polymerase I is used to synthesize a complementary strand, where the double-stranded region serves as primer. Finally, DNA ligase seals the nick. After introduction of this double-stranded molecule into an *Escherichia coli* host, DNA mismatch repair processes will lead to the occurrence of a mixed population of molecules that consists of 50% wild-type and 50% site-specifically mutagenized clones (because the repair system uses both the original as well as the mutated strand as template). Oligo-mismatch mutagenesis is a way of site-specific mutagenesis.

P

Paracentric inversion: Any segment of DNA that is reversed in orientation relative to the rest of the chromosome, but does not involve the centromere.

Paramutation: Any heritable change in the activity of one allele induced by the corresponding allele on the homologous chromosome. Such changes are brought about by modifications of chromatin structure or cytosine methylation (i.e., are epigenetic) and do not result in, or depend, on changes of the underlying DNA sequence.

PCR–ligation–PCR mutagenesis (PLP mutagenesis): A technique for the generation of fused genes, site-directed mutagenesis, or introduction of specific deletions, insertions, or point mutations into target DNA. For example, the fusion of two (or more) genes starts with the amplification of each gene in a separate PCR. The amplification products are then phosphorylated using T4 polynucleotide kinase and ligated with T4 DNA ligase, creating different combinations of joined fragments. The fused gene is then specifically PCR-amplified out of this heterogeneous mixture with a primer directed to the 5′ end of the upstream gene and a primer complementary to the 3′ end of the downstream gene. The resulting amplification product is then subcloned into the original target sequence, creating a type of insertion mutation. PLP mutagenesis relies on a DNA polymerase with exonuclease (i.e., proofreading) activity, so that the blunt-ended fragments match exactly with the primer sequence.

Perfect match (PM): The complete correspondence of two (or more) bases in two (or more) strands of a DNA molecule. Perfect matches are only possible by Watson–Crick base pairing of AT and GC pairs, respectively. Any other combination inevitably leads to a mismatch.

Permutation: Any permanent mutation in a gene without phenotypic consequences. Permutations predispose the carrier for further mutation(s).

Plasmid-enhanced PCR-mediated mutagenesis (PEP mutagenesis): A variant of the splice overlap extension PCR that allows us to introduce mutations (e.g., deletions or insertions) into a target DNA. In short, the target DNA is first amplified as two parts using two primer pairs, designed to introduce the mutation and two restriction sites (which are incorporated into the most distal primers). The internal 5′-phosphorylated primers permit an efficient blunt-end ligation of the two parts. For a deletion mutation, the targeted sequence is simply omitted; for an insertion, it is incorporated into one of the primers. The two parts together with the cloning plasmid are then digested with the two restriction endonucleases, ligated, and used to transform bacterial host cells. The efficiency and orientation of the blunt-end ligation process is controlled by sequence-specific overlapping interactions with the plasmid.

Plastome mutation: Any mutation in plastid DNA (chloroplast DNA).

Point mutation (microlesion; micromutation): A mutation involving a chemical change in only a single nucleotide.

Polar mutation (dual-effect mutation): Any gene mutation of the nonsense or frameshift type in an operon producing two effects – the repression of nonmutated genes located farther downstream and the failure to express the gene subsequent to the mutation site.

Polymerase chain reaction mutagenesis (PCR mutagenesis): A variant of conventional oligonucleotide-directed mutagenesis that allows us to introduce insertions, deletions, or point mutations in a target DNA with its concomitant amplification using PCR techniques.

Premutation: Any mutation in a gene that does not lead to phenotypic consequences, but a predisposition for a disease in the next generation. For example, a normal transmitting male carries a premutation in the *FMR1* gene on the distal long arm of the X chromosome. This premutation consists of an increased number of CGG repeats in the 5'-untranslated region of the *FMR1* gene (repeat numbers in normal individuals: 5–44; premutation: 55–200). CGG alleles with intermediate numbers of repeats are considered intermediate alleles (also called "gray zone" alleles). A further expansion of the CGG repeat leads to an inhibition of the transport of the 40S ribosomal subunit from the nucleus and therefore to the suppression of translation. Repeat numbers beyond 200, accompanied by aberrant methylation of cytidyl residues (full mutation), generally cause clinical symptoms of the full-blown fragile X syndrome in males, whereas females with the full mutation are less affected.

Primer-specific and mispair extension analysis (PSMEA): A technique for the detection of single nucleotide variations (e.g., deletion, insertion, transition, transversion) between two DNA templates. The method exploits the highly efficient $3' \rightarrow 5'$ exonuclease proofreading activity of *Pyrococcus furiosus* DNA polymerase that prevents the extension of a primer when an incomplete set of deoxynucleotide triphosphates is present and a mismatch occurs at the initiation site of DNA synthesis (i.e., the 3' end of the primer). For example, in the presence of only dCTP and dGTP, primer 3'-CTCTG...5' can easily be extended on template A (5'-**G**AGAC....3'), because the crucial nucleotide (in bold face) matches. The same primer cannot be extended on template B (5'-**A**AGAC...3'), so that genome A can be discriminated from genome B (presence/absence of an extension product). In contrast, the use of dTTP and dGTP allowed the extension of the primer on template B, not on template A. PSMEA therefore allows genotyping of organisms that differ in only one (or few) nucleotide pairs at the 5' end of the primer-binding site.

Programmed DNA deletion: The programmed destruction of both single-copy and moderately repetitive DNA sequences ("deletion elements") from several hundred base pairs to more than 20 kb in size and specific for the micronucleus in *Tetrahymena thermophila*. In short, this ciliated protozoon contains one germinal nucleus (micronucleus) and one somatic nucleus (macronucleus) per cell. During sexual conjugation, the micronucleus goes through a series of events to produce a zygotic nucleus that divides and differentiates into the new macro- and micronucleus of the progeny cell. The old macronucleus is destroyed. Formation of the new macronucleus involves extensive genome-wide DNA rearrangements. Thousands of specific DNA segments (about 15% of the genome) are deleted, and the remaining DNA is fragmented and endoduplicated about 23-fold to form the somatic genome responsible for all transcriptional activities during growth. The programmed DNA deletion is triggered and guided by dsRNA transcribed from germline sequences during conjugation.

Promoter mutation: Any mutation that occurs within the promoter sequence of a gene. For example, in so-called *aphakia* (*ak*) mouse mutants, which do not form any lens or pupil in their otherwise normal embryonic development, the underlying gene *Pitx3* on chromosome 19 is absolutely identical to the wild-type gene. However, two deletions of 652 and 1423 bp, respectively, in the promoter and in the transcription initiation region lead to an almost complete silencing of the *Pitx3* gene. As a consequence, the encoded homeobox transcription factor is not functional and the eye development does not occur.

Promoter polymorphism: An imprecise term for any sequence polymorphism that occurs in a promoter of a gene. Usually such polymorphisms are caused by small deletions, insertions, or, most frequently, SNPs. Promoter polymorphisms may be neutral (i.e., without effect on the transcription of the adjacent gene) or inhibit the binding of specific proteins necessary for accurate transcription (e.g., transcription factors) and account for differences in the expression of a distinct gene between two (or more) individuals.

Promoter single nucleotide polymorphism (promoter SNP (pSNP)): Any SNP that occurs in the promoter sequence of a gene. If a pSNP prevents the binding of a transcription factor to its recognition sequence in the promoter, the promoter becomes partly dysfunctional.

Promoter-up mutant (up-promoter mutant; up mutant): Any mutant with a mutation in the promoter of one of its genes that leads to a higher rate of expression of this gene. In gene technology such up-promoters are used for the overexpression of genes encoding useful proteins.

R

Random elongation mutagenesis: A technique for random mutagenesis of protein-encoding DNA sequences that capitalizes on the ligation of partially randomized oligonucleotides ahead of the stop codon of a target sequence such that short chains of about 16 amino acids are introduced into the carboxy ends of the encoded proteins. This leads to the generation of mutant populations, where each mutant carries a different peptide tail, which expands the protein sequence space. This expansion may produce mutants with favorable properties. For example, random elongation mutagenesis generated catalase mutants with increased thermostability of the enzyme. The technique therefore allows *in vitro* evolution of proteins with new properties (evolutionary molecular engineering).

Random mutagenesis: Any nondirected (random) introduction of mutations into a DNA molecule (e.g., a plasmid, PCR fragment, chromosome). Usually, a target DNA is randomly mutagenized by error-prone PCR with, for example, *Taq* DNA polymerase, that introduces a distinct spectrum of mutations into the template. For example, *Taq* polymerase induces the following transitions or transversions (in decreasing efficiency): A → T, T → A, A → G, T → C, G → A, C → T, A → C, T → G, G →

T, C → A, G → C, and C → G. Such mutations can be generated at a low, medium, or high frequency depending on the input template concentration. The mutagenized DNA is then introduced into an *Escherichia coli* strain deficient in several of the primary DNA repair pathways such that the mutation spectrum in the target DNA is maintained.

Reduction-of-function mutation: Any mutation within a gene that does not abolish the function of the encoded protein, but only reduces its catalytic or regulatory properties.

Reference single nucleotide polymorphism (refSNP; rsSNP; rsID; SNP ID): Any SNP at a specific site of a genome (or part of a genome, e.g., a bacterial artificial chromosome clone) that serves as reference point for the definition of other SNPs in its neighborhood. The rsID number ("tag") is assigned to each refSNP at the time of its submission to the databanks.

Regulatory single nucleotide polymorphism (regulatory SNP (rSNP)): A relatively rare SNP that affects the expression of a gene (or several genes). Usually this SNP is located in the promoter of the gene. For example, an A/G-SNP ("A allele") in a promoter may allow the binding of a cognate protein (e.g., a transcription factor), whereas a C/T-SNP ("C allele") may reduce the binding affinity of that protein such that no transcription of the adjacent gene is possible. Other regulatory SNPs are located in enhancers or silencers.

Repair nuclease: Any one of a series of nucleases that functions in DNA repair, either by recognizing and removing an incorrect base or an otherwise damaged site. Repair nucleases may act as an endonuclease or as an exonuclease, removing nucleotides on one strand from the end of the duplex.

Repairosome: The protein complex catalyzing DNA repair.

Repeat-associated polymorphism (RAP): Any nucleotide polymorphism that is generated by an elevated mutation frequency around repeated sequences (e.g., variable number of tandem repeats). For example, in the so-called control region of mitochondrial DNA (with the origin of replication and transcription) a series of repeated sequences are located in tandem, which undergo expansion or contraction as a result of insertion or deletion of repeat units (by e.g., slipped strand mispairing during mitochondrial DNA replication). These processes actually lead to the generation of mitochondrial DNA length variants. Now, the nucleotides surrounding the original site of the repeat are also mutated. Since genome expansion and contraction events occur independently in germ cells of different individuals, unique RAPs are created that serve to identify individuals, demes, and whole populations.

Repeat expansion detection (RED): A method to detect trinucleotide repeats and their multimers in genomic DNA, which is heat-denatured, and oligonucleotide(s) complementary to either strand of the repeat target is (are) hybridized at a temperature close to the melting point. Then thermostable DNA ligase is added and will covalently join only those adjacent oligonucleotides that are not separated by gaps.

Then the ligation products are separated from their template sequences by denaturation. As a consequence, a mixed population of ssDNA molecules is generated. The ligation step is then repeated from 180 to 400 times. The single-stranded multimers are size-separated in denaturing polyacrylamide gels, subsequently blotted onto hybridization membranes, and hybridized to radiolabeled complementary oligonucleotides. The RED technique can also be modified to include several different oligonucleotide repeats simultaneously (multiplex RED). In order to differentiate between the (RED) products formed by each type of repeat, oligonucleotides of different length have to be used (e.g., $[CCG]_{11}$, $[TAG]_{12}$, $[CTG]_{13}$, $[CAT]_{14}$, etc.). This technique can be used, for example, to detect the expansion of trinucleotide repeats in human diseases (e.g., fragile X, myotonic dystrophy, Huntington's disease, spinobulbal muscular atrophy, etc.).

Repeat-induced point mutation (RIP; originally: rearrangement induced premeiotically): A mechanism to inactivate repetitive sequences, to reduce their copy numbers, or both, by introducing transition mutation(s), predominantly affecting GC pairs (and some GC pairs more than others), that are replaced by AT pairs in *Neurospora crassa* and some other fungi. RIP involves enzymatic deamination of cytosines or 5-methylcytosines to uracil or thymine, respectively, and occurs after fertilization, specifically in the subsequent mitoses prior to nuclear fusion.

Restriction fragment length polymorphism (RFLP; DNA polymorphism): The variation(s) in the length of DNA fragments produced by a specific restriction endonuclease from genomic DNAs of two or more individuals of a species. RFLPs are generated by rearrangements or other mutations that either create or delete recognition sites for the specific endonuclease. They may also be due to the presence of repetitive DNA in different copy numbers on a specific chromosomal region (for example variable number of tandem repeats). RFLPs are now widely used to localize specific genes in complex genomes (linkage analysis), to discriminate between closely related individuals (fingerprint tailoring) and to establish genome maps.

Retrotransposon-based insertion polymorphism (RBIP): Any sequence difference between two (or more) genomes that is based on either the presence or absence of a retrotransposon at a specific chromosomal locus, which can be detected by its site-specific amplification in a conventional PCR. This PCR employs genomic DNA as template and three different primers (triplex PCR), two targeting at the genomic DNA flanking a retrotransposon at its 5′ and 3′ side, respectively, and a third primer complementary to sequences within the retrotransposon. The flanking primers define a specific genomic locus, the internal primer in combination with one of the flanking primers allows us to amplify part of the retrotransposon sequence. The polymorphism is based on the presence (amplification product detectable) or absence (no amplification product) of the retrotransposon. Therefore, RBIP markers are codominant (i.e., different allelic states at a locus can be revealed). Since this type of polymorphism is based on either presence or absence of a PCR product, there is no need for gel-electrophoretic separation of bands. Instead, a simple dot-blot hybridization assay is sufficient and amenable to high-throughput automation. Since retrotransposon insertions are frequent events in eukaryotic genomes, many different loci can be targeted each with locus-specific primers and therefore many different codominant markers be generated.

Reverse mutation (back mutation; reversion): The restoration of the original nucleotide sequence of a gene previously mutated by a "forward" mutation.

Robertsonian translocation: A special type of translocation that involves breaks in the heterochromatic regions of a chromosome close to the centromere in the short and long arm, respectively, of two acrocentric chromosomes and the reciprocal translocation of the intermediates. The products of this centric fusion during a Robertsonian translocation are a long-arm metacentric chromosome and a very small metacentric chromosome that may even be lost without any genetic damage to the carrier.

S

Sequence-based amplified polymorphism detection (SBAP detection): A technique to discover sequence polymorphisms between two (or more) genomes, generally DNA sequences, that combines the multiplexing capacity of amplified fragment length polymorphism and the simplicity of random amplified polymorphic DNA methods. In short, genomic DNA is first amplified with 17-nucleotide primers, each consisting of a fixed GC-rich sequence of 14 nucleotides at its 5′ terminus and three selective bases at its 3′ terminus, using conventional PCR techniques. The second amplification is primed by 19-nucleotide primers, each consisting of 16 nucleotides at its 5′ terminus complementary to an AT-rich template sequence and three selective bases at its 3′ end. One of the primers is end-labeled using polynucleotide kinase and [γ-^{32}P]ATP. After PCR, the amplification products are separated in denaturing polyacrylamide gels, and the bands detected by autoradiography.

Sequence saturation mutagenesis (SeSaM): A random mutagenesis technique that allows us to exchange any nucleotide in a DNA sequence for another one, without any interference by the DNA polymerase used. In short, the technique consists of four basic steps. In a first step, a pool of DNA fragments with a random length distribution is generated. To that end, the target gene (or target DNA sequence) is amplified in a conventional PCR with dNTPs, phosphorothioate nucleotides, and primers complementary to the upper strand of the dsDNA. The phosphorothioate nucleotides are statistically incorporated and linked with the neighboring nucleotides via phosphorothioester bonds, which are then cleaved by alkaline ethanolic iodine. Cleavage leads to the generation of a series of fragments with free 3′-OH groups. In a second step, the various fragments are tailed at their 3′ termini with universal bases (e.g., deoxyinosine (dI), N^6-methoxy-2,6-diaminopurine [K] and 6-(2-deoxy-β-D-ribofuranosyl)-3,4-dihydroxy-8H-pyrimido-[4,5]-C [1,2] oxazine-7-one [P]), using terminal transferase. In a third step, the elongated DNA fragments are extended to the full-length genes by PCR, using a single-stranded template, a reverse primer, and *Taq* DNA polymerase. The longer complementary strand is selectively separated by gel electrophoresis. In the last step, the universal bases are replaced with standard nucleotides. First, a reverse primer is annealed to the 3′ end of the full-length gene strand and PCR-extended to produce a complementary strand. During the extension process, standard nucleotides are incorporated at positions where universal nucleotides are present on the template strand. Finally, the mutated genes are amplified by PCR to produce a library.

Sequence-specific amplification polymorphism (S-SAP): A technique that combines amplified fragment length polymorphism with sequence-specific PCR to identify DNA polymorphisms between related organisms. As a sequence-specific primer an oligonucleotide complementary to the conserved 5′ terminus of long terminal repeats (LTRs) of retrotransposons (e.g., *Ty1 copia* and others) is used. In short, genomic DNA is first restricted with a rarely and a frequently cutting restriction endonuclease (e.g., *Eco*RI and *Mse*I, respectively), then *Eco*RI and *Mse*I adaptors are ligated to the fragments, which are then preamplified with adaptor-homologous primers in conventional PCR. The preamplified products are then selectively amplified with a ^{33}P- or ^{32}P-labeled oligonucleotide complementary to the end of the LTR of a suitable retrotransposon and either a rare or frequent site adapter-homologous oligonucleotide. The resulting fragments are denatured and separated on sequencing gels. The detected polymorphisms may result from sequence variation at flanking or internal restriction sites of the genomic DNA or from variation(s) at the 5′ terminus of the LTR of the targeted retrotransposon, or also from Indel events within the retrotransposable element. In, for example, barley, S-SAP reveals more polymorphisms than amplified fragment length polymorphism alone.

Sequencing error: Any base introduced into a newly synthesized DNA strand that does not match the complementary base on the corresponding template strand. Sequencing errors are a source for misinterpretations of sequence data (e.g., can suggest the presence of a SNP at a particular locus that does not exist in reality) and can only be corrected by resequencing.

Shooter mutant: A mutant of *Agrobacterium tumefaciens* in which the auxin genes have been partially or completely deleted. Therefore, shooter mutants induce an enhanced cytokinin level in transformed plant tissue. In some plant species, transformation with a shooter mutant leads to a "shooty" phenotype (i.e., the appearance of many shoots) so that these mutants are preferably used to induce regeneration of species with a low regeneration capacity.

Short patch repair: The excision of about 20 nucleotides around and including a site of DNA damage (e.g., a missing base or a thymine dimer) and the repair of the resulting gap by DNA polymerase that uses the undamaged strand as template. The genes involved in this type of repair (*uvrA*, *uvrB*, *uvrC*, and *uvrD*) are constitutively expressed in *Escherichia coli*, but their expression is enhanced by the induction of the SOS repair system.

Signature-tagged mutagenesis (STM; signature-tagged transposon mutagenesis): A transposon mutagenesis technique in which transposons carrying unique sequence tags are used for the isolation of bacterial virulence genes. In short, each transposon is first tagged with a unique DNA sequence tag, composed of a variable 40-bp central region flanked by arms of 10- to 20-bp constant sequences (common to all tags). The central region can be varied to yield a huge diversity of transposon tags. These double-stranded tags are then ligated into a vector, transformed into *Escherichia coli* (producing a pool of *E. coli* mutants, each of which carries another tag), and transferred into the bacterium under investigation (the pathogen) by conjugation. Transposition leads to the stable and random single-copy integration of the transposon together with its specific tag into the target bacterial genome. Then transposon-tagged mutant exconjugants are grown in 96-well microtiter plates. These arrays are replica-plated onto two membranes. The 96 mutants of each microtiter plate are then pooled (input pool), an aliquot retained for DNA extraction, and the rest injected into an animal (usually mouse) susceptible to the disease caused by the test bacterium. In the case of mice, the spleens are removed after 3 days (symptoms of disease visible) and mutants that multiplied within this organ recovered by plating of spleen homogenates onto growth medium. Some 10 000 colonies are combined (recovered pool) and DNA is isolated. Then both the tags of the original pool and the recovered pool are amplified with primers complementary to the invariable arm regions of the tag, and PCR radiolabeled. After release of the arms by *Hin*dIII digestion, the tags are used to probe the replica colony blots from the microtiter plate. Colonies hybridizing to the probe from the injected pool but not to the probe from the recovered pool are mutants with attenuated virulence (carrying mutated virulence genes).

Silent mutation (same sense mutation; silent site mutation): Any gene mutation that is of no consequence for the phenotype of the organism (e.g., a point mutation in the third position of a codon that does not change the amino acid specificity of the mutated codon such as a change of UAU to UAC, which both code for tyrosine).

Silent single nucleotide polymorphism (silent SNP): Any SNP that occurs in the coding region of a gene, but does not change the amino acid sequence of the encoded protein. A silent SNP may nevertheless change the folding of the corresponding protein, thereby changing its conformation. For example, a synonymous SNP in the human multidrug resistance 1 gene leads to a change in the structure of the substrate interaction site of the encoded P-glycoprotein and reduced function of the protein.

Silent site: A nucleotide sequence in a genome where silent mutations occur (i.e., nucleotide substitutions that leave the encoded amino acid unchanged).

Simple-sequence length polymorphism (SSLP): The variation(s) in the length of DNA fragments containing simple repetitive sequences in genomic DNA from two or more individuals of a species. SSLPs are used to discriminate between closely related individuals and to detect relationships in population genetics.

Simple-sequence length polymorphism DNA fingerprinting (SSLP fingerprinting): A variant of the DNA amplification fingerprinting technique that aims at detecting amplified fragment length polymorphisms based on the presence of a variable number of simple repetitive sequences at specific loci (simple-sequence length polymorphism). The synthetic oligodeoxynucleotide primers (amplimers) used in this procedure are complementary to genomic DNA flanking simple repetitive sequences. DNA from two or more individuals of a species is compared.

Single base-pair replacement: A technique for the substitution of a correct base pair in a target DNA for an incorrect one. In short, the double-stranded target DNA region is first nicked and digested by exonuclease such that a single-stranded protrusion (i.e., single-stranded section) is generated. Then this single-stranded overhang is repaired with DNA polymerase in the presence of only three bases (e.g., dATP, dGTP, and dCTP). This forces the polymerase to insert a mismatching base whenever a thymidine occurs in the template strand. In the next replication round this mismatched base is complemented, resulting in a replacement of the original base pair for a new one.

Single-hit single nucleotide polymorphism (single-hit SNP): Any SNP for which each allele is present in only one sample from a distinct population.

Single nucleotide amplified polymorphism (SNAP): A technique for the reliable detection of SNPs in specific alleles. Normally, allele-specific primers are designed such that the 3′ nucleotide of a primer corresponds to the site of the target SNP. Therefore, the allele-specific primer perfectly matches with one allele (the specific allele), but carries a 3′ mismatch with the nonspecific allele. Now mismatched 3′ termini are extended by DNA polymerases with much lower efficiency than perfectly matched termini. Therefore, the allele-specific primer preferentially amplifies the specific allele. However, a single base-pair change at the 3′ end of the nonspecific allele is not sufficient to reliably discriminate between the two alleles. Therefore, in the SNAP procedure a second mismatch is introduced into the primer within the last four bases prior to the 3′ terminus, preferentially substituting T with G or C with A. The creation of this extra mismatch in combination with the natural mismatch significantly reduces the amplification of the nonspecific allele, but leaves the amplification of the specific allele unabated. This SNAP primer design allows SNPs to be typed by the presence or absence of PCR-amplified products on standard agarose gels.

Single nucleotide divergence (SND): The difference in quantity and location of SNPs in the genomes of two related organisms that occurred in the time after their divergence from a common ancestor.

Single nucleotide polymorphism (SNP; pronounce "snip"): Any polymorphism between two genomes that is based on a single nucleotide exchange, small deletion, or insertion. Statistically, an SNP will occur every kilobase in the human genome (soybean: 3–4 SNPs/kb). The genomic distribution of SNPs is biased towards genic DNA (1 SNP/2000 bp) as compared to extragenic DNA (1 SNP/500 bp). The number of SNPs in the human genome is estimated to 3–30 million. This relatively good genome coverage and the distribution of SNPs throughout the genome in both coding and noncoding regions makes SNPs highly informative markers for mapping procedures. Since specific SNPs correlate with increased risk for a particular disease, these polymorphisms are diagnostic ("disease-associated SNPs"). The basic difference between point mutations and SNPs lies in their different frequencies (point mutation: 1%; SNP: 1% of a population).

Single nucleotide polymorphism (SNP) typing: The scanning of specific regions (e.g., genes) in two (or more) genomes for SNPs to detect a SNP composition typical for each of the genomes. SNP typing generates a specific SNP profile of each genome, in which associations of a specific SNP (or SNPs) with a particular phenotype (e.g., a disease) have diagnostic value.

Single nucleotide polymorphism chip (SNP chip): Any microarray onto which specific PCR-amplified regions of genes known to contain one (or more) SNPs are spotted. Using primer extension or quantitative PCR techniques, SNPs can be detected on the chip.

Single nucleotide polymorphism database (dbSNP): A database for SNPs (i.e., single base exchanges), short deletions, and insertion polymorphisms. Web site: http://www.ncbi.nlm.nih.gov/SNP. HGVbase maintains an extensive list of other SNP databases at: http://hgvbase.cgb.ki.se/databases.htm.

Single nucleotide polymorphism density (SNP density): The frequency of SNPs per unit length of genomic DNA (usually SNPs/100 kb). SNP density is different for different regions of genome. For example, regional heterogeneity of SNP density is characteristic for different chromosomes of *Anopheles gambiae*, possibly caused by the introgression of divergent Mopti and Savanna cytotypes (chromosomal forms). SNPs are therefore distributed along the chromosomes in a bimodal way: one mode contains about one SNP/10 kb, the other one about one SNP/200 bp. SNP density is high in intergenic and intronic regions, as compared with genic SNPs. Since introgression is excluded from the X chromosome, its SNP density is lower and not bimodal. In the human genome, SNP density is about 12.13 SNPs/10 kb; in the mouse genome SNP density is only 0.821 SNPs/10 kb.

Single nucleotide polymorphism island (SNP island): Any region of a genome where SNPs are clustered.

Single nucleotide polymorphism mass spectrometry (SNP-MS): A technique for the detection of SNPs between two (or more) genomes. In short, a SNP (e.g., an AC mismatch) is first localized in a specific region of a genome (e.g., within a gene), and flanking primers used to amplify this region in a conventional PCR. Then the amplification product is single-stranded, a probe oligonucleotide annealed immediately 5′ adjacent to the SNP, and a primer extension reaction started with one dideoxynucleotide complementary to the matching nucleotide in the wild-type strand (here, ddTTP). The matching nucleotide can be incorporated and the probe be extended by one nucleotide, whereas the extension is not possible on the other strand carrying the mismatch (here, C). Therefore both products differ by one base and this difference in mass can be detected by mass spectrometry.

Single nucleotide polymorphism scanning (SNP scanning): The *in silico* search for SNPs in a sequenced stretch of DNA (e.g., a bacterial artificial chromosome clone, a genomic segment, or, in extreme cases, whole genomes).

Single nucleotide polymorphism scoring (SNP scoring): The search for SNPs in two (or more) DNA sequences (in extreme cases, genomes), and their characterization and use for genome analysis (e.g., establishment of a SNP map).

Single nucleotide polymorphism simple tandem repeat (SNPSTR; pronounce "snipster"): Any simple tandem repeat (STR) locus closely (i.e., within less than 500 bp) linked to one (or more) SNP loci. A SNPSTR system is generally characterized by lower levels of homoplasy than are STR loci alone. Therefore, SNPSTR systems provide good estimates of intraspecific population divergence times.

Single primer mutagenesis: An *in vitro* technique to introduce mutations into a target DNA. An oligodeoxynucleotide containing a single base mismatch ("mutagenic oligonucleotide") is first annealed to the single-stranded template DNA and then extended by the Klenow fragment of *Escherichia coli* DNA polymerase I in the presence of T4 DNA ligase. As a result, a double-stranded covalently closed circular DNA (cccDNA) molecule is generated that contains a single base-pair mismatch at the mutated site. Subsequently, this heteroduplex is transformed into a suitable host cell. Segregation of the two strands of the heteroduplex leads to a mixed progeny containing either wild-type or mutant DNA.

Single-strand conformation analysis (SSCA; single-strand conformation polymorphism analysis (SSCPA); polymerase chain reaction single-strand conformation polymorphism (PCR-SSCP) analysis): The detection and characterization of single-strand conformation polymorphisms in genomic DNAs from different organisms that is based on subtle differences of electrophoretic mobility between a nonmutated single-stranded sequence and its mutated counterpart. Such differences arise from mutation-induced changes in the three-dimensional structure of the target DNA. In short, the PCR primers (amplimers) are first radioactively labeled and annealed to the target sequence. Then PCR is started, and the amplified sequences are denatured and separated in thin nondenaturing polyacrylamide gels (denaturing gradient gel electrophoresis). The single-stranded PCR products and the mutation-induced shift in mobility of one such product can then be detected by autoradiography.

Singleton polymorphism: A single base difference between otherwise completely monomorphic (i.e., identical) alleles, genes, genomes, or, more generally, DNA sequences.

Site-specific mutagenesis (site-directed mutagenesis; directed mutagenesis; targeted mutagenesis; localized mutagenesis): The introduction of single base-pair mutations (point mutations) at specific sites in a target DNA.

Slipped-strand mispairing (SSM; replication slippage): A nonreciprocal intrahelical process that leads to the mispairing of complementary bases at the site of short, tandemly repeated sequence motifs (microsatellites) in viral, bacterial, and eukaryotic

DNA. In short, during replication local denaturation (unwinding) of the DNA duplex, displacement of the two strands in the region of tandem repeats, and a slip during renaturation may lead to nonpaired single-stranded loops that are target sites for excision and repair. Therefore, replication slippage leads to a reassociation of the strands in a misaligned configuration. If, for example, the primer strand containing the first newly synthesized direct repeat dissociates from the template strand and then misaligns ("slipping forward") at the second direct repeat, continued DNA synthesis leads to the deletion of one of the direct repeats and the intervening sequence between the two direct repeats. Alternatively, if the primer strand containing the newly synthesized second repeat dissociated from the template strand and then misaligns ("slipping backward") at the first direct repeat, then continued DNA synthesis leads to the insertion of one of two direct repeats together with the intervening sequence. These processes can therefore lead to one or more duplications, deletions or insertions of tandem repeat units. Slipped-strand mispairing occurs in only one allele and is regarded as the driving force in the expansion of simple repetitive sequences in eukaryotic genomes.

SNP discovery: The detection of new SNPs in a genome.

SNP frequency: The frequency with which SNPs occur along a defined stretch of DNA or a chromosome (usually expressed as SNP/kb). SNP frequency varies between genomic regions in the same individual and between related individuals. For example, in some regions of the maize genome, SNP frequency is about 1/65 bp, in other regions 1/85 bp. Generally, SNP frequency is much higher than the frequency of insertions or deletions (1/250 bp).

SNP ID: See Reference SNP.

SNP identification technology (SNP-IT™): A proprietary technology for the detection of SNPs in a gene or genome that is based on primer extension. In short, the genomic target region is first amplified by conventional PCR technology, then so-called SNP-IT primers, designed to specifically bind to their complementary target sequences immediately adjacent to the SNP site of interest, are added and extended by DNA polymerase with a fluorescently labeled dideoxynucleotide as terminator ("single-base extension"). The fluorochrome of the labeled dideoxynucleotide on the extended primer is then detected after its excitation. The technique can also be expanded into a high-throughput format ("ultra-high throughput" (UHT)), that combines multiplexed SNP detection assays with high-density oligonucleotide microarrays. In one single UHT, up to 4608 SNPs can be screened simultaneously.

SNP image map: A graphical depiction of the distribution and number of confirmed and candidate SNPs along a stretch of DNA, a bacterial artificial chromosome clone, or a chromosome. Web site: http://lpg.nci.nih.gov/html-cgap/validated.html

SNPing: Laboratory slang term for the detection of SNPs.

SNP map: A linear array of SNP sites along a particular DNA molecule (e.g., a chromosome). Since most genomes contain many such sites (e.g., human genome:

3–30 million SNPs), SNP maps are highly saturated – the distance between two neighboring markers very short (i.e., 1–10 cM). Clustering of SNPs on such a map indicates genomic DNA (1 SNP/500 bp), random distribution of SNPs is characteristic for extragenomic DNA (1 SNP/2000 bp).

SNP minisequencing: A technique for the detection of SNPs. In short, the genomic target region is first amplified with appropriate forward and reverse primers and conventional PCR, and the amplification product purified (e.g., by separation of the amplicon from primers and dNTPs). Then a primer is hybridized to the denatured target DNA such that it ends exactly one base 3′ upstream of the SNP ("SNP primer"), and extended in the presence of dideoxynucleoside triphosphates labeled with different fluorochromes (e.g., ddATP-TAMRA, ddCTP-Cy5, ddGTP-Cy3, and ddTTP-fluorescein). The primer extension reaction allows the incorporation of only the nucleotide matching the template, but is then blocked (hence the name SNP minisequencing). The product is then denatured, electrophoresed in capillaries (capillary electrophoresis), and the type of fluorescence analyzed. SNP minisequencing allows to identify the target locus as, for example, homozygous wild-type, homozygous mutant, or diallelic heterozygote.

SNPper (pronounce: "snipper"): A web-based tool package (snpper.chip.org) that allows to screen public databases for SNPs. Users can search for SNPs in defined positions of a chromosome or in genes and SNPper records all SNPs of the corresponding region, with information about each SNP, the number of SNPs, alleles, the SNP position, the genomic sequences surrounding a SNP, and links to the relevant pages in the public dbases. The "Save SNPset" command allows storage the SNPset on the server so that it can be reloaded later on. Web site: http://bio.chip.org/biotools.

SNP primer: Any primer oligonucleotide that matches a SNP site at its 3′ or 5′ flank, hybridizes to this site, and can be extended by one fluorescently labeled dideoxynucleotide.

SNP profile: The pattern of distribution of SNPs in a distinct gene or genome (generally DNA). SNP profiles are unique for individual organisms (e.g., individuals of a population). SNP profiles have potential for pharmacogenomics, since clinical trials with newly developed drugs are expected to create less adverse reactions and more targeted responses in patients defined by their specific SNP patterns ("individualized medicine").

SNP profiling: The establishment of a profile of a particular SNP in a whole population.

SNP-rich segment: Any one of a series of genomic regions in which SNPs occur at a considerably higher frequency than in the rest of the genome. For example, in the mouse genome such SNP-rich segments contain about 40 SNPs per 10 kb, whereas other regions contain much less SNPs (e.g., the intermitting sequences harbor about 0.5 SNPs per 10 kb).

Somatic mutation: Any mutation occurring in somatic, but not germline cells. Somatic mutations lead to the formation of chimeric tissues if the mutant cell undergoes mitoses.

Spontaneous mutation ("background mutation"): Any mutation that is not experimentally induced but occurs naturally. The spontaneous mutation rate varies from genome to genome (e.g., in bacteriophages: 7×10^{-5} to 1×10^{-11}; in bacteria: 2×10^{-6} to 4×10^{-10}; in fungi: 2×10^{-4} to 3×10^{-9}; in plants: 1×10^{-5} to 1×10^{-6}; in insects: 1×10^{-4} to 2×10^{-5}; in humans: 1×10^{-5} to 2×10^{-6}). It is caused by errors in DNA replication, DNA repair, and also free radicals generated by, for example, the mitochondrial metabolism.

Stop codon mutation: Any change in the base sequence of the stop codon of a gene that abolishes its transcription termination function. As a consequence, RNA polymerase II(B) transcribes beyond the original termination signal, generating an abnormally long mRNA. The encoded protein either does not function correctly or loses its function. Stop codon mutations cause human diseases. For example, the thanatophoric dwarfism type 1, the most common form of lethal neonatal chondrodysplasia with shortened limbs and extremities, a narrow thorax, enlarged head, and an abnormal lobulation of the brain, is caused by a stop codon mutation in the fibroblast growth factor receptor 3 (*FGFR3*) gene. Therefore, the coding region continues to be transcribed for 423 bp beyond the original stop codon until another in-frame stop codon is reached. The expressed protein is elongated by 141 amino acids.

Structural variant (SV): Any one of two (or more) genomes that differ in large segments of around 3 kb to more than around 50 kb, generated by deletions, insertions, and inversions. Such SVs are more significant for phenotypic variation than SNPs, and are involved in gene expression variation between two individuals, susceptibility to HIV infection, female fertility, systemic autoimmunity, and disorders (e.g., the Williams–Beuren syndrome, velocardiofacial syndrome). SVs can be detected by paired-end mapping.

Structural variation (SV): Any large-scale and substantial changes (i.e., copy-number variations or copy-number polymorphisms, deletions, duplications, insertions, inversions, or large tandem repeats translocations) that occur in one of two comparable human (but also animal and plant) genomes, and therefore contribute to human diversity and disease susceptibility. Such SVs are frequent components of repetitive elements in the human genome (e.g., *Alu*I sequences, L1- and L2-LINES), but also hit genes (more frequently genes encoding proteins involved in interaction with the environment, e.g., immune reactions and odor perceptions, and less frequently genes encoding proteins functioning in metabolic processes). About 4000 SVs, each more than 50 kb in length, responsible for about 5% variation in the human genome (involving more than 800 independent genes), influence both disease susceptibility and the normal functioning and appearance of the body. Color-blindness, increased risk of prostate cancer, and susceptibility to some forms of cardiovascular disease result from deletions of particular genes or parts of genes.

Extra copies of a gene known as *CC3L1* reduce a person's susceptibility to HIV infection and progression to AIDS. Lower than normal quantities of other genes can lead to intestinal or kidney diseases. These SVs span thousands or even millions of base pairs, are frequently clustered ("hotspots"), and to some extent conserved throughout evolution. Their existence make the human genome a highly dynamic structure that shows significant large-scale variation from the currently published genome reference sequence. Moreover, SVs between human and chimpanzee genomes were driving lineage-specific evolution, based on their potential for dramatic and irreversible mutations. Most of the SVs are products of nonhomologous end-joining, retrotransposition events, and nonallelic homologous recombination.

Substitution: Any mutation in DNA that leads to the replacement of one base pair by another base pair.

Substitution mutagenesis: Any change in the base sequence of a DNA duplex molecule caused by the replacement of one base with another base (e.g., $C \rightarrow T$).

Superimposed substitution (multiple hit): The occurrence of two (or more) base substitutions at the same site in a genome. Principally, any nucleotide can be substituted by any other nucleotide with equal probability (Jukes–Cantor model), but there may be some nucleotide sites that substitute more frequently than others (i.e., evolve more rapidly).

Suppressor mutation (suppression; second site mutation): A secondary mutation that totally or partially restores function(s) lost by a primary mutation in a defined DNA sequence. The site of the secondary mutation is distinctly different from the site of the primary mutation so that a suppressor mutation does not eliminate the original mutation (as is the case in classical reverse mutations). Frequently a suppressor mutation corrects a previous reading frameshift mutation.

Suppressor-mutator element (Spm; suppressor-mutator system): A transposable element of *Zea mays* that is almost identical to the related enhancer element (En) and encodes four alternatively spliced transcripts translated into four proteins (TnpA, TnpB, TnpC, and TnpD). Complete Spm elements transpose autonomously and additionally trans-activate internally deleted, transposition-defective derivatives (dSpm). This trans-activation is catalyzed by TnpA and TnpD. The Spm element can be inactivated through methylation at cytosine residues at its 5' end close to the transcription start site. Selective TnpA-mediated reactivation of Spm leads to a demethylation of these sequences. TnpA binds to a specific 12-bp recognition sequence that is repeated several times in direct and inverted orientation at the 5' and 3' end of the transcription start site, bringing the ends of Spm together. TnpD acts as a specific endonuclease and generates the transposition complex. In addition to its function in transposition and transactivation of the methylated Spm promoter, TnpA also acts as a repressor of the unmethylated constitutive Spm promoter, but as activator of the methylated promoter.

Suppressor-sensitive mutant: Any mutant organism that is only viable in the presence, not the absence of a suppressor.

Synonymous sequence change: Any alteration in the nucleotide sequence of a coding region that does not change the amino acid sequence of the encoded protein.

Synonymous single nucleotide polymorphism (synonymous SNP; synSNP):   Any SNP that occurs in an exon, but does not change the amino acid composition of the encoded protein.

Synthetic mutant library: Any collection of *Escherichia coli* cells where each cell harbors a plasmid with an insert carrying a different, experimentally introduced mutation (or mutations) into the sequence of a specific gene. Such libraries provide sufficient genetic diversity of the target gene and the expression of each mutant sequence produces a pool of proteins that is screened for a protein with novel or desirable properties. Multiple mutations in the underlying sequence generate cumulative effects that are superior over single mutations. Therefore, many rounds of mutagenesis, expression of the mutated sequence, and selection ("directed evolution") eventually lead to a protein with the desirable properties (e.g., an enzyme with better substrate binding). Mutations can be introduced by a series of different techniques as, for example, error-prone PCR (generating random mutations within a gene template during amplification by e.g., *Taq* DNA polymerase), oligo mismatch mutagenesis, or DNA shuffling (i.e., the random recombination of homologous gene sequences). Synthetic mutant libraries are used for the selective modification and improvement of particular properties of a specific protein.

T

Tag single nucleotide polymorphism (tag SNP; haplotype tagging SNP):   Any one of two (or more) strongly associated (i.e., commonly inherited) SNP loci that are characteristic for a specific haplotype. For example, in a specific genomic region usually many SNPs are present at a frequency specific for this region (e.g., on average about one SNP per 300 bases in the human genome; however, there is much variation in SNP frequency across the genome). Determination of all SNPs in the selected region usually produces a SNP distribution profile, which requires extensive sequencing in many different genotypes. The selection of only few distinct SNPs from this region (the so-called tag SNPs) for genotyping predicts the remainder of the common SNPs in this region and only the tag SNPs need to be known to identify each of the common haplotypes in a population.

T-complex (transfer complex):   A stable complex between the T-strand of the *Agrobacterium tumefaciens* Ti-plasmid and proteins encoded by the *vir* region genes *E* and *D*. The T-complex mediates T-DNA transfer from *Agrobacterium* to plant cells. Vir E2 proteins possibly protect the T-strand from endonucleolytic attack during the transfer process, and vir D2 proteins guide the complex from the bacterium to the plant cell nucleus (pilot protein).

T-DNA (transferred DNA):   A part of the Ti-plasmid in virulent *Agrobacterium tumefaciens* or the Ri-plasmid in virulent *Agrobacterium rhizogenes* that has been transferred from the bacterium to the nuclear genome of plant host cells. There it is stably integrated and expressed, causing permanent proliferation of the host cell(s) into tumors (crown gall tumors, hairy root disease). T-DNA is flanked by direct repeats – the so-called T-DNA borders. Proliferation is induced by the expression of genes 1, 2, and 4 of the T-DNA. Genes 1 and 2 encode enzymes that convert tryptophan to indoleacetamide (tryptophan monooxygenase) and indole acetamide to indole acetic acid (indoleacetamidehydrolase). Gene 4 codes for an isopentenyl transferase catalyzing the synthesis of the plant hormone cytokinin of the zeatin type. The T-DNA, integrated into the plant genome, differs in tumors induced by different *A. tumefaciens* strains with regard to size and copy number as well as the genes encoding enzymes catalyzing the synthesis of different opines. For example, in so-called nopaline tumors the T-DNA is present as contiguous stretch of about 23 kb encoding at least 13 genes, among them a nopaline synthase, that produces the abnormal amino acid nopaline. In octopine tumors (producing the abnormal amino acid octopine) the T-DNA is divided into a 12-kb T_L-DNA (left) and a 7-kb T_R-DNA (right) with an intervening plant sequence. The T_L-DNA is responsible for tumorigenesis and is transcribed into eight transcripts, the T_R-DNA codes for five distinct transcripts. Do not confuse with T-DNA, a term used for the T segment (transferred segment) in DNA topoisomerase II-catalyzed reaction.

T-DNA border: One of two nearly perfect 25-bp direct repeat sequences flanking the T-region of *Agrobacterium tumefaciens* Ti-plasmids. These borders are essential for the *vir*-induced excision of the T-strand after contact of the bacterium with wounded plant cells, because they contain recognition sites for a border-specific, *vir* region-encoded endonuclease that nicks the bottom strand within the border. Border sequence: 5'-TCTTTCTTTTAGGTTTACCCGCCAATATATCCTGTCAAACACAACA-3' (excision site). Only the right border in its natural orientation seems to be essential for T-DNA transfer and possibly its integration into the host cell genome. Advanced plant transformation vector systems contain only the border sequences flanking foreign DNA of up to 50 kb which replaces the T-DNA.

T-DNA insertion mutant (T-DNA insertion line): Any plant into whose nuclear genome a complete or truncated T-DNA from the *Agrobacterium tumefaciens* Ti-plasmid has been inserted. This insertion may be unique (i.e., at only one single locus) or several insertion events occur at different loci in the target genome. T-DNA insertion mutants of, for example, *Arabidopsis thaliana* are available that exhibit a variety of different phenotypes. The genes underlying these phenotypes can be identified, since the T-DNA can be localized on genomic DNA by either hybridization with labeled complementary probes (Southern blotting) or by its conventional PCR amplification, using primers complementary to the T-DNA.

T-DNA tagging (T-DNA gene tagging): A method to isolate a gene that has been mutated by the insertion of a T-DNA sequence. In short, T-DNA is integrated into the genomes of plant protoplasts, the transformants regenerated to complete plants, and these plants screened for mutant phenotypes (e.g., a change in growth behavior due to loss-of-function or gain-of-function of an interesting gene). Then a genomic library is constructed from a T-DNA-induced mutant and screened with a radiolabeled T-DNA as probe. The T-DNA-containing clones are sequenced and the gene into which the T-DNA inserted, can be isolated directly.

Temperature gradient gel electrophoresis (TGGE): An electrophoretic technique for the analysis of conformational transitions and sequence variations of DNA and RNA, and protein–nucleic acid interactions. The TGGE combines gel electrophoresis (separation of macromolecules by size, charge, and conformation) with a superimposed temperature gradient perpendicular to the electrical field (separation by differing thermal stabilities of the different macromolecules).

Temperature-modulated heteroduplex analysis (TMHA): A technique to scan genomes for the presence of mutations (e.g., SNPs) and, more generally, sequence polymorphisms. In short, wild-type and mutant DNA are first amplified in a touchdown PCR with *Pfu* DNA polymerase (thus minimizing misextension by proofreading), using two primers flanking the region of interest. DNA polymerase is inactivated by ethylenediaminetetraacetic acid and heating to 95 °C. The mixture is then allowed to cool to 25 °C over 1 h. During this time, a mixture of heteroduplexes (harboring base mismatches) and homoduplexes is formed that can be resolved by denaturing high-performance capillary electrophoresis as a function of temperature. For example, under nondenaturing conditions (such as necessary for separating DNA fragments), homo- and heteroduplexes have identical retention times. As the temperature increases, the heteroduplexes start to denature in the region flanking the mismatched bases and emerge as separate peaks ahead of the still intact homoduplexes. At still higher temperatures, the homoduplexes also begin to denature, with the AT wild-type homoduplex being earlier denatured than the GC homoduplex ("temperature modulation"). In TMHA, DNA from homozygous wild-type individuals forms one homoduplex species.

Temperature-sensitive mutant (ts mutant): Any mutant with an upper limit of temperature tolerance that is lower than that of the wild-type organism (heat-sensitive mutant) or, alternatively, that is inactivated by lower temperatures (cold-sensitive mutant).

Temperature-sensitive mutation (ts mutation): Any conditional lethal mutation that becomes apparent only above (or below) a certain temperature threshold. The gene affected by such a mutation encodes a protein that is unstable above (or below) a certain temperature. The mutant therefore behaves like the wild-type at permissive temperatures, but exhibits the mutant phenotype at nonpermissive (restrictive) temperature.

Temperature sweep gel electrophoresis (TSGE): A variant of the temperature gradient gel electrophoresis (TGGE) technique for the detection of point mutations in DNA. In contrast to TGGE, the temperature of the gel plate is raised gradually and uniformly from 45 to 63 °C during electrophoresis. TSGE is based on the same principles as TGGE, which exploits the decrease in electrophoretic mobility of a DNA fragment that occurs if localized regions begin to melt. The melting point of such regions is shifted by a point mutation so that single base substitutions in these first-denatured regions can be detected.

Template-directed correction: The reversion of stress-induced mutations in a specific plant gene (e.g., the hothead gene *hth* of *Arabidopsis thaliana*) possibly guided by an RNA cache, a correction template independent of the genome, that represents an ancestral wild-type copy of the gene. These caches may direct sequence-specific methylation of cytosines in the gene, recruit a DNA repair enzyme rather than a DNA methyltransferase, and reverse the mutation(s).

Tetra-allelic single nucleotide polymorphism (tetra-allelic SNP): Any SNP of which four different alleles are present in a population.

Thionucleotide (α-thionucleotide; NTPαS; thiophosphate): Any nucleoside triphosphate that carries a sulfur atom at the 5'-α (or also 5'-γ) position instead of an oxygen atom. Thionucleotides are more stable against nucleases than their normal counterparts and are used for site-directed mutagenesis. These compounds can also be labeled with ^{35}S in the α or γ position, and possess a relatively long half-life ($r_{1/2} = 87.1$ days).

Thionucleotide mutagenesis: A variant of the oligo-mismatch mutagenesis technique for the introduction of mutations in DNA that is based on the use of thionucleotides. In short, the mutagenic oligonucleotide is hybridized to the single-stranded template DNA, and extended by the Klenow fragment of DNA polymerase in the presence of a DNA ligase and a thionucleotide, producing a mutated phosphothio heteroduplex. The phosphothio strand of the heteroduplex cannot be digested by certain restriction endonucleases. The corresponding strand can be removed by exonucleases, whereas the mutated strand remains intact. It then serves as template for the *in vitro* synthesis of a complementary new strand and the mutated duplex DNA is generated.

Ti-plasmid (tumor-inducing plasmid; pTi): A large conjugative plasmid of about 200 MDa in size that is found in all virulent strains of the Gram-negative soil bacterium *Agrobacterium tumefaciens*. It contains genes for replication (*oriV*), plasmid transfer (*tra* genes), phage exclusion (*Ape*), incompatibility (*Inc*), virulence (*vir* region), root and shoot induction in host plants (*Roi* and *Shi*), opine synthesis in host plants (*Nos, Ocs,* and *Ags*), opine catabolism (*Noc, Occ, Agc,* and *Arc*), and catabolism of phosphorylated sugars (*Psc*). The genes for root and shoot induction and opine synthesis are clustered in a specific segment of the Ti-plasmid (T-region) and flanked by 25-bp border sequences. These borders are the recognition sites for a Ti-plasmid-encoded endonuclease ("border endonuclease") that excises one strand of this region (T-strand) which is packaged into a T-complex and transported into a recipient plant cell. There it is integrated into the nuclear genome and causes the cell's permanent proliferation into a tumor.

Tnt1 (transposon: *Nicotianatabacum*): Any member of a large superfamily of autonomous and active *Ty1 copia*-like retrotransposons of solanaceous plants (e.g., tobacco, tomato, and other Solanaceae), that carry structural features of viral retroelements including two perfect, 610-bp long terminal repeats (LTRs), known to contain promoter regions (e.g., a putative TATAA-box), a linker (a noncoding region with a polypurine track), a repeat RNA region flanked by a unique 3′ RNA region (U3) and a unique 5′ RNA region (U5), a ribonuclease H gene, and a single open reading frame (ORF) typically containing *gag–pol* domains, and comprise 5.3 kb. Elements in tobacco (and several other Solanaceae) are each present in greater than 100 copies/haploid genome and fall into at least three major subfamilies (*Tnt1A*, *B*, and *C*), that differ in their U3 sequences, but share conserved flanking coding and LTR regions. The tomato elements, called *Retrolyc1* (retrotransposon *Lycopersicum peruvianum*), comprise two subfamilies, *Retrolyc1A* and *B*, and share extensive nucleotide similarities to *Tnt1* elements, except in the regulatory U3 region. Several wild and cultivated *Solanum* genotypes from South America contain the retrotransposon *Retrosol*. *Tnt1* retrotransposons found in various Solanaceae species are characterized by a high level of variability in the LTR sequences involved in transcription and evolved by gaining new expression patterns, mostly associated with diverse stress conditions. *Tnt1A* insertions into genic regions are initially favored, but are subsequently counterselected, while insertions into repetitive DNA are maintained (from four to 40 insertions per plant). The transcriptional BII regulatory element in *Tnt1* retrotransposons tightly controls their activation by restricting expression to responses upon stress (e.g., protoplast isolation and culture or bacterial and fungal attack). *Tnt1* activation by microbial elicitors is followed by DNA amplification, which generates genetic diversity in plants. The *Tnt1A* promoter is activated by various biotic and abiotic stimuli, but transcripts appear only in few tissues (e.g., in roots, but not leaves). In addition to tight transcriptional control, *Tnt1A* retrotransposons self-regulate their activity through gradual generation of defective copies with reduced transcriptional activity.

Transcription-based mutation: The change of nucleotides in the coding strand of a gene that is single-stranded locally in the so-called transcription bubble (a melted portion of the transcribed gene). The single-stranded region is highly exposed to mutagens.

Transcription-coupled repair (TCR; transcription-coupled nucleotide excision repair (TC-NER); transcription-coupled DNA repair): A somewhat misleading term for the repair of DNA damage (e.g., UV-induced cyclobutane dimers, thymine glycols, 8-oxo-guanine, or other mutated bases) involving the recognition of the damaged site by the transcriptional complex. During the transcription of a gene, the elongating DNA-dependent RNA polymerase arrests at an injury on the template strand, causing to stall the transcription complex and leading to a transcriptional collapse. The stalled RNA polymerase II complex harbors, among others, CSA, CSB, transcription factor IIH, XPG, and probably other proteins, that process the arrested RNA polymerase, thereby making the damaged site accessible for repair proteins. Processing involves ubiquitination and subsequent degradation of the polymerase. The damage itself is then repaired by lesion-specific or lesion-independent repair systems. The term CS derives from the so-called Cockayne syndrome (CS), a complex human disease, whose carriers suffer from severe growth defects, mental retardation, microencephaly, and skeletal and retinal abnormalities. At least five complementation groups exist: CSA, CSB, XPB, XPD, and XPG. The protein encoded by the CSA gene belongs to the so-called WD family, which is involved in many pathways, including signal transduction, transcription, mRNA modification, and cell division. CSB codes for a helicase, which may induce DNA unwinding and thereby disengage stalled RNA polymerase II for subsequent repair of the lesion. The repair process therefore occurs more quickly in the transcribed rather the nontranscribed DNA strand. In bacteria, TCR is mediated by the transcription repair coupling factor (TRCF), that disrupts the stalled RNA polymerase and recruits the DNA excision repair machinery, and is encoded by the *mfd* gene. The TRCF protein consists of a compact arrangement of eight structural domains (D1a, D1b, and D2–D7), where D1a/D2/D1b form the UvrB homology module, D4 comprises an RNA polymerase interacting domain (RID), and D5/D6 harbors the translocation module (translocation domains TD1 and TD2). The different domains are connected by 12–25-amino-acid long linkers. TRCF relaxes the transcription-dependent inhibition of nucleotide excision repair (NER) by recognition and ATP-dependent removal of the stalled polymerase on the damaged DNA, and stimulation of DNA repair by recruitment of the Uvr(A)BC endonuclease. TRCF uses the ATPase motor to translocate on dsDNA upstream of the transcription bubble, inducing forward translocation of the polymerase, bubble collapse, and transcript release.

Transition (transition mutation; transitional mutation; base-pair substitution): The replacement of one purine base by another purine or a pyrimidine base by another pyrimidine in a DNA duplex molecule, leading to a transition mutation. The result of a transition finally is the exchange of a GC pair with an AT pair or vice versa.

Transition single nucleotide polymorphism (transition SNP): Any SNP originating from a transition mutation where a purine base is replaced by another purine or a pyrimidine base by another pyrimidine. The majority of SNPs in the human genome are transition SNPs.

Transposase: An enzyme encoded by a gene of class II transposons that catalyzes the excision, transfer, and insertion of the transposable element that carries its gene.

Transposon (Tn; transposable element (TE); mobile element; mobile genetic element; "jumping gene;" selfish genetic element (SGE)): The usage of the term "transposon" is partly contradictory. In a strict sense it designates only prokaryotic transposable elements, while eukaryotic transposable elements are called "transposon-like elements." It is, however, also used synonymously to "transposable element," as a name for both eukaryotic and prokaryotic mobile sequences. Accordingly, the following definitions are used:

  a. Generally, all segments of DNA that can change their location within a genome. Transposable elements are flanked by short inverted repeat sequences and encode enzymes that catalyze their excision from their original site, and their transfer to and insertion into a new site (transposition). Transposons can be used for the construction of transposon-based cloning vectors, for transposon mutagenesis, and for transposon tagging.

b. Mobile DNA sequences of bacteria, bacteriophages, or plasmids, flanked by terminal repeat sequences and typically harboring genes for transposition functions (transposase, resolvase). They insert at random and independently of the host cell recombination system into plasmids or chromosomes. Transposons can be broadly categorized into compound (class I) and complex (class II) transposons. Class I transposable elements are characterized by a drug resistance gene flanked by insertion sequences either as direct or, more frequently, as inverted repeats. The IS elements provide the transposition functions and transpose the intervening drug resistance gene(s) in concert. The IS elements may also transpose independently. Examples for class I transposons are *Tn5* and *Tn10*. Class II transposons contain genes encoding transposition functions and drug resistance(s) flanked by short, inverted repeats of 30–40 bp in length. A copy of the transposon is retained at the original location. The transposon's insertion at a new target site generates a direct repeat of 3–11 bp within the target DNA borders. A class II transposon is, for example, *Tn3*. Transposons of both classes are used as source for drug resistance genes and as mutagens.

Transposon insertion display (TID; transposon display TD): A high-resolution technique for the simultaneous visualization of individual members of transposable element families in high-copy-number lines, the analysis of element copy numbers, insertion frequencies and transpositional activities of the elements, and the isolation of transposon-tagged genes and sequences flanking transposable elements in plants (e.g., *En/Spm*, *Mu*1 and the non-long-terminal-repeat retrotransposon *Cin*4 in *Zea mays*). TID is a variant of the conventional amplified fragment length polymorphism technique. In short, genomic DNA is first isolated, restricted with a frequently cutting restriction enzyme (e.g., *Bbs*I and *Mse*I for *Cin*4TID, *Bst*XI and *Mse*I for *Mu*1-TID), then splinkerette-like linkers ligated to the fragments, and biotinylated primers complementary to the linkers used to amplify the fragments in a conventional PCR. The amplified products are then purified from the primers and bound to streptavidin-coated beads. Then a second PCR with a nested 5' or 3' end transposon-specific and labeled primer (e.g., labeled with ^{32}P or ^{33}P) and a linker-primer is performed, the products electrophoretically separated on a denaturing polyacrylamide gel and detected by autoradiography (display of only transposon-specific fragments). Fragments of interest are then isolated and amplified for sequencing.

Transposon mutagenesis: A method to introduce insertion mutations at random within a target DNA using transposons. Basically, two approaches can be followed – in the nontargeted approach the transposon inserts at random, in the targeted approach it inserts into a particular gene sequence.

Transposon tagging: A method to isolate a gene that has been mutated by the insertion of a transposon. Basically, two approaches can be followed. In the nontargeted (random) approach the transposon is randomly integrated into the target genome. Insertion into, or very close to the gene of interest may alter the phenotype of the host organism, which makes selection of interesting transformants easier. The targeted approach is based on the insertion of a transposon into a cloned gene and the transformation of a suitable host with this construct. The host incorporates the construct into its chromosome by homologous recombination. In both cases the target DNA can be identified by hybridization of genomic restriction digests or genomic clones to a radiolabeled probe completely, or only partly identical to the transposable element.

Transversion (transversion mutation; base-pair substitution): The substitution of a purine by a pyrimidine, or a pyrimidine by a purine base in duplex DNA. The result of a transversion finally is the exchange of an AT pair with a TA pair or CG pair. Transversions are less frequent than transitions.

Transversion single nucleotide polymorphism (transversion SNP): Any SNP originating from a transversion mutation where a purine is exchanged for a pyrimidine, and vice versa.

Triple-knockout mutation ("knock-knock-knock mutation"): Any mutation that simultaneously occurs in three genes within the same genome and knocks out the functions of all three.

T-strand: A linear, ssDNA molecule arising through a strand-specific nick within the so-called border sequences flanking the T-region of *Agrobacterium tumefaciens* Ti-plasmids. The excision of the T-strand is catalyzed by a *vir*-region-encoded endonuclease that is induced in a cascade of events following the contact of an *Agrobacterium* with wounded plant cells. In most plants, wounding induces the synthesis of phenolic compounds, some of which (e.g., acetosyringone) serve as signal substances for Agrobacteria (i.e., are recognized by specific *vir*-encoded membrane chemoreceptors, vir A proteins, and are transmitted to the *vir* region through specific *vir*-encoded transfer or activator proteins, vir G proteins). After the excision of the T-strand it is packaged into a T-complex – a protective coat of vir E2 proteins (single-strand specific DNA-binding proteins) – and transported into the recipient plant cell by a vir D2 protein (pilot protein).

U/V

Unique mutation: Any mutation that is present only once in a genome or a plasmid.

Unstable mutation: Any mutation with a high frequency of reversion (reverse mutation; e.g., a mutation caused by the insertion of a transposon that has a high frequency of transposition).

Variable number of tandem repeats (VNTR): A set of tandemly repeated, short (11- to 60-bp) oligonucleotide sequences with a conserved core sequence 5'-GGGCAGGAXG-3' The number of these repeats within a given DNA region of the human genome varies from one individual to another. The diminution or amplification of the number of such tandem repeats may be due to a high frequency of unequal crossing-over events at the VNTR recombinational hotspots. VNTRs thus are responsible for considerable DNA sequence polymorphisms in the human genome (e.g., since the length of a restriction fragment carrying VNTRs is a function of the copy number of tandem repeats present within the fragment, restriction fragment length polymorphisms may be due to the presence of VNTRs). VNTRs can be detected, for example, with ligated oligonucleotide probes.

W/X/Y/Z

Whole-genome duplication: The amplification of a whole genome (or also parts of it) during evolution, such that the present-day genomes of many organisms consist of duplicated regions. For example, the duplicated regions of the *Arabidopsis thaliana* genome constitute 68 Mb, or 60% of the genome. The number of homologous genes in the duplicated regions vary considerably, ranging from 20 to 50%, which is either a consequence of tandem duplications or gene loss after segmental duplication. Whole-genome duplications are a cause for polyploidization.

Wild-type (standard type):

a. The most frequently occurring phenotype (strain, organism) in natural breeding populations.
b. The genetic constitution of an organism at the onset of recombinant DNA experiments with its genome or its plasmid(s). Thus, this "wild-type" is an arbitrarily specified genotype used as a basis for comparison in genetic programs.

Zero-cycle artifact: Any one of the nonspecifically primed amplification products that arises from *Taq* polymerase-catalyzed, low-stringency priming events prior to the first amplification cycle in a conventional PCR.

Index

a

adaptor-ligation method   18, 23, 26, 27, 311
agarose gel   132, 134, 137, 145, 310, 312
– electrophoresis   134
– purifying DNA   310
Agrobacterium tumefaciens   31, 35
– T-DNA integration   32
amino acid sequences   169, 175
amplified fragment length polymorphism
    (AFLP)   180
analysis of variance (ANOVA) method   235
animals, mutations and natural
    polymorphisms   131
antiauxin resistant1-1 (aar1-1)   12
Apium graveolens   132
Arabidopsis thaliana   3, 5, 14, 19, 31–33, 84, 94,
    105, 233
– dry seeds   9, 10
– – survival curves   9
– genome   27
– T-DNA lines   33, 83, 94
– transposon-tagged mutant resources,
    databases   19
association mapping   226, 231–233, 241.
    See also linkage disequilibrium mapping
– applications   238–242
– – limitations   241, 242
– – QTL mapping *vs.* association
    mapping   240, 241
– germplasm collections   234
– methods   233–238
– protocols   233–238
– – association mapping, population   233,
    234
– – genotyping   234
– – phenotyping   234, 235
– – statistical procedures   235–238

– studies, types   232, 241
azimuthally varying field (AVF)   8

b

backcross breeding   247–252, 259
– dominant mutation   251
– drawbacks   250
– goal   249
– marker-assisted selection   259
– molecular markers, use   259
– objective   248
– recessive mutation   251, 252
bacterial artificial chromosome (BAC)
– yeast artificial chromosome (YAC)
    clones   106
bacteriophage T4 DNA polymerase   74
barley chromosome   228
– distal portion, graphical genotypes   228
barley genes, comparison   226
barley mutant   217–221
– breeding, applications   220–222
– classes   218, 220
– review   217–221
barley yellow mosaic virus (BaYMV)
    complex   227
Beckman polyallomer tubes   309
binary vector pAC161, plasmid map   35
biological model systems   289
blunt-ended fragments   108
bound oligomers, optical detection   349
Bowman mutant collection   223
breeding methods   247–249
– alternatives   249
– factors   248
– methods   249–258
– – backcross breeding   247, 249–252
– – forward breeding   247, 252, 253

The Handbook of Plant Mutation Screening. Edited by Günter Kahl and Khalid Meksem
Copyright © 2010 WILEY-VCH Verlag GmbH & Co. KGaA, Weinheim
ISBN: 978-3-527-32604-4

– protocols 249–258
– – supplementary protocols 253–258
bulk segregant analysis 228

c

candidate-gene approach, use 242
candidate-gene association mapping
 projects 234
cauliflower mosaic virus (CaMV) 35
– 35S promoter 271
CCD camera 151
– detection system 299, 349
cDNA molecules 301, 302
– libraries 107
– microarrays 301
– synthesis 107
cetyltrimethylammonium (CTAB)-DNA
 preparation protocol 37
chemical cleavage of mismatch (CCM)
 method 121, 122
– applications 127
– duplex formation 126
– liquid-phase CCM analysis 126
– schematic presentation 122, 123
– solid-phase CCM analysis 126, 127
chemical mutagens 3, 218
– ethylmethane sulfonate (EMS) 3
chemical synthesis, cycle 296
chip-immobilized DNA template 326
CLAVATA1-like receptor kinase gene 12
cleavable fluorescent (CF)
– ddNTPs 329
– – terminators ddNTP-N_3-fluorophores 332
– NRTs 325, 331
– – design synthesis, and characterization
 324–326, 331–333
– – DNA chip construction 326
– – four-color DNA SBS 323, 326–329
– – overview 323, 324
– structures 325, 332
cluster amplification 107, 111
complete TLS assay 68
– calculation 70
– DNA
– – polymerase reactions 69
– – P/T design considerations 68
– polymerase reaction products analysis 69
corn breeding programs 247
– applications 258, 259
– – breeding with natural mutation 258
– – breeding with transgene 259
– perspectives 259, 261
– – doubled haploids 260
– – marker-assisted selection 259

Coulter counter principle 340
crops, genetic diversity 231
crude celery juice extract (CJE) 132
– enzyme activity test, agarose gel image 134
cytochrome coxidase subunit I (COI) 317

d

database 24
– construction/sequence analysis 24
ddNTP-N_3-fluorophores 333, 334
denaturing high-performance liquid
 chromatography (DHPLC) 200
de novo sequencing reaction 327, 349
diacylglycerol acyltransferase (DGAT) 240
2,4-dichlorophenoxyacetic acid (2,4-D),
 synthetic auxin 12
digital gene expression (DGE) analysis 290
diploid Bell peppers, LunaProbe™ chlorophyll
 genotyping 161
direct gene sequencing methods 200
DISTILLING database 223
DNA 55, 73, 124, 190, 264, 265, 296, 303, 341,
 344, 347
– alterations 5
– analysis methods 320
– backbone 341
– bulking 190
– chemical modification 73
– chip, four-color SBS data 328
– damage 55, 75, 78
– four-color DNA sequencing 330
– fractionation 106–109, 168, 265, 289, 295,
 309, 310, 320
– hybridization 348
– metabolic pathways 56
– mismatches, classes 122
– modification 347–349
– molecules 290, 292, 340, 341, 347, 348
– – challenges 292
– – pore probes 340
– pellets 37, 153
– physical properties 341
– pools, polymerase chain reaction (PCR)
 screening 84
– polymerase 67, 78, 322, 324, 333
– primer 39
– – plates 154
– replication 320
– sequence 307, 319–321, 329, 330, 335, 344,
 345, 350
– sequencing-by-synthesis (SBS) 319, 321
– sugar-phosphate backbone 122
– templates, primer 323, 324
– translocation 347

DNA markers 194
– EcoTILLING bands, use 194
– EcoTILLING polymorphisms, use 194
DNA polymerase error 55, 79
– applications 73–78
–– alkylation damage, mutational processing by DNA polymerases 75
–– DNA lesion discrimination mechanisms 75–78
–– *in vitro* genetic assay, features 73
–– polymerase accuracy in absence of DNA damage 74
– discrimination 79
– examination method 55
– genetic assay 56
– methods/protocols 56–73
– perspectives 78
DNA polymerase reaction 55, 74, 61, 62, 69, 74, 321, 326, 355
– classification 55
– DNA modification by alkylating agents 61
– heteroduplex plasmid formation 63
– primer/template (P/T) DNA substrates preparation 61
– small fragment (SF) isolation 62
dNTP base-stacking interactions 77
doubled haploid technology, use 260
double-strand breaks (DSBs) 5, 6, 132
double-stranded DNA (dsDNA) bridge 110
Dover systems 307
Ds/genomic DNA junction 24
– PCR amplification 24–26
Ds-transposed plants, selection 20, 21
– *Ds*-transposed lines 20
–– applications 26–28
–– construction 20
dual-end-labeling strategy 132
dual-tag sequencing method 303

e

EcoTILLING 188, 193, 195, 200
– application 188
– gel 195
– strategy 193
– use for rapid forward genetics in rice 193
– use to identify causal mutation 195
electron energy loss spectroscopy (EELS) 350
electropherograms 167, 169, 170, 172–174, 176, 180, 181
ELOSSM code program 8
enzymatic fragmentation method 316
enzymatic incorporation reactions 323
enzymatic mismatch cleavage method 132, 135, 136

– application 143
– CJE preparation 136
– detection 132
– fluorescent detection 137, 138
– nuclease cleavage 139, 140
– PCR amplification 138, 139
– primer design 138
– testing enzyme activity 137
enzyme
– extraction procedure 134
– free sequencing chemistry 316
ethylmethane sulfonate (EMS) 31, 192
– mutations 201, 207, 220
– plants, viability/fertility 192
– screening 210
European Research Area in Plant Genomics (ERA-PG) project 223, 228
– BARCODE project 223
exonuclease method 351
– KF polymerase 77
expressed sequence tag (EST) sequences 225

f

fabrication methods 345
false discovery rate (FDR) 238
flanking sequence tags (FSTs) 21, 84
– data, sequence analysis 26
– high-throughput analysis 21–23
– locus 96, 97
– production strategy 38
– T-DNA populations 33
flavanoid biosynthesis 219
fluorescence scanner, four-color images 335
forward breeding 248, 252–254, 260
– dominant mutation 252, 253
– goal 252
– marker-assisted selection 260
– objective 248
– recessive mutation 253
four-color sequencing-by-synthesis (SBS) 330, 334
– comparison 330
– data 328
– hybrid DNA 333, 334

g

gapped duplex (GD) molecules 56, 74
– formation 60
– preparation, ssDNA purification 58
gap-filling reaction 78
gel electrophoresis 153, 182
GenBank files 172, 174

gene-based genetic mapping technologies 226
gene expression 290, 302
– approaches 302
– quantitative assessment 302
– serial analysis 302
– whole-transcriptome analysis 290
gene function analysis 263
GeneMarker software 180–182
– applications 181
– mutation detection with DNA fragments 180
– TILLING analysis 181, 182
general public license (GPL) 305
gene-specific primers (GSPs) 33, 95
– PCR 201
gene targeting 263–265, 277
– application 263
– development 264
– efficiency 265
– methods/protocols 266–279
– vector designs 265
genetic assay 55, 78
– adaptation 78
genome alignment method 316, 317
genome analyzer 105, 115, 116
– instrument control computer 112
genome scan calculation 237
Genome Sequencer (GS) 193, 200, 203, 290
– FLX platform 200
– – library preparation 203–205
– – sequencing 205
– – titanium platform 211
– 454 Genome Sequencer 177
– fragment analysis 167
– mutation detection software 167
genome-wide association (GWA) mapping 232, 240
genomic DNA 23, 106, 291, 308
– adaptor, restriction enzyme-mediated ligation 23, 24
– fragments 104
– purification 22
grafting experiment 12
green fluorescent protein (GFP) 258

h

hairpin DNA molecules 348, 350
– proof-of-concept study 351
haploid inducer line 260
Helicos BioSciences 289, 301
Helicos-driven digital gene expression (DGE) 301, 302
Helicos flow cell 295

Helicos genetic analysis system 300, 304, 305
– performance specifications 304
Helicos true single molecule sequencing (tSMS^{TM}) technology 291, 297, 298
Heliscope single molecule sequencer 291, 295–297, 300, 304
– design 297
– optics configuration 298
herpes simplex virus type 1 thymidine kinase (HSV-*tk*) assay 65, 73, 75
– forward mutation assay, DNA modification 56
– gene 56, 73
– – MluI-EcoRV target region 73
– *in vitro* assay, schematic presentation 57
– mutant 74
– – frequency calculation/determination 65
– mutations 63
– – bacterial selection for plasmids 63
– – heteroduplex molecules electroporation 64
– – VBA selective media preparation 64
– polymerase errors, mutational spectra 76
– use 75
heteroduplex DNA 122
– modification 122
– use 122
heterozygous mutations, robust detection 135
hexachlorocarboxyfluorescein (HEX) 124
high-resolution DNA melting 150
– applications 159
– – Bell pepper multiplex genotyping 161
– – potato tetraploid genotyping 161, 162
– – SNP heterozygote detection 159, 160
– – 5′-untranslated region (UTR) 160
– LightScanner® instrument 151
– – Lunaprobe^{TM} genotyping 156–159
– – melting analysis 155
– – PCR 153–155
– – PCR optimization 152, 153
– – scanning primer design 151, 152
– methods 150–159
– oligonucleotides 150
– protocols 150–159
high-resolution melt curve analysis 122
high-speed sequencing technologies, benefits 339
high-throughput scheme 21
– DNA sequencing system 319
– instrumentation 150
– mutation/polymorphism discovery technique 199
– scanning 150

– single molecule sequencing-by-synthesis (SBS)   300
– steps   21
Hordeum vulgare   227
Human Genome Project   305, 319, 339
Human Genome Research Institute   339
hybrid DNA sequencing-by-synthesis (SBS)   329
– four-color   333–335
– overview   329–331
hybridization   314, 315
– assisted nanopore sequencing   347
– selection   105

i

Illumina genome analyzer system   103
– adaptor ligation   109
– applications   116–118
– data analysis   117
– DNA clusters preparation   110–112
– DNA fragments, library preparation   107, 108
– overview   103, 104
– PCR   110
– resequencing strategies   104–107
– – sequencing transcriptomes   107
– – targeted genome selection   105–107
– – whole genome   105
– sequencing forward strand   112–115
– sequencing process, stages   104
– sequencing reverse strand   115, 116
– size selection and gel purification   110
Illumina Solexa Genome Analyzer   175, 177
Indels   170
– detection   178
– discovery   132
infrared dye (IRD)-labeled primers   187, 188
intermating   252, 256, 257
International Maize and Wheat Improvement Center   258
internode length, classes   218
ion beam irradiation methods   8, 9, 350
– dose determination   9
– schematic diagram   8
ion beam mutagenesis   5, 6
– applications   11–14
– – forward genetics   12
– – limitations   13, 14
– – plant breeding   13
ionic current blockage sequencing   344, 345
in silico mutation methods, detection   167
in situ rolling-circle amplification   308, 313
in vitro DNA synthesis assays   56
in vitro genetic assays   78

j

jellyfish protein, GFP   259
JelMarker™ software   182

k

KeyPoint technology   199, 200, 208, 210, 211
– screening   208
Kolmogorov–Smirnov test   236

l

LI-COR DNA analyzer   134
– fluorescence detection   134
– slab gel electrophoresis   134
LI-COR electrophoresis system   191
LI-COR gels, preparation   141, 142
LightScanner® instrument   152, 159
– master mix   157
– primer design software   151, 156
– scanning data   160
– SNP scanning   159
linear energy transfer (LET)   4, 5
– conceptual diagram   4
linkage disequilibrium (LD)   232, 239
– mapping   231
linkage drag, definition   250
liquid-phase analysis   124
long terminal repeat (LTR) retrotransposons   83
loss of heterozygosity (LOH)   168, 180
LunaProbe™ genotyping   156

m

maize kernels   240
– oleic acid content, ANOVA   240
manifold methods   278
Mann–Whitney test   236
marker-assisted breeding programs   168
marker gene   41, 48, 265, 266, 271, 276, 278
Markov chain Monte Carlo (MCMC) parameters   236
matrix-assisted laser desorption/ionization-time of flight mass spectrometry (MALDI-TOF MS)   326
Medicago truncatula   83, 84, 95
– application   95–98
– – FST sequence in *Tnt1* database   97
– – gene to work   97
– FSTs identification   96
– *in vitro* transformation   83
– line with mutant phenotype   96
– methods/protocols   84–95
– – *Tnt1* insertion sites identification   84–94
– – *Tnt1*-tagged collection   96, 98
– – use   96

microsatellite instability (MSI)   168, 180
microsatellite markers   181, 194
molecular markers   258–260
– development   258
mRNA   107, 168, 301, 302, 317
multidimensional pooling strategy   210
mutagenesis technique   3, 14
– ion beam irradiation   14
mutant population   207
– KeyPoint analysis results   207
mutant resource development methods   18–26
– protocols   18–26
– stages   18
mutation   5, 55, 247, 257
– classification   5
– pleiotropic effects   257, 258
– score, factors   172
– specificity determination   66
– spectra generation   66
Mutation Surveyor®   168–175
– detection of variations   170
– electropherogram generation   172
– GBK file editor   174
– graphic analysis display   171
– heterozygous Indel detection tool   173
– project   172
– reviewing/exporting analysis   175
– Sanger sequencing data analysis   168, 172
– – mutation detection   168
– – procedure   172

n

nanopore-based sequencing methods   339, 342–344, 349
– attraction   343
– biological nanopore history   341–343
– current developments   344
– for mutation detection   339
– important parameters   341
– improving biological nanopores   344
– improving solid-state nanopores   345
– resequencing applications   349, 350
– solid-state nanopore history   343
nanopore concept   339–342, 346
– detection principle   340
– development   339
– promise   343
– sensing region   342
– sensing resolution   341
National Center for Biotechnology Information (NCBI) Entrez Gene   174
nested polymerase chain reaction (PCRs)   86, 92
– adaptor/oligonucleotides   86
– border amplification   93
– fragment characterization   93
– genomic DNA digestion   92
– I-PCR 1/2 program   92, 93
– ligation   92
– oligonucleotides   92
next-generation genome analyzers   168, 176
next generation sequence technology   167, 175, 176
– mutation detection   175–180
– steps   176
NextGENe™ software   175, 176, 179, 180
– charts   180
– condensation   178
– mutation detection   175–180
– mutation output   179
– reporting options   179
– sequence alignment   177
Nicotiana tabacum   5
Nordic Gene Bank   223
nucleotide polymorphism   131
– genotyping   131
– polymorphism detection   350
nucleotide reversible terminators (NRTs)   321, 329
numerical aperture (NA)   298

o

3′-O-allyl-dATP, structures   329
3′-O-allyl-dCTP, structures   329
3′-O-allyl-dGTP, structures   329
3′-O-allyl-dNTPs, structures   327–329
3′-O-modified dNTPs   330
3′-O-modified NRTs   330
3′-O-N$_3$-dATP
– ddNTP-N$_3$-fluorophore solution   335
– structures   331
open reading frame (ORF)   34
Oxford Nanopore Technologies Ltd   345

p

paired-end sequencing   303
paired-read sequencing   303
photolithographic methods   344
Physcomitrella patens   263, 264, 273, 280
– gene duplications   272
– gene targeting   264, 280
– materials used   273
Physcomitrella patens tissue culture methods   266
– basic media and solutions   266
– materials   268
– routine propagation   267

–– materials   267
–– procedures   267
– sexual propagation   269
–– materials   269
–– procedures   269
–– solutions and media   269
– sterilization of spores   270
– storage of cultures   270
physical mutagens   3
– electrons   3
– γ-rays   3
– X-rays   3
plants   5, 10, 36, 131, 149
– effective irradiation dose   10
– high-resolution DNA melting   149
–– genotyping   149
–– mutation scanning   149
– high-resolution melting, applications   150
– ion beams, mutational effects   5–7
– mutagenesis, gene targeting tool   263
– mutation detection methods   121, 122, 131
–– applications   143, 144
–– CCM protocols   124
–– data analysis   142, 143
–– enzymatic mismatch cleavage   131, 133
–– LI-COR gels   141
–– polymerase chain reaction (PCR)   121
–– Sanger sequencing   121
–natural polymorphisms   131
– nomenclature   36
– radiation sensitivity   10, 11
– research tagged lines   17
–– T-DNA-tagged lines   17
–– transposon-tagged lines   17
plant ion beam mutagenesis   7
– facilities   7
– methods   7–11
– protocols   7–11
–– ion beam irradiation   8
polonator, see Dover systems
polyacrylamide gel electrophoresis (PAGE)   132, 308
polyethylene glycol (PEG) molecules   326, 342
– mediated DNA   278
polymerase chain reaction (PCR) techniques   25, 85, 105, 121, 149, 188, 201, 264, 277, 311, 320
– amplicon, electrophoretic scan   125
– amplification   168
– amplified DNA   122
– based genetic fingerprinting technique   180
– based protocols   84
– cleanup   311
– emulsion   320

– ligated and purified template   311, 312
– melting temperature   150
– optimization   152, 153
– products   95, 277
– reactions   277
– reverse transcription   278
– schematic representation   25, 85
– sequencing primers   25
Polymerase dissociation assay   70–73
– calculation   72
– dissociation experiments   72
– DNA P/T preparation   71
polyT oligonucleotides, characteristic   295
polyvinylpyrrolidone (PVP)   37
population-based association mapping methods   236
postligation restriction enzyme   38
potato   162
– DNA   161
– tetraploid bialleleic genotyping, LunaProbe assay   162
PpCOL2 gene   278
principal component analysis (PCA)   237
pro-anthocyanidin pathway   219
pyrosequencing   321
– data, comparison   330

q

QIA-quick Gel Extraction Kit   310, 312
– DNA purification   312
quality protein maize (QPM)   258
– breeding programs   258
– development   258
quantitative trait locus (QTLs)   260
– mapping   231, 240

r

rat DNA polymerase   324
– ternary complexes, three-dimensional structure   324
recessive mutations   248, 255, 256
– genotyping   255
– progeny testing   255, 256
reduced rachis internodes, benefits   218
RefSeq databases   174
reporter gene   18, 55, 56, 271, 277
– β-glucuronidase gene   271
resistive pulse sensing method   340
reverse genetic approach   94, 135, 187, 188
– FST sequencing   94
– role   187
– schematic illustration   95
– screening DNA pools   94
– tools   188

rice   190–193
– EMS mutagenesis   191, 192
– forward genetics   193
– TILLING, bulking of DNA   190
– transposon-tagged mutant resources, databases   19
RNA
– isoforms   107
– polymerase run-on assays profiling   118

s

Saccharomyces cerevisiae, profiling   302
sample preparation protocol   292
– blocking   294
– DNA
– – fragments, size selection   292
– – polyA tailing   294
– fill-and-lock   294
– flow cell, load samples   294
– genomic DNA, ultrasonic shearing   292
Sanger dideoxy chain termination method   320, 321
Sanger sequencing method   167–175, 207, 320, 321, 323
– mutation detection software   167
Scandinavian mutation program   219
scanning tunneling microscopy (STM)   346
seeds   21, 36
– nomenclature   36
– sterilization, chlorine gas   21
sequencing-based screening method   199, 202, 206
– applications   206–210
– – EMS mutation screening/validation   206–208
– – natural polymorphism screening/validation   208–210
– methods   202–206
– protocols   202–206
– TILLING/EcoTILLING   199
sequencing-by-synthesis (SBS)   106, 290, 320, 322, 323
short sequence repeats (SSRs)   168
short tandem repeats (STRs)   168
shotgun sequencing-by-hybridization (SBH)   308
signal-to-noise ratios   140, 172, 297, 298, 347, 351
single-marker methods   235
single-stranded DNA (ssDNA) molecules   56, 60, 74, 79, 110, 308, 320, 342
– background mutation frequency, determination   65
– carbon nanotube binding   346

– nanopore sequencing   320
– proportions   60
single nucleotide polymorphism (SNPs)   104, 140, 150, 159, 168, 170, 199, 209, 217, 223–229, 232, 234, 238, 290, 345
– barley, two/six-row locus   224–227
– detection   188, 321
– disease resistance locus, graphical genotyping   227
– enzyme concentration optimization   140
– genotypes   150
– haplotype   238
– heterozygote detection   159
– high-throughput genotyping, protocol   227–229
– marker associations, genome scan   225
– mutation scanning   159
– resources   223, 224
– scanning   159
slab gel electrophoresis   134
SleIF4E gene   204, 209, 210
– haplotypes   210
– SNPs   210
– wild-type sequence   209
Softgenetics software   167
– SoftGenetics' NextGENe software   178, 180
Solanum lycopersicum   201
solid-state nanopore   345–347, 351
– tunneling electrodes   346
SOLiD system   175, 177
simple sequence repeat (SSR) markers   195
statistical probability calculation methods   201
Staudinger reaction   331
step-by-step genome analyzer protocol   103
STRAT method models   237
surface functionalization methods   345

t

TAIL-PCR approach   18, 26, 89, 90
– fragment characterization   91
– primers   90
– TAIL-PCR1 program   90
Takasaki Ion Accelerators for Advanced Radiation Application (TIARA)   3
targeting experiment   272
– plating   275
– – alternative procedure   275
– – materials   275
– – procedures   275
– – solutions and media   275
– protoplast isolation   272
– – materials   272
– – procedures   273

– – solutions and media   272
– selection of transformants   275
– – materials   276
– – procedures   276
– – solutions and media   276
– transformation   274
– – materials   274
– – procedures   274
– – solutions and media   274
targeting induced local lesions in genomes (TILLING) technique   105, 168, 181, 187–189, 190, 192, 196, 200, 220
– application   188, 279, 280
– DNA fragments, PCR procedure   189
– EcoTILLING   181
– methods/protocols   188–196
– mutation frequency evaluation   192
– perspectives   196, 280
– platform, mutation detection in rice   188
– use, for reverse/forward genetics   187
targeting vectors   270
– construction   270
– homology length   271
– reporter genes   271
– selectable marker genes   271
TASSEL   237
T-DNA insertion collection   33, 34
– methods   34–47
– protocols   34–47
– – *Agrobacterium* culture   35
– – FST production   37–40
– – growth conditions   34
– – plant material   34
– – plant transformation   35
– – plasmid design   34, 35
– – sulfadiazine-selected T1 plants, DNA preparation   36, 37
– – T1 seed harvesting   35
TD-PCR   85–89
– border amplification   88
– fragment characterization   89
– genomic DNA digestion   87
– ligation   87
– specific oligonucleotides   87
– TD-PCR1/2 program   88
three-dimensional pooling strategy   94
TimeLogic DeCypher system   205
Tnt1-containing plants, transposon display analysis   86
translesion synthesis (TLS)   55, 67, 77
– biochemical TLS assay   67

– KF polymerase   77
tobacco, nitrate reductase (NiaD) gene   84
tomato, GS FLX KeyPoint runs results   206
total internal reflection fluorescence microscopy (TIRFM), advantage   297
transformants analysis   277–279
transgenic plants   258, 259
– maize   259
transmission electron microscopy (TEM)   341
– based fabrication methods   343
– tomography, three-dimensional   350
transposon display technology   84
tris(2-carboxy-ethyl) phosphine (TCEP)   331
true single molecule sequencing (tSMS™)   289, 291, 297, 298
– applications   301–304
– – DNA sequencing applications   303
– – single molecule DGE/RNA-Seq   301–303
– – single molecule sequencing techniques under development   303, 304
– methods/protocols/technical principles   291–301
– – cyclic SBS   295
– – data analysis   301
– – flow cell surface architecture   294, 295
– – growing strands, optical imaging   297
– – mechanical operation   300
– – single molecule sequencing technical challenges/solutions   291
– – system components   300
– perspectives   304
– SBS reactions   296, 300
– – HeliScope single molecule sequencer   300

u
United Kingdom spring barley certified seed production, percentage   222
5′-untranslated region (UTR)   160

v
vector DNA, transformation   264
viral genomes   316
virtual terminator nucleotide mixture   295

w
West European malting market   222
whole-genome mutagenesis   247
wild-type genomic fragment   98

y
yeast artificial chromosome (YAC)   105